Combinatorial Optimization: Algorithms and Complexity

CHRISTOS H. PAPADIMITRIOU
University of California-Berkeley

KENNETH STEIGLITZ
Princeton University

DOVER PUBLICATIONS, INC.
Mineola, New York

To the memory of our fathers
Harilaos Papadimitriou
Irving Steiglitz

Copyright

Bibliographical Note

This Dover edition, first published in 1998, is a corrected, unabridged republication of the work originally published in 1982 by Prentice-Hall, Inc., New Jersey. It contains a new preface by the authors written especially for this Dover edition.

Library of Congress Cataloging-in-Publication Data

Papadimitriou, Christos H.
Combinatorial optimization : algorithms and complexity / Christos H. Papadimitriou, Kenneth Steiglitz.
p. cm.
Originally published: Englewood Cliffs, N.J. : Prentice Hall, © 1982. With new pref.
Includes bibliographical references and index.
ISBN 0-486-40258-4 (pbk.)
1. Mathematical optimization. 2. Combinatorial optimization. 3. Computational complexity. I. Steiglitz, Kenneth, 1939– II. Title.
QA402.5.P37 1998
519.3—dc21 98–21476
CIP

Manufactured in the United States by Courier Corporation
40258411
www.doverpublications.com

Contents

Preface to the Dover Edition

During the fifteen years since *Combinatorial Optimization* first appeared, its authors have often discussed the possibility of a second edition. In some sense a second edition seemed very appropriate, even called for: Many exciting new results had appeared that would merit inclusion, while not quite so many and so exciting that the basic premises, style, and approach of the book would need to be reworked.

The current republication of the book by Dover gives us an interesting opportunity to look critically at the book, and the field it recounts, fifteen years later. In retrospect, we are now happy with our decision (if you can call it that) not to proceed with a second edition. This was a book about two fields, at a moment when they had reached a degree of joint maturity and cross-fertilization that can inspire and justify a book; this feeling of novelty and timeliness would have been lost in a second edition. It is so much more appropriate (and, quite frankly, fun) to contemplate the interim developments from the armchair of a preface-writer.

To take the subjects of the book in order, the ellipsoid algorithm for linear programming (which had just made our publication deadline) did not have as much impact in practice as it did in theory, but the interior point algorithm developed in 1984 by Karmarkar has changed the practice of solving linear programs—at least by providing simplex with the kind of competition that leads to faster implementations. As for simplex and its variants, there were some very interesting algorithms due to Megiddo and Tardos, and the polynomial average-case result by Borgwardt; however, the quest for a strongly polynomial algorithm for linear programming is still very much the major open problem in the field. The questions of integrality and total unimodularity are also much better understood now. All these important advances are covered in an excellent more recent book:

A. Schrijver, *Theory of Linear and Integer Programming,* New York: Wiley-Interscience, 1986.

There have been substantial improvements in the running times of virtually all network and graph optimization algorithms discussed in the book: Network flow, minimum cost flow, shortest path, minimum spanning tree, matching, etc. These improvements were occasionally precipitated by an ingenious new way of looking at the problem, but more often they were the result of the introduction of interesting new data structures, the innovative use of randomization, of scaling, of methods for dealing with density, and other algorithmic techniques. For a snapshot ca. 1993 see:

R. K. Ahuja, T. L. Magnanti, J. B. Orlin, *Network Flows,* Englewood Cliffs, N.J.: Prentice-Hall, 1993.

As for complexity, the central question of P vs. NP is now as unanswered as ever. If anything, a solution seems somehow farther away today than then, because many more avenues of attack have been explored by now with interesting but markedly inconclusive results. There has been progress and new insights in complexity, albeit in aspects such as space complexity, interaction, and randomization, that are not central to the book (but see the next paragraph regarding the complexity of approximation). Parallel algorithms and complexity came and went. For a recent book on the subject see:

C. H. Papadimitriou, *Computational Complexity,* Reading, Mass.: Addison-Wesley, 1994.

Two of the last topics of the book, approximation algorithms and local search, have since exploded into two very active fields. The more impressive theoretical advances were made in approximation—we even have now an approximation scheme for the Euclidean traveling salesman problem. Many of these exciting approximation algorithms were in fact based on mathematical programming concepts and methods (primal dual, fixed-dimension integer programming, semidefinite programming). On the other hand, a sequence of unexpected results in complexity culminated in a proof, in 1992, that several of these problems cannot have a polynomial approximation scheme, unless of course P = NP. For a compilation of surveys of these topics see:

D. Hochbaum (ed.), *Approximation Algorithms for NP-hard problems,* Boston, Mass.: PWS Publishing, 1996.

Finally, in the past fifteen years we have seen the development of many families of heuristics for optimization problems, typically inspired by metaphors from physics or biology (and sometimes referred to collectively as *new age algorithms*), which can be considered as clever variants of local search: Simulated annealing, tabu search, genetic algorithms, Boltzmann machines, neural networks, and so on. For a review of some of these approaches, see:

C. R. Reeves (ed.), *Modern Heuristic Techniques for Combinatorial Problems,* New York: J. Wiley, 1993.

Although very little has been rigorously established about the performance of such algorithms, they often seem to do remarkably well on certain problems. Developing the mathematical methodology for explaining and predicting the performance of these and other heuristics is one of the most important challenges facing the fields of optimization and algorithms today.

Christos H. Papadimitriou
Kenneth Steiglitz

Preface

The goal of this book is to bring together in one volume the important ideas of computational complexity developed by computer scientists over the past fifteen years, and the foundations of mathematical programming developed by the operations research community. The first seven chapters comprise a self-contained treatment of linear programming and duality theory, with an emphasis on graph and network flow interpretations. Chapter 8 is a transition chapter which introduces the techniques for analyzing the complexity of algorithms. Modern, fast algorithms for flow, matching, and spanning trees, as well as the general setting of matroids, are described in Chapters 9–12. The next two chapters, 13 and 14, treat integer linear programming, including Gomory's cutting-plane algorithm. Chapters 15–16 take up the relatively new ideas of the theory of *NP*-completeness and its ramifications. The last three chapters, 17–19, describe practical ways of dealing with intractable problems—approximation algorithms, branch-and-bound, dynamic programming, and local (or neighborhood) search.

The book can be used as a text in many ways, depending on the background of the students. At Princeton, for example, computer science students with a background in the theory of algorithms have covered chapters 1–10, 13, 14, 18, 19 in a one-semester graduate course. On the other hand, students with a background in operations research might start with Chapter 8 (after reviewing 1–7) and cover the rest of the book. Chapter 8, we might point out, contains the new Soviet ellipsoid algorithm, which provided a missing link to this book while it was being written.

We take the liberty here of mentioning several topics which we feel are treated in novel ways: The theory of the simplex algorithm avoids the awkwardness of unbounded polyhedra by "boxing in" the feasible set with bounds derived from the input data. The finiteness issue is settled by the new and elegant algorithm of R.G. Bland, with the lexicographic method introduced later for integer programming. And throughout the linear programming sections we have given geometric interpretations of the basic mathematical objects and operations.

The reader will also find unusual emphasis given to the primal-dual algorithm. It is used to derive Dijkstra's algorithm, the Ford and Fulkerson max-flow algorithm, all the min-cost flow algorithms, and the weighted matching algorithms. In this way the primal-dual idea is used to link general linear programming with its combinatorial applications, and to unify several algorithms for combinatorial problems.

NP-completeness theory is developed with minimal recourse to Turing machine theory, and the class of problems *NP* is defined by using the idea of "certificate," avoiding thereby the notion of nondeterminism. New and hopefully more transparent reductions are used for many proofs of *NP*-completeness, and in Chapter 13 a new and simple proof is given that integer linear programming is in *NP*, based on a discrete form of Farkas' Lemma.

The final chapter surveys some very successful applications of a widely used approach to intractable problems—local search. We come full circle here, showing that it is a simplex-like algorithm.

A book such as this one owes a great deal, of course, to many previous workers and writers. Without trying to be complete, we would like to mention the very important succession of books by Ford and Fulkerson; Dantzig; Simonnard; T.C. Hu; Lawler; Aho, Hopcroft, and Ullman; and Garey and Johnson; which have influenced our work at many points.

The authors also owe thanks to many co-workers and colleagues for comments on the work in progress. Among them are B.D. Dickinson, R. Ginosar, K. Lieberherr, A. Mirzaian, W.P. Niedringhaus, J. Orlin, F. Sadri, S. Toueg, J. Valdes, M. Yannakakis, and N. Zadeh.

Finally, we thank L. Furman, G. Pecht, R. D'Arcangelo, and C. Cole for unfailing secretarial help through many drafts of the manuscript.

Christos H. Papadimitriou
Kenneth Steiglitz

1

Optimization Problems

1.1 Introduction

Many problems of both practical and theoretical importance concern themselves with the choice of a "best" configuration or set of parameters to achieve some goal. Over the past few decades a hierarchy of such problems has emerged, together with a corresponding collection of techniques for their solution. At one end of this hierarchy is the general *nonlinear programming problem:* Find x to†

$$\begin{aligned} &\text{minimize } f(x) \\ &\text{subject to } g_i(x) \geq 0 \qquad i = 1, \ldots, m \\ &\qquad\qquad h_j(x) = 0 \qquad j = 1, \ldots, p \end{aligned}$$

where f, g_i, and h_j are general functions of the parameter $x \in R^n$. The techniques for solving such problems are almost always iterative in nature, and their convergence is studied using the mathematics of real analysis.

When f is convex, g_i concave, and h_j linear, we have what is called a *convex programming problem.* This problem has the most convenient property

†See the appendix at the end of this chapter for a résumé of terminology and notation.

that local optimality implies global optimality. We also have conditions for optimality that are *sufficient*, the *Kuhn-Tucker conditions*.

To take the next big step, when f and all the g_i and h_j are linear, we arrive at the *linear programming problem*. Several striking changes occur when we restrict attention to this class of problems. First, any problem in this class reduces to the selection of a solution from among a *finite* set of possible solutions. The problem is what we can call *combinatorial*. The finite set of candidate solutions is the set of vertices of the convex polytope defined by the linear constraints.

The widely used *simplex algorithm* of G.B. Dantzig finds an optimal solution to a linear programming problem in a finite number of steps. This algorithm is based on the idea of improving the cost by moving from vertex to vertex of the polytope. Thirty years of refinement has led to forms of the simplex algorithm that are generally regarded as very efficient—problems with hundreds of variables and thousands of constraints are solved routinely. It is also true, however, that there are specially devised problems on which the simplex algorithm takes a disagreeably exponential number of steps.

Soviet mathematicians, in a relatively recent development, have invented an ellipsoid algorithm for linear programming that is guaranteed to find an optimal solution in a number of steps that grows as a *polynomial* in the "size" of the problem—a state of affairs that we shall come to regard as very favorable in the course of this book. At the time of this writing, it is uncertain whether refinements analogous to those of the simplex algorithm will render the ellipsoid algorithm competitive with it. The nesting of the problems mentioned so far—general nonlinear, convex, and linear programs—is indicated in Figure 1–1.

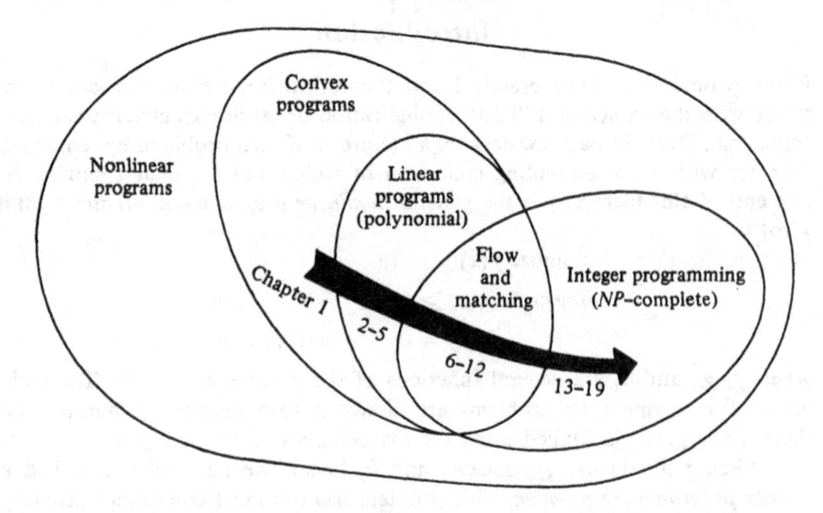

Figure 1–1 The classes of problems considered in this book and the path followed by the chapters.

Certain linear programs, the *flow and matching problems*, can be solved much more efficiently than even general linear programs. On the other hand, these problems are also closely related to other problems that are apparently intractable! As an example, the point-to-point shortest-path problem in a graph is in our class of flow and matching problems, and in fact has an $O(n^2)$ algorithm for its solution, where n is the number of nodes in the graph. In contrast, the *traveling salesman problem*, which asks for the shortest closed path that visits every node exactly once, is in the class of *NP-complete problems*, all of which are widely considered unsolvable by polynomial algorithms. This fine line between "very easy" and "very hard" problems is a recurrent phenomenon in this book and has naturally attracted the attention of algorithm designers.

Another way of looking at the flow and matching problems is as special cases of *integer linear programs*. These come about when we consider linear programs and ask for the best-cost solution with the restriction that it have *integer-valued* coordinates. These problems, like linear programs, have a finite algorithm for solution. But there the resemblance stops: The general integer linear programming problem is itself *NP*-complete. The complete state of affairs is shown in Figure 1–1, together with an indication of the general path that this book will take through these classes of problems.

We shall begin with the most fundamental and easily accessible facts about convex programs. We then take up linear programming in earnest, studying the simplex algorithm, its geometry, and the algorithmic implications of duality. We shall stress graph-theoretic interpretations of the algorithms discussed, and that will lead naturally to the flow and matching problems. It is an easy step from there to a consideration of complexity issues and the ellipsoid algorithm and to a study of the *NP*-complete problems that have become representative of difficult combinatorial optimization problems. The last part of the book is concerned with approaches toward the practical solution of *NP*-complete problems of moderate size: approximation, enumerative techniques, and local search.

1.2 Optimization Problems

Optimization problems seem to divide naturally into two categories: those with *continuous* variables, and those with *discrete* variables, which we call *combinatorial*. In the continuous problems, we are generally looking for a set of real numbers or even a function; in the combinatorial problems, we are looking for an object from a finite, or possibly countably infinite, set—typically an integer, set, permutation, or graph. These two kinds of problems generally have quite different flavors, and the methods for solving them have become quite divergent. In our study of combinatorial optimization we start—in some sense—at its boundary with continuous optimization.

Linear programming plays a unique role in optimization theory; it is in one sense a continuous optimization problem, but, as we mentioned above, it can also be considered combinatorial in nature and in fact is fundamental to the study of many strictly combinatorial problems. We shall therefore give a definition of an *optimization problem* general enough to include linear programming (and almost any other optimization problem).

Definition 1.1

An *instance of an optimization problem* is a pair (F, c), where F is any set, the domain of feasible points; c is the cost function, a mapping

$$c: F \longrightarrow R^1$$

The problem is to find an $f \in F$ for which

$$c(f) \leq c(y) \quad \text{for all} \quad y \in F$$

Such a point f is called a *globally optimal* solution to the given instance, or, when no confusion can arise, simply an *optimal* solution. □

In many examples the cost function will take on only nonnegative integer values.

Definition 1.2

An *optimization problem* is a set I of instances of an optimization problem. □

We have been careful to distinguish between a *problem* and an *instance* of a problem. Informally, in an *instance* we are given the "input data" and have enough information to obtain a solution; a *problem* is a collection of instances, usually all generated in a similar way. Thus, in the following example, an instance of the traveling salesman problem has a given distance matrix; but we speak in general of the traveling salesman problem as the collection of all instances associated with all distance matrices.

Example 1.1 (Traveling Salesman Problem (TSP))

In an instance of the TSP we are given an integer $n > 0$ and the distance between every pair of n cities in the form of an $n \times n$ matrix $[d_{ij}]$, where $d_{ij} \in Z^+$. A *tour* is a closed path that visits every city exactly once. The problem is to find a tour with minimal total length. We can take

$$F = \{\text{all cyclic permutations } \pi \text{ on } n \text{ objects}\}$$

A cyclic permutation π represents a tour if we interpret $\pi(j)$ to be the city visited after city j, $j = 1, \ldots, n$. Then the cost c maps π to

$$\sum_{j=1}^{n} d_{j\pi(j)} \quad \square$$

Example 1.2 (Minimal Spanning Tree (MST))

As above, we are given an integer $n > 0$ and an $n \times n$ symmetric distance matrix $[d_{ij}]$, $d_{ij} \in Z^+$. The problem is to find a spanning tree on n vertices that has minimal total length of its edges. In our definition of an instance of an optimization problem, we choose

$$F = \{\text{all spanning trees } (V, E) \text{ with } V = \{1, 2, \ldots, n\}\}$$
$$c\colon (V, E) \longrightarrow \sum_{[i,j] \in E} d_{ij}$$

(By a *spanning tree* we mean an undirected graph (V, E) that is connected and acyclic. See the appendix at the end of this chapter.) □

Example 1.3 (Linear Programming (LP))

Let m, n be positive integers, $b \in Z^m$, $c \in Z^n$, and A an $m \times n$ matrix with elements $a_{ij} \in Z$. An instance of LP is defined by

$$F = \{x\colon x \in R^n, \quad Ax = b, \quad x \geq 0\}$$
$$c\colon x \longrightarrow c'x$$

Stated as such, linear programming is a continuous optimization problem, with, in fact, an uncountable number of feasible points $x \in F$. To see how it can be considered combinatorial in nature, consider the simple instance defined by $m = 1, n = 3$ and

$$A = (1 \quad 1 \quad 1)$$
$$b = (2)$$

Figure 1–2 shows the feasible set F in this instance, the intersection of a plane with the first octant in R^3.

The problem is to minimize the value of the linear function $c'x = c_1x_1 +$

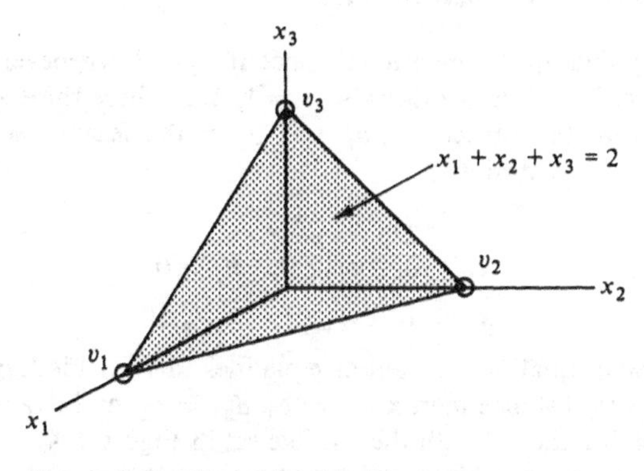

Figure 1–2 The feasible set F for an instance of LP.

$c_2x_2 + c_3x_3$ on this triangle. It is not hard to see intuitively that a minimum will always occur at one of the corners v_1, v_2, or v_3 indicated in the figure. If we grant this, we can solve this instance by finding all the vertices and evaluating $c'x$ at each one. This may be a formidable task in a larger instance, but the point is that it is a *finite* one. It is in this sense that LP is combinatorial. □

In many cases it is possible to do the opposite: express a purely combinatorial problem as an LP.

Example 1.3 (Continued)

Consider an instance of the MST problem with $n = 3$ points. There are three spanning trees of these points, shown in Figure 1–3. They can also be

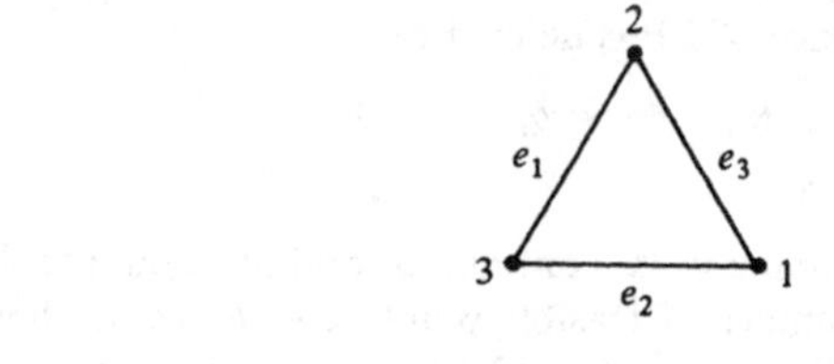

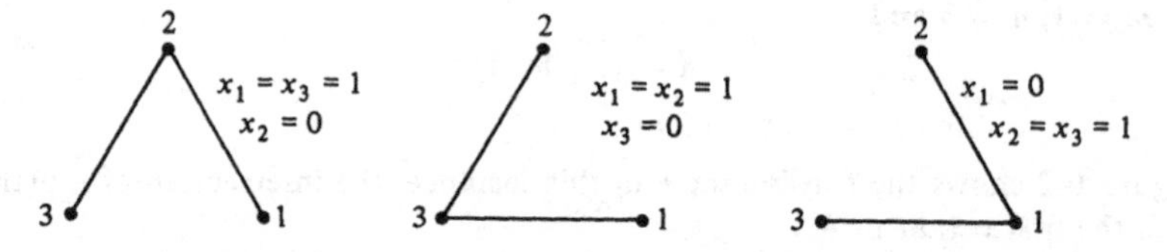

Figure 1–3 Three nodes and their three spanning trees, thought of as points in 3-dimensional space.

thought of as points in 3-dimensional space if $x_j = 1$ whenever edge e_j is in the tree considered, and zero otherwise, $j = 1, 2, 3$. These three spanning trees then coincide with the vertices v_1, v_2, and v_3 of the feasible set F in Figure 1–4 defined by the constraints

$$x_1 + x_2 + x_3 = 2$$

$$x_1 \geq 0, \quad x_2 \geq 0, \quad x_3 \geq 0$$

$$x_1 \leq 1, \quad x_2 \leq 1, \quad x_3 \leq 1$$

(We shall allow inequalities as well as equalities in LP.) Finding the minimal spanning tree with distance matrix $d_{12} = c_3$, $d_{23} = c_1$, and $d_{31} = c_2$ is exactly the same as solving the LP with the feasible set in Figure 1–4.

Thus this purely combinatorial problem can, in principle, be solved by LP. This point of view will be very useful later for developing algorithms for certain combinatorial problems. □

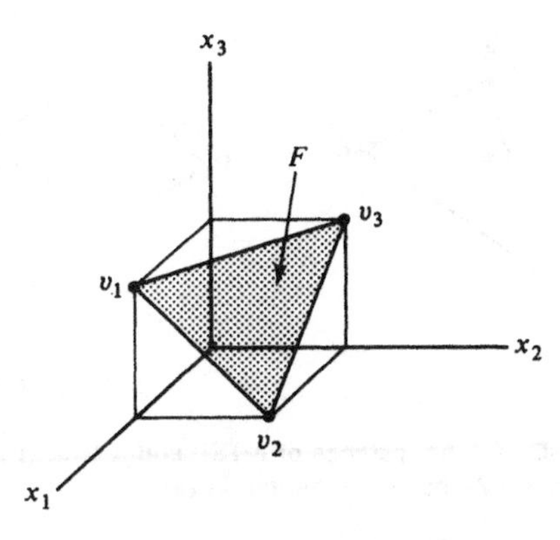

Figure 1–4 The feasible set F for the simple spanning tree problem.

1.3 Neighborhoods

Given a feasible point $f \in F$ in a particular problem, it is useful in many situations to define a set $N(f)$ of points that are "close" in some sense to the point f.

Definition 1.3

Given an optimization problem with instances (F, c), a *neighborhood* is a mapping

$$N: F \longrightarrow 2^F$$

defined for each instance. □

If $F = R^n$, the set of points within a fixed Euclidean distance provides a natural neighborhood. In many combinatorial problems, the choice of N may depend critically on the structure of F.

Example 1.4 [Lin1]

In the TSP we may define a neighborhood called 2-*change* by

$N_2(f) = \{g : g \in F$ and g can be obtained from f as follows: remove two edges from the tour; then replace them with two edges$\}$

Figure 1–5 shows an example of a tour f and another tour $g \in N_2(f)$ for an instance of the TSP with seven cities and a distance matrix determined by Euclidean distance between points in the plane.

This neighborhood can be generalized in the obvious way to N_k, called

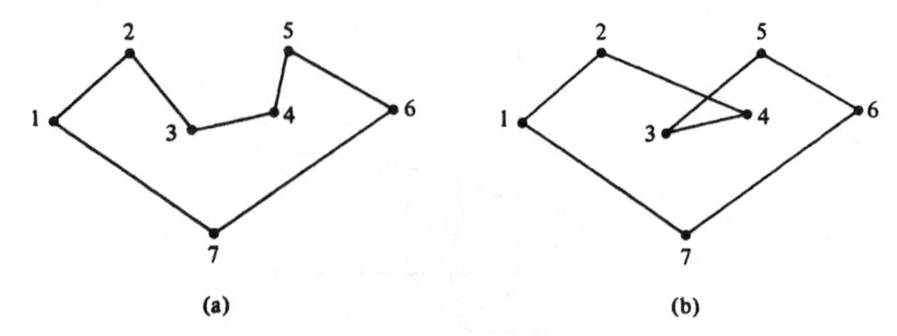

Figure 1–5 (a) An instance of a TSP and a tour. (b) Another tour which is a 2-change of the tour in (a).

k-change, where at most k links are replaced. Such neighborhoods lead to very effective heuristics for the TSP. □

Example 1.5

In the MST, an important neighborhood is defined by

$N(f) = \{g : g \in F$ and g can be obtained from f as follows: add an edge e to the tree f, producing a cycle; then delete any edge on the cycle$\}$ □

Example 1.6

In LP, we can define a neighborhood by

$$N_\epsilon(x) = \{y : Ay = b, \quad y \geq 0, \quad \text{and} \quad \|y - x\| \leq \epsilon\}$$

This is simply the set of all feasible points within Euclidean distance ϵ of x, for some $\epsilon > 0$. □

1.4 Local and Global Optima

Finding a globally optimal solution to an instance of some problems can be prohibitively difficult, but it is often possible to find a solution f which is best in the sense that there is nothing better in its neighborhood $N(f)$.

Definition 1.4

Given an instance (F, c) of an optimization problem and a neighborhood N, a feasible solution $f \in F$ is called *locally optimal with respect to N* (or simply *locally optimal* whenever N is understood by context) if

$$c(f) \leq c(g) \quad \text{for all} \quad g \in N(f) \quad \square$$

Example 1.7

Consider the instance of an optimization problem (F, c) defined by

$$F = [0, 1] \subseteq R^1$$

and the cost function c sketched in Fig. 1–6.

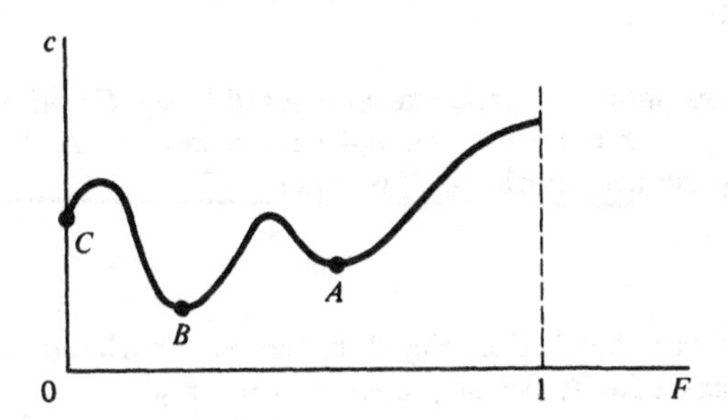

Figure 1–6 A 1-dimensional Euclidean optimization problem.

Further, let the neighborhood be defined simply by closeness in Euclidean distance for some $\epsilon > 0$.

$$N_\epsilon(f) = \{x : x \in F \quad \text{and} \quad |x - f| \leq \epsilon\}$$

Then if ϵ is suitably small, the points A, B, and C are all locally optimal, but only B is globally optimal. ☐

Example 1.8

In the TSP, solutions locally optimal with respect to the k-change neighborhood N_k are called *k-opt* [Linl]. To find a k-opt tour in an instance of the TSP, define the function improve (t), where $t \in F$, as follows:

$$\text{improve } (t) = \begin{cases} \text{any } s \in N_k(t) \text{ such that } c(s) < c(t), \text{ if such an } s \text{ exists} \\ \text{'no' otherwise} \end{cases}$$

That is, improve (t) searches $N_k(t)$ for a better tour s. If one is found, it returns the improved tour; otherwise it returns the value 'no.' An algorithm for finding a k-opt tour is then†

```
procedure k-opt
begin
   t := some initial tour;
   while improve(t) ≠ 'no' do
      t := improve(t);
   return t
end
```

†Algorithms are written in an informal notation called *pidgin algol*. See the appendix at the end of this chapter.

Because we are generally interested in finding a global optimum and because many algorithms can do no more than search for local optima, it is important to know whether a local optimum is or is not global. This depends, of course, on the neighborhood N. The following terminology describes the happy situation when every local optimum is also a global optimum.

Definition 1.5

Given an optimization problem with feasible set F and a neighborhood N, if whenever $f \in F$ is locally optimal with respect to N it is also globally optimal, we say the neighborhood N is *exact*. □

Example 1.9

In the instance sketched in Fig. 1–6, the neighborhood N_ϵ is exact when $\epsilon \geq 1$ but not exact for sufficiently small $\epsilon > 0$. □

Example 1.10

In the TSP, N_2 is not exact; but N_n, where n is the number of cities, is exact. (See Problem 2.) □

Example 1.11

In the MST, the neighborhood described in Example 1.5 is exact. (See Problem 3.) □

1.5 Convex Sets and Functions

We now turn our attention to the class of problems where $F \subseteq R^n$. In particular, we should like very much to find classes of problems where N_ϵ is exact for every $\epsilon > 0$, for in such problems we can be assured that any local optimum found is a global one. Such a property is enjoyed by the class of convex programming problems, of which linear programming is a special case. We start with some important definitions.

Definition 1.6

Given two points $x, y \in R^n$, a *convex combination* of them is any point of the form

$$z = \lambda x + (1 - \lambda)y, \qquad \lambda \in R^1 \quad \text{and} \quad 0 \leq \lambda \leq 1$$

If $\lambda \neq 0, 1$, we say z is a *strict* convex combination of x and y. □

Definition 1.7

A set $S \subseteq R^n$ is *convex* if it contains all convex combinations of pairs of points $x, y \in S$. □

Example 1.12

The entire set R^n is convex, as is the empty set $\varnothing$ and any singleton set. □

Example 1.13

In R^1, any interval is convex and any convex set is an interval. □

Example 1.14

In R^2, convex sets, loosely speaking, are those without indentations. Thus, in Fig. 1–7, set A is convex but B is not. □

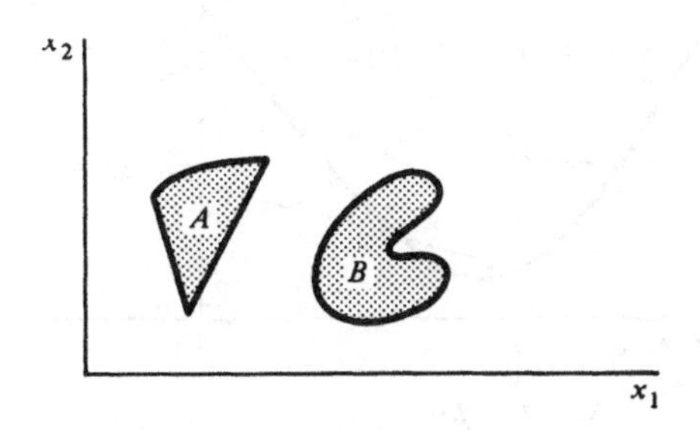

Figure 1–7 A convex set A and a nonconvex set B.

An important property of convex sets is expressed in the following lemma.

Lemma 1.1 *The intersection of any number of convex sets S_i is convex.*

Proof If x and y are two points in $\cap S_i$, they are in every S_i. Any convex combination of them is then in every S_i and therefore in $\cap S_i$. □

We now introduce the idea of a convex function defined on a convex set.

Definition 1.8

Let $S \subseteq R^n$ be a convex set. The function

$$c: S \longrightarrow R^1$$

is *convex in S* if for any two points $x, y \in S$

$$c(\lambda x + (1-\lambda)y) \leq \lambda c(x) + (1-\lambda)c(y), \qquad \lambda \in R^1 \text{ and } 0 \leq \lambda \leq 1$$

If $S = R^n$, we say simply that c is *convex.* □

Example 1.15

Any linear function is convex in any convex set S. □

Example 1.16

Intuitively, a convex function is one that "bends up." Figure 1–8 shows a sketch of a convex function in $[0, 1] \subseteq R^1$.

$$c\colon [0, 1] \longrightarrow R^1$$

The convexity condition implies that chords always lie above the function. □

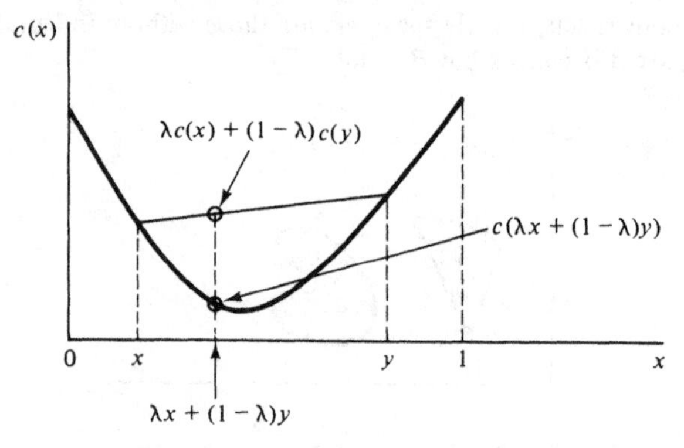

Figure 1–8 A function c convex in [0, 1].

The set of points where a convex function is less than or equal to a given value is a convex set. More precisely, we have the following.

Lemma 1.2 *Let $c(x)$ be a convex function on a convex set S. Then for any real number t, the set*

$$S_t = \{x : c(x) \leq t, \quad x \in S\}$$

is convex.

Proof Let x and y be two points in S_t. Then the convex combination $\lambda x + (1-\lambda)y$ is in S and

$$\begin{aligned} c(\lambda x + (1-\lambda)y) &\leq \lambda c(x) + (1-\lambda)c(y) \\ &\leq \lambda t + (1-\lambda)t \\ &\leq t \end{aligned}$$

which shows that the convex combination $\lambda x + (1-\lambda)y$ is also in S_t. □

Example 1.17

When c is defined on R^2, the boundaries of the sets S_t are level contours such as are drawn on topographic maps. Such a map is shown in Fig. 1–9. □

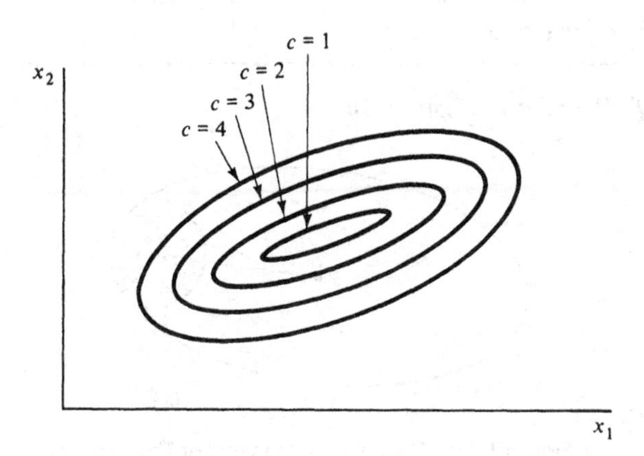

Figure 1–9 The level contours of a convex function defined on R^2.

Finally, functions that in some sense behave oppositely to convex functions are called *concave functions*.

Definition 1.9

A function c defined in a convex set $S \subseteq R^n$ is called *concave* if $-c$ is convex in S. □

Example 1.18

Every linear function is concave as well as convex. Loosely speaking, a linear function bends neither down nor up, and so walks the line between convexity and concavity. □

1.6 Convex Programming Problems

An important class of optimization problems concerns the minimization of a convex function on a convex set. These problems have the convenient property (mentioned above) that local optima are global. More precisely, we establish the following fact.

Theorem 1.1 *Consider an instance of an optimization problem (F, c), where $F \subseteq R^n$ is a convex set and c is a convex function. Then the neighborhood defined by Euclidean distance*

$$N_\epsilon(x) = \{y : y \in F \text{ and } \|x - y\| \leq \epsilon\}$$

is exact for every $\epsilon > 0$.

Proof We refer to Figure 1–10.

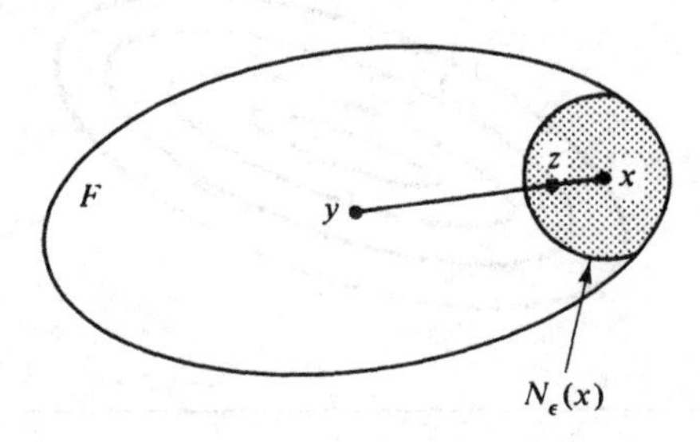

Figure 1–10 The points in the proof of Theorem 1.1.

Let x be a local optimum with respect to N_ϵ for any fixed $\epsilon > 0$ and let $y \in F$ be any other feasible point, not necessarily in $N_\epsilon(x)$. We can always choose a λ sufficiently close to 1 that the strict convex combination

$$z = \lambda x + (1 - \lambda)y, \qquad 0 < \lambda < 1$$

lies within the neighborhood $N_\epsilon(x)$. Evaluating the cost function c at this point, we get, by the convexity of c,

$$c(z) = c(\lambda x + (1 - \lambda)y) \leq \lambda c(x) + (1 - \lambda)c(y)$$

Rearranging, we find that

$$c(y) \geq \frac{c(z) - \lambda c(x)}{1 - \lambda}$$

But since $z \in N_\epsilon(x)$

$$c(z) \geq c(x)$$

so

$$c(y) \geq \frac{c(x) - \lambda c(x)}{1 - \lambda} = c(x)$$

Note that we have made no extra assumptions about the function c; it need not be differentiable, for example. □

For our purposes, the convex feasible region will always be defined by a set of inequalities involving concave functions. Such problems are conventionally known as *convex programming problems.*

Definition 1.10

An instance of an optimization problem (F, c) is a *convex programming problem* if c is convex and $F \subseteq R^n$ is defined by

$$g_i(x) \geq 0, \qquad i = 1, \ldots, m$$

where

$$g_i : R^n \longrightarrow R^1$$

are *concave* functions. □

It is not hard to see that the set F defined in this way is in fact convex.

Lemma 1.3 *The feasible set F in a convex programming problem is convex.*

Proof The functions $-g_i$ are convex, so by Lemma 1.2, the sets

$$F_i = \{x : g_i(x) \geq 0\}$$

are convex. Hence, by Lemma 1.1,

$$F = \bigcap_{i=1}^{m} F_i$$

is also convex. □

With this, we have shown the following theorem.

Theorem 1.2 *In a convex programming problem, every point locally optimal with respect to the Euclidean distance neighborhood N_ϵ is also globally optimal.*

Example 1.19

A convex function $c(x)$ defined on $[0, 1] \subseteq R^1$ can have many local optima but all must be global, as illustrated in Fig. 1–11 □

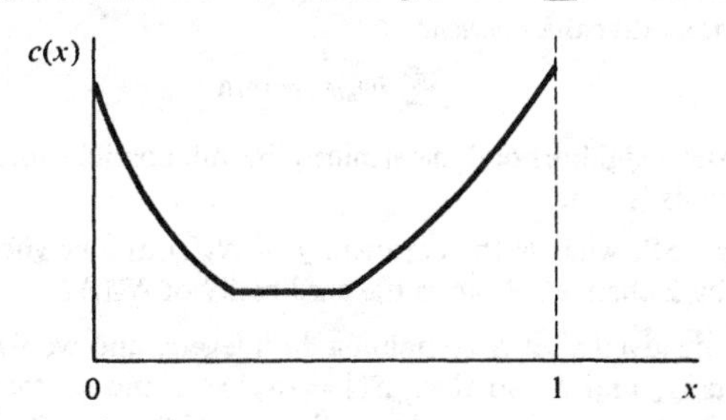

Figure 1–11 A convex programming problem with many local optima, all of which are global.

Example 1.20

Every instance of LP is a convex programming problem, because linear functions are both convex and concave. Thus a local optimum of an instance of LP must also be global. □

PROBLEMS

1. Formulate the following as optimization problem instances, giving in each case the domain of feasible solutions F and the cost function c.
   (a) Find the shortest path between two nodes in a graph with edge weights representing distance.
   (b) Solve the *Tower of Hanoi* problem, which is defined as follows: We have 3 diamond needles and 64 gold disks of increasing diameter. The disks have holes at their centers, so that they fit on the needles, and initially the disks are all on the first needle, with the smallest on top, the next smallest below that, and so on, with the largest at the bottom. A legal move is the transfer of the top disk from any needle to any other, *with the condition that no disk is ever placed on top of one smaller than itself.* The problem is to transfer, by a sequence of legal moves, all the disks from the first to the second needle. (There is a story that the world will end when this task is completed [Kr], which may be an optimistic expectation.) Generalize to n gold disks.
   (c) Win a game of chess. How many instances of this problem are there?
   (d) Find a cylinder with a given surface area A that has the largest volume V.
   (e) Find a closed plane curve of given perimeter that encloses the largest area.

*2. Show by example that 2-change does not define an exact neighborhood for the TSP. Repeat for 3-change, and for $(n - 3)$-change, where n is the number of cities.

*3. Show that the neighborhood defined in Example 1.5 for the MST is exact.

4. The *moment problem* is that of finding a permutation π of n weights w_i, $i = 1, \ldots, n$, so that the *moment*

$$\sum_{i=1}^{n} i w_{\pi(i)} = \min$$

   Show that the neighborhood determined by all possible interchanges of two adjacent weights is exact.

5. In the n-city TSP, what is the cardinality of $N_2(t)$, the neighborhood of tour t determined by 2-change? What is the cardinality of $N_3(t)$?

6. Suppose we are given a set S containing $2n$ integers, and we wish to partition it into two sets S_1 and S_2 so that $|S_1| = |S_2| = n$ and so that the sum of the numbers in S_1 is as close as possible to the sum of those in S_2. Let the neighbor-

*Throughout this book an asterisk means that a problem is relatively difficult.

hood N be determined by all possible interchanges of two integers between S_1 and S_2. Is N exact?

7. Is the product of two convex functions convex? If yes, prove it; if not, give a counterexample.

8. Let $f(x)$ be convex in R^n. Is $f(x + b)$, where b is a constant, convex in R^n?

9. Let $f(x)$ be convex in R^n. Fix $x_2, \ldots, x_n$ and consider the function $g(x_1) = f(x_1, \ldots, x_n)$. Is g convex in R^1?

10. Let $f(x_i)$ be a convex function of the single variable x_i. Then $g(x) = f(x_i)$ can also be considered as a function of $x \in R^n$. Is $g(x)$ convex in R^n?

11. Show that the sum of two convex functions is convex.

12. Justify the inclusion of integer linear programming within the class of nonlinear programs in Figure 1–1.

13. The following is a very useful criterion for determining if a function is convex [SW, vol. I, p. 152]:

Let C be an open convex set in R^n and let f have continuous second partial derivatives in C. Then f is convex in C if and only if the matrix of second partial derivatives

$$H(x) = \left[\frac{\partial^2 f}{\partial x_i \, \partial x_j}\right]_{ij}$$

is positive semidefinite for all x in C.

Determine whether or not the following functions are convex in the indicated domains.

(a) $f = x_1 x_2, \quad C = R^2$

(b) $f = e^{x_1 + x_2}, \quad C = R^2$

(c) $f = x_1^4 + x_2^4 - x_1^2 x_2^2, \quad C = R^2$

(d) $f = x_1^3 + x_2^3, \quad C = \{x \in R^2 : x > 0\}$

(e) $f = \tan x_1, \quad C = \{x_1 : 0 < x_1 < 1\}$

14. Consider the nonlinear program

$$\min f = x_1 x_2$$
$$\text{such that } g = (x_1 - 1)^2 + (x_2 - 1)^2 = 1$$

Find all the global and local minima.

***15.** Formulate the minimal spanning tree problem for an n-node graph as a linear program with $\binom{n}{2}$ variables, one for each edge, thus generalizing Example 1.3. (*Hint:* this may require many constraints.)

NOTES AND REFERENCES

Further discussion of nonlinear programming problems and optimality conditions can be found in

[FM] Fiacco, A. V., and G. P. McCormick, *Nonlinear Programming: Sequential*

Unconstrained Minimization Techniques. New York: John Wiley & Sons, Inc., 1968.

[Had1] HADLEY, G., *Nonlinear and Dynamic Programming*. Reading, Mass.: Addison-Wesley Publishing Co., Inc., 1964.

[Ao] AOKI, M., *Introduction to Optimization Techniques*. New York: Macmillan, Inc., 1971.

[SW] STOER, J., and C. WITZGALL, *Convexity and Optimization in Finite Dimensions*. Berlin: Springer-Verlag, 1970.

The idea of a k-change neighborhood for the TSP is due to

[Lin1] LIN, S., "Computer Solutions to the Traveling Salesman Problem," *BSTJ*, 44, no. 10 (1965), 2245–69.

The origin of the Tower of Hanoi problem (Problem 1) is described in

[Kr] KRAITCHIK, M., *Mathematical Recreations*. New York: W. W. Norton and Co., 1942.

APPENDIX
TERMINOLOGY AND NOTATION

A.1 Linear Algebra

The *real number line* is denoted by R (or sometimes R^1) and the *n-dimensional real vector space*, the set of ordered n-tuples of real numbers, by R^n. Other fixed sets are the set of nonnegative reals, R^+; the set of integers, Z; the set of nonnegative integers, Z^+; and the set of ordered n-tuples of integers, Z^n.

A *set* of elements $s_1, s_2, s_3, \ldots$ is written explicity as

$$S = \{s_1, s_2, s_3, \ldots\}$$

and a set defined to contain all elements x for which a condition P is true is defined by writing

$$S = \{x : P(x)\}$$

For example, Z^+ can be defined by

$$Z^+ = \{i : i \in Z \quad \text{and} \quad i \geq 0\}$$

The size of a finite set S is denoted by $|S|$. A mapping μ from set S to set T is written

$$\mu: S \longrightarrow T$$

and 2^S stands for the set of all subsets of S.

An $m \times n$ matrix with element a_{ij} in Row i and Column j is written

$$A = [a_{ij}]$$

The n-vector consisting of the ith row of A is denoted by a_i and the m-vector which is the jth column of A by A_j. All vectors x (no prime) are *column* vectors; vectors x' (prime) are *row* vectors. Thus the matrix equation

$$Ax = b$$

is equivalent to the set of scalar equations

$$a_i'x = b_i \qquad i = 1, \ldots, m$$

where b is an m-vector and b_i, its ith component, is a scalar. The *transpose* of a matrix A is written A^T and the *determinant* of a square matrix is denoted by det (A).

The *unit square matrix* is denoted by I, where its dimension is usually understood by context. It is defined by

$$I_{ij} = \begin{cases} 1 & \text{if } i = j \\ 0 & \text{otherwise} \end{cases}$$

Similarly, 0 denotes either the zero scalar, vector, or matrix, depending on the context.

To construct an n-vector x with ith component x_i, we write

$$x = \text{col}\,(x_1, \ldots, x_n)$$

or, when there is no danger of confusion,

$$x = (x_1, \ldots, x_n)$$

To construct an $(n + m)$-vector z if its first n components are those of x and next m components those of y, we write

$$z = (x \mid y)$$

A.2 Graph Theory

A *graph* G is a pair $G = (V, E)$, where V is a finite set of *nodes* or *vertices* and E has as elements subsets of V of cardinality two called *edges*. The vertices of V are usually called $v_1, v_2, \ldots$. For example, the graph

$$G = (\{v_1, v_2, v_3, v_4\}, \{[v_1, v_2], [v_2, v_3], [v_3, v_4], [v_4, v_1], [v_1, v_3]\})$$

is shown in Fig A–1. (Notice that we denote edges using brackets.) It is occasionally useful to consider *multigraphs*, that is, graphs with repeated edges (see Figure A–2).

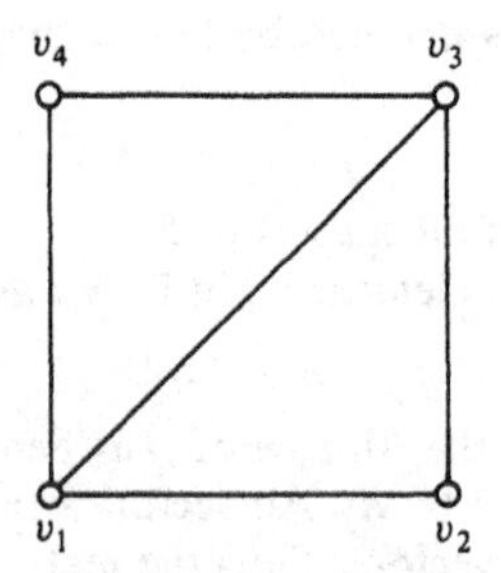

Figure A–1 A graph.

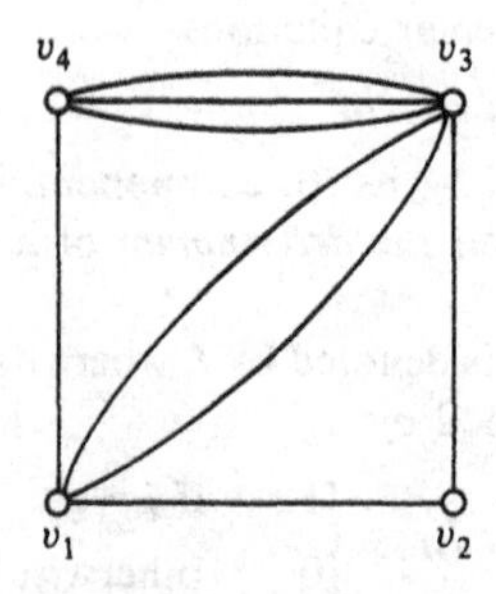

Figure A–2 A multigraph.

A *directed graph*, or *digraph*, is a graph with directions assigned to its edges. Formally, a digraph D is a pair $D = (V, A)$ where V is again a set of vertices and A is a set of *ordered pairs* of vertices called *arcs;* that is, $A \subseteq V \times$

V. In Figure A–3, we have drawn the digraph $D = (\{v_1, v_2, v_3, v_4\}, \{(v_1, v_2), (v_2, v_3), (v_4, v_3), (v_4, v_1), (v_1, v_4), (v_1, v_3)\})$.

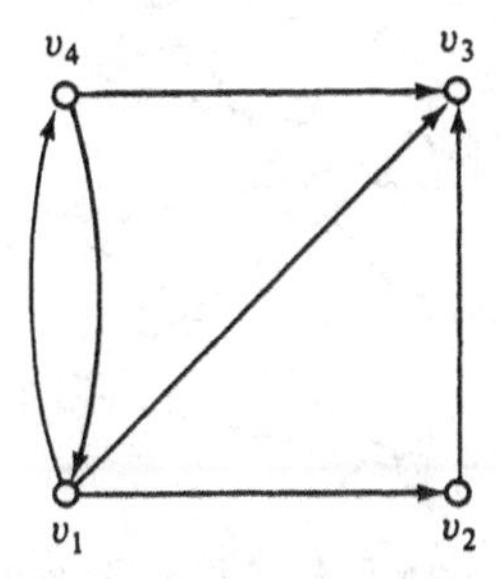

Figure A–3 A digraph.

If $G = (V, E)$ is a graph and $e = [v_1, v_2] \in E$, then we say that v_1 is *adjacent* to v_2 (and vice-versa) and that e is *incident* upon v_1 (and v_2). The *degree* of a vertex v of G is the number of edges incident upon v. So, for the graph of Figure A–1, the degree of v_1 is 3. A *walk* in G is a sequence of nodes $w = [v_1, v_2, v_3, \ldots, v_k]$, $k \geq 1$, such that $[v_j, v_{j+1}] \in E$ for $j = 1, \ldots, k - 1$. The walk is *closed* if $k > 1$ and $v_k = v_1$. A walk without any repeated nodes in it is called a *path*; a closed walk with no repeated nodes other than its first and last one is called a *circuit* or *cycle*. For example, in Figure A–1, $[v_1]$, $[v_1, v_2, v_3, v_1, v_4, v_1]$, $[v_1, v_2, v_3, v_1]$, and $[v_2, v_3, v_4]$ are all walks; the second and third are closed, the first and fourth are paths, and the third is a cycle. The *length* of the *path* $[v_1, \ldots, v_k]$ is $k - 1$; the *length* of the *cycle* $[v_1, \ldots, v_k = v_1]$ is $k - 1$.

In a digraph $D = (V, A)$ the *indegree* of a node v is the number of arcs of the form (u, v) that are in A; similarly, the *outdegree* of v is the number of arcs of A that have the form (v, u). We can readily extend the above definitions to digraphs. A *directed walk* $w = (v_1, v_2, \ldots, v_k)$ of G is a sequence of nodes in V such that $(v_j, v_{j+1}) \in A$ for $j = 1, \ldots, k - 1$. Furthermore, if $k > 1$ and $v_k = v_1$, then w is *closed*. A *directed path* in G is a walk without repetitions. A *directed circuit* or *cycle* is a closed directed path. The *length* of a directed path and cycle are defined by analogy to the undirected case. (Notice that we always use brackets for the edges of a graph and parentheses for the arcs of a digraph.)

Suppose that $B = (W, E)$ is a graph that has the following property. The set of vertices W can be partitioned into two sets, V and U, and each edge in E has one vertex in V and one vertex in U (Figure A–4). Then B is called a *bipartite* graph and is usually denoted by $B = (V, U, E)$. Not all graphs have such a partition. The precise conditions under which they do are the following:

Proposition 1 A graph is bipartite iff it has no circuit of odd length.

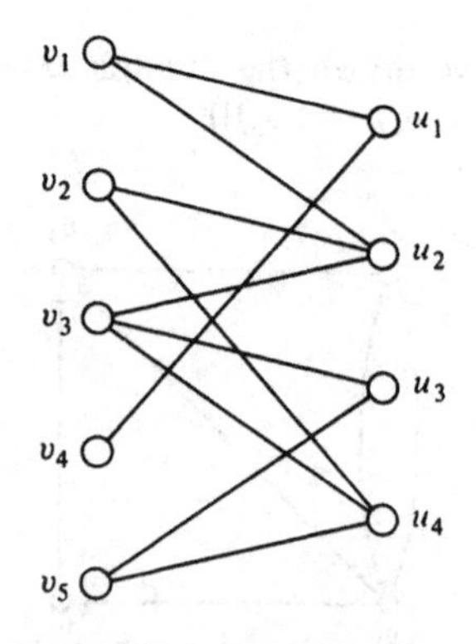

Figure A–4 A bipartite graph.

Another interesting class of graphs is the class of *trees*. A graph is *connected* if there is a path between any two nodes in it. A *tree* $G = (V, T)$ is a connected graph without cycles. For example, Fig. A–5 shows a tree. A *forest*

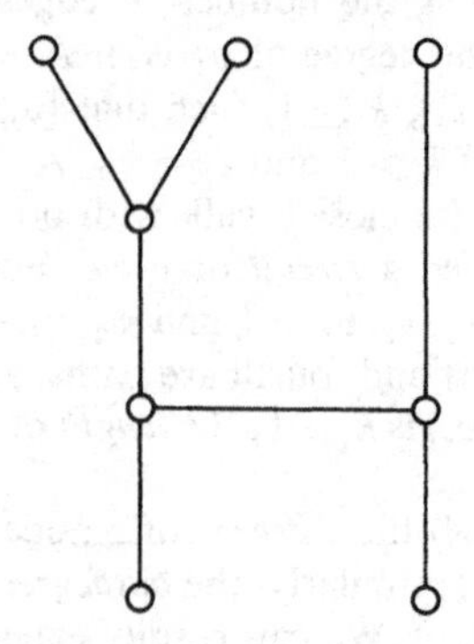

Figure A–5 A tree.

is a set of node-disjoint trees $F = \{(V_1, T_1), \ldots, (V_k, T_k)\}$ (Fig. A–6). We usually say that a tree (V, T) *spans* its set of nodes V or is a *spanning tree* of V. Similarly, a forest $F = \{(V_1, T_1), \ldots, (V_k, T_k)\}$ spans $V_1 \cup V_2 \cup \cdots \cup V_k$.

Proposition 2 Let $G = (V, E)$ be a graph. Then the following are equivalent.

1. G is a tree.
2. G is connected and has $|V| - 1$ edges.
3. G has no cycles, but if an edge is added to G, a unique cycle results.

We shall frequently consider *weighted* graphs, that is, graphs $G = (V, E)$ together with a function w from E to Z (usually just Z^+; it can also be R^+

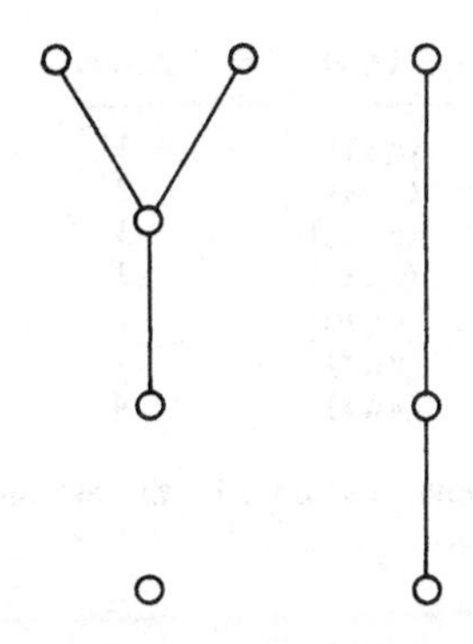

Figure A–6 A forest.

when, for example, the weights are Euclidean distances). In certain cases we shall use more mnemonic names for weights, such as c (for costs) or d (for distances). We denote the weight of the edge $[u, v]$ by $w[u, v]$, or w_{uv}. Notice that a symmetric $n \times n$ *distance matrix* $[d_{ij}]$ (recall Examples 1.1 and 1.2) can also be thought of as a *weighted complete graph* $G = (\{v_1, \ldots, v_n\}, K_n)$, where $K_n = \{[v_i, v_j]: 1 \leq i < j \leq n\}$.

A *network* $N = (s, t, V, A, b)$ is a digraph (V, A) together with a *source* $s \in V$ with 0 indegree, a *terminal* $t \in V$ with 0 outdegree, and with a *bound* (or *capacity*) $b(u, v) \in Z^+$ for each $(u, v) \in A$ (Fig. A–7). A *flow* f in N is a

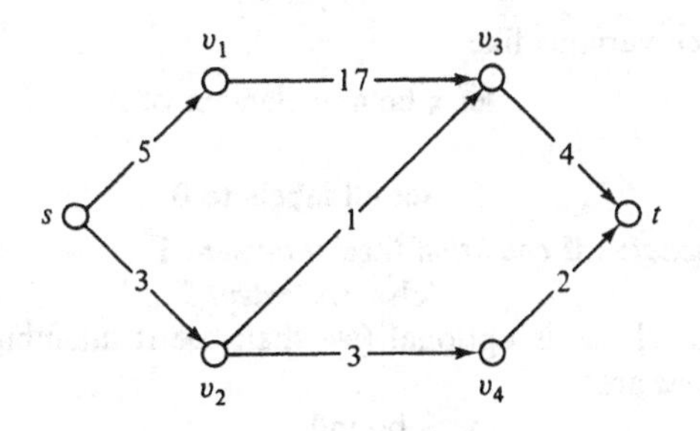

Figure A–7 A network.

vector in $R^{|A|}$ (one component $f(u, v)$ for each arc $(u, v) \in A$) such that:

1. $0 \leq f(u, v) \leq b(u, v)$ for all $(u, v) \in A$.
2. $\sum_{(u,v)\in A} f(u, v) = \sum_{(v,u)\in A} f(v, u)$ for all $v \in V - \{s, t\}$.

The *value* of f, sometimes denoted by $|f|$, is the following quantity: $|f| = \sum_{(s,u)\in A} f(s, u)$. For example, in Figure A–8, we show a legitimate flow for N; its value is $|f| = 5$.

(u, v)	$f(u, v)$
(s, v_1)	3
(s, v_2)	2
(v_1, v_3)	3
(v_2, v_3)	1
(v_2, v_4)	1
(v_4, t)	1
(v_3, t)	4

Figure A–8 A flow of value 5 for the network in Figure A-7.

A.3 Pidgin Algol†

We express most algorithms in this book in terms of our version of *pidgin algol.* For the reader familiar with PASCAL, ALGOL, or PL/I, algorithms in this language should be trivial to understand. For others, skimming this subsection and exercising some care in the beginning should be more than enough. Pidgin algol should be viewed as an informal *notation,* rather than a high-level programming language.

The basic unit of a pidgin algol algorithm is the *statement.* A statement can be of different kinds.

1. *Assignment*: *variable*:= *expression*
   We allow things like

   Q := {s}

   and even variants like

   let x be any element of S

   or

   set all labels to 0

2. *Conditional*: **if** *condition* **then** *statement* 1
   else *statement* 2
   The **else** clause is optional (we shall use it unambiguously). Typical *conditions* are

   x > bound

   v = s **and** done = "yes"

   (*Note*: We write the *reserved words* of pidgin algol in boldface.)

3. *For Statement*: **for** *list* **do** *statement* 1
   Here *list* contains a list of parameters for which *statement* 1 is to be repeated. Examples are

   for j := 1, 2, . . . , n **do** A[j] := A[j + 1]

†The term *pidgin algol* appears to have been introduced in A. V. Aho, J. E. Hopcroft, and J. D. Ullman, *The Design and Analysis of Computer Algorithms* (Reading, Mass.: Addison-Wesley Publishing Co., Inc., 1974).

and

for all v ∈ V such that [v, u] ∈ E **do** rank [v]: = 0

4. *While Statement*: **while** *condition* **do** *statement* 1
   Statement 1 is executed repeatedly, as long as *condition* holds.
5. *Go-to Statement*: **go to** *label*
   For example, **go to** loop means that the (unique) statement prefixed by the label "loop:" should be executed next.
6. *Compound Statement*:

```
begin
   statement 1;
   statement 2;
      .
      .
      .
   statement k − 1;
   statement k
end
```

Traditionally, this is considered to be the tricky part for newcomers to the ALGOL family, but it is really easy. Remember that in a conditional statement, **then** must be followed by a single statement. But what happens if we want to do something complicated (that is, several statements) if *condition* holds? We put these statements together, we separate them with semicolons, and surround them with a **begin-end** pair. To make this more readable, we indent appropriately. Keep in mind that *statement* 1 through *statement k* could be of any of the sorts 1 through 6. If they are all assignments or **go-to**'s, we sometimes write a compound statement without **begin-end** and with the statements separated by commas. For example,

j := j + 1, A[j] := i, **go to** loop

stands for

```
begin
   j := j + 1;
   A[j] := i;
   go to loop
end
```

7. *Comments*: These are of the form (**comment**: this is a comment).
8. *Miscellaneous statements*: We shall allow practically anything, as long as it is readable and reasonably unambiguous. Some extreme examples are

   Construct the auxiliary network AN(f).
   Augment the current matching using the path p.

2

The Simplex Algorithm

2.1 Forms of the Linear Programming Problem

The linear programming problem defined in the last chapter was not the most general one; we could have had some inequality as well as equality constraints and we could have had some variables unconstrained in sign, as well as variables restricted to be nonnegative. We now define the most general form of a linear program as follows.

Definition 2.1

Given an $m \times n$ integer matrix A with rows a_i', let M be the set of row indices corresponding to equality constraints, and let $\bar{M}$ be those corresponding to inequality constraints. Similarly, let $x \in R^n$ and let N be the column indices corresponding to constrained variables and $\bar{N}$ those corresponding to unconstrained variables. Then an instance of the *general LP* is defined by

$$\begin{array}{lll}\min c'x & & \\ a_i'x = b_i & i \in M & \\ a_i'x \geq b_i & i \in \bar{M} & (2.1) \\ x_j \geq 0 & j \in N & \\ x_j \gtreqless 0 & j \in \bar{N} & \end{array}$$

where b is an m-vector of integers and c an n-vector of integers. □

Example 2.1 (Diet Problem)

One of the first problems ever formulated as an LP is the *diet problem* [Sti]. We consider the problem faced by a homemaker when buying food. He has a choice of n foods, and each food has some of each of m nutrients. Suppose

a_{ij} = amount of ith nutrient in a unit of the jth food, $i = 1, \ldots, m$, $j = 1, \ldots, n$

r_i = yearly requirement of ith nutrient, $i = 1, \ldots, m$

x_j = yearly consumption of the jth food, $j = 1, \ldots, n$, in units.

c_j = cost per unit of the jth food, $j = 1, \ldots, n$.

A yearly diet is represented by a choice of a vector $x \geq 0$. That such a diet satisfies the minimal nutritional requirements is expressed by

$$Ax \geq r$$

If we want to find the least expensive diet that is nutritionally adequate, we then need to consider the LP

$$\begin{array}{ll}\min c'x & \\ Ax \geq r & (2.2) \\ x \geq 0 \quad \square & \end{array}$$

The form of LP obtained in the diet problem and the form given in Chapter 1 are common enough to be given special names. We use the following terminology.

Definition 2.2

An LP in the form of (2.2) is said to be in *canonical form*. An LP in the form of (2.3)

$$\begin{array}{ll}\min c'x & \\ Ax = b & (2.3) \\ x \geq 0 & \end{array}$$

is said to be in *standard form*. Finally, an LP in the form of (2.1) is said to be in *general form*. □

We next prove that *the canonical, standard, and general forms are all equivalent*. By this we mean that an instance in one form can be converted to one in another form by a simple transformation, in such a way that the two instances have the same solution. The canonical and standard forms are both already in general form, so we need show only that a general-form problem can be put in canonical and standard forms.

1. To put a general-form problem in canonical form, we need to eliminate any equality constraints and unconstrained variables. Given an equality constraint in the general-form program

$$\sum_{j=1}^{n} a_{ij}x_j = b_i$$

we can replace this with two inequality constraints

$$\sum_{j=1}^{n} a_{ij}x_j \geq b_i$$

and

$$\sum_{j=1}^{n} (-a_{ij})x_j \geq (-b_i)$$

Given an unconstrained variable x_j in the general-form program

$$x_j \gtreqless 0$$

we create two variables x_j^+ and x_j^- in the canonical-form program and write

$$x_j = x_j^+ - x_j^- \quad \text{where} \quad x_j^+ \geq 0, \quad x_j^- \geq 0$$

2. To put a general-form problem in standard form, we need to eliminate inequality constraints; unconstrained variables can be eliminated as above. Given an inequality constraint in the general-form program

$$\sum_{j=1}^{n} a_{ij}x_j \geq b_i$$

introduce the variable s_i in the canonical problem and write

$$\sum_{j=1}^{n} a_{ij}x_j - s_i = b_i, \qquad s_i \geq 0$$

The variable s_i introduced in this transformation is called a *surplus* variable; it represents the amount by which the left-hand side of the inequality exceeds the right-hand side. If, when formulating an LP, we get an inequality of the form

$$\sum_{j=1}^{n} a_{ij}x_j \leq b_i$$

we can introduce a variable s_i and write

$$\sum_{j=1}^{n} a_{ij}x_j + s_i = b_i, \qquad s_i \geq 0$$

Such a variable is called a *slack* variable.

2.2 Basic Feasible Solutions

Our goal now is to develop the simplex algorithm for solving LP's, and it is convenient to assume we are given an LP in *standard form*

$$\begin{aligned} &\min c'x \\ &Ax = b \qquad (A \text{ is an } m \times n \text{ matrix of integers, and } m < n) \\ &x \geq 0 \end{aligned}$$

which we do without loss of generality by the results in the previous section.

We argued intuitively in Example 1.3 that there should always be an optimal "corner" of the convex feasible set F of an LP. There are two ways to define such "corners" precisely—one geometric, and one algebraic. For our algebraic definition we need the following assumption, which we shall see later is hardly restrictive:

Assumption 2.1 There are m linearly independent columns A_j of A. That is, A is of rank m.

Definition 2.3

A *basis* of A is a linearly independent collection $\mathcal{B} = \{A_{j_1}, \ldots, A_{j_m}\}$. We can alternatively think of $\mathcal{B}$ as an $m \times m$ nonsingular matrix $B = [A_{j_i}]$. The *basic solution* corresponding to $\mathcal{B}$ is a vector $x \in R^n$ with

$$\begin{aligned} x_j &= 0 \text{ for } A_j \notin \mathcal{B} \\ x_{j_k} &= \text{the } k\text{th component of } B^{-1}b, \qquad k = 1, \ldots, m. \quad \square \end{aligned}$$

Thus a basic solution x can be found by the following procedure:

1. Choose a set $\mathcal{B}$ of linearly independent columns of A.
2. Set all components of x corresponding to columns not in $\mathcal{B}$ equal to zero.
3. Solve the m resulting equations to determine the remaining components of x. These are the *basic variables*.

Example 2.2

Let us consider the LP

$$\begin{array}{lllllllll} \min & 2x_2 + x_4 + 5x_7 \\ x_1 + & x_2 + x_3 + x_4 & & & & = 4 \\ x_1 & & + x_5 & & & = 2 \\ & x_3 & & + x_6 & & = 3 \\ & 3x_2 + x_3 & & & + \; x_7 & = 6 \\ x_1, & x_2, \quad x_3, & x_4, \quad x_5, & x_6, & x_7 & \geq 0 \end{array}$$

One basis is certainly $\mathcal{B} = \{A_4, A_5, A_6, A_7\}$, which corresponds to the matrix $B = I$. The corresponding basic solution is $x = (0, 0, 0, 4, 2, 3, 6)$. Another basis is $\mathcal{B}' = \{A_2, A_5, A_6, A_7\}$, with basic solution $x' = (0, 4, 0, 0, 2, 3, -6)$. Notice that x' is *not* a feasible solution, since $x'_7 < 0$. □

One can bound from above the absolute value of the components of any basic solution by using our assumption that the entries of A, b, and c are integers.

Lemma 2.1 *Let $x = (x_1, \ldots, x_n)$ be a basic solution. Then*

$$|x_j| \leq m!\,\alpha^{m-1}\beta$$

where

$$\alpha = \max_{i,j} \{|a_{ij}|\}$$

and

$$\beta = \max_{j=1,\ldots,m} \{|b_j|\}$$

Proof This is true for x_j not a basic variable, because then $x_j = 0$. For basic variables, recall that x_j is the sum of m products of elements of B^{-1} by elements of b. Now, each element of B^{-1} is, by definition of the inverse, equal to an $(m - 1) \times (m - 1)$ determinant divided by a nonzero $m \times m$ determinant. By integrality, the denominator is of absolute value at least 1. The determinant of the numerator is the sum of $(m - 1)!$ products of $m - 1$ elements of A; therefore it has absolute value no greater than $(m - 1)!\,\alpha^{m-1}$. Because each x_j is the sum of m elements of B^{-1} multiplied by an element of b, we have

$$|x_j| \leq m!\,\alpha^{m-1}\beta. \qquad \square$$

This bound will be used many times in future arguments.

Definition 2.4

If a basic solution x is in F, then x is a *basic feasible solution* (bfs). □

For example, in the LP of Example 2.2, $x = (0, 0, 0, 4, 2, 3, 6)$ is a bfs. Basic feasible solutions play a very central role in both the theory and the computing practice of LP. One aspect of their importance is expressed in the following lemma, stating that *all* basic feasible solutions are potential uniquely optimal solutions of the corresponding LP.

Lemma 2.2 *Let x be a bfs of*

$$Ax = b$$
$$x \geq 0$$

corresponding to the basis $\mathcal{B}$. Then there exists a cost vector c such that x is the unique optimal solution of the LP

$$\min c'x$$
$$Ax = b$$
$$x \geq 0$$

Proof Consider the cost vector c defined by

$$c_j = \begin{cases} 0 & \text{if } A_j \in \mathcal{B} \\ 1 & \text{if } A_j \notin \mathcal{B} \end{cases}$$

The cost of the bfs x is $c'x = 0$; obviously, x is optimal because all c_j's are nonnegative. Furthermore, if any other feasible solution y is also going to have zero cost, it must be the case that $y_j = 0$ for all $A_j \notin \mathcal{B}$. Hence y must be equal to x, and x is uniquely optimal. □

It is not at all certain, however, that all LP's have bfs's. For example, if $F = \varnothing$, naturally there can be no bfs. It is convenient, however, to exclude this pathological case at this point. We shall come back, in time, to see how one can remove this assumption.

Assumption 2.2 The set F of feasible points is not empty.

We can now show that bfs's do exist.

Theorem 2.1 *Under Assumptions 2.1 and 2.2, at least one bfs exists.*

Proof Assume that F contains a solution x with $t > m$ nonzero components, and in fact that x is the solution in F with the largest number of zero components. Without loss of generality, we have

$$x_1, \ldots, x_t > 0; \qquad x_{t+1}, \ldots, x_n = 0$$

Consider the first t columns of A. They obviously satisfy

$$A_1 x_1 + \cdots + A_t x_t = b \tag{2.4}$$

Let r be the rank of the matrix of these t columns; $r > 0$, because if $r = 0$ the bfs $x = 0$ is in F. Also, $r \leq m < t$. We may thus assume that the matrix

$$\begin{bmatrix} a_{11} & a_{12} & \cdots & a_{1r} \\ a_{21} & a_{22} & \cdots & a_{2r} \\ \cdot & \cdot & & \cdot \\ \cdot & \cdot & & \cdot \\ \cdot & \cdot & & \cdot \\ a_{r1} & a_{r2} & \cdots & a_{rr} \end{bmatrix}$$

is nonsingular. Therefore we can solve Eq. 2.4 to express $x_1, \ldots, x_r$ in terms of $x_{r+1}, \ldots, x_t$. In other words

$$x_j = \beta_j + \sum_{i=r+1}^{t} \alpha_{ij} x_i, \qquad j = 1, \ldots, r$$

Now, let θ be the quantity

$$\theta = \min \{x_{r+1}, \theta_1\}$$

where

$$\theta_1 = \min_{\alpha_{r+1,i} > 0} \left\{ \frac{x_i}{\alpha_{r+1,i}}, i = 1, \ldots, r \right\}$$

Construct a new feasible solution $\hat{x}$ by

$$\hat{x}_j = \begin{cases} x_j - \theta & \text{if } j = r + 1 \\ x_j & \text{if } j > r + 1 \\ \beta_j + \sum_{i=r+1}^{t} \alpha_{ij} \hat{x}_i & \text{if } j < r + 1 \end{cases}$$

Then, for $j \leq r$, $\hat{x}_j = x_j - \alpha_{r+1,j}\theta$. If $\theta = x_{r+1}$, then $\hat{x}_{r+1} = 0$; if $\theta = \theta_1 = x_k/\alpha_{r+1,k}$ for some $k \leq r$, then $\hat{x}_k = 0$. In either case, $\hat{x}$ is a feasible solution with one more zero component than x, which is a contradiction.

This argument shows that there is a solution x with $t \leq m$ nonzero components, and furthermore that the corresponding columns can be assumed to be linearly independent. This set of columns can then be augmented to a basis for x because A is of rank m. □

One final convenient assumption, which we shall also show to be unnecessary later: We shall assume that the LP has a finite minimum value of the objective function $c'x$.

Assumption 2.3 The set of real numbers $\{c'x : x \in F\}$ is bounded from below.

Even though the cost $c'x$ of an LP is bounded from below, the feasible set may still extend infinitely far in some directions. We conclude this section by showing that, under Assumption 2.3, we can nevertheless restrict our attention to LP's with *bounded* feasible sets F. More precisely, F can be assumed to lie within a suitably large hypercube.

Theorem 2.2 *Let Assumptions 2.1 through 2.3 hold for the LP*

$$\begin{aligned} &\min c'x \\ &Ax = b \\ &x \geq 0 \end{aligned} \tag{LP}$$

Then LP is equivalent, in the sense that it has the same optimal value of its cost function:*

$$\begin{aligned} &\min c'x \\ &Ax = b \\ &x \geq 0 \\ &x \leq M \end{aligned} \tag{LP*}$$

where

$$\begin{aligned} M &= (m+1)!\, \alpha^m \beta \\ \alpha &= \max \{|a_{ij}|, |c_j|\} \\ \beta &= \max \{|b_i|, |z|\} \end{aligned}$$

and z is the greatest lower bound of the set $\{c'x\colon Ax = b, x \geq 0\}$.

Proof Consider the set of real numbers

$$G = \{c'x\colon Ax = b, x \geq 0\}$$

It is not hard to show that G is closed (see Problem 2.10). Therefore there is a point x that is feasible in LP and achieves the greatest lower bound z. Next consider the set of points that satisfy the constraints

$$\begin{aligned} c'x &= z \\ Ax &= b \\ x &\geq 0 \end{aligned} \tag{2.5}$$

This set is then nonempty, and in fact consists of all the *optimal* feasible solutions to LP.

Assume first that the equations in (2.5) are of rank $m + 1$. Then Theorem 2.1 implies that (2.5) has a bfs, and Lemma 2.1 shows that its components satisfy the desired bound. Hence the constraints $x \leq M$ do not change the optimal solution of LP.

We have left to consider the case in which the equations in (2.5) are of rank m. In that case c' can be written as a linear combination $\sum d_i a_i'$ of the rows of A, and the cost $c'x = \sum d_i b_i$ is a constant for all feasible points of LP. Therefore LP has an optimal bfs, and its components by Lemma 2.1 are bounded by a number no larger than M. ☐

From now on we shall use this result to assume that F is *always bounded.*

2.3 The Geometry of Linear Programs

We shall now give some important definitions and results pertaining to an alternative *geometric* way of viewing LP.

2.3.1 Linear and Affine Spaces

Consider the vector space R^d. A (*linear*) *subspace* S of R^d is a subset of R^d closed under vector addition and scalar multiplication. Equivalently, a subspace S of R^d is the set of points in R^d that satisfy a set of homogenous linear equations:

$$S = \{x \in R^d: a_{j1}x_1 + \cdots + a_{jd}x_d = 0, \quad j = 1, \ldots, m\} \tag{2.6}$$

It is well known that every subspace S has a *dimension*, dim(S), equal to the maximum number of linearly independent vectors in it. Equivalently, dim(S) = d − rank ($[a_{ij}]$), where $[a_{ij}]$ is the matrix of the coefficients in (2.6) above.

An *affine subspace* A of R^d is a linear subspace S translated by a vector u: $A = \{u + x: x \in S\}$. The *dimension* of A is that of S. Equivalently, an affine subspace A of R^d is the set of all points satisfying a set of (inhomogeneous) equations

$$A = \{x \in R^d: a_{j1}x_1 + \cdots + a_{jd}x_d = b_j; \quad j = 1, \ldots, m\}$$

The dimension of *any subset* of R^d is the smallest dimension of any affine subspace which contains it. For example, any line segment has dimension 1; any set of k points, $k \leq d + 1$, has dimension at most $k - 1$. The dimension of the set F defined by the LP (satisfying Assumptions 2.1 and 2.2)

$$\begin{aligned} &\min c'x \\ &Ax = b \qquad A \text{ an } m \times d \text{ matrix} \\ &x \geq 0 \end{aligned}$$

is therefore at most $d - m$.

2.3.2 Convex Polytopes

An affine subspace of R^d of dimension $d - 1$ is called a *hyperplane*. Alternatively, a hyperplane is a set of points x satisfying

$$a_1x_1 + a_2x_2 + \cdots + a_dx_d = b$$

with not all a's equal to zero. A hyperplane defines two (*closed*) *halfspaces*, namely the sets of points satisfying, respectively,

$$a_1x_1 + \cdots + a_dx_d \geq b$$
$$a_1x_1 + \cdots + a_dx_d \leq b$$

A halfspace is a convex set. Therefore the intersection of halfspaces is also convex. The intersection of a finite number of halfspaces, when it is bounded and nonempty, is called a *convex polytope*, or simply a *polytope*.

We shall henceforth be interested only in convex polytopes that are included in the nonnegative orthant; in other words, by *convention*, d of the halfspaces defining a polytope will always be $x_j \geq 0, j = 1, \ldots, d$.

Example 2.3

The 3-dimensional polytope P of Figure 2–1 is the intersection of the halfspaces indicated by the inequalities in (2.7). As required, P is bounded, because it can easily be shown to be totally contained in the cube $0 \leq x_1, x_2, x_3 \leq 3$.

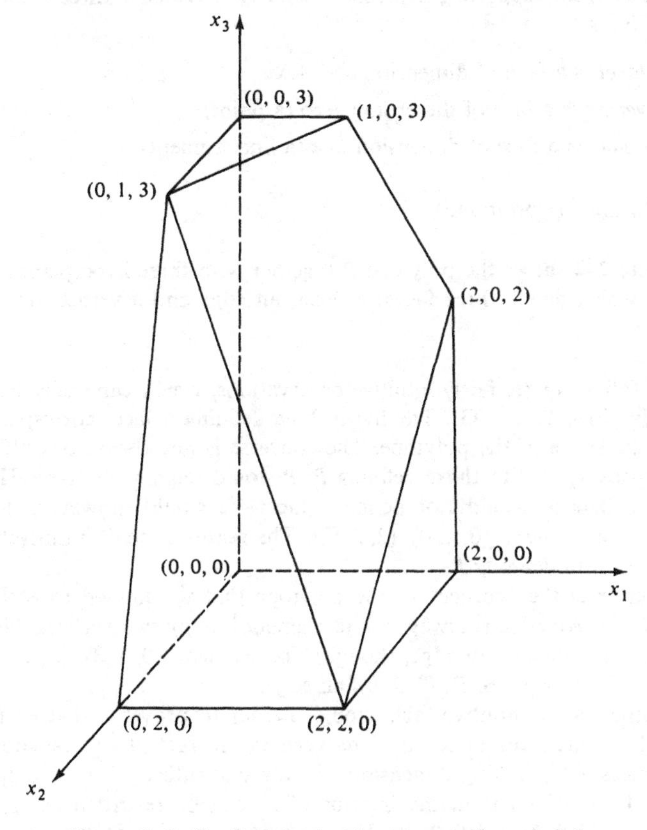

Figure 2–1 The 3-dimensional polytope in Example 2.3.

$$\begin{aligned} x_1 + x_2 + x_3 &\leq 4 \\ x_1 \qquad\qquad &\leq 2 \\ x_3 &\leq 3 \\ 3x_2 + x_3 &\leq 6 \\ x_1 \qquad\qquad &\geq 0 \\ x_2 \qquad &\geq 0 \\ x_3 &\geq 0 \quad \square \end{aligned} \tag{2.7}$$

Let P be a convex polytope of dimension d and let HS be a halfspace defined by hyperplane H. If the intersection $f = P \cap HS$ is a subset of H—in other words, P and HS just "touch in their exteriors"—then f is called a *face* of P and H is the *supporting hyperplane defining* f. We have three distinguished kinds of faces.

A *facet* is a face of dimension $d - 1$.

A *vertex* is a face of dimension zero (a point).

An *edge* is a face of dimension one (a line segment).

Example 2.3 (Continued)

Figure 2–2 shows the polytope P together with three hyperplanes H_1, H_2, and H_3, which define three faces: a facet, an edge and a vertex, respectively. □

The following are fairly intuitive observations, which can easily be proved rigorously [Gru, Roc, YG]. The hyperplane defining a facet corresponds to a defining halfspace of the polytope. The converse is not always true: If we add the halfspace $x_2 \leq 2$ to those defining P, P would remain the same. However, the new halfspace would not define a facet—it would, however, define an edge, the line segment [(0, 2, 0), (2, 2, 0)]. The reason is that, intuitively, $x_2 \leq 2$ is *redundant* in defining P.

A vertex is the "corner" of the polytope that we alluded to earlier with less precision. An edge is always a line segment joining two vertices. Not every pair of vertices defines an edge, though: The segment [(0, 0, 3), (2, 2, 0)] is *not* an edge; neither is [(1, 0, 3), (2, 2, 0)] an edge.

Another fairly intuitive fact, though harder to prove, is that every point in P is the convex combination of its vertices—in fact, it can be shown that four vertices ($d + 1$ for dimension d) always suffice. For example, the point (1, 1, 1), which is in the interior of P, can be rewritten as $(1, 1, 1) = \frac{1}{2}(2, 2, 0) + \frac{1}{3}(0, 0, 3) + \frac{1}{6}(0, 0, 0)$. We now state a general theorem to this effect.

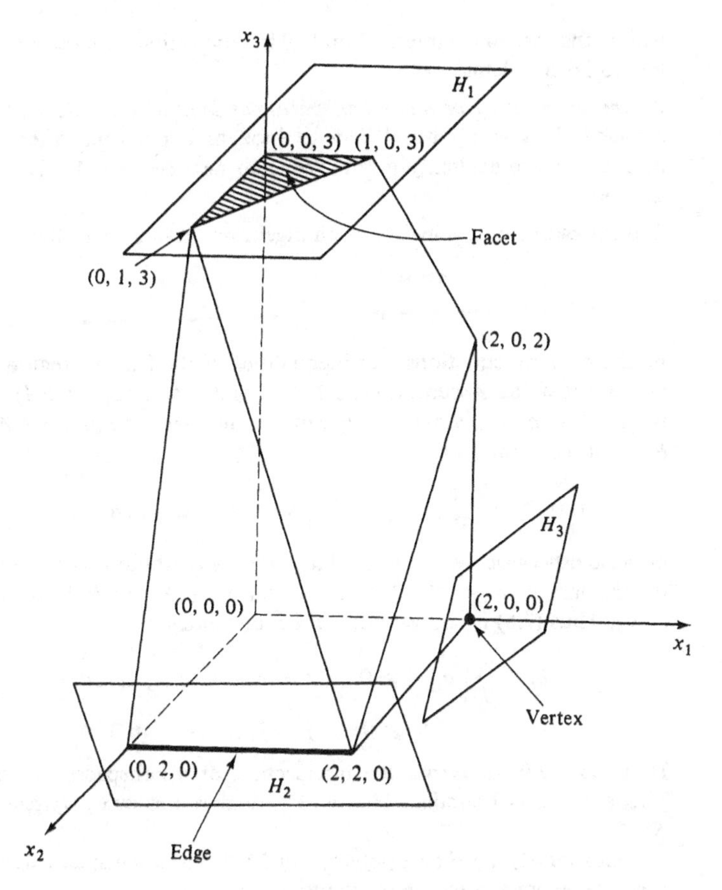

Figure 2–2

Theorem 2.3 **[Gru, Roc, YG]**

(*a*) *Every convex polytope is the convex hull of its vertices.*

(*b*) *Conversely, if V is a finite set of points, then the convex hull of V is a convex polytope P. The set of vertices of P is a subset of V.*

2.3.3 Polytopes and LP

By Theorem 2.3, a polytope P can be thought of in several different ways.

1. *As the convex hull of a finite set of points.* This point of view is fairly convenient when we are given only the vertices of the polytope. This

will be the case in Chapters 13 and 19 in connection with certain combinatorial problems.

2. *As the intersection of many halfspaces, as long as this intersection is bounded.* This is a natural way to look at a polytope when these inequalities are explicitly given. We shall next see that LP is such a situation.

3. A third aspect of a polytope is an algebraic version of 2. above. Let

$$\begin{aligned} Ax &= b \\ x &\geq 0 \end{aligned} \tag{2.8}$$

be the defining equations and inequalities of the feasible region F of an LP satisfying Assumptions 2.1, 2.2, and 2.3. Since rank$(A) = m$, where A is an $m \times n$ matrix, we can assume that the equations $Ax = b$ are of the form

$$x_i = b_i - \sum_{j=1}^{n-m} a_{ij}x_j, \qquad i = n - m + 1, \ldots, n \tag{2.8'}$$

because otherwise we can find a basis B of A (without loss of generality the last m rows of A) and premultiply (2.8) by B^{-1} to obtain (2.8′). Thus (2.8) is equivalent to the inequalities

$$\begin{aligned} b_i - \sum_{j=1}^{n-m} a_{ij}x_j \geq 0 \qquad & i = n - m + 1, \cdots, n \\ x_j \geq 0 \qquad & j = 1, \ldots, n - m \end{aligned} \tag{2.9}$$

However, (2.9) describes the intersection of n halfspaces, which by Theorem 2.2 is bounded. Hence (2.9) defines a *convex polytope* $P \subseteq R^{n-m}$.

Conversely, let P be a polytope in R^{n-m}. The n halfspaces defining P can be expressed by the inequalities

$$h_{i1}x_1 + \cdots + h_{i,n-m}x_{n-m} + g_i \leq 0 \qquad i = 1, \ldots, n \tag{2.10}$$

By our *convention*, we may assume that the first $n - m$ inequalities in (2.10) are of the form

$$x_i \geq 0 \qquad i = 1, \ldots, n - m$$

Let H be the matrix of the coefficients of the remaining inequalities. We can introduce m slack variables $x_{n-m+1}, \ldots, x_n$ to obtain

$$\begin{aligned} Ax &= b \\ x &\geq 0 \end{aligned}$$

where the $m \times n$ matrix $A = [H|I]$ and $x \in R^n$. Thus any polytope (satisfying our convention) can be alternatively viewed as the feasible region F of an LP via a simple transformation. Furthermore, any point $\hat{x} = (x_1, \ldots, x_{n-m}) \in P$ can be transformed to $x = (x_1, \ldots, x_n) \in F$ by defining

$$x_i = -g_i - \sum_{j=1}^{n-m} h_{ij} x_j \qquad i = n - m + 1, \ldots, n \tag{2.11}$$

Conversely, any $x = (x_1, \ldots, x_n) \in F$ can be easily transformed to $\hat{x} = (x_1, \ldots, x_{n-m}) \in P$ by simply truncating the last m coordinates of x.

We can now show how these three points of view affect our notion of a "corner."

Theorem 2.4 *Let P be a convex polytope, $F = \{x: Ax = b, x \geq 0\}$ the corresponding feasible set of an LP, and $\hat{x} = (x_1, \ldots, x_{n-m}) \in P$. Then the following are equivalent.*

(*a*) *The point $\hat{x}$ is a vertex of P.*

(*b*) *If $\hat{x} = \lambda\hat{x}' + (1 - \lambda)\hat{x}''$, with $\hat{x}', \hat{x}'' \in P$, $0 < \lambda < 1$, then $\hat{x}' = \hat{x}'' = \hat{x}$ (in other words, $\hat{x}$ cannot be the* strict *convex combination of points of P).*

(*c*) *The corresponding vector x in F defined by (2.11) is a bfs of F.*

Proof (a) $\Rightarrow$ (b) Suppose that $\hat{x}$ is a vertex and yet there are points $\hat{x}', \hat{x}'' \in P$ different from $\hat{x}$ such that, for $0 < \lambda < 1$, $\hat{x} = \lambda\hat{x}' + (1 - \lambda)\hat{x}''$. Since $\hat{x}$ is a vertex, there is a halfspace $HS = \{\hat{x} \in R^{n-m}: h'\hat{x} \leq g\}$ such that $HS \cap P = \{\hat{x}\}$. Thus $\hat{x}', \hat{x}'' \notin HS$, and hence $h'\hat{x}' > g$ and $h'\hat{x}'' > g$. It follows that $h'\hat{x} = h'(\lambda\hat{x}' + (1 - \lambda)\hat{x}'') > g$ and $\hat{x} \notin HS$, a contradiction.

(b) $\Rightarrow$ (c) Suppose that $\hat{x}$ has Property (b), and consider the corresponding element x of F. Consider the subset $\mathcal{B}$ of the columns of A defined by $\mathcal{B} = \{A_j: x_j > 0, 1 \leq j \leq n\}$. We wish first to show that this is a linearly independent set of columns. Suppose it is not. Then there are integers d_j, not all 0, such that

$$\sum_{A_j \in \mathcal{B}} d_j A_j = 0 \tag{2.12}$$

Since $x \in F$ we have

$$\sum_{A_j \in \mathcal{B}} x_j A_j = b, \tag{2.13}$$

and also

$$x_j \geq 0 \qquad j = 1, \ldots, n$$

Now multiply (2.12) by some number θ and add and subtract from (2.13).

$$\sum_{A_j \in \mathcal{B}} (x_j \pm \theta\, d_j) A_j = b$$

Since $x_j > 0$ for $A_j \in \mathcal{B}$, we can choose a positive and sufficiently small θ such that

$$x_j \pm \theta\, d_j \geq 0 \text{ for all } A_j \in \mathcal{B}$$

Thus we have found two points defined by

$$x'_j = \begin{cases} x_j - \theta\, d_j & A_j \in \mathcal{B} \\ 0 & A_j \notin \mathcal{B} \end{cases}$$

and

$$x''_j = \begin{cases} x_j + \theta\, d_j & A_j \in \mathcal{B} \\ 0 & A_j \notin \mathcal{B} \end{cases}$$

such that $x', x'' \in F$ or $\hat{x}', \hat{x}'' \in P$, and yet $\hat{x} = \frac{1}{2}\hat{x}' + \frac{1}{2}\hat{x}''$, a contradiction.

We have shown that the set of columns $\mathcal{B}$ is linearly independent, and so $|\mathcal{B}| \leq m$. Since we have assumed that there are m linearly independent columns of A, we can always augment the set $\mathcal{B}$ so that it is linearly independent and has m vectors. These then form basic columns, which render x a bfs.

(c) $\Rightarrow$ (a) If $y = (y_1, \ldots y_n)$ is a bfs of $Ax = b, x \geq 0$, then, by Lemma 2.2, there exists a cost vector c such that y is the unique vector $x \in R^n$ satisfying

$$c'x \leq c'y$$
$$Ax = b$$
$$x \geq 0$$

It is easy to see, however, that this means that $\hat{y} = (y_1, \ldots, y_{n-m})$ is the unique point in R^{n-m} satisfying

$$d'\hat{x} \leq d'\hat{y} \qquad \hat{x} \in P$$

where

$$d_i = c_i - \sum_{j=1}^{m} h_{n-m+j,i}c_{n-m+j} \qquad i = 1, \ldots, n-m$$

Hence $\hat{y}$ is indeed a vertex of P, with supporting hyperplane defined by $d'\hat{x} = d'\hat{y}$. □

In Sec. 2.9 we shall derive a very similar characterization of the *edges* of a polytope P.

By Theorem 2.4 we have a correspondence between vertices of P and bases of A. Given two different vertices of P, u and u', the corresponding bases $\mathcal{B}$ and $\mathcal{B}'$ must be different, because a basis uniquely determines a bfs and hence a vertex. However, *two different bases $\mathcal{B}$ and $\mathcal{B}'$ may correspond to the same bfs x.*

Example 2.4

Recall the LP and polytope of Figure 2–1. The matrix A is

$$A = \begin{bmatrix} 1 & 1 & 1 & 1 & 0 & 0 & 0 \\ 1 & 0 & 0 & 0 & 1 & 0 & 0 \\ 0 & 0 & 1 & 0 & 0 & 1 & 0 \\ 0 & 3 & 1 & 0 & 0 & 0 & 1 \end{bmatrix} \quad \text{and} \quad b = \begin{bmatrix} 4 \\ 2 \\ 3 \\ 6 \end{bmatrix}$$

Consider the bases $\mathcal{B} = \{A_1, A_2, A_3, A_6\}$ and $\mathcal{B}' = \{A_1, A_2, A_4, A_6\}$. Both have

$$B^{-1}b = B'^{-1}b = (2, 2, 0, 0, 0, 3, 0)$$

A look at Figure 2–1 explains what has happened. To calculate the vertex corresponding to $\mathcal{B}$, we first set $x_4 = x_5 = x_7 = 0$, which means that the corresponding three inequalities must be satisfied by equality, determining the vertex (2, 2, 0) by the intersection of three facets. Now in $\mathcal{B}'$ we replace the constraint $x_1 + x_2 + x_3 \leq 4$ by $x_3 \geq 0$. But $x_3 = 0$ also happens to pass through the same vertex (2, 2, 0), and so nothing has changed. Thus a vertex like this must lie on more than $n - m = 3$ facets; equivalently, the bfs must have more than $n - m = 3$ zeros. □

Definition 2.5

A bfs (and the corresponding vertex) is called *degenerate* if it contains more than $n - m$ zeros. □

We now give the essential result of the above discussion.

Theorem 2.5 *If two distinct bases correspond to the same bfs x, then x is degenerate.*

Proof Suppose that $\mathcal{B}$ and $\mathcal{B}'$ both determine the same bfs x. Then x has zeros in the $n - m$ columns not in $\mathcal{B}$; it also must have zeros in the columns in $\mathcal{B} - \mathcal{B}' \neq \varnothing$. Hence it is degenerate. □

We can now show the following, which is tantamount to showing that LP can be solved in a finite number of steps.

Theorem 2.6 *There is an optimal bfs in any instance of LP. Furthermore, if q bfs's are optimal, their convex combinations are also optimal.*

Proof By Theorem 2.4 and its proof, this is equivalent to proving that there is an optimal vertex of P and that if q vertices of P are optimal, their convex combinations are also, where the linear cost is $d'x$. The set P is closed and bounded, so the linear function d achieves its minimum in P. Let x_0 be an optimal solution and let $x_1, \ldots, x_N$ be the vertices of P. We know from Theorem 2.3 that x_0 can be written

$$x_0 = \sum_{i=1}^{N} \alpha_i x_i$$

where

$$\sum_{i=1}^{N} \alpha_i = 1, \qquad \alpha_i \geq 0$$

Let j be the index corresponding to the vertex with lowest cost. Then

$$d'x_0 = \sum_{i=1}^{N} \alpha_i d'x_i \geq d'x_j \sum_{i=1}^{N} \alpha_i = d'x_j$$

which shows that x_j is optimal.

For the second part of the result, assume that vertices $x_{j_1}, \ldots, x_{j_q}$ are optimal, and let y be a convex combination of these vertices. Then y is optimal, because

$$d'y = d' \sum_{i=1}^{q} \alpha_i x_{j_i} = \sum_{i=1}^{q} \alpha_i (d'x_{j_i}) = d'x_{j_1} \qquad \square$$

We have thus established that an instance of LP can be solved in a finite number of steps: We need examine the cost only at each vertex of the polytope P. Furthermore, all vertices of P (in fact, all bfs's) can be generated systematically by taking each set of m columns, inverting the corresponding matrix B, and rejecting those that have a negative component of $B^{-1}b$. This is hardly a practical algorithm in a reasonably sized instance, however, since there are just too many possible vertices. With the geometric picture we have of the polytope P and its vertices, we are now in a position to develop the simplex algorithm, in which we move from vertex to vertex in a systematic way, thus avoiding an enumeration of all vertices.

2.4 Moving from bfs to bfs

Let x_0 be a bfs of an instance of LP with matrix A, corresponding to the ordered set of indices of basic columns

$$\mathcal{B} = \{A_{B(i)}: i = 1, \ldots, m\}$$

If the basic components of x_0 are x_{i0}, $i = 1, \ldots, m$, then

$$\sum_{i=1}^{m} x_{i0} A_{B(i)} = b, \quad \text{where} \quad x_{i0} \geq 0 \tag{2.14}$$

where as usual we use $A_j \in R^m$ to represent the jth column of A. The set of basic column vectors $\mathcal{B}$ is linearly independent, by definition, so we can write any nonbasic column $A_j \in R^m$, $A_j \notin \mathcal{B}$ as a linear combination of the basic columns as follows:

$$\sum_{i=1}^{m} x_{ij} A_{B(i)} = A_j \tag{2.15}$$

If we now multiply Eq. 2.15 by a scalar $\theta > 0$ and subtract from Eq. 2.14, we get a most important equation:

$$\sum_{i=1}^{m} (x_{i0} - \theta x_{ij}) A_{B(i)} + \theta A_j = b \tag{2.16}$$

Assume for the moment that x_0 is nondegenerate; then all the $x_{i0} > 0$, and as we increase θ from zero, we move from the bfs to feasible solutions with $m + 1$ strictly positive components. How far can we move θ and still remain feasible? Until the first component $(x_{i0} - \theta x_{ij})$ becomes zero, which occurs at the value

$$\theta_0 = \min_{\substack{i \\ \text{such that} \\ x_{ij} > 0}} \frac{x_{i0}}{x_{ij}} \tag{2.17}$$

Example 2.5

Consider the LP with the constraints of Example 2.2 (or 2.4). The basis $\mathcal{B} = \{A_1, A_3, A_6, A_7\}$ given by $B(1) = 1$, $B(2) = 3$, $B(3) = 6$, $B(4) = 7$ has bfs $x = (2, 0, 2, 0, 0, 1, 4)$. We can write the nonbasic column $A_5 = \text{col}\,(0, 1, 0, 0)$ as

$$\begin{aligned} A_5 &= x_{15}A_1 + x_{25}A_3 + x_{35}A_6 + x_{45}A_7 \\ &= 1 \cdot A_1 - 1 \cdot A_3 + 1 \cdot A_6 + 1 \cdot A_7 \end{aligned}$$

Then Eq. 2.16 becomes

$$(2 - \theta)A_1 + (2 + \theta)A_3 + (1 - \theta)A_6 + (4 - \theta)A_7 + \theta A_5 = b$$

A look at Fig. 2.1 shows that this family of feasible points

$$(2 - \theta, 0, 2 + \theta, 0, \theta, 1 - \theta, 4 - \theta)$$

moves from the vertex (2, 0, 2)—and the bfs (2, 0, 2, 0, 0, 1, 4)—to the vertex (1, 0, 3) and the bfs (1, 0, 3, 0, 1, 0, 3) as θ increases from zero to 1. Equation 2.17 yields $\theta_0 = 1$, and the new basis becomes $\mathcal{B}'$ with $B'(1) = 1$, $B'(2) = 3$, $B'(3) = 5$ and $B'(4) = 7$. □

We now take up two special conditions that might prevail at the bfs x_0.

Special Case 1 If x_0 is degenerate because some $x_{i0} = 0$ and the corresponding x_{ij} is positive, then $\theta_0 = 0$ by (2.17), and we do not move in R^n. We stay at the same vertex, but can think of ourselves as moving to the *new basis* with column j replacing column $B(i)$. We sometimes say in such a case that variable x_j has entered the basis *at zero level.*

Special Case 2 If all the x_{ij}, $i = 1, \ldots, m$ are nonpositive, we can move arbitrarily far without becoming infeasible. In such a case F is unbounded, violating our taking F bounded after Theorem 2.2.

It remains to show that the new point reached by the above process is in fact a bfs.

Theorem 2.7 *Given a bfs x_0 with basic components x_{i0}, $i = 1, \ldots, m$ and basis $\mathcal{B} = \{A_{B(i)}: i = 1, \ldots, m\}$, let j be such that $A_j \notin \mathcal{B}$. Then the new* *feasible solution determined by*

$$\theta_0 = \min_{\substack{i \\ \text{such that} \\ x_{ij} > 0}} \frac{x_{i0}}{x_{ij}} = \frac{x_{l0}}{x_{lj}} \tag{2.18}$$

$$x'_{i0} = \begin{cases} x_{i0} - \theta_0 x_{ij} & i \neq l \\ \theta_0 & i = l \end{cases} \tag{2.19}$$

is a bfs with basis $\mathcal{B}'$ defined by

$$B'(i) = \begin{cases} B(i) & i \neq l \\ j & i = l \end{cases} \tag{2.20}$$

When there is a tie in the min operation of (2.18), the new bfs is degenerate.

Proof We need to show that x'_0 with components given by Eq. 2.19 is basic, since it is a feasible solution by the discussion surrounding Eqs. 2.16 and 2.17. Thus, we must show that the set of basic columns $\mathcal{B}'$ is linearly independent.

Suppose then that for some constants d_i we have

$$\sum_{i=1}^{m} d_i A_{B'(i)} = d_l A_j + \sum_{\substack{i=1 \\ i \neq l}}^{m} d_i A_{B(i)} = 0 \tag{2.21}$$

Substituting

$$A_j = \sum_{i=1}^{m} x_{ij} A_{B(i)} \tag{2.22}$$

this becomes

$$\sum_{\substack{i=1 \\ i \neq l}}^{m} (d_l x_{ij} + d_i) A_{B(i)} + d_l x_{lj} A_{B(l)} = 0 \tag{2.23}$$

This is a linear combination of the original basis vectors, so all the coefficients must be zero; in particular $d_l x_{lj} = 0$, and hence $d_l = 0$. Equation 2.21 then implies that the remaining d_i are zero and hence that the new basis is in fact linearly independent.

We conclude the proof by noting that if a tie occurs in the min operation of Eq. 2.18, the corresponding entries in x'_0 become zero by Eq. 2.19, which means x'_0 is degenerate. □

This method of moving from one bfs to another is called *pivoting;* we say column $B(l)$ *leaves* the basis and column j *enters* the basis.

2.5 Organization of a Tableau

In the last section we assumed that we always had available to us the representation of any nonbasic Column A_j in terms of the basic columns, as in Eq. 2.22. It is crucial that we have the coefficients x_{ij} at our fingertips if we are to pursue an algorithm that does a great deal of moving from vertex to vertex. We do this by keeping our set of equations diagonalized with respect to the basic variables.

Suppose that at any stage we keep an $m \times (n+1)$ array of numbers that represents the information in the original equality constraints $Ax = b$. Thus we represent the equations

$$\begin{aligned} 3x_1 + 2x_2 + x_3 \qquad\quad &= 1 \\ 5x_1 + x_2 + x_3 + x_4 &= 3 \\ 2x_1 + 5x_2 + x_3 + x_5 &= 4 \end{aligned}$$

by

	x_1	x_2	x_3	x_4	x_5
1	3	2	1	0	0
3	5	1	1	1	0
4	2	5	1	0	1

separating the right-hand sides of the equations with a vertical bar and considering them as Column 0. We can multiply a row by a nonzero constant or add a multiple of any row to any other without changing the information in these equations; these are usually called *elementary row operations*. If we have a basis $\mathcal{B}$ available, we can perform elementary row operations until the basic columns form an identity submatrix:

$$A_{B(i)} = e_i = \begin{pmatrix} 0 \\ \cdot \\ \cdot \\ \cdot \\ 1 \\ \cdot \\ \cdot \\ \cdot \\ 0 \end{pmatrix} \longleftarrow i\text{th row} \qquad i = 1, \ldots, m$$

where we conventionally use e_i to represent the m-vector with a 1 in the ith row and zero elsewhere. Thus, in our example, if $\mathcal{B} = \{A_3, A_4, A_5\}$, we can multiply Row 1 by -1 and add it to Rows 2 and 3, yielding

	x_1	x_2	x_3	x_4	x_5
1	③	2	1	0	0
2	2	−1	0	1	0
3	−1	3	0	0	1

(2.24)

Column 0 now gives the values of the basic variables $x_{B(i)} = x_{i0}$, $i = 1, \ldots, m$ obtained by setting the nonbasic variables to zero. Notice also that the nonbasic columns contain precisely the numbers x_{ij}; for example,

$$A_1 = 3A_3 + 2A_4 - A_5$$
$$= \sum_{i=1}^{m} x_{i1} A_{B(i)}$$

The calculations necessary to change the basis can therefore be carried out immediately. Suppose, for example, that we wish to bring Column $j = 1$ into the basis; then by Eq. 2.18

$$\theta_0 = \min_{\substack{i \\ \text{such that} \\ x_{ij} > 0}} \left(\frac{x_{i0}}{x_{ij}}\right) = \frac{1}{3} \text{ for } i = l = 1$$

We now need to introduce a unit vector in Column $j = 1$, with the 1 in Row $l = 1$. We do this by dividing Row 1 by 3, adding to Row 3, and then multiplying by -2 and adding to Row 2. We usually represent this operation by circling the "pivot" element x_{lj} in the tableau, as in (2.24). The new tableau is

	x_1	x_2	x_3	x_4	x_5
$\frac{1}{3}$	1	$\frac{2}{3}$	$\frac{1}{3}$	0	0
$\frac{4}{3}$	0	$-\frac{7}{3}$	$-\frac{2}{3}$	1	0
$\frac{10}{3}$	0	$\frac{11}{3}$	$\frac{1}{3}$	0	1

The new basis is $\mathcal{B}' = \{A_1, A_4, A_5\}$, corresponding to the bfs $x_1' = \frac{1}{3}$, $x_4' = \frac{4}{3}$ and $x_5' = \frac{10}{3}$. In general, if x_{ij} and x_{ij}' are the old and new tableaux, respectively; $\mathcal{B}$ and $\mathcal{B}'$ the old and new basic sets, respectively; and the pivot element is x_{lj}, then

$$
\begin{aligned}
x_{lq}' &= \frac{x_{lq}}{x_{lj}} && q = 0, \ldots, n \\
x_{iq}' &= x_{iq} - x_{lq}' x_{ij} && i = 1, \ldots, m; \quad i \neq l \\
& && q = 0, \ldots, n
\end{aligned}
\tag{2.25}
$$

$$B'(i) = \begin{cases} B(i) & i \neq l \\ j & i = l \end{cases}$$

Now that we know how to move from bfs to bfs, we need to investigate the effect of such moves on the cost.

2.6 Choosing a Profitable Column

The cost of a bfs x_0 with basis $\mathcal{B}$ is

$$z_0 = \sum_{i=1}^{m} x_{i0} c_{B(i)}$$

Now consider the process of bringing Column A_j into the basis: We write

A_j in terms of the basis columns as

$$A_j = \sum_{i=1}^{m} x_{ij} A_{B(i)} \tag{2.26}$$

This can be interpreted as meaning that for every unit of the variable x_j that enters the basis, an amount x_{ij} of each of the variables $x_{B(i)}$ must leave. Thus a unit increase in the variable x_j results in a net change in the cost equal to

$$c_j - \sum_{i=1}^{m} x_{ij} c_{B(i)}$$

The quantity on the right is important enough to be assigned its own symbol, z_j; and we call the difference

$$\bar{c}_j = c_j - z_j$$

the *relative cost* of Column j. It is then profitable to bring Column j into the basis exactly when $\bar{c}_j < 0$. Furthermore, when for all j, $\bar{c}_j \geq 0$, we are at a local optimum, which is also a global optimum. We prove all this in detail in the following.

First we introduce some vector and matrix notation. For any tableau X, let B be the $m \times m$ matrix comprised of the columns of A corresponding to the basis in X, and let c_B be the m-vector of costs corresponding to these basic variables. Then because the tableau X is obtained by diagonalizing the basic columns of A, we can write the tableau X as

$$X = B^{-1}A$$

and the vector $z = \text{col}\,(z_1, \ldots, z_n)$ from its definition as

$$z' = c'_B X = c'_B B^{-1} A$$

We use this matrix terminology again and again in what follows.

Theorem 2.8 (Optimality Criterion) *At a bfs x_0, a pivot step in which x_j enters the basis changes the cost by the amount*

$$\theta_0 \bar{c}_j = \theta_0 (c_j - z_j) \tag{2.27}$$

If

$$\bar{c} = c - z \geq 0 \tag{2.28}$$

then x_0 is optimal.

Proof From Eq. 2.19 in Theorem 2.7, the new solution is

$$x'_{i0} = \begin{cases} x_{i0} - \theta_0 x_{ij} & i \neq l \\ \theta_0 & i = l \end{cases}$$

so the new cost is

$$\begin{aligned} z'_0 &= \sum_{\substack{i=1 \\ i \neq l}}^{m} (x_{i0} - \theta_0 x_{ij}) c_{B(i)} + \theta_0 c_j \\ &= z_0 + \theta_0 (c_j - z_j) \end{aligned}$$

which establishes Eq. 2.27.

To show that $\bar{c} \geq 0$ implies that x_0 is optimal, let y be any feasible vector whatsoever, not necessarily basic. That is,

$$Ay = b$$

and

$$y \geq 0$$

Since $\bar{c} = c - z \geq 0$, the cost of y is

$$c'y \geq z'y = c'_B B^{-1}Ay = c'_B B^{-1}b = c'x_0$$

which shows that the cost of y can never be less than that of x_0. □

Since the values $\bar{c}_j$ tell us when a column can profitably enter the basis, we would like to keep them as part of the tableau. This is usually done in Row 0, as follows. Write the cost equation as

$$0 = -z + c_1 x_1 + \cdots + c_n x_n \tag{2.29}$$

Now the relative cost $\bar{c}_j$ associated with a *basis* column j is

$$\bar{c}_j = c_j - z_j = c_j - \sum_{i=1}^{m} x_{ij} c_{B(i)} = 0$$

since x_{ij} is a unit vector with a 1 where $B(i) = j$. If we consider Eq. 2.29 the zeroth row of our tableau, we can make its components over basis columns zero by multiplying the ith row by $-c_{B(i)}$ and adding the result to the zeroth row. This yields in a nonbasic column the quantity

$$\bar{c}_j = c_j - \sum_{i=1}^{m} x_{ij} c_{B(i)}$$

and on the left-hand side of the zeroth equation

$$-z_0 = -\sum_{i=1}^{m} x_{i0} c_{B(i)}$$

The zeroth row therefore becomes

$$-z_0 = -z + \sum_{\substack{j=1 \\ A_j \notin \mathcal{B}}}^{n} \bar{c}_j x_j \tag{2.30}$$

If we now think of the tableau as a diagonal form in terms of the $m + 1$ variables $x_{B(1)}, x_{B(2)}, \ldots, x_{B(m)}$ and $-z$, we see that the same pivoting rules apply to the zeroth row as to Rows 1 to m. Hence we can carry along the relative costs $\bar{c}_j$ by keeping one more row in the tableau. There is, of course, no need to keep a column for the variable $-z$.

If we ignore for now the problem of degeneracy, then every pivot yields $\theta_0 > 0$, and we have a finite algorithm for LP, the *simplex algorithm*: If any $\bar{c}_j < 0$, pivot on Column j; when finally $\bar{c}_j \geq 0$ for all j, we have reached an optimal bfs. We never return to a previously visited bfs, because the cost decreases monotonically. Since there are a finite number of bfs's, we must terminate in a finite number of steps. We postpone the question of degeneracy to the next section, and conclude this section with an informal program (Fig. 2–3) and an example of the simplex algorithm.

```
procedure simplex
begin
  opt:= 'no', unbounded:= 'no';
  (comment: when either becomes 'yes' the algorithm terminates)
  while opt = 'no' and unbounded = 'no' do
    if c̄_j ≥ 0 for all j then opt:= 'yes'
      else begin
          choose any j such that c̄_j < 0;
          if x_ij ≤ 0 for all i then unbounded:= 'yes'
          else
            find θ_0 = min_{i, x_ij>0} [x_i0 / x_ij] = x_k0 / x_kj
            and pivot on x_kj
          end
end
```

Figure 2–3 The simplex algorithm.

Example 2.6

We consider the LP with the constraints of Sec. 2.5 and the cost function

$$z = x_1 + x_2 + x_3 + x_4 + x_5$$

The tableau of the original problem is therefore

	x_1	x_2	x_3	x_4	x_5
0	1	1	1	1	1
1	3	2	1	0	0
3	5	1	1	1	0
4	2	5	1	0	1

To start, we need a bfs, and we need to make zero the $\bar{c}_j$'s corresponding to the basic columns. We know from Sec. 2.5 that Columns 3, 4, and 5 yield a bfs. Subtracting Row 1 from Rows 2 and 3 and then subtracting the resulting Rows 1, 2, and 3 from Row 0 yields

		x_1	x_2	x_3 (basis)	x_4 (basis)	x_5 (basis)
$-z =$	−6	−3	−3	0	0	0
$x_3 =$	1	3	②	1	0	0
$x_4 =$	2	2	−1	0	1	0
$x_5 =$	3	−1	3	0	0	1

This represents the bfs indicated by the variables on the left, with cost $z = 6$. We have in Row 0, Columns 1 and 2, $\bar{c}_1 = -3$ and $\bar{c}_2 = -3$, respectively; so it is profitable for Column 1 or 2 to enter the basis. Choosing Column 2, we find

$$\theta_0 = \tfrac{1}{2} \quad \text{for} \quad l = 1$$

and we pivot on the element $x_{12} = 2$, which is circled. The resulting tableau is

		x_1	x_2	x_3	x_4	x_5
$-z =$	$-\frac{9}{2}$	$\frac{3}{2}$	0	$\frac{3}{2}$	0	0
$x_2 =$	$\frac{1}{2}$	$\frac{3}{2}$	1	$\frac{1}{2}$	0	0
$x_4 =$	$\frac{5}{2}$	$\frac{7}{2}$	0	$\frac{1}{2}$	1	0
$x_5 =$	$\frac{3}{2}$	$-\frac{11}{2}$	0	$-\frac{3}{2}$	0	1

which is optimal, with cost $z = \frac{9}{2}$. □

2.7 Pivot Selection and Bland's Anticycling Algorithm

There is a certain amount of uncertainty in the simplex algorithm as we described it: We have not said how to choose which column j (with $c_j - z_j < 0$) enters the basis; and we have not said how to resolve ties in the calculation of θ_0, which determines the row l and the variable $x_{B(l)}$ to leave the basis.

We first take up the question of column selection. Unfortunately, there is no theory to guide us here, and we must rely on empirical observations. The oldest and most widely used criterion is simply to choose the $\bar{c}_j < 0$ which is most negative. As we established above, a unit increase in the variable x_j entering the basis results in a change of $\bar{c}_j$ in the cost, so $\bar{c}_j$ can be thought of as the *derivative* of the cost with respect to distance in the *space of nonbasic variables*. Choosing the most negative $\bar{c}_j$ then corresponds to a kind of steepest descent policy, which is called the *nonbasic gradient* method [KQ]. By no means does this ensure, however, that the actual decrease in cost, $\theta_0 \bar{c}_j$, will be as large as possible, since we do not know θ_0 until we compute the ratios for row selection. This suggests another policy: Choose the column that results in the largest decrease in cost. This method, called the *greatest increment* method, carries with it an additional computational burden at each pivot step but offers the possibility of reaching optimality after a fewer number of pivots than the nonbasic gradient method.

A unit increase in the nonbasic variable x_j changes the entire vector x by

$$x_k = \begin{cases} +1 & k = j \\ -x_{ij} & k = B(i), \quad i = 1, \ldots, m \\ 0 & \text{otherwise} \end{cases}$$

We therefore can compute the derivative of the cost with respect to distance in the *space of all variables*,

$$\frac{\bar{c}_j}{\sqrt{1 + \sum_{i=1}^{m} x_{ij}^2}}$$

and the column selection policy corresponding to this derivative is called the *all-variable gradient* method.

Kuhn and Quandt [KQ] report the results of extensive computer experiments with these and other methods. The results, on problems with up to 25 rows, indicate that the all-variable gradient method converges in fewer pivots than the nonbasic gradient or greatest increment methods and is also faster. Goldfarb and Reid [GR] have described a fast way to compute the all-variable derivative and report good results with the all-variable gradient policy. The reader should view these results with some caution, however. First, the computation times reported by Kuhn and Quandt show at best no more than an improvement factor of two, and such improvements in running time can often result from changes in programming details. Second, the random class of LP's used for the tests may not reflect anybody's "typical" LP. Last, the nonbasic gradient method has the important advantage of simplicity and is still the most popular method actually programmed.

We now turn next to the resolution of ties in the row selection procedure. Here the possibility of degeneracy presents a certain danger: If we pivot during the simplex algorithm on element $x_{ij} > 0$ and the component x_{i0} of the bfs is zero, then $\theta_0 = 0$ and the cost increment $\theta_0(c_j - z_j) = 0$. That is, the cost z does not decrease, even though we choose a Column j with $c_j - z_j < 0$. It is further possible that we go through a sequence of such pivots, returning to our starting point. This means that the algorithm will loop indefinitely (assuming that the choices of column and row are made deterministically), and that would be a most undesirable situation. This phenomenon is called *cycling*.

Example 2.7 [Be1]

Consider the tableau

	x_1	x_2	x_3	x_4	x_5	x_6	x_7
3	$-\frac{3}{4}$	$+20$	$-\frac{1}{2}$	$+6$	0	0	0
0	$\textcircled{\frac{1}{4}}$	-8	-1	9	1	0	0
0	$\frac{1}{2}$	-12	$-\frac{1}{2}$	3	0	1	0
1	0	0	1	0	0	0	1

Let us pivot from this bfs with the following tie-breaking rules.

(a) Always select the nonbasic variable with the most negative $\bar{c}_j$ to enter the basis.

(b) In case of a tie, always select the basic variable with the smallest subscript to leave the basis.

We obtain the following sequence of tableaux (pivots are circled)

3	0	−4	$-\frac{7}{2}$	33	3	0	0
0	1	−32	−4	36	4	0	0
0	0	④	$\frac{3}{2}$	−15	−2	1	0
1	0	0	1	0	0	0	1

3	0	0	−2	18	1	1	0
0	1	0	⑧	−84	−12	8	0
0	0	1	$\frac{3}{8}$	$-\frac{15}{4}$	$-\frac{1}{2}$	$\frac{1}{4}$	0
1	0	0	1	0	0	0	1

3	$\frac{1}{4}$	0	0	−3	−2	3	0
0	$\frac{1}{8}$	0	1	$-\frac{21}{2}$	$-\frac{3}{2}$	1	0
0	$-\frac{3}{64}$	1	0	($\frac{3}{16}$)	$\frac{1}{16}$	$-\frac{1}{8}$	0
1	$-\frac{1}{8}$	0	0	$\frac{21}{2}$	$\frac{3}{2}$	−1	1

3	$-\frac{1}{2}$	16	0	0	−1	1	0
0	$-\frac{5}{2}$	56	1	0	②	−6	0
0	$-\frac{1}{4}$	$\frac{16}{3}$	0	1	$\frac{1}{3}$	$-\frac{2}{3}$	0
1	$\frac{5}{2}$	−56	0	0	−2	6	1

3	$-\frac{7}{4}$	44	$\frac{1}{2}$	0	0	−2	0
0	$-\frac{5}{4}$	28	$\frac{1}{2}$	0	1	−3	0
0	$-\frac{1}{6}$	−4	$-\frac{1}{6}$	1	0	($\frac{1}{3}$)	0
1	0	0	1	0	0	0	1

3	$-\frac{3}{4}$	+20	$-\frac{1}{2}$	6	0	0	0
0	$\frac{1}{4}$	−8	−1	9	1	0	0
0	$\frac{1}{2}$	−12	$-\frac{1}{2}$	3	0	1	0
1	0	0	1	0	0	0	1

Then, after six pivots, *we arrive at the same bfs with which we started.* All intermediate pivots introduced new basic variables at zero level, and there was no change in the cost. We say that simplex (with this particular pivot rule) has *cycled.* □

We can view our problem at this point as one of resolving the uncertainties in the simplex algorithm in such a way as to prevent cycling. It is sometimes reported that cycling simply does not occur in practice, even though artificial examples can be constructed. This is contradicted by a recent report [KS]. In addition, some LP formulations of combinatorial problems are highly degenerate, and it is not at all clear that we can trust to luck to avoid cycling, aesthetic considerations aside.

The simplest way to avoid cycling is to resolve ties in a random way—with probability 1 we shall escape from any loop. This complicates the programming of the ratio test, however, and is not as intellectually satisfying as a deterministic rule that guarantees finiteness of the simplex algorithm. It does not seem to be a popular policy.

In one standard approach to cycle avoidance, any choice of column is allowed, and ties in the θ_0 calculation are resolved in such a way as to ensure that the zeroth row increases lexicographically, thus ensuring that no basis is ever repeated. This has the advantage of allowing any column selection policy. We postpone a description of this method until Chapter 14. We shall describe here a relatively recent algorithm that prevents cycling, due to R.G. Bland [Bl], which is remarkable in its simplicity. We first need the following lemma.

Lemma 2.3 *Let $\bar{c}'$ be the relative cost row for any tableau X_1 with a unit basis, not necessarily corresponding to a feasible solution. (That is, some x_{i0} in the zeroth column can be negative.) Let y be any solution to the constraints $Ay = b$, not necessarily corresponding to a feasible solution. (That is, some y_j can be negative.) Let f be the cost associated with X_1 and g with y. Then*

$$\bar{c}'y = g - f$$

Proof A direct calculation yields

$$\bar{c}'y = (c' - z')y = c'y - z'y = g - c'_B B^{-1}Ay = g - c'_B B^{-1}b = g - f$$

since $B^{-1}b$ is the zeroth column of tableau X_1. □

Theorem 2.9 (Bland's anticycling algorithm [Bl]) *Suppose in the simplex algorithm we choose the column to enter the basis by*

$$j = \min\{j: c_j - z_j < 0\}$$

(choose the lowest numbered favorable column), and the row by

$$B(i) = \min\left\{B(i): x_{ij} > 0 \ \text{ and } \ \frac{x_{i0}}{x_{ij}} \le \frac{x_{k0}}{x_{kj}} \ \text{ for every } k \text{ with } x_{kj} > 0\right\}$$

(choose in case of tie the lowest numbered column to leave the basis). Then the algorithm terminates after a finite number of pivots.

***Proof*†** We find a contradiction from the assumption that a cycle exists. For a cycle to occur, there must be a finite sequence of pivots that returns to a bfs. The cost z remains constant during this cycle, and the x_{i0} associated with each pivot must be zero, for otherwise $\theta_0 > 0$, which would imply that z decreases. This in turn implies that the zeroth column x_{i0}, $i = 1, \ldots, m$ remains constant during the cycle.

Throw away those rows and columns not containing pivots during the cycle, yielding a new program which still cycles and has all $x_{i0} = 0$ and constant z during the cycle.

Now let q be the *largest* index of a variable entering the basis during the cycle, and consider two tableaux: T_1, the tableau when x_q is about to enter the basis; and T_2, the tableau when x_q is about to leave the basis (see Fig. 2–4). Denote the entries in T_1 by x_{ij}, with basis $\mathcal{B}$, and in T_2 by $\hat{x}_{ij}$, with basis $\hat{\mathcal{B}}$, and let column p be the column entering in T_2. We now apply Lemma 2.3 by

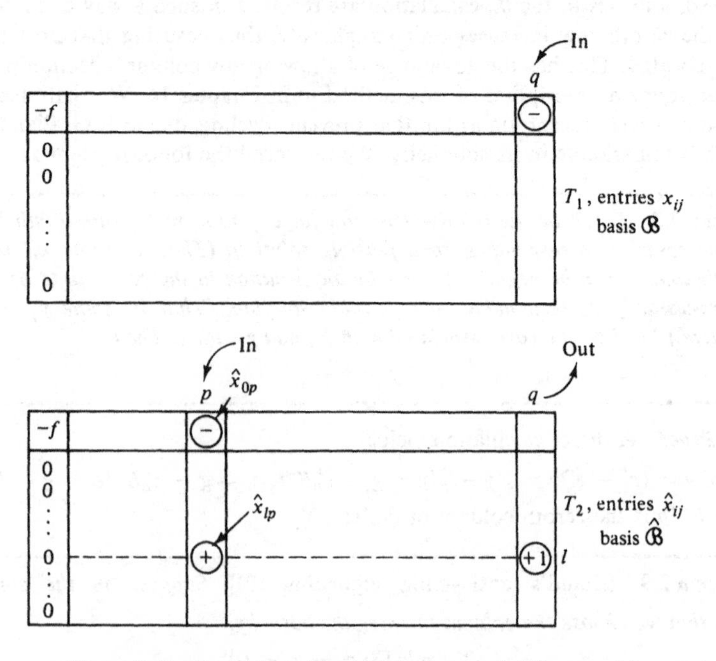

Figure 2–4 The tableaux T_1 and T_2 in the proof of Theorem 2.9. The variable x_q is about to enter in T_1 and leave in T_2.

†This proof is a simplification of Bland's proof due to H. W. Kuhn [Ku].

constructing two solutions: In T_1 we use simply x_0, the bfs, and identify T_1 with X_1. From T_2, we define a solution y by

$$y_j = \begin{cases} 1 & \text{if } j = p \\ -\hat{x}_{ip} & \text{if } A_j \in \hat{\mathcal{B}} \\ 0 & \text{otherwise} \end{cases}$$

Notice that y is neither basic nor feasible but is a solution of $Ay = b$, and hence it satisfies the requirements of Lemma 2.3. Furthermore, the cost of y is $f + \hat{x}_{0p}$, so the conclusion of Lemma 2.3 gives

$$\bar{c}'y = \hat{x}_{0p} < 0$$

The inequality follows since Column p is entering in T_2 and therefore must have negative relative cost $\hat{x}_{0p}$.

Now by the choice of pivot column in T_1,

$$\bar{c}_j \begin{cases} \geq 0, & j < q \\ < 0, & j = q \end{cases}$$

and by the choice of pivot row in T_2

$$y_j = \begin{cases} -\hat{x}_{ip} < 0, & j = q \\ 0, 1, \text{ or } -\hat{x}_{ip} \geq 0, & j < q \end{cases}$$

Therefore

$$\bar{c}'y = \sum_{j<q} \bar{c}_j y_j + \bar{c}_q y_q \geq \bar{c}_q y_q > 0$$

which is a contradiction. □

2.8 Beginning the Simplex Algorithm

We are left with only one detail: How do we obtain an initial bfs with which to start the simplex algorithm? Sometimes, of course, we may inherit a bfs as part of the problem formulation. For example, we may begin with inequalities of the form $Ax \leq b$, in which case the slack variables constitute a bfs. If we are not so lucky, we may use the *artificial variable*, or *two-phase*, method. In this method we simply append new, "artificial" variables x_i^a, $i = 1, \ldots, m$ to the left of the tableau as follows.

$$\begin{array}{c|ccc|ccc|}
 & x_1^a & \cdots & x_m^a & x_1 & \cdots & x_n \\
\hline
 & 1 & & 0 & & & \\
b & & \ddots & & & A & \\
 & 0 & & 1 & & & \\
\hline
\end{array}
\qquad
\begin{array}{ll}
x_j \geq 0 & j = 1, \ldots, n \\
x_i^a \geq 0 & i = 1, \ldots, m
\end{array}$$

We multiplied some of the original equations by -1 when necessary to make $b \geq 0$. We then have a bfs $x_i^a = b_i$.

In *Phase I*, we minimize the cost function

$$\xi = \sum_{i=1}^{m} x_i^a$$

subject to the above constraints, using the simplex algorithm. There are three possible outcomes.

Case 1 We reduce ξ to zero, and all the x_i^a are driven out of the basis; in this case we now have a bfs to the original problem.

Case 2 We reach optimality with $\xi > 0$, in which case we know that the original problem violates Assumption 2.2—that there is some feasible solution. (If there were a feasible solution to the original problem, it would show that the minimum value of ξ is zero.)

Case 3 We reduce ξ to zero, but some artificial variables remain in the basis at *zero level.*

In Case 1 the columns corresponding to the artificial variables can be dropped and we can continue directly with *Phase II*: The ordinary simplex algorithm using the original cost function $z = c'x$. It is sometimes convenient to start with two cost rows, one for ξ and one for z. When we switch from Phase I to II, we simply change from the first cost row to the second. In Case 2, of course, we must simply stop.

In Case 3 suppose that the ith column of the basis at the end of Phase I is the column corresponding to an artificial variable, and $x_{i0} = 0$. We may pivot on any nonzero (not necessarily positive) element x_{ij} of Row i corresponding to a non-artificial variable. Since θ_0 will be zero, no infeasibility or change in cost ξ will result. This is not exactly pivoting, since it might be the case that $x_{ij} < 0$ or $\bar{c}_j > 0$; we simply say that we are *driving the artificial variable out of the basis.* We repeat this until we obtain a feasible basis with the original variables. The only way that this can fail is that a row can be zero in all the columns corresponding to non-artificial variables. But this means that we have arrived at a zero row in the original matrix by elementary row operations, which contradicts Assumption 2.1 and shows that A was not of full rank m. We can delete such zero rows and continue in Phase II with a basis of lower dimension.

Conversely, if the original set of equations is not of full rank m, we cannot reach Case 1. We shall therefore reach Case 2 if the problem has no feasible solution, or an artificial variable will remain in the basis at zero level at the end of Phase I and, in fact, at the end of Phase II.

Example 2.8

In Example 2.6 we knew a set of basic columns a priori. To use the two-phase method, we would begin with the tableau

		x_1^a	x_2^a	x_3^a	x_1	x_2	x_3	x_4	x_5	
$-z=$	0	0	0	0	1	1	1	1	1	row 0′
$-\xi=$	0	1	1	1	0	0	0	0	0	row 0
	1	1	0	0	3	2	1	0	0	
	3	0	1	0	5	1	1	1	0	
	4	0	0	1	2	5	1	0	1	

We subtract Rows 1, 2, and 3 from the ξ cost row, Row 0, to begin with zero relative costs for our original basis x_1^a, x_2^a, and x_3^a; this yields

		x_1^a	x_2^a	x_3^a	x_1	x_2	x_3	x_4	x_5
$-z=$	0	0	0	0	1	1	1	1	1
$-\xi=$	−8	0	0	0	−10	−8	−3	−1	−1
$x_1^a=$	1	1	0	0	③	2	1	0	0
$x_2^a=$	3	0	1	0	5	1	1	1	0
$x_3^a=$	4	0	0	1	2	5	1	0	1

The successive pivots and tableaux in Phase I are shown below.

		x_1^a	x_2^a	x_3^a	x_1	x_2	x_3	x_4	x_5
$-z=$	$-\frac{1}{3}$	$-\frac{1}{3}$	0	0	0	$\frac{1}{3}$	$\frac{2}{3}$	1	1
$-\xi=$	$-\frac{14}{3}$	$\frac{10}{3}$	0	0	0	$-\frac{4}{3}$	$\frac{1}{3}$	−1	−1
$x_1=$	$\frac{1}{3}$	$\frac{1}{3}$	0	0	1	($\frac{2}{3}$) circled	$\frac{1}{3}$	0	0
$x_2^a=$	$\frac{4}{3}$	$-\frac{5}{3}$	1	0	0	$-\frac{7}{3}$	$-\frac{2}{3}$	1	0
$x_3^a=$	$\frac{10}{3}$	$-\frac{2}{3}$	0	1	0	$\frac{11}{3}$	$\frac{1}{3}$	0	1

		x_1^a	x_2^a	x_3^a	x_1	x_2	x_3	x_4	x_5
$-z=$	$-\frac{1}{2}$	$-\frac{1}{2}$	0	0	$-\frac{1}{2}$	0	$\frac{1}{2}$	1	1
$-\xi=$	−4	4	0	0	2	0	1	−1	−1
$x_2=$	$\frac{1}{2}$	$\frac{1}{2}$	0	0	$\frac{3}{2}$	1	$\frac{1}{2}$	0	0
$x_2^a=$	$\frac{5}{2}$	$-\frac{1}{2}$	1	0	$\frac{7}{2}$	0	$\frac{1}{2}$	①	0
$x_3^a=$	$\frac{3}{2}$	$-\frac{15}{6}$	0	1	$-\frac{11}{2}$	0	$-\frac{3}{2}$	0	1

		x_1^a	x_2^a	x_3^a	x_1	x_2	x_3	x_4	x_5
$-z =$	-3	0	-1	0	-4	0	0	0	1
$-\xi =$	$-\frac{3}{2}$	$\frac{7}{2}$	1	0	$\frac{11}{2}$	0	$\frac{3}{2}$	0	-1
$x_2 =$	$\frac{1}{2}$	$\frac{1}{2}$	0	0	$\frac{3}{2}$	1	$\frac{1}{2}$	0	0
$x_4 =$	$\frac{5}{2}$	$-\frac{1}{2}$	1	0	$\frac{7}{2}$	0	$\frac{1}{2}$	1	0
$x_3^a =$	$\frac{3}{2}$	$-\frac{5}{2}$	0	1	$-\frac{11}{2}$	0	$-\frac{3}{2}$	0	①

		x_1^a	x_2^a	x_3^a	x_1	x_2	x_3	x_4	x_5
$-z =$	$-\frac{9}{2}$	$\frac{5}{2}$	-1	-1	$\frac{3}{2}$	0	$\frac{3}{2}$	0	0
$-\xi =$	0	1	1	1	0	0	0	0	0
$x_2 =$	$\frac{1}{2}$	$\frac{1}{2}$	0	0	$\frac{3}{2}$	1	$\frac{1}{2}$	0	0
$x_4 =$	$\frac{5}{2}$	$-\frac{1}{2}$	1	0	$\frac{7}{2}$	0	$\frac{1}{2}$	1	0
$x_5 =$	$\frac{3}{2}$	$-\frac{5}{2}$	0	1	$-\frac{11}{2}$	0	$-\frac{3}{2}$	0	1

At the end of Phase I, $\xi = 0$, and the resulting tableau is in fact optimal for Phase II as well. (The final tableau for variables x_1 to x_5 agrees with the final, optimal tableau in Example 2.6.) □

The final two-phase algorithm is shown in Fig. 2–5. Notice that we have obviated Assumptions 2.1–2.3: (1) If the matrix A of the original problem is

```
           procedure two-phase
           begin
             infeasible:= 'no', redundant:= 'no';
             (comment: Phase I may set these to 'yes')
Phase I:     introduce an artificial basis, x_i^a;
             call simplex with cost ξ = Σ x_i^a;
             if ξ_opt > 0 in Phase I then infeasible:= 'yes'
               else begin
                   if an artificial variable is in the basis and
                     cannot be driven out then redundant:= 'yes',
                     and omit the corresponding row;
Phase II:          call simplex with original cost
                   end
           end
```

Figure 2–5 The final two-phase algorithm.

not of rank m, we learn so at the end of Phase I and can continue; (2) if the original problem is infeasible, we also learn that at the end of Phase I; and (3) if the problem is of unbounded cost, we learn about it in Phase II.

2.9 Geometric Aspects of Pivoting

Let us solve the LP of Example 2.2 by simplex and trace the sequence of bfs's obtained on the corresponding polytope. The resulting sequence of tableaux is shown in Fig. 2–6, together with the sequence of vertices of the polytope corresponding to the bfs's produced. We observe that simplex simply traces a path along edges of the polytope. We shall next prove formally that this is always the case.

−34	−1	−14	−6	0	0	0	0	
4	1	1	1	1	0	0	0	
2	①	0	0	0	1	0	0	①
3	0	0	1	0	0	1	0	
6	0	3	1	0	0	0	1	

−32	0	−14	−6	0	1	0	0	
2	0	1	①	1	−1	0	0	
2	1	0	0	0	1	0	0	②
3	0	0	1	0	0	1	0	
6	0	3	1	0	0	0	1	

−20	0	−8	0	6	−5	0	0	
2	0	①	1	1	−1	0	0	
2	1	0	0	0	1	0	0	③
1	0	−1	0	−1	1	1	0	
4	0	2	0	−1	1	0	1	

−4	0	0	8	14	−13	0	0	
2	0	1	1	1	−1	0	0	
2	1	0	0	0	1	0	0	④
3	0	0	1	0	0	1	0	
0	0	0	−2	−3	③	0	1	

-4	0	0	$-\frac{2}{3}$	1	0	0	$\frac{13}{3}$
2	0	1	$\frac{1}{3}$	0	0	0	$\frac{1}{3}$
2	1	0	$\textcircled{\frac{2}{3}}$	1	0	0	$-\frac{1}{3}$
3	0	0	1	0	0	1	0
0	0	0	$-\frac{2}{3}$	-1	1	0	$\frac{1}{3}$

⑤

-2	1	0	0	2	0	0	4
1	$-\frac{1}{2}$	1	0	$-\frac{1}{2}$	0	0	$\frac{1}{2}$
3	$\frac{3}{2}$	0	1	$\frac{3}{2}$	0	0	$-\frac{1}{2}$
0	$-\frac{3}{2}$	0	0	$-\frac{3}{2}$	0	1	$\frac{1}{2}$
2	1	0	0	0	1	0	0

⑥

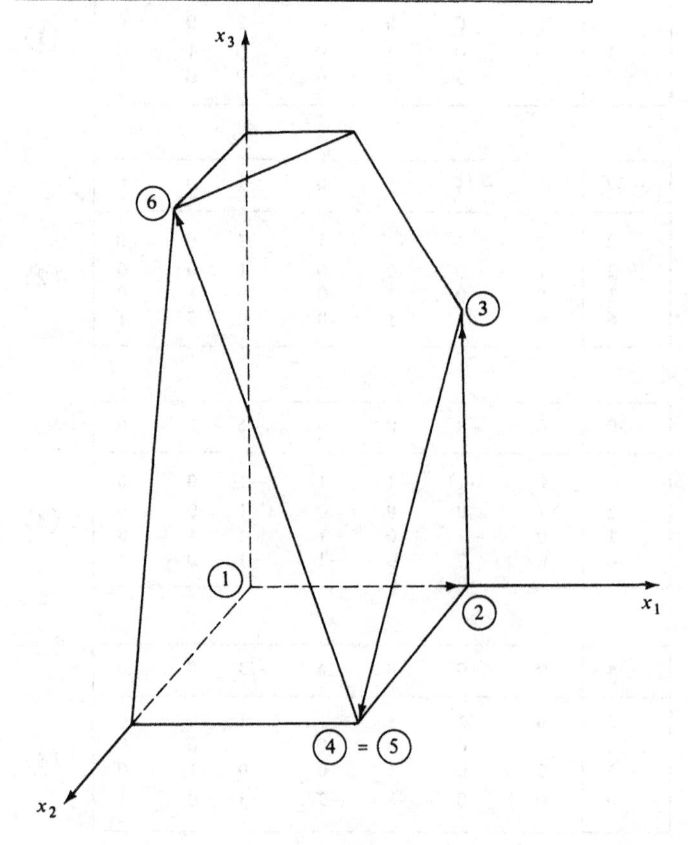

Figure 2–6

Definition 2.6

Two vertices $\hat{x}$ and $\hat{y}$ of a polytope are called *adjacent* if the line segment $[\hat{x}, \hat{y}]$ is an edge of the polytope. Two distinct bfs's x and y of an LP $Ax = b, x \geq 0$ are called *adjacent* if there exist bases $\mathcal{B}_x$, $\mathcal{B}_y$ such that $\mathcal{B}_y = (\mathcal{B}_x - \{A_j\}) \cup \{A_k\}$ and $x = B_x^{-1}b, y = B_y^{-1}b$. □

Thus simplex proceeds by replacing one bfs with another adjacent one, having no greater cost, until the optimal bfs is obtained.

We can now prove the following extension of Theorem 2.4 to edges.

Theorem 2.10 *Let P be a polytope, $F = \{x: Ax = b, x \geq 0\}$ the corresponding feasible set, and $\hat{x} = (x_1, \ldots, x_{n-m})$, $\hat{y} = (y_1, \ldots, y_{n-m})$ be distinct vertices of P. Then the following are equivalent.*

(a) *The segment $[\hat{x}, \hat{y}]$ is an edge of P.*

(b) *For every $\hat{z} \in [\hat{x}, \hat{y}]$, if $\hat{z} = \lambda\hat{z}' + (1 - \lambda)\hat{z}''$ with $0 < \lambda < 1$ and $\hat{z}'$, $\hat{z}'' \in P$, then $\hat{z}', \hat{z}'' \in [\hat{x}, \hat{y}]$.*

(c) *The corresponding vectors x, y of F are adjacent bfs's.*

Proof (a) ⟹ (b) If $[\hat{x}, \hat{y}]$ is an edge of P, then there is a supporting hyperplane H with equation, say, $h'\hat{x} = g$. Every $\hat{z} \in [\hat{x}, \hat{y}]$ therefore satisfies $h'\hat{z} = g$. Now, assume that $\hat{z} = \lambda\hat{z}' + (1 - \lambda)\hat{z}''$ with $1 < \lambda < 0, \hat{z}', \hat{z}'' \in P$ but not both in $[\hat{x}, \hat{y}]$. Thus $h'\hat{z}' \leq g$, $h'\hat{z}'' \leq g$, and one inequality is strict. Therefore, $h'\hat{z} = h'(\lambda\hat{z}' + (1 - \lambda)\hat{z}'') < g$, a contradiction.

(b) ⟹ (c) Assume that bfs's $x, y \in F$ correspond to points in P with Property (b), but are nonadjacent.

Let $\mathcal{M}_x$ and $\mathcal{M}_y$ be the sets of columns corresponding to nonzero components of x and y, respectively. Now it is easy to see that there is a bfs $w \neq x, y$ with nonzero components only in $\mathcal{M}_x \cup \mathcal{M}_y$. Otherwise, we could have a cost vector

$$c_j = \begin{cases} 0 & A_j \in \mathcal{M}_y \\ 1 & A_j \in \mathcal{M}_x - \mathcal{M}_y \\ nM & \text{otherwise} \end{cases}$$

where M is a suitably large number, say the one defined in Lemma 2.1. Then y is uniquely optimal, and any feasible solution with nonzero components out of $\mathcal{M}_x \cup \mathcal{M}_y$ has cost more than x. Thus simplex started at x would fail to discover a sequence of adjacent bfs's with nonincreasing cost leading to the optimum, which is absurd. So such a $w \neq x, y$ does exist, and, furthermore, $\hat{w}$ does not lie on $[\hat{x}, \hat{y}]$ because the points $\hat{w}, \hat{x}, \hat{y}$ correspond to distinct vertices of the polytope P.

Now let $z = \frac{1}{2}(x + y)$ and consider the difference

$$d = z - w$$

It is nonzero only for columns in $\mathcal{M}_x \cup \mathcal{M}_y$, and hence there exists a positive number θ such that

$$u_1 = z + \theta d$$

and

$$u_2 = z - \theta d$$

are feasible. Hence $z = \frac{1}{2}(u_1 + u_2)$, where $\hat{u}_1$ and $\hat{u}_2$ do not lie on $[\hat{x}, \hat{y}]$; this contradicts Property (b).

(c) ⇒ (a) Let $\mathcal{B}_x, \mathcal{B}_y$ be the bases corresponding to x and y, respectively, with $\mathcal{B}_y = \mathcal{B}_x \cup \{A_j\} - \{A_k\}$ for some columns A_j, A_k. Let us construct a cost vector c by

$$c_j = \begin{cases} 0 & \text{if } A_j \in \mathcal{B}_y \cup \mathcal{B}_x \\ 1 & \text{otherwise} \end{cases}$$

All feasible solutions that are convex combinations of x and y are optimal. Furthermore, these are the only optimal solutions. To show this suppose that z is optimal. Then z is, by Theorem 2.3, a convex combination of bfs's, and, in particular, of bfs's with bases subsets of $\mathcal{B}_x \cup \mathcal{B}_y$; however, x and y are the only such bfs's.

It follows that only convex combinations w of x and y satisfy $Aw = b$, $w \geq 0$ and $c'w \leq c'x$. Therefore, in P, only points $\hat{w}$ on the segment $[\hat{x}, \hat{y}]$ satisfy

$$d'\hat{w} \leq d'\hat{x}$$

where d is defined, as in the proof of Theorem 2.4, to be

$$d_l = c_l - \sum_{j=1}^{m} h_{n-m+j,l}c_{n-m+j}$$

Hence $[\hat{x}, \hat{y}]$ is the intersection of a halfspace with P and is therefore an edge. □

A final comment on simplex: By our discussion of Chapter 1, LP is a convex programming problem, and so the Euclidean neighborhood N_ϵ is exact. That is, if we search in the neighborhood of all points in F that are within ϵ of some $x_0 \in F$ and find no solution better than x_0, then x_0 is globally optimal (see Figure 2–7(a)).

The simplex algorithm has revealed another exact neighborhood, combinatorially and computationally much more meaningful. First, we do not have to consider all of the (uncountably infinite) set F, but just the finite set of basic feasible solutions. Furthermore, within this set of bfs's, we have the neighborhood

$$N_A(x_0) = \{y : y \text{ is a bfs adjacent to } x_0\}$$

Then Theorem 2.8 tells us that N_A is *exact* for LP (see Figure 2–7(b)). What is more, $N_A(x_0)$ contains a few (at most $n - m$) bfs's and can be searched very fast—in fact, just by looking at the signs of the $\bar{c}_j$'s. Thus simplex can be viewed

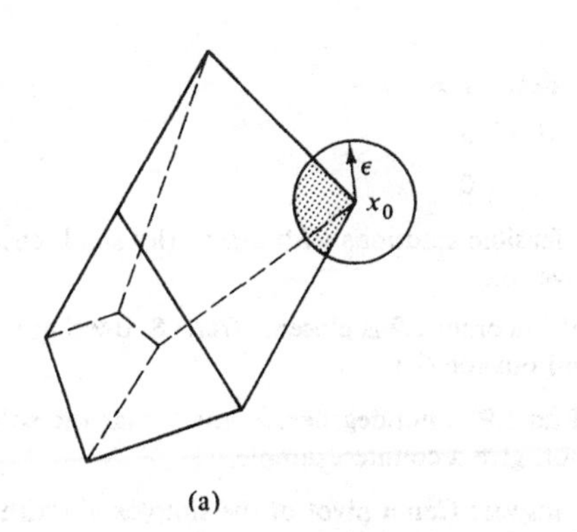

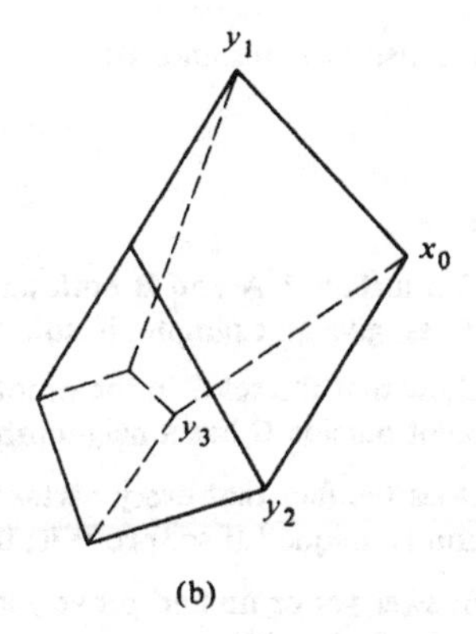

Figure 2–7 (a) The exact neighborhood N_ϵ. (b) The exact neighborhood $N_A(x_0) = \{y_1, y_2, y_3\}$.

as just a clever implementation of the general neighborhood search scheme for the exact neighborhood structure N_A.

PROBLEMS

1. Show that the converse of Theorem 2.5 is not true; that is, that there can exist a degenerate vertex whose corresponding basis is unique.
2. Show that a polytope F defined by an instance of LP is a closed set.
3. Suppose in an instance of LP, we have n variables that are unconstrained in sign. Show how they can be replaced by $n + 1$ variables that are constrained to be nonnegative.
4. Check the statement in the proof of Theorem 2.4 that the set $\mathcal{B}$ can be augmented to a basis, and the similar statement in the proof of Theorem 2.1.

*5. Show that the optimality criterion of Theorem 2.8 is not necessary at an optimal vertex.

*6. Show that the condition $\theta_0 = 0$ for every possible pivot in the simplex algorithm does not imply optimality.

*7. Show that a linear program cannot cycle unless we have at least two basic variables that are zero.

8. Show that the set of optimal points of an instance of LP is a convex set.
9. We are given the following instance A of LP in standard form:

$$\min c'x$$
$$Ax = b$$
$$x \geq 0$$

We also have instance **B**:

$$\begin{aligned} \min\ & -c'x \\ Ax &= b \\ x &\geq 0 \end{aligned}$$

Can instances A and B both have feasible solutions with arbitrarily small cost? If yes, give an example; if not, prove so.

10. Show that the set G in the proof of Theorem 2.2 is closed. (*Hint:* Show that any point outside G has a neighborhood outside G.)
11. Does the fact that every vertex of an LP is nondegenerate imply that the solution is unique? If so, prove it; if not, give a counterexample.
12. Answer yes or no and prove your answer: Can a pivot of the simplex algorithm move the feasible point a positive distance in R^n while leaving the cost unchanged?
13. Can a vector which has just left the basis in the simplex algorithm reenter on the very next pivot?
14. The following fragment of FORTRAN code calculates $\bar{c}_j$ for pricing in the simplex algorithm and decides whether to pivot in order to bring Column j into the basis:

```
      .
      .
      .
      CBAR = C(J)
      DO 1  I = 1,M
    1 CBAR = CBAR - C(BASIS(I))*X(I,J)
      IF(CBAR.LT.O.)GO TO 3
      .
      .
      .
```

Assume the variables *C*, *BASIS*, and *X* are defined appropriately at this point in the program. Give a reason why this will not work well in practice, and suggest a simple alteration which will. (The issue here is not language dependent.)

15. (Programming project) Write a computer program that implements the two-phase simplex algorithm for an LP in standard form. The input should be the vectors b, and c, and the matrix A. The program should terminate in one of the following four ways.
    1. Unbounded solution found in Phase I. This is impossible (why?), but should be a logical branch in the program as an error check.
    2. Optimal solution found in Phase I with positive cost. This means the original problem is infeasible.
    3. Unbounded solution found in Phase II. This means that the original problem has unbounded cost.
    4. Optimal solution found in Phase II. This means that the original problem has been solved.

Your program should print out:

(a) The problem data;

(b) The row, column, and cost after each pivot in both phases;

(c) A message after Phase I and after Phase II if entered;

(d) The final basis, tableau, and cost, regardless of the termination point.

Test your program on problems that terminate in as many ways as possible.

16. Prove the following: If F is a k-dimensional face of a convex polytope P in R^d, then F is also a convex polytope, and furthermore every vertex of F is also a vertex of P.

17. Prove: If an LP is unbounded, then there is a rational vector α such that (*a*) $c'\alpha < 0$, and (*b*) if x is feasible and $k > 0$, then $x + k\alpha$ is also feasible.

NOTES AND REFERENCES

The simplex algorithm was invented in 1947 by G. B. Dantzig, and we cannot recommend too highly his comprehensive text

[Da1] DANTZIG, G. B., *Linear Programming and Extensions.* Princeton, N.J.: Princeton University Press, 1963.

The reader will find there a detailed and first-hand account of the origins of linear programming, as well as a development of the simplex algorithm and its variations. Besides this, there are many other excellent texts devoted to linear programming. Among them are

[BJ] BAZARAA, M. S., and J. J. JARVIS, *Linear Programming and Network Flows.* New York: John Wiley & Sons, Inc., 1977.

[CS] COOPER, L., and D. STEINBERG, *Methods and Applications of Linear Programming.* Philadelphia: W.B. Saunders, 1974.

[Gal] GALE, D., *The Theory of Linear Economic Models.* New York: McGraw-Hill Book Company, 1960.

[Gas] GASS, S. I., *Linear Programming* (4th ed.). New York: McGraw-Hill Book Company, 1975.

[Had2] HADLEY, G., *Linear Programming.* Reading, Mass.: Addison-Wesley Publishing Co., Inc., 1962.

[Hu] HU, T. C., *Integer Programming and Network Flows.* Reading, Mass.: Addison-Wesley Publishing Co., Inc., 1970.

[Si] SIMONNARD, M., *Linear Programming* (translated from the French by W. S. Jewell). Englewood Cliffs, N.J.: Prentice-Hall, Inc., 1966.

[YG] YUDIN, D. B., and E. G. GOL'SHTEIN, *Linear Programming* (translated from the Russian by Z. Lerman). Jerusalem: Israel Program for Scientific Translations, 1965.

Dantzig [Da1] attributes the formulation of the diet problem to

[Sti] STIGLER, G. J., "The Cost of Subsistence," *J. Farm Econ.*, 27, no. 2 (May 1945), 303–14.

He also gives the first publication of the simplex algorithm as

[Da2] DANTZIG, G. B., "Programming of Interdependent Activities, II, Mathematical Model," pp. 19–32, in *Activity Analysis of Production and Allocation*, ed. T. C. Koopmans. New York: John Wiley & Sons, Inc., 1951. Also in *Econometrics* 17, nos. 3 and 4 (July–Oct. 1949), 200–11.

Theorem 2.3 can be considered a special case of the general fact that any closed, bounded convex set is the convex hull of its extreme points. For much more about polytopes, see

[Gru] GRÜNBAUM, B., *Convex Polytopes*. New York: John Wiley & Sons, Inc., 1967.

[Roc] ROCKAFELLAR, R. T., *Convex Analysis*. Princeton, N.J.: Princeton University Press, 1970.

Cycling in practical problems is described in

[KS] KOTIAH, T. C. T., and D. I. STEINBERG, "On the Possibility of Cycling with the Simplex Method," *OR*, 26, no. 2 (March–April 1978), 374–6.

The anticycling rule given in Section 2.7 is due to

[Bl] BLAND, R. G., "New Finite Pivoting Rules," Discussion Paper 7612, Center for Operations Research and Econometrics (CORE), Université Catholique de Louvain, Heverlee, Belgium, June 1976 (revised January 1977).

The proof given is after

[Ku] KUHN, H. W., Class Notes, Princeton University, 1976.

Computational experiments comparing different column selection rules are described in

[KQ] KUHN, H. W., and R. E. QUANDT, "An Experimental Study of the Simplex Method," pp. 107–24, in *Proceedings of Symposia on Applied Mathematics*, vol. XV, ed. N. Metropolis and others. American Mathematical Society, Providence, R.I.; 1963.

An all-variable steepest-descent method is described in

[GR] GOLDFARB, D., and J. K. REID, "A Practicable Steepest-Edge Simplex Algorithm, *Math. Prog.*, 12, no. 3 (June 1977), 361–71.

The cycling example is from

[Be1] BEALE, E. M. L., "Cycling in the Dual Simplex Algorithm," *Naval Research Logistics Quarterly*, 2, no. 4 (1955), 269–75.

3

Duality

3.1 The Dual of a Linear Program in General Form

If the simplex algorithm were all there were to linear programming, that would be useful enough. But there are also many interesting theoretical aspects to the subject, especially relating to combinatorial problems. All of these are related in one way or another to the idea of duality, to which we next turn our attention.

Consider an LP in general form:

$$\begin{aligned} &\min c'x \\ &a_i'x = b_i \qquad i \in M \\ &a_i'x \geq b_i \qquad i \in \bar{M} \\ &x_j \geq 0 \qquad j \in N \\ &x_j \gtrless 0 \qquad j \in \bar{N} \end{aligned} \tag{3.1}$$

We wish to use the optimality criterion of Theorem 2.8, so we convert this to standard form. For each inequality in $\bar{M}$, create a surplus variable x_i^s, $i \in \bar{M}$; for each unconstrained variable x_j, $j \in \bar{N}$, create two new nonnegative variables by $x_j = x_j^+ - x_j^-$, and replace column A_j by two columns A_j and $-A_j$. This yields the LP

$$\begin{aligned} \min\ & \hat{c}'\hat{x} \\ & \hat{A}\hat{x} = b \\ & \hat{x} \geq 0 \end{aligned} \tag{3.2}$$

where

$$\hat{A} = \left[A_j, j \in N \middle| (A_j, -A_j),\ j \in \bar{N} \middle| \begin{matrix} 0, i \in M \\ \hline -I, i \in \bar{M} \end{matrix} \right]$$

and

$$\hat{x} = \text{col}\,(x_j, j \in N | (x_j^+, x_j^-),\ j \in \bar{N} | x_i^s, i \in \bar{M})$$
$$\hat{c} = \text{col}\,(c_j, j \in N | (c_j, -c_j),\ j \in \bar{N} | 0)$$

We then know from the optimality criterion $\bar{c} \geq 0$ and the simplex algorithm that if there is an optimal solution $\hat{x}_0$ to (3.2), then there exists a basis $\hat{\mathcal{B}}$ for the LP in Eq. 3.2 such that

$$\hat{c}' - (\hat{c}_B'\hat{B}^{-1})\hat{A} \geq 0$$

Thus, $\pi' = \hat{c}_B'\hat{B}^{-1}$ is a feasible solution to the linear constraints

$$\pi'\hat{A} \leq \hat{c}' \tag{3.3}$$

where $\pi \in R^m$, and m is the number of rows in the original A. These inequalities have three parts, depending on which set of columns of $\hat{A}$ is involved. The first set yields simply

$$\pi' A_j \leq c_j, \qquad j \in N \tag{3.4}$$

The next set corresponds to the unconstrained x_j, $j \in \bar{N}$, and comes in pairs:

$$\begin{aligned} \pi' A_j &\leq c_j, \\ -\pi' A_j &\leq -c_j \end{aligned} \qquad j \in \bar{N}$$

which is equivalent to

$$\pi' A_j = c_j, \qquad j \in \bar{N} \tag{3.5}$$

The final set corresponds to the inequalities $i \in \bar{M}$:

$$-\pi_i \leq 0 \qquad i \in \bar{M}$$

or

$$\pi_i \geq 0 \qquad i \in \bar{M} \tag{3.6}$$

Equations 3.4, 3.5, and 3.6 define the constraints of a new LP, called the *dual* of the starting LP; the starting LP is called the *primal*. The value $\pi' = \hat{c}_B'\hat{B}^{-1}$ is feasible in the dual. If we define the cost function of the dual as max $\pi'b$, then π' is not only feasible, but optimal! We summarize this in the next definition and theorem.

Definition 3.1

Given an LP in general form, called the *primal*, the *dual* is defined as follows:

Primal		Dual
$\min c'x$		$\max \pi'b$
$a_i'x = b_i$	$i \in M$	$\pi_i \gtrless 0$
$a_i'x \geq b_i$	$i \in \bar{M}$	$\pi_i \geq 0$
$x_j \geq 0$	$j \in N$	$\pi'A_j \leq c_j$
$x_j \gtrless 0$	$j \in \bar{N}$	$\pi'A_j = c_j$ ☐

Theorem 3.1 *If an LP has an optimal solution, so does its dual, and at optimality their costs are equal.*

Proof Let x and π be feasible solutions to the primal and dual, respectively. Then

$$c'x \geq \pi'Ax \geq \pi'b \tag{3.7}$$

That is, the cost in the primal always dominates the cost in the dual. Since we assume the primal has a feasible solution, the dual cannot have a solution unbounded in cost. The dual has the feasible solution π' discussed above, so by the simplex algorithm, it has an optimum. We note that the cost of this π' is

$$\pi'b = \hat{c}_B'\hat{B}^{-1}b = \hat{c}_B'\hat{x}_0$$

which is the optimal cost in the primal. Therefore, by (3.7), this π' is optimal in the dual. ☐

An important feature of duality is the symmetry expressed in the following theorem.

Theorem 3.2 *The dual of the dual is the primal.*

Proof Write the dual as

$$\begin{aligned} &\min \pi'(-b) \\ &(-A_j')\pi \geq -c_j \qquad j \in N \\ &(-A_j)\pi = -c_j \qquad j \in \bar{N} \\ &\pi_i \geq 0 \qquad i \in \bar{M} \\ &\pi_i \gtrless 0 \qquad i \in M \end{aligned}$$

and consider it the primal. Then Def. 3.1 produces the dual of the dual as

$$\begin{aligned} &\max x'(-c) \\ &x_j \geq 0 \qquad && j \in N \\ &x_j \gtrless 0 \qquad && j \in \bar{N} \\ -&a_i'x \leq -b \qquad && i \in \bar{M} \\ -&a_i'x = -b \qquad && i \in M \end{aligned}$$

which will be recognized as the original primal. □

A linear program always falls into exactly one of three categories: Either (1) it has a finite optimum, (2) it has a solution of unbounded cost, or (3) it has no feasible solution. Thus a primal and its dual have nine combinations as shown in Fig. 3–1.

Primal \ Dual	Finite optimum	Unbounded	Infeasible
Finite optimum	(1)	X	X
Unbounded	X	X	(3)
Infeasible	X	(3)	(2)

Figure 3–1 Possible categories of a primal–dual pair.

By Theorems 3.1 and 3.2, we have already eliminated all entries on the first row and column except the case corresponding to both primal and dual with finite optima. The eliminated cases are indicated with crosses. The matrix is filled in as follows.

Theorem 3.3 [vN, Ga, GKT] *Given a primal-dual pair, exactly one of three situations occurs, as indicated in Fig. 3–1.*

Proof By Eq. 3.7, if either the primal or dual has unbounded cost, the other cannot have a feasible solution. We have left only two cases, indicated by Cases 2 and 3 in Fig. 3–1. Simple examples show that these are both possible.

Case 2 occurs when neither program is feasible. Consider the infeasible primal

$$\min x_1$$
$$x_1 + x_2 \geq 1$$
$$-x_1 - x_2 \geq 1$$
$$x_1 \gtrless 0$$
$$x_2 \gtrless 0$$

The dual is

$$\max \pi_1 + \pi_2$$
$$\pi_1 - \pi_2 = 1$$
$$\pi_1 - \pi_2 = 0$$
$$\pi_1 \geq 0$$
$$\pi_2 \geq 0$$

which is also infeasible.

Restrict x_1, $x_2 \geq 0$ in the primal and it remains infeasible; but the dual becomes unbounded, providing an example of Case 3. □

Example 3.1 (The Dual of the Diet Problem)

Let us return to the diet problem discussed in Example 2.1. The dual is

$$\max \pi' r$$
$$\pi' A \leq c'$$
$$\pi' \geq 0$$

This has the following illuminating interpretation. A pill-maker wishes to market pills containing each of the m nutrients, at a price π_i per unit of nutrient i. He wishes to be competetive with the price of real food, while at the same time maximizing the cost of an adequate diet. The constraints

$$\sum_{i=1}^{m} \pi_i a_{ij} \leq c_j \qquad j = 1, \ldots, n$$

in fact express the fact that the cost in pill form of all the nutrients in the jth food is no greater than the cost of the jth food itself; the cost function $\pi' r$ is simply the cost of an adequate diet. By Theorem 3.1, the optimal cost of the homemaker's primal is equal to the optimal cost of the pill-maker's dual: They are really two ways of expressing the same problem. □

3.2 Complementary Slackness

If we examine the definition of dual, we see a kind of tug-of-war exists between primal and dual: The more severe a constraint in one, the looser its counterpart in the other. The ultimate expression of this balance is the necessary and sufficient condition for a pair x, π to be respective optima in a primal-dual pair, known as the *complementary slackness* condition.

Theorem 3.4 (Complementary Slackness) *A pair x, π respectively feasible in a primal-dual pair is optimal if and only if*

$$u_i = \pi_i(d_i'x - b_i) = 0 \qquad \text{for all } i \tag{3.8}$$

$$v_j = (c_j - \pi' A_j)x_j = 0 \qquad \text{for all } j \tag{3.9}$$

Proof We first note that $u_i \geq 0$ for all i and $v_j \geq 0$ for all j by the duality relations. Define

$$u = \sum_{\text{all } i} u_i \geq 0$$

$$v = \sum_{\text{all } j} v_j \geq 0$$

Then $u = 0$ if and only if Eq. 3.8 holds and $v = 0$ if and only if Eq. 3.9 holds. Next, note that

$$u + v = c'x - \pi'b$$

because the terms involving both x and π cancel if we add Eqs. 3.8 and 3.9 for all i and j. Thus, Eqs. 3.8 and 3.9 hold if and only if $u + v = 0$, or

$$\pi'b = c'x$$

which is necessary and sufficient for x and π both to be optimal, by Eq. 3.7. □

Theorem 3.4 has important ramifications. Notice that it implies that if an inequality constraint in a dual is not binding, then the corresponding variable in the primal must be zero if the pair considered is at optimality. Symmetrically, if a nonnegative variable is strictly positive, the corresponding inequality must be binding.

Example 3.2

The linear program used in Example 2.6 has the dual

$$\begin{aligned}
&\max \pi_1 + 3\pi_2 + 4\pi_3 \\
&3\pi_1 + 5\pi_2 + 2\pi_3 \leq 1 \\
&2\pi_1 + \pi_2 + 5\pi_3 \leq 1 \\
&\pi_1 + \pi_2 + \pi_3 \leq 1 \\
&\pi_2 \leq 1 \\
&\pi_3 \leq 1 \\
&\pi_i \gtrless 0 \qquad \text{for all } i
\end{aligned}$$

Because the primal is in standard form, the complementary slackness condition in (3.8) is automatically taken care of. Condition 3.9 becomes

$$c_2 - \pi' A_2 = 0$$
$$c_4 - \pi' A_4 = 0$$
$$c_5 - \pi' A_5 = 0$$

because the optimal primal has positive x_2, x_4, and x_5. Thus the second, fourth, and fifth inequalities in the dual must be binding:

$$\begin{aligned} 2\pi_1 + \pi_2 + 5\pi_3 &= 1 \\ \pi_2 \qquad &= 1 \\ \pi_3 &= 1 \end{aligned}$$

This is seen to be satisfied by

$$\pi_1 = -\tfrac{5}{2}$$
$$\pi_2 = 1$$
$$\pi_3 = 1$$

which corresponds to a dual cost of $\frac{9}{2}$, the optimal primal cost. □

3.3 Farkas' Lemma

Farkas' lemma is a fundamental fact about vectors in R^n that in a sense captures the essense of duality. It could have been used to derive the results earlier in this chapter, where we relied instead on the constructive aspect of the simplex algorithm. At this point, we are in a position to prove Farkas' lemma as a consequence of what we already know about linear programming.

We first introduce a useful definition.

Definition 3.2

Given a set of vectors $a_i \in R^n$, $i = 1, \ldots, m$, the *cone generated by the set* $\{a_i\}$, denoted by $C(a_i)$, is

$$C(a_i) = \{x \in R^n : x = \sum_{i=1}^{m} \pi_i a_i, \quad \pi_i \geq 0, \quad i = 1, \ldots, m\} \quad □$$

Example 3.3

Figure 3–2 shows two vectors in R^2 and their corresponding cone. □

Now suppose we are given a set of vectors $\{a_i\}$ and another vector $c \in R^n$, and suppose further that the following property ties the $\{a_i\}$ and c together: whenever a vector $y \in R^n$ has a nonnegative projection on the a_i's, it also has a nonnegative projection on c. Farkas' lemma states that this condition is equivalent to c being in the cone generated by the a_i's. In Fig. 3–2, this means

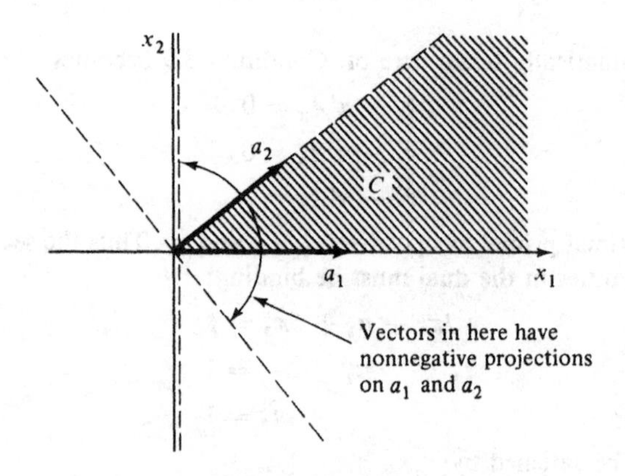

Figure 3–2 An example of a cone.

that if every vector in the indicated arc has a nonnegative projection on c, then $c \in C(a_i)$; and conversely, if $c \in C(a_i)$, then every vector in the arc has a nonnegative projection on c.

We now see this result as an immediate consequence of Theorem 3.1.

Theorem 3.5 (Farkas' Lemma) [Fa] *Given vectors* $a_i \in R^n$, $i = 1, \ldots, m$ *and another* $c \in R^n$, *then*

$$(y'a_i \geq 0 \text{ for all } i \Longrightarrow y'c \geq 0) \Longleftrightarrow c \in C(a_i)$$

Proof We dispose first of the *if* part, which is really trivial. If

$$c = \sum_{i=1}^{m} \pi_i a_i, \qquad \pi_i \geq 0$$

then

$$y'c = \sum_{i=1}^{m} \pi_i(y'a_i) \geq 0$$

if the $y'a_i \geq 0$.

The *only if* part is the heart of the matter. Consider the LP

$$\begin{aligned} &\min c'y \\ &a_i'y \geq 0 \qquad i = 1, \ldots, m \\ &y \gtrless 0 \end{aligned}$$

This program is feasible, because $y = 0$ is a feasible point. It is also bounded by the hypothesis that $a_i'y \geq 0$ for all i implies $y'c \geq 0$. Therefore the dual

$$\begin{aligned} &\max 0 \\ &\pi' A_j = c_j \qquad j = 1, \ldots, n \\ &\pi \geq 0 \end{aligned}$$

has a feasible solution, where

$$A_j = \text{col}\,(a_{ij},\quad i = 1, \ldots, m) \in R^m$$

if

$$a_i = \text{col}\,(a_{ij},\quad j = 1, \ldots, n) \in R^n$$

Thus there is a π such that

$$c = \sum_{i=1}^{m} \pi_i a_i, \quad \pi_i \geq 0$$

□

3.4 The Shortest-Path Problem and Its Dual

Definition 3.3

Given a directed graph $G = (V, E)$ and a nonnegative weight $c_j \geq 0$ associated with each arc $e_j \in E$, an instance of the *shortest-path problem* (SP) is the problem of finding a directed path from a distinguished source node s to a distinguished terminal node t, with the minimum total weight. □

As an optimization problem, the feasible set is

$$F = \{\text{sequences } P = (e_{j_1}, \ldots, e_{j_k}) : \text{this sequence is a directed path from } s \text{ to } t \text{ in graph } G\}$$

and cost mapping

$$c(P) = \sum_{i=1}^{k} c_{j_i}$$

We can formulate an instance of SP as an LP by first defining the (node-arc) incidence matrix $A = [a_{ij}]$ of the graph G by

$$a_{ij} = \begin{cases} +1 & \text{if arc } e_j \text{ leaves node } i \\ -1 & \text{if arc } e_j \text{ enters node } i \\ 0 & \text{otherwise} \end{cases} \quad \begin{pmatrix} i = 1, \ldots, |V| \text{ and} \\ j = 1, \ldots, |E| \end{pmatrix}$$

Example 3.4

Figure 3–3 shows a directed graph and corresponding arc weights that define an instance of SP. The corresponding incidence matrix is:

$$A = \begin{array}{c} \\ s \\ t \\ a \\ b \end{array} \begin{array}{c} \begin{array}{ccccc} e_1 & e_2 & e_3 & e_4 & e_5 \end{array} \\ \begin{bmatrix} +1 & +1 & 0 & 0 & 0 \\ 0 & 0 & 0 & -1 & -1 \\ -1 & 0 & +1 & +1 & 0 \\ 0 & -1 & -1 & 0 & +1 \end{bmatrix} \end{array}$$

□

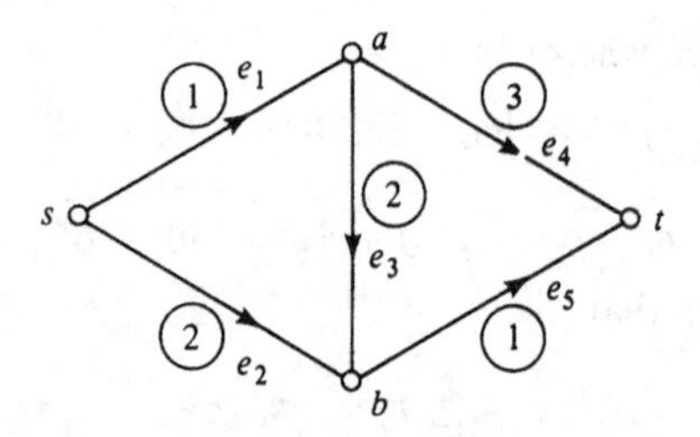

Figure 3–3 A weighted, directed graph for SP; the weights are circled.

To continue with the general LP formulation, associate a variable f_j with arc e_j to represent a flow of some imaginary commodity through the arc in the direction of orientation. Then flow conservation at a node i is expressed by the equation

$$a_i' f = 0$$

where f is the column vector with ith component f_i and, as usual, a_i' is the ith row of A. A path from s to t can be thought of as a flow of 1 unit leaving s and entering t; such a flow must satisfy

$$Af = \begin{bmatrix} +1 \\ -1 \\ 0 \\ 0 \\ \cdot \\ \cdot \\ \cdot \\ 0 \end{bmatrix} \begin{matrix} \text{Row } s \\ \text{Row } t \\ \\ \\ \\ \\ \\ \\ \end{matrix}$$

Of course, there is no reason why f cannot in general take on noninteger values, but if we consider the minimum-cost flow problem

$$\min c'f$$

such that

$$Af = \begin{bmatrix} +1 \\ -1 \\ 0 \\ \cdot \\ \cdot \\ \cdot \\ 0 \end{bmatrix} \begin{matrix} \text{Row } s \\ \text{Row } t \\ \\ \\ \\ \\ \\ \end{matrix} \qquad (3.10)$$

$$f \geq 0$$

we see intuitively that there is an optimal solution in which each f_i is zero or one, representing a unit flow along a shortest path from s to t in G. (This will be proved later, in a more general context, in Chapter 13. Also, see Problem 17.)

Notice next that the $|V|$ equations in our problem are redundant—flow conservation at any $|V| - 1$ nodes implies conservation at the remaining node. Thus we can leave out any one equation.

Example 3.4 (Continued)

Let us omit the Row t equation; this has the advantage of leaving a nonnegative zeroth column. This yields the tableau

		f_1	f_2	f_3	f_4	f_5
$-z =$	0	1	2	2	3	1
	1	1	1	0	0	0
	0	-1	0	1	1	0
	0	0	-1	-1	0	1

Adding Row 1 to Row 2 yields a bfs comprised of Columns 1, 4, and 5. Making the relative costs over these basis columns zero yields the tableau

		f_1	f_2	f_3	f_4	f_5
$-z =$	-4	0	-1	0	0	0
$f_1 =$	1	1	1	0	0	0
$f_4 =$	1	0	(1)	1	1	0
$f_5 =$	0	0	-1	-1	0	1

A basis represents in general a set of $|V| - 1$ arcs, a subset of which corresponds to a path from s to t with the given cost. The nonpath arcs represent degenerate components of the basis: In this example the current basis represents the path (e_1, e_4) with cost 4 and the degenerate arc e_5.

A pivot bringing the column corresponding to e_2 into the basis yields the optimal tableau shown next (the pivot element is circled above).

		f_1	f_2	f_3	f_4	f_5
$-z =$	-3	0	0	1	1	0
$f_1 =$	0	1	0	-1	-1	0
$f_2 =$	1	0	1	1	1	0
$f_5 =$	1	0	0	0	1	1

This corresponds to the shortest path (e_2, e_5) with cost 3 and the degenerate arc e_1. Figure 3–4 illustrates the pivot step; one arc is added and one is subtracted from the basis to create a new path with lower cost. □

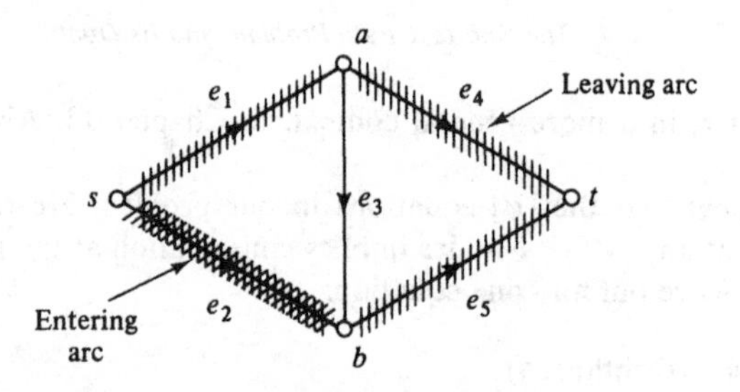

Figure 3–4 The basis (shaded arcs) and the entering column (thatched arc).

Returning to the general LP formulation in Eq. 3.10, we can write the dual of an instance of SP by assigning a variable π_i to each node i:

$$\begin{aligned} &\max \pi_s - \pi_t \\ &\pi' A \leq c' \\ &\pi \gtrless 0 \end{aligned} \tag{3.11}$$

Because of the way the incidence matrix is defined, the inequalities in the dual can be written simply as

$$\pi_i - \pi_j \leq c_{ij} \qquad \text{for each } (i, j) \in E \tag{3.12}$$

The complementary slackness conditions in Theorem 3.4 have a simple interpretation in this problem. A path f and an assignment of variables π are jointly optimal if and only if (a) each arc in the shortest path (which corresponds to a positive f_j in the primal) corresponds to equality in the corresponding inequality of (3.12), and (b) each strict inequality in (3.12) corresponds to an arc not in the shortest path.

Example 3.4 (Continued)

Figure 3–5 shows an optimal choice of π corresponding to the shortest path (s, b), (b, t). Notice that there is really no π corresponding to the node t, since the equation for node t was left out.

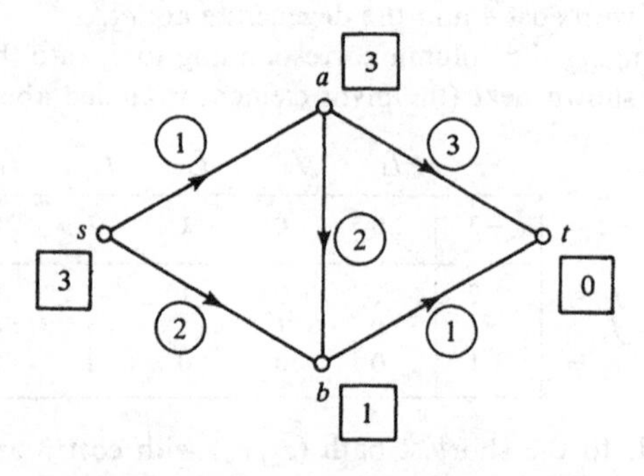

Figure 3–5 An optimal π for the dual of the shortest-path problem (shown in squares).

By complementary slackness, however, it can be supplied easily by subtracting the cost of the final arc in the shortest path (b, t) from π_b (as shown). □

3.5 Dual Information in the Tableau

It should come as no surprise that the final tableau of the simplex algorithm gives us the optimal dual solution as well as the optimal primal. It is most convenient to assume that we begin with an identity matrix at the left of the tableau; this usually corresponds to artificial or slack variables. Figure 3–6 shows the initial tableau with this assumption.

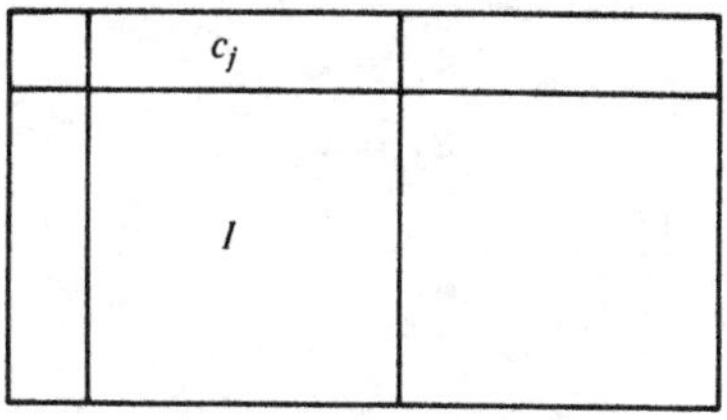

Figure 3–6 Assumed initial tableau.

At termination of the simplex algorithm at an optimal solution, we have in effect multiplied the tableau below the cost row by B^{-1}, where B is the set of columns in the original tableau corresponding to the optimal bfs. Furthermore, the cost row at optimality becomes

$$\bar{c}_j = c_j - \pi' A_j \geq 0 \tag{3.13}$$

where π is an optimal solution to the dual, as we saw in the proof of Theorem 3.1. In Columns 1 to m, where we assumed we started with an identity matrix, A_j is the unit vector e_j, so that

$$\bar{c}_j = c_j - \pi_j \qquad j = 1, \ldots, m \tag{3.14}$$

Hence, we may obtain an optimal dual from the final tableau by

$$\pi_j = c_j - \bar{c}_j \qquad j = 1, \ldots, m \tag{3.15}$$

We also note that in the final tableau the position of the initial identity matrix is occupied by B^{-1}, as shown in Fig. 3–7. This fact will be used to great advantage in the next chapter.

Example 3.5

In Example 2.8, the initial tableau has $c_j = 0$ in the (actual z) cost row. Thus, in the final tableau,

$$\pi_j = -\bar{c}_j$$

	$c_j - \pi_j$	
	B^{-1}	

Figure 3–7 The final tableau.

which yields

$$\pi_1 = -\tfrac{5}{2}$$
$$\pi_2 = 1$$
$$\pi_3 = 1$$

checking Example 3.2. □

Example 3.6

In Example 3.4, the shortest-path example, we did not begin with the required identity matrix. If we want to get an optimal dual solution, it is convenient to start with such a matrix in the tableau; if we had, we would have obtained the values shown in Fig. 3–5 in its place in the final tableau (see Problem 4). □

3.6 The Dual Simplex Algorithm

The optimality criterion $\bar{c} \geq 0$ can be thought of as expressing feasibility of the dual variable $\pi' = c'_B B^{-1}$, because

$$\begin{aligned} \bar{c}' &= c' - z' \\ &= c' - c'_B B^{-1} A \\ &= c' - \pi' A \end{aligned} \tag{3.16}$$

Thus we can think of the simplex algorithm as maintaining a primal feasible solution and working towards dual feasibility. Such an algorithm is called a *primal* algorithm. In the same way, we can start with a dual feasible solution and work towards primal feasibility. This is called the *dual simplex algorithm.*

In more detail, assume we have a tableau with a basic (but not primal feasible) solution and a feasible dual solution (cost row $x_{0j} \geq 0$). We select a row, such as Row r, corresponding to an $x_{r0} < 0$, an infeasible component of the primal. The possible pivots lie in Row r. We consider only those entries with

$$x_{rj} < 0$$

because when we pivot, we wish to increase z (decrease $-z$). This follows because we stay dual feasible and hence below optimal cost; we are really doing a maximization problem. The new elements in the cost row after pivoting about element x_{rs} are

$$x'_{0j} = x_{0j} - \left(\frac{x_{0s}}{x_{rs}}\right)x_{rj}$$

We want these elements to remain nonnegative to maintain dual feasibility. Hence, for $x_{rj} < 0$, we must have

$$\frac{x_{0j}}{x_{rj}} \leq \frac{x_{0s}}{x_{rs}}$$

The choice of pivot column is therefore determined by

$$\max_{\substack{j \text{ such that} \\ x_{rj}<0}} \left[\frac{x_{0j}}{x_{rj}}\right]$$

Notice the beautiful symmetry with respect to the primal simplex algorithm—in the dual simplex algorithm we choose the row first and then find the column to enter the basis. The ratio test is a maximum over negative entries of ratios between Row 0 and Row r, instead of a minimum over positive entries of ratios between Column 0 and Column s.

In the absence of degeneracy, we move from (primal infeasible) basic solution to (primal infeasible) basic solution, increasing the cost. Hence we must run out of basic solutions and therefore terminate. In the presence of degeneracy, we must use some rule, such as Bland's (see Problem 2).

Example 3.7

Return to the shortest-path problem of Example 3.4, and choose the basis in the original tableau consisting of Columns 2, 3, and 4. From Fig. 3–3, this is clearly primal infeasible, because e_2, e_3, e_4 does not include a directed s-t path. Pivoting to obtain this basis yields the tableau

		f_1	f_2	f_3	f_4	f_5
$-z =$	-3	1	0	0	0	0
$f_2 =$	1	1	1	0	0	0
$f_3 =$	-1	-1	0	1	0	(-1)
$f_4 =$	1	0	0	0	1	1

This represents a primal-infeasible, dual-feasible point, and we can apply the dual simplex algorithm, yielding the pivot element x_{25} as shown. The next tableau is

$-z =$	-3	1	0	0	0	0
$f_2 =$	1	1	1	0	0	0
$f_5 =$	1	1	0	-1	0	1
$f_4 =$	0	-1	0	1	1	0

which is optimal; arc e_3 has been replaced by arc e_5, yielding the optimal path (e_2, e_5). □

3.7 Interpretation of the Dual Simplex Algorithm†

The dual simplex algorithm is so symmetrically related to the primal simplex algorithm that the reader may well suspect that the former is simply the latter applied to the dual. We now verify this suspicion.

At any point in a primal simplex algorithm, we can interpret the basic variables as slack variables $u \in R^m$ and write the primal as

$$\begin{aligned} \min\ & c'x \\ & Ax + u = b \\ & x, u \geq 0 \end{aligned} \tag{3.17}$$

This is equivalent to the program

$$\begin{aligned} \max\ & -c'x \\ & Ax \leq b \\ & x \geq 0 \end{aligned} \tag{3.18}$$

with dual

$$\begin{aligned} \min\ & \pi'b \\ & \pi'A \geq -c' \\ & \pi' \geq 0 \end{aligned} \tag{3.19}$$

†We thank F. Sadri for help with this section.

Introducing surplus variables $s \in R^n$, we can write this in standard form as follows.

$$\begin{aligned} &\min \pi' b \\ &\pi' A - s' = -c' \\ &\qquad \pi', s' \geq 0 \end{aligned} \tag{3.20}$$

or as the standard form program

$$\begin{aligned} &\min b' \pi \\ &(-A')\pi + s = c \\ &\qquad \pi, s \geq 0 \end{aligned} \tag{3.21}$$

We now have two standard form programs, (3.17) and (3.21), which can be interpreted as a primal-dual pair. The primal problem has basic variables u and nonbasic variables x; the dual has basic and nonbasic variables s and π, respectively. (These two programs can, in fact, be represented in the same tableau, with the identity basis elements omitted.)

Let us keep the two programs in (3.17) and (3.21) in our usual standard form and write them side by side, as shown in Fig. 3–8.

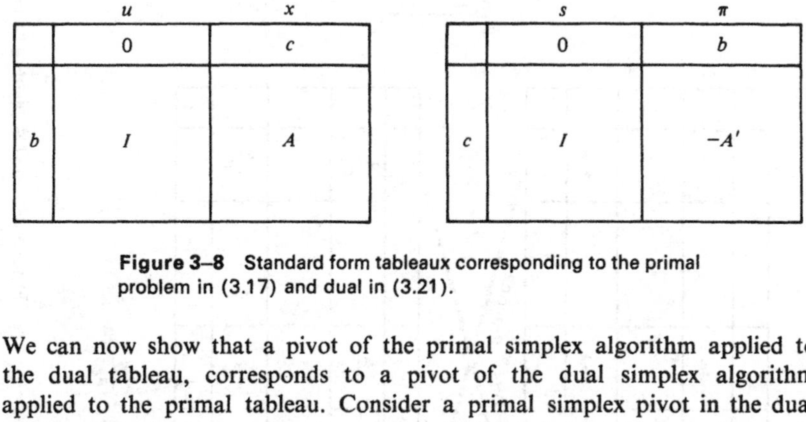

	u	x
	0	c
b	I	A

	s	π
	0	b
c	I	$-A'$

Figure 3–8 Standard form tableaux corresponding to the primal problem in (3.17) and dual in (3.21).

We can now show that a pivot of the primal simplex algorithm applied to the dual tableau, corresponds to a pivot of the dual simplex algorithm applied to the primal tableau. Consider a primal simplex pivot in the dual tableau: We begin by choosing a $b_r < 0$. The choice of pivot element is determined by

$$\min_{\substack{j \text{ such that} \\ (-a_{rj}) > 0}} \left[\frac{c_j}{(-a_{rj})} \right] = - \max_{\substack{j \text{ such that} \\ a_{rj} < 0}} \left[\frac{c_j}{a_{rj}} \right]$$

which checks the choice of pivot in the dual simplex algorithm. Figure 3–9 then shows a typical pivot in both tableaux, side by side for comparison. The reader can verify that the pivot operation in the dual tableau corresponds exactly

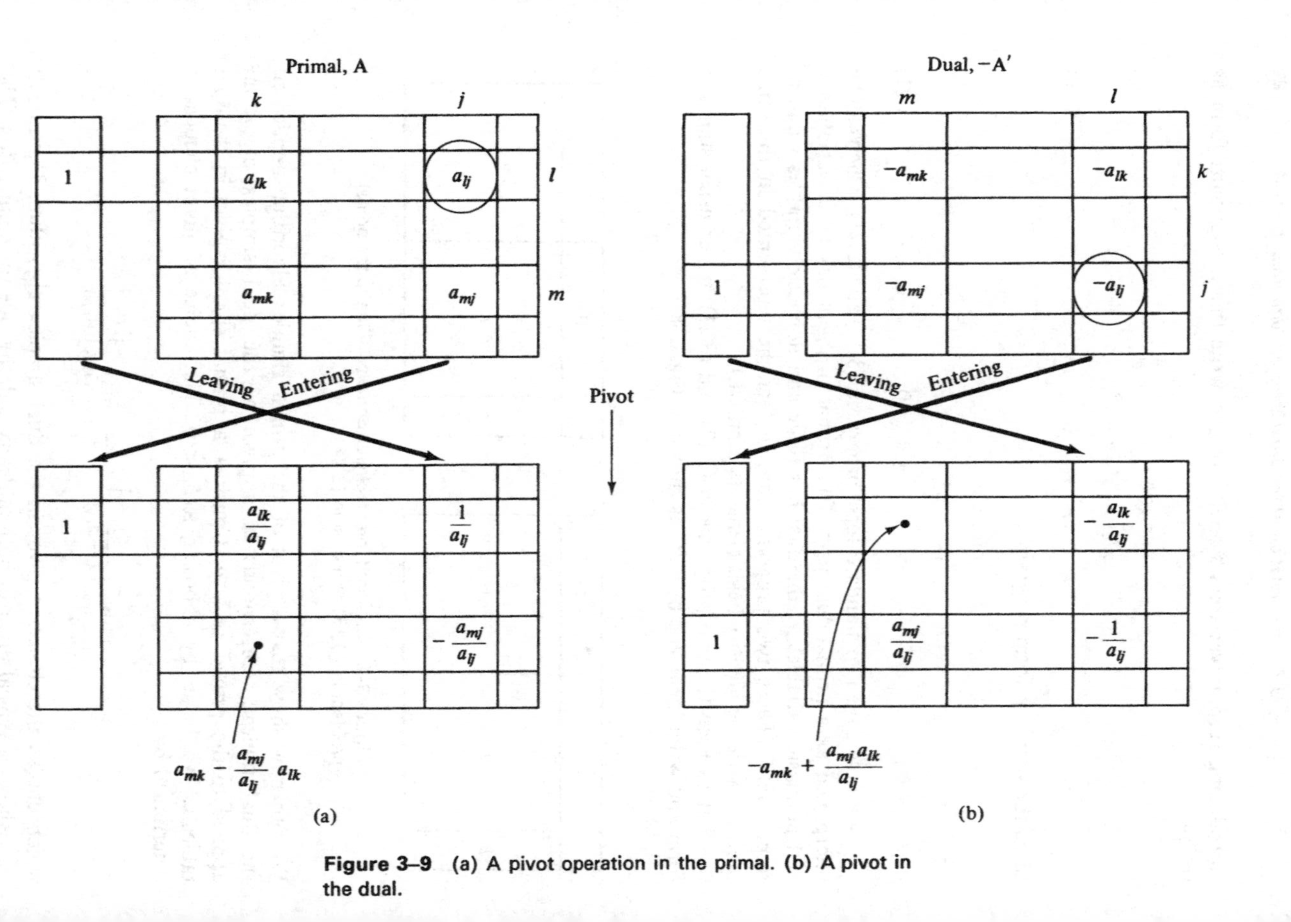

Figure 3–9 (a) A pivot operation in the primal. (b) A pivot in the dual.

to that in the primal tableau, provided that the new nonbasic column is exchanged with the new basic column in each tableau (see Problem 1).

PROBLEMS

1. (F. Sadri) Verify that a pivot of the primal simplex algorithm applied to the dual tableau of Fig. 3–8 corresponds exactly to a pivot of the dual simplex algorithm applied to the corresponding primal tableau, where the new nonbasic column is exchanged with the new basic column in each tableau.
2. (F. Sadri) From the result in the previous problem, derive an anticycling rule for the dual simplex algorithm which is analogous to Bland's rule for the primal simplex algorithm. Show that your rule works.
3. Justify carefully the inequalities in Eq. 3.7 in the proof of Theorem 3.1.
4. Find an optimal dual solution to the shortest-path problem in Example 3.4 by starting the simplex algorithm with the appropriate unit matrix as suggested in Example 3.6, thus verifying the dual solution shown in Fig. 3–5.
5. A physicist takes measurements of a variable $y(x)$; the results are in the form of pairs (x_i, y_i). The physicist wishes to find the straight line that fits this data best in the sense that the maximum vertical distance between any point (x_i, y_i) and the line is as small as possible. Formulate this problem as an LP. Why might you decide to solve the dual?

*6. [YG] Prove that if a standard form LP has an optimal solution but no optimal solutions that are degenerate, then the dual problem has a unique optimal solution.

*7. Consider an instance of LP in standard form. Suppose the primal has a degenerate optimal solution. Is the optimal solution to the dual necessarily not unique? If yes, prove it. If not, give a counterexample.

8. Show that a cone is a convex set.

*9. Show that a cone is a closed set. (Watch out for this one! A solution can be found in [YG].)

10. Check that Farkas' lemma holds when the set $\{a_i\}$ is empty.
11. Answer *true* or *false*:
    (a) If a linear program is infeasible, the dual must be unbounded.
    (b) If a linear program is unbounded, the dual must be infeasible.
12. Show that the dual of a canonical form LP is also in canonical form.
13. A network problem is formulated for a directed graph $G = (V, E)$ using the node-arc incidence matrix, as in Example 3.4. Show that a set of $|V| - 1$ columns is linearly independent if and only if the corresponding arcs, considered as undirected edges in the undirected version of G, is a tree. (Thus a basis corresponds to a tree.) Interpret the pivot step in the light of this fact.

14. Consider the following LP, which we call Problem P:

$$\min x_1 + x_3$$
$$x_1 + 2x_2 \leq 5$$
$$x_2 + 2x_3 = 6$$
$$x_1, x_2, x_3 \geq 0$$

(a) Solve P by the simplex algorithm.

(b) Write the dual D of P by inspection.

(c) Write the complementary slackness conditions for this problem and use them to solve the dual D. Check your answer by evaluating the optimal costs of P and D.

15. Repeat Problem 14 with the inequality in P replaced by $x_1 + 2x_2 \leq -5$.

16. Suppose we allow negative arc weights in the shortest-path problem, and furthermore, we consider an example in which there is a directed cycle that has negative total weight. What will happen when we apply simplex? Construct such an example. What can we say about the dual in such a circumstance?

17. Prove there is an optimal solution to the LP for the shortest-path problem in which each $f_i = 0$ or 1. When are there also optimal solutions which violate this condition?

18. Consider the $n \times n$ node-arc incidence matrix of a directed graph with exactly n nodes and n arcs. Show that its determinant is zero.

19. Suppose a standard form LP has a unique optimal solution. Does it follow that the dual has a nondegenerate optimal solution? Does the converse hold?

20. Suppose an LP in standard form has two columns that are proportional with a positive constant of proportionality. Construct an equivalent LP with one fewer column.

21. Another form of Farkas' lemma and the duality theorem is as follows: A system of linear inequalities $Ax \leq b$ is called *inconsistent* if there exists a y such that $y'A = 0$, $y'b < 0$, and $y \geq 0$. Show that the system $Ax \leq b$ has no solutions if and only if it is inconsistent.

NOTES AND REFERENCES

[vN] J. von Neumann is credited by D. Gale with being the first to state the duality theorem, Theorem 3.3, in privately circulated notes at least as early as 1947. See

[Ga] GALE, D., *The Theory of Linear Economic Models.* New York: McGraw-Hill Book Company, 1960.

Gale cites the first proof, based on von Neumann's notes, in

[GKT] GALE, D., H. W. KUHN, and A. W. TUCKER, "On Symmetric Games," in *Contributions to the Theory of Games*, ed. H. W. Kuhn and A. W. Tucker, *Ann. Math Studies*, no. 24. Princeton, N.J.: Princeton University Press, 1950.

Farkas' lemma is from

[Fa] FARKAS, J., "Theorie der Einfachen Ungleichungen," *J. Reine und Angewandte Math.*, 124 (1902), 1–27.

An elementary proof for Problem 9 is given in

[YG] YUDIN, D. B., and E. G. GOL'SHTEIN, *Linear Programming* (trans. from the Russian by Z. Lerman). Jerusalem: Israel Program for Scientific Translations, 1965.

Problem 21 is due to

[Ku] KUHN, H. W., "Solvability and Consistency for Linear Equations and Inequalities," *Amer. Math. Monthly*, 63 (1956), 217–32.

and was suggested by the following interesting paper of Chvátal.

[Ch] CHVÁTAL, V., "Some Linear Programming Aspects of Combinatorics," Report STAN-CS-75-505, Computer Science Department, Stanford University (September 1975).

4

Computational Considerations for the Simplex Algorithm

4.1 The Revised Simplex Algorithm

If we were to implement the simplex method as described in Chapter 2, we would update the entire $(m + 1) \times (n + 1)$ tableau at each iteration. It turns out we can avoid this by keeping a smaller, $(m + 1) \times (m + 1)$ matrix updated, and generating relative costs $\bar{c}_j$ and columns X_j as we need them. This has important practical implications, which we shall discuss after describing the method. We shall also see how the method, called the *revised simplex method* (or sometimes the *inverse matrix method*), can be applied to the important max-flow problem, and how it leads to a decomposition principle for problems of a certain form.

We saw in the last chapter that if we begin with an identity matrix in the initial tableau, then at any later iteration with a basis B, the matrix B^{-1} resides in its place. Let us assume for now that we begin with such an identity at the left end of the tableau, with a zero cost row, and keep only Rows and Columns 0 through m. At iteration l, we call this $(m + 1) \times (m + 1)$ matrix $CARRY^{(l)}$; the initial $CARRY^{(0)}$ is shown below.

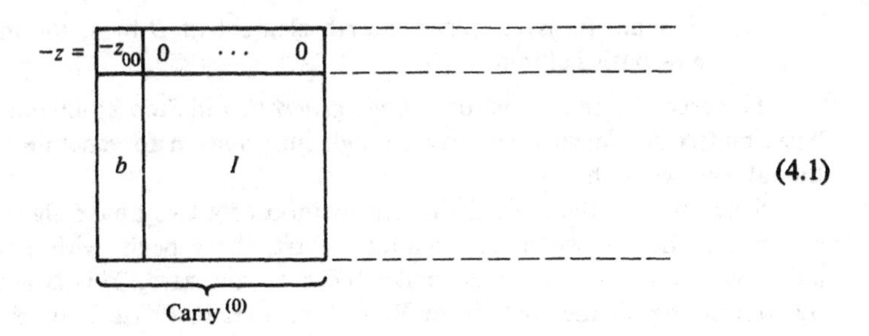

$$(4.1)$$

After iteration l in the simplex algorithm we shall have

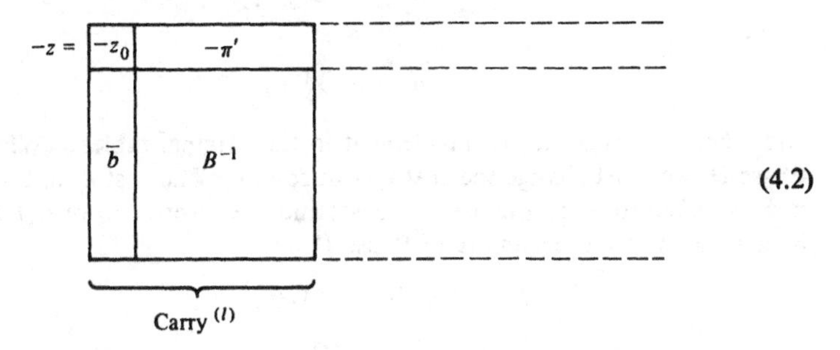

$$(4.2)$$

We also keep track of the current basic variables, the ordered index set B.

The original tableau A, the current $CARRY^{(l)}$, and the basis set B are sufficient to carry on with the primal simplex algorithm. The steps are as follows.

1. (Pricing Operation) Generate the relative costs

$$\bar{c}_j = c_j - \pi' A_j \tag{4.3}$$

one at a time, until we either find one that is negative—say for $j = s$—or we terminate with an optimal solution.

2. (Column Generation) Generate the column X_s as it would exist in the lth tableau by

$$X_s = B^{-1} A_s \tag{4.4}$$

Then determine the pivot element, say x_{rs}, by the usual ratio test

$$\min_{\substack{i \text{ such that} \\ x_{is} > 0}} \left(\frac{\bar{b}_i}{x_{is}} \right), \tag{4.5}$$

or discover that the cost is unbounded.

3. (Pivot) Update $CARRY^{(l)}$ to obtain $CARRY^{(l+1)}$ by operating on it as if we were pivoting on x_{rs} and $CARRY^{(l)}$ were the left end of the tableau. This uses the generated column X_s.

4. (Update Basis) Replace the rth element of B by s, the index of the new basis column.

The crucial step here is the *column generation* in Step 2; having the inverse basis matrix B^{-1} means we have enough information to generate any part of the tableau we wish.

When we use the revised simplex method as a two-phase algorithm, there are certain details we must attend to. First, if we begin with artificial variables with unit cost, we must make the cost row zero. This is accomplished by subtracting all the rows from Row 0, so that in Phase I the $\bar{c}_j$ in Eq. 4.3 are calculated by

$$\bar{c}_j = d_j - \pi' A_j$$

where (4.6)

$$d_j = -\sum_{i=1}^{m} a_{ij}$$

and where, as usual, a_{ij} is the element in the original tableau. When we enter Phase II, we must change the cost row as follows. The cost d_j in Eq. 4.6 should now be taken to be c_j, the original cost; and the zeroth row of *CARRY* should be generated at the beginning of Phase II by

$$-\pi' = -c'_B B^{-1}$$

4.2 Computational Implications of the Revised Simplex Algorithm [Be, OH, Las]

At first glance, it would seem that updating an $(m + 1) \times (m + 1)$ matrix instead of an $(m + 1) \times (n + 1)$ tableau would provide a direct computational savings. Examine the pricing operation of the revised simplex method, however: It requires the calculation of the inner products $\pi' A_j$ between the dual variable π and the column A_j in the original problem tableau. If this is carried out for every nonbasic column, this requires $m \times (n - m)$ multiplications. But this is not significantly less than the number of multiplications needed to update the tableau in the ordinary simplex method!

The real advantages of the revised method are somewhat more subtle, but very important nonetheless. First, we need not compute the relative costs of every nonbasic column; we can take the first column with negative cost, as in Bland's rule (Sec. 2.7). This will reduce the computational burden to some fraction of that required for pricing every column—a fraction determined by the average number of columns that must be examined until one with negative relative cost is found.

A second real advantage of the revised method stems from the fact that the pricing operation uses the columns A_j of the *original* tableau. As we saw in the shortest-path problem and shall see in many similar problems of a combinatorial nature, the original tableau is often very sparse. If it is the node-arc incidence matrix of a graph, for example, it contains only $2n$ nonzero elements. This implies not only that the pricing operation can be done quickly, but also that the original tableau can be stored in a very compact way and perhaps not generated explicitly at all. We shall illustrate these advantages in the next section.

We should mention a refinement of the revised simplex method in which the inverse basis matrix is stored in *product form*, rather than explicitly as an $(m + 1) \times (m + 1)$ matrix. Each pivot operation can be represented as premultiplication by a matrix P defined to be the $(m + 1) \times (m + 1)$ identity matrix except for Column r, which contains the vector

$$\eta = \begin{bmatrix} \frac{-x_{1s}}{x_{rs}} \\ \frac{-x_{2s}}{x_{rs}} \\ \cdot \\ \cdot \\ \cdot \\ \frac{1}{x_{rs}} \\ \cdot \\ \cdot \\ \cdot \\ \frac{-x_{m+1,s}}{x_{rs}} \end{bmatrix} \begin{matrix} \\ \\ \\ \\ \\ \ldots \text{Row } r \\ \\ \\ \\ \\ \end{matrix}$$

At any stage l, then, the current $CARRY^{(l)}$ can be written

$$CARRY^{(l)} = P_l \cdots P_1 \cdot CARRY^{(0)}$$

and each P can be stored very compactly, because it is determined only by the index r and the vector η. When the sequence of η's becomes too long, it can be replaced by a shorter sequence, using a process called *reinversion*, in which an equivalent, but shorter, sequence of pivots is found to arrive at the current basis. Such techniques can greatly reduce the storage and time required to perform the simplex algorithm, especially if special attention is paid to reducing the number of nonzero elements in the η-sequence. The reader is referred to [OH, Las] for a detailed discussion of these techniques.

4.3 The Max-Flow Problem and Its Solution by the Revised Method

In some network problems that can be formulated as LP's, the constraint matrix can be derived directly from the graph underlying the problem, and, as we suggested in the previous section, need not be generated explicitly at all.

We illustrate this situation by a formulation of the max-flow problem, a fundamental problem that will appear again later on.

Definition 4.1

Given a flow network $N = (s, t, V, E, b)$, with $n = |V|$ nodes and $m = |E|$ arcs, an instance of the max-flow problem (MFP) is defined as the problem of finding an s-t flow $f \in R^m$ with maximum value v. □

Example 4.1

Figure 4–1 shows a flow network with $n = 4$ nodes and $m = 5$ arcs. The maximum s-t flow value is 2 by inspection. □

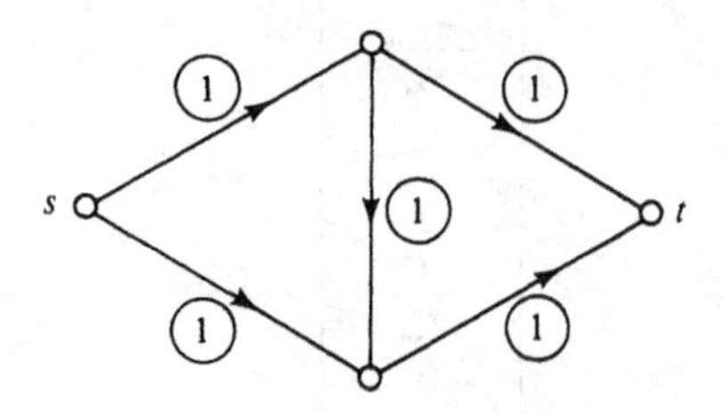

Figure 4–1 A flow network illustrating an MFP. The circled numbers are capacities.

We now formulate this problem as an LP in a somewhat surprising way: Let the arcs be numbered $e_1, \ldots, e_m$ as usual; let $C_1, \ldots, C_p$ be an enumeration of *every chain* (i.e., path) from s to t. Our LP will begin with m rows, one for each arc, and p columns, one for each s-t chain C_j. Of course, in any reasonably large problem, there will be an enormous number of s-t chains; but, as we suggested above, we do not intend to write down the constraint matrix explicitly. Now define the *arc-chain incidence matrix* $D = [d_{ij}]$ by

$$d_{ij} = \begin{cases} 1 & \text{if } e_i \text{ is in } C_j \\ 0 & \text{otherwise} \end{cases} \qquad i = 1, \ldots, m; \quad j = 1, \ldots, p$$

The capacity constraints then become simply

$$Df \leq b$$

The problem is to maximize the sum of all the flows in all the chains, so we want

$$\min c'f$$

where

$$c = (-1, \ldots, -1) \in R^p$$

Thus the complete formulation is

$$\min c'f$$
$$Df \leq b$$
$$f \geq 0$$

To convert this LP to standard form, we introduce a slack vector $s \in R^m$ and define an augmented flow vector

$$\hat{f} = (f \mid s)$$

a corresponding cost vector

$$\hat{c} = (c \mid 0)$$

and a new constraint matrix

$$\hat{D} = (D \mid I)$$

The problem in standard form is then

$$\begin{aligned} \min z &= \hat{c}'\hat{f} \\ \hat{D}\hat{f} &= b \\ f &\geq 0 \end{aligned} \tag{4.7}$$

Each slack variable s_i represents the difference between the flow in arc i and the capacity b_i, $i = 1, \ldots, m$.

Let us now assume we are using the revised simplex algorithm to solve this problem, and at some stage we have a bfs and the corresponding dual variable $\pi \in R^m$. (We note that there is no need for a Phase I, since an initial feasible solution to the problem of Eq. 4.7 is provided by $f = 0$ and $s = b$, which corresponds to zero flow.) The criterion for a column (chain C_j) to enter the basis profitably is then, by Eq. 4.3,

$$\bar{c}_j = c_j - \pi' D_j < 0$$

where D_j is the jth column of the arc-chain incidence matrix D. We may write this criterion as

$$(-\pi')D_j < +1$$

because $c_j = -1$ for every $j = 1, \ldots, p$.

This has the following interpretation: the vector $-\pi$ is a weight vector on the arcs, and $-\pi' D_j$ is the cost of the chain C_j under this weight. A chain can then enter the basis profitably if its cost is less than 1.

We now come to the main point: To find a profitable column, we need find only the *shortest chain* from s to t under the weight $-\pi$. If that shortest chain, say C_j, has cost no less than 1, then the optimality criterion is satisfied; if not, we introduce C_j into the basis. The calculation therefore requires only the maintenance of an $(m+1) \times (m+1)$ *CARRY* matrix and the repeated solution of the shortest-path problem.

The shortest-path problem is (as we shall see later) much easier if the edge costs are nonnegative; this can be ensured as follows: If any $-\pi_i$ is negative, then we simply insert the corresponding slack variable s_i into the basis.

Example 4.1 (Continued)

We return to the simple max-flow problem in Fig. 4–1, number the arcs as shown in Fig. 4–2, and write the first *CARRY* matrix as

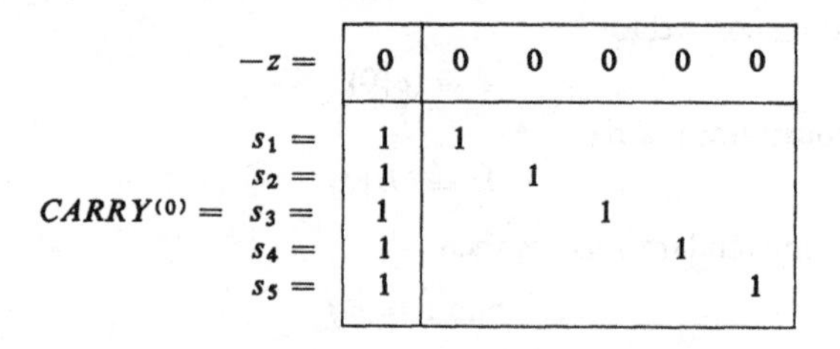

$CARRY^{(0)} =$

$-z =$	0	0	0	0	0	0
$s_1 =$	1	1				
$s_2 =$	1		1			
$s_3 =$	1			1		
$s_4 =$	1				1	
$s_5 =$	1					1

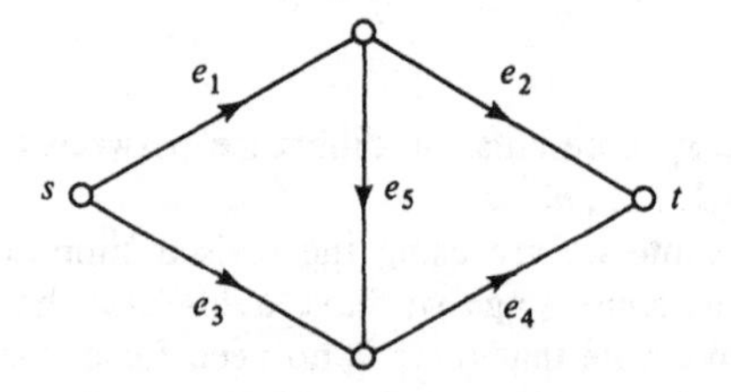

Figure 4–2 Numbering the arcs in the example.

The initial arc weights are all zero, so we may begin by introducing any shortest chain (of length zero) into the basis. Let us be perverse and start with a chain that is not in the optimal solution, say $C_1 = (e_1, e_5, e_4)$. The corresponding column of the tableau is

$$\begin{array}{|c|}\hline -1 \\ \hline \textcircled{1} \\ 0 \\ 0 \\ 1 \\ 1 \\ \hline \end{array}$$

and when we perform the indicated pivot we get the next *CARRY*:

$CARRY^{(1)} =$

$-z =$	1	1				
$C_1 =$	1	1				
$s_2 =$	1		1			
$s_3 =$	1			1		
$s_4 =$		-1			1	
$s_5 =$		-1				1

The weighted network corresponding to this solution is shown in Fig. 4–3.

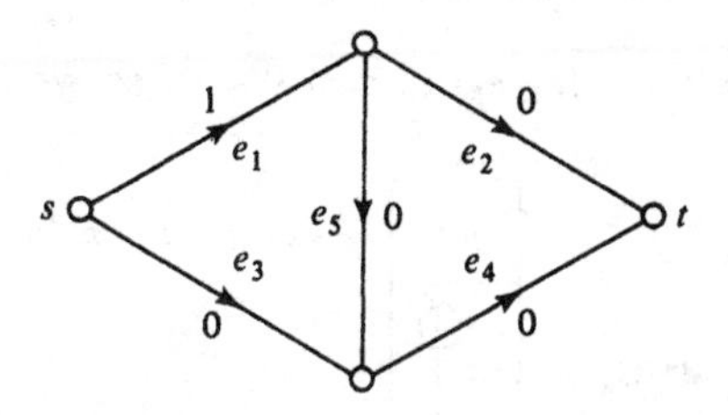

Figure 4–3 The shortest-path problem corresponding to the second pivot.

The shortest path is chain $C_2 = (e_3, e_4)$ with length $0 < 1$, so we bring C_2 into the basis. Before we do this, we must generate the current column by

$B^{-1}C_2 =$

-1	$\longleftarrow \bar{c}_j = -1 - \pi' D_j$
0	
0	
1	
①	
0	

After the indicated pivot, we get

$CARRY^{(2)} =$

$-z =$	1				1	
$C_1 =$	1	1				
$s_2 =$	1		1			
$s_3 =$	1	1		1	-1	
$C_2 =$		-1			1	
$s_5 =$		-1				1

The new weighted network is shown in Fig. 4–4.

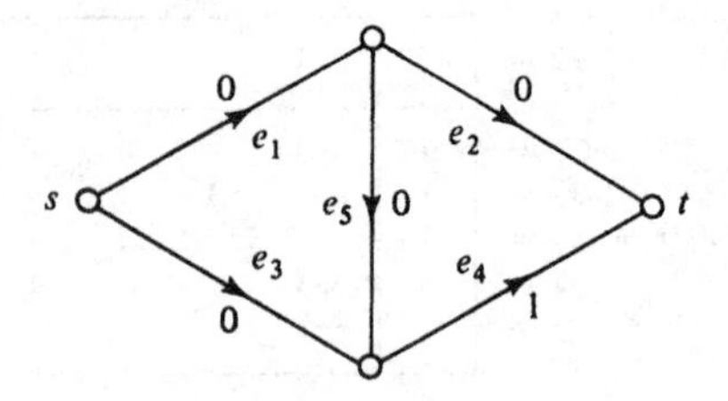

Figure 4–4 The shortest-path problem for the third pivot.

The shortest path is $C_3 = (e_1, e_2)$ of cost $0 < 1$, so we generate the current column

$$\begin{array}{c} \boxed{-1} \leftarrow \bar{c}_j = -1 - \pi' D_j \\ B^{-1}C_3 = \begin{array}{|c|} \hline \textcircled{1} \\ 1 \\ 1 \\ -1 \\ -1 \\ \hline \end{array} \end{array}$$

and pivot as shown, yielding

	$-z =$	2	1			1	
	$C_3 =$	1	1				
	$s_2 =$		-1	1			
$CARRY^{(3)} =$	$s_3 =$				1	-1	
	$C_2 =$	1				1	
	$s_5 =$	1					1

with $z = -2$ (corresponding to a flow of value 2). The new shortest-path problem is shown in Fig. 4–5.

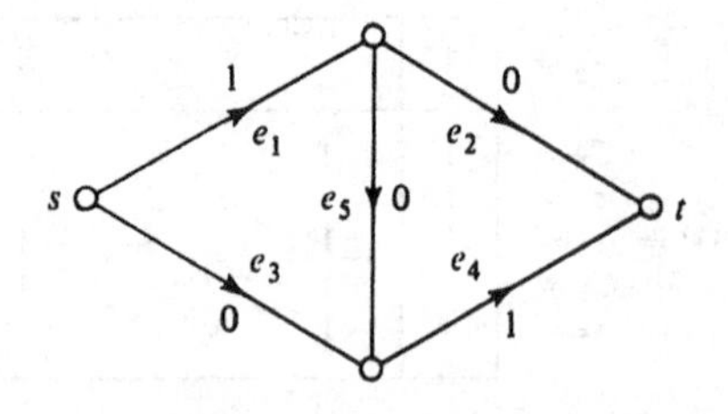

Figure 4–5 The shortest-path problem at optimality.

The shortest path now has length $1 \nless 1$, so we have reached an optimal solution. □

Before going on, we observe that we started with one problem—max-flow—and generated a sequence of related and simpler problems, shortest-path problems. This is a recurrent theme throughout optimization theory, one that will appear many times in this book.

4.4 Dantzig-Wolfe Decomposition [DW]

It often happens that a large linear program is really a collection of smaller linear programs that are largely independent of each other. As an example, suppose we have some LP where the constraint matrix is the node-arc incidence matrix of a very large graph, but the graph has the form shown in Fig. 4–6.

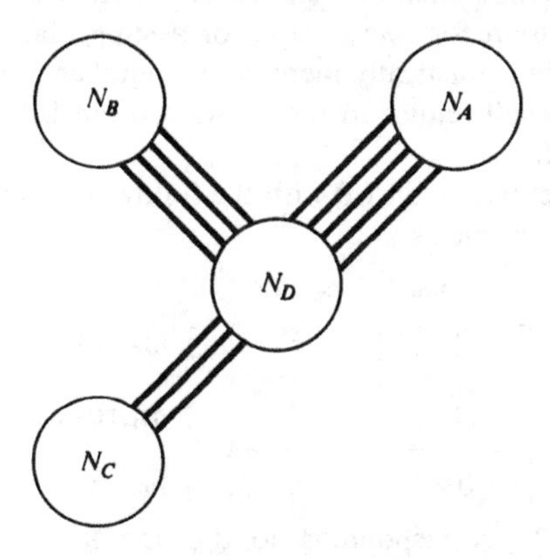

Figure 4–6 A graph with an easily decomposable node-arc incidence matrix.

Arcs exist only between nodes of the same sets and between the set N_D and the others. The node-arc incidence matrix of such a graph can be written as follows.

N_D {	D_1	D_2	D_3
N_A {	A	0	0
N_B {	0	B	0
N_C {	0	0	C

The decomposition method is designed to take advantage of this special structure by allowing us to solve the entire problem by iteratively solving problems the sizes of A, B, and C.

We already have the material from which to construct the decomposition method. First, the revised form of the simplex algorithm allows us to increase the number of columns indefinitely without increasing the size of the working matrix (what we called the *CARRY* matrix).

A second idea comes from our formulation of the MFP in arc-chain form: We reformulated the problem to introduce an enormous number of columns, but we counted on the fact that in the revised method we never need to write out all the columns explicitly. In that arc-chain formulation, the pricing operation of a column turned out to be the calculation of the price of a chain under a cost determined by the dual variable π (sometimes called, in fact, a *price* vector). The pricing of each column, one at a time, was then avoided by finding only the *shortest* chain.

In the decomposition method, this device is carried to an extreme. We shall introduce a column for *every vertex* of a subproblem. This will reduce the number of rows but drastically increase the number of columns. However, the pricing operation will reduce to the solution of an LP the size of the subproblem we broke off.

In more detail, let us use an LP with the following constraint matrix as an example with two subproblems:

$$\begin{array}{cc} n_1 \text{ cols.} & n_2 \text{ cols.} \\ \left[\begin{array}{c|c} D & F \\ \hline A & 0 \\ \hline 0 & B \end{array}\right] & \begin{array}{l} \} \; m_0 \text{ rows} \\ \} \; m_1 \text{ rows} \\ \} \; m_2 \text{ rows} \end{array} \end{array} \tag{4.8}$$

with variables $x \in R^{n_1}$ corresponding to the first n_1 columns and $y \in R^{n_2}$ corresponding to the next n_2 columns. We write the complete LP (in standard form) as

$$\begin{aligned} \min z = c'x + d'y & \\ Dx + Fy &= b_0 \\ Ax \qquad &= b_1 \\ By &= o_2 \\ x, y &\geq 0 \end{aligned} \tag{4.9}$$

We call the first m_0 equations the *coupling equations*; the problems associated with the succeeding sets of rows we call *Subproblems* A and B, respectively. In particular, consider the constraints of Subproblem A:

$$\begin{aligned} Ax &= b_1 \\ x &\geq 0 \end{aligned} \tag{4.10}$$

By Theorem 2.3, any feasible point in this subproblem can be written as a convex combination of vertices of the feasible set. Call these vertices $x_1, \ldots,$ x_p, and write

$$x = \sum_{j=1}^{p} \lambda_j x_j \tag{4.11}$$

where

$$\lambda_j \geq 0$$

and

$$\sum_{j=1}^{p} \lambda_j = 1$$

Similarly, we write

$$y = \sum_{j=1}^{q} \mu_j y_j \tag{4.12}$$

$$\mu_j \geq 0$$

$$\sum_{j=1}^{q} \mu_j = 1$$

where the y_j are the vertices of Subproblem B.

Replacing x and y by their representations in (4.11) and (4.12), Eq. 4.9 becomes the LP in the variables λ_i and μ_i shown below:

variables:	$\lambda_1 \ldots \lambda_p$	$\mu_1 \ldots \mu_q$		
m_0 rows {	$\Delta_1 \ldots \Delta_p$	$\Phi_1 \ldots \Phi_q$	b_0	
1 row {	$1 \ldots 1$	$0 \ldots 0$	1	(4.13)
1 row {	$0 \ldots 0$	$1 \ldots 1$	1	

where

$$\begin{aligned} \Delta_j &= Dx_j \qquad j = 1, \ldots, p \\ \Phi_j &= Fy_j \qquad j = 1, \ldots, q \end{aligned} \tag{4.14}$$

and the cost is

$$\min z = \xi'\lambda + \delta'\mu \tag{4.15}$$

where

$$\begin{aligned} \xi_j &= c'x_j \qquad j = 1, \ldots, p \\ \delta_j &= d'y_j \qquad j = 1, \ldots, q \end{aligned} \tag{4.16}$$

and

$$\lambda, \mu \geq 0 \tag{4.17}$$

are the new variables, in R^p and R^q, respectively. The problem in this form is usually called the *master problem*.

In any reasonable example, we now have an astronomical number of columns, one for each vertex of each of the two subproblems. But the number of rows has been reduced from $m_0 + m_1 + m_2$ to $m_0 + 2$ and the revised method can be implemented with a *CARRY* matrix of size $(m_0 + 3) \times (m_0 + 3)$. All other considerations aside, this means we can fit much larger problems into fast-access storage. We now require $(m_0 + 3)^2$ storage locations for *CARRY*, as opposed to $(m_0 + m_1 + m_2 + 1)^2$ in the original formulation. If $m_0 = m_1 = m_2$, for example, we have reduced the main storage requirement by almost a factor of nine.

Assume now that we are using the revised simplex method; we next must see how to negotiate the pricing operation. At any stage, *CARRY* will have a set of prices in Row 0, which we shall write as a partitioned vector

$$\text{col}\,(\pi, \alpha, \beta)$$

where $\pi \in R^{m_0}$ corresponds to the first m_0 rows and $\alpha, \beta \in R^1$ to the next two rows of the master problem in (4.13). The relative cost of the λ_j-column is therefore

$$\begin{aligned} \bar{c}_j &= \xi_j - (\pi, \alpha, \beta)' \begin{bmatrix} \Delta_j \\ 1 \\ 0 \end{bmatrix} \\ &= \xi_j - \pi'\Delta_j - \alpha \qquad j = 1, \ldots, p \end{aligned} \tag{4.18}$$

so the criterion for a column to be brought profitably into the current basis is

$$\xi_j - \pi'\Delta_j < \alpha \qquad j = 1, \ldots, p \tag{4.19}$$

If this is true for any vertex of Subproblem A, we have found a profitable column among the first p columns.

We now come to the point. As promised, we do not have to evaluate the left-hand side of Eq. 4.19 for every vertex of Subproblem A; this would clearly be impractical for a problem of reasonable size. Instead, we search for the smallest value of the left-hand side of (4.19) over all the vertices of the subproblem. That is, we find

$$\min_{1 \leq j \leq p} (\xi_j - \pi'\Delta_j) \tag{4.20}$$

This is equivalent to

$$\min_{\substack{\text{all vertices} \\ x_j \text{ of} \\ \text{Subproblem A}}} (c' - \pi' D)x_j \tag{4.21}$$

But searching for the minimum value of a linear function at all the vertices of the feasible set of an LP is simply an LP! Thus the pricing operation for the first p columns can proceed by solving the LP

$$\begin{aligned} &\min\,(c' - \pi'D)x \\ &Ax = b_1 \\ &x \geq 0 \end{aligned} \tag{4.22}$$

Similarly, we can determine if there is a favorable column among the last q columns by solving

$$\begin{aligned} &\min\,(d' - \pi'F)y \\ &By = b_2 \\ &y \geq 0 \end{aligned} \tag{4.23}$$

and comparing the minimum cost to β, by an argument analogous to that surrounding Eq. 4.18. The complete algorithm is shown in Fig. 4–7.

```
procedure decomposition
begin
  opt := 'no';
  set up CARRY with zero row (π, α, β);
  while opt = 'no' do
  begin solve the LP
     min (c'−π'D)x = z0
        Ax = b1
         x≥0;
     if z0 < α then generate the column corresponding to the solution with this cost,
                    and pivot in CARRY
     else begin     solve the LP
                    min (d'−π'F)y = z0
                       By = b2
                        y≥0;
                    if z0<β then generate the column corresponding to the solution
                                  with this cost, and pivot in CARRY
                       else opt := 'yes'
          end
  end
end
```

Figure 4–7 The decomposition algorithm.

We may describe the operation of the decomposition method in the following terms. The master problem, based on its overall view of the entire situation, sends a price to Subproblem A. This subproblem then responds with a solution (called a *proposal*) for possibly improving the overall problem, based on its local information and the price. The master problem then weighs the cost (z_0) of this proposal against its criterion α for Subproblem A. If the proposal is cheaper than α, it is implemented by bringing it into the basis. If not, Subproblem B is sent a price and asked for a proposal. As long as a subproblem can produce a favorable proposal, the master problem can find a favorable pivot. When neither subproblem can come up with a favorable proposal, we have reached an optimal solution of the entire problem. (The reader is referred to [Dal] for an entertaining dramatization of an example.)

What we have done for two subproblems can clearly be done for any number of subproblems, in which case the constraint matrix will take the form shown below.

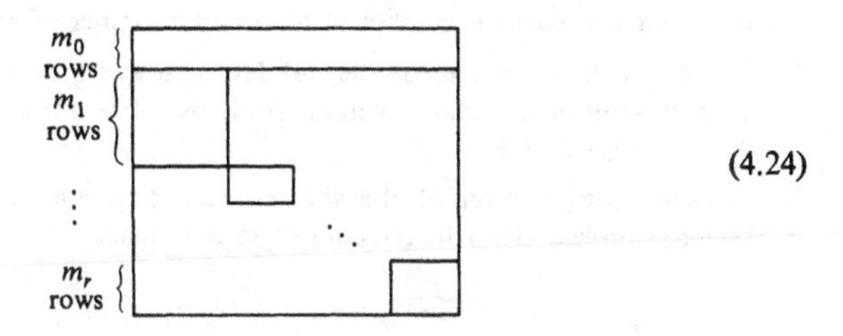

This problem has $\sum_{i=0}^{r} m_i$ rows, as opposed to $m_0 + r$ in the master problem of the decomposition algorithm.

Example 4.2

A little arithmetic shows the effectiveness of the approach in fitting large problems into a computer. Suppose we have 100 subproblems, each with 100 rows, and 100 coupling equations. Then the original problem has 10,100 rows; *CARRY* will have $10{,}101 \times 10{,}101 \approx 10^8$ entries, which cannot fit in the fast storage of any computer known to the authors. On the other hand, the master problem will have a *CARRY* with $201 \times 201 \approx 4 \times 10^4$ entries, which is practical to store in the fast memory of any reasonably large computer. □

The clear advantage of the decomposition algorithm is its effect on the space requirements; we cannot say much about the time requirements, because we do not know how many times the subproblems must be solved before optimality is reached.

PROBLEMS

1. Show that the number of *s-t* chains in a flow network can grow exponentially as a function of the number of vertices.
2. From the solution of the max-flow problem by the simplex algorithm, prove the following: In any max-flow problem for a network with m arcs, there is an optimal flow that can be decomposed into the sum of flows along no more than m chains.
3. Verify that Bland's anticycling rule can be followed easily in the revised simplex method.

*4. Investigate the details of implementing a two-phase version of the decomposition method using the revised algorithm. In particular, what if a subproblem is infeasible? Redundant? Unbounded? Is it easy to follow Bland's rule and thereby to prove finiteness?

5. (Programming project) Repeat Problem 15 in Chapter 2 for the revised simplex algorithm. Compare the regular and revised methods on various problems by collecting statistics on number of pivots and number of multiplications.

*6. Prove or give a counterexample: (a) The inverse basis matrix during the revised simplex solution of max-flow never contains entries other than 0 or ± 1. (b) The π_i are always 0 or 1.

7. Consider the solution of the shortest-path problem in node-arc form using revised simplex. Describe the column selection step.

8. Let $p_i = x_{rs}$ be the ith pivot element in an application of revised simplex. Show that after k pivots, the determinant of the basis matrix B is given by $\prod_{i=1}^{k} p_i$.
9. If we solve a linear programming problem that is highly degenerate, we may have many tied rows at the beginning of the application of simplex. Explain how we may encounter severe numerical problems if we resolve these ties without regard for the size of the potential pivots. (*Hint:* Consider Problem 8.) Suggest a remedy. (*Hint:* Ignore the problem of cycling and consider how Gauss elimination is usually implemented.)
10. If we use revised simplex to solve a linear programming problem where all the data consists of q-bit numbers and the original tableau is $m \times n$, how many bits do we need to keep for each word in our calculations to obtain an answer of infinite precision?

NOTES AND REFERENCES

More on the practical computational aspects of the simplex and revised simplex algorithms can be found in

[Be] BEALE, E. M. L., *Mathematical Programming in Practice.* New York: John Wiley & Sons, Inc., 1968.

[OH] ORCHARD-HAYS, W., *Advanced Linear-Programming Computing Techniques.* New York: McGraw-Hill Book Company, 1968.

[Las] LASDON, L. S., *Optimization Theory for Large Systems.* London: MacMillan, Inc., 1970.

The decomposition method was first published in

[DW] DANTZIG, G. B., and P. WOLFE, "Decomposition Principle for Linear Programming," *OR*, 8, no. 1 (1960), 101–11.

Dantzig has dramatized the idea that the decomposition method is a mathematical expression of decentralized planning in a four-act play with the characters "Staff," the coordinator of a shipping operation, and "F. M. Dalks," a fictitious (but composite) consultant. See

[Dal] DANTZIG, G. B., *Linear Programming and Extensions.* Princeton, N.J.: Princeton University Press, 1963.

5

The Primal-Dual Algorithm

5.1 Introduction

This chapter is devoted to the primal-dual algorithm, which is a general algorithm for solving LP's. The method was actually developed from a less general algorithm that was devised for certain network problems, and, as we shall see in this and succeeding chapters, provides the key idea for generating specialized algorithms for many graph-related problems. We begin with an informal outline of the method.

Starting with an LP in standard form, which we call the primal problem P,

$$\begin{aligned} \min z &= c'x \\ Ax &= b \geq 0 \\ x &\geq 0 \end{aligned} \tag{P}$$

construct its dual D:

$$\begin{aligned} \max w &= \pi' b \\ \pi' A &\leq c' \\ \pi' &\gtrless 0 \end{aligned} \tag{D}$$

Notice that we have multiplied the equalities in P by -1 where necessary to make $b \geq 0$. Recall the *complementary slackness* conditions (Theorem 3.4): If x is feasible in P and π in D, a necessary and sufficient condition for both to be optimal is that

$$\pi_i(a_i'x - b_i) = 0 \qquad \text{for all } i \tag{5.1}$$

$$(c_j - \pi' A_j)x_j = 0 \qquad \text{for all } j \tag{5.2}$$

The set of relations in (5.1) will be satisfied by any x that is feasible in P since P is in standard form, so we focus attention on the set in (5.2).

Now suppose that we have a π feasible in the dual D. If we could somehow find an x feasible in P that satisfied

$$x_j = 0$$

whenever $c_j - \pi' A_j > 0$, this x (and indeed this π) would be optimal. The *primal-dual algorithm* is derived from the idea of searching for such an x, given a π. We search for such an x by solving an auxiliary problem, called the *restricted primal* (RP), determined by the π we are working with. If our search for the x is unsuccessful, we nevertheless obtain information from the dual of RP, which we call DRP, that tells us how to improve the particular π with which we started. Iterating in this way, we converge to optimality in a finite number of steps (see Sec. 5.3 for a discussion of finiteness). Fig. 5–1 illustrates the method in schematic form.

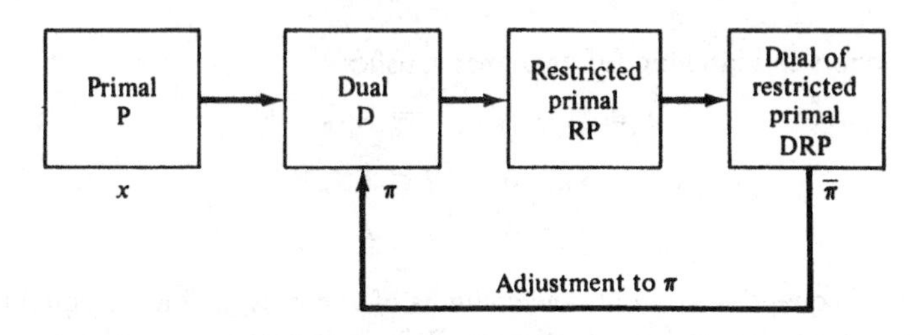

Figure 5–1 An outline of the primal-dual method.

5.2 The Primal-Dual Algorithm [DFF]

The primal-dual algorithm is properly called a *dual* algorithm, because we begin with a π feasible in D and maintain dual feasibility throughout. In the case that $c \geq 0$, we may take $\pi = 0$ immediately as an initial dual feasible point. When c is not nonnegative, we may nevertheless find a feasible π easily using a device attributed to Beale [DFF], among others. Introduce the variable x_{n+1} to the primal problem P and add the constraint

$$x_1 + x_2 + \cdots + x_n + x_{n+1} = b_{m+1} \tag{5.3}$$

where b_{m+1} is taken larger than the sum of any possible solution values $x_1, \ldots, x_n$ to P (for example, n times the M of Lemma 2.1) and the cost $c_{n+1} = 0$. Clearly, Eq. 5.3 does not change the solution to P. The dual of the new primal then has one new variable π_{m+1} and one new equation as follows:

$$\max w = \pi' b + \pi_{m+1} b_{m+1}$$
$$\pi' A_j + \pi_{m+1} \leq c_j \qquad j = 1, \ldots, n$$
$$\pi_{m+1} \leq 0$$

A feasible solution to this LP is simply

$$\pi_i = 0 \qquad i = 1, \ldots, m$$
$$\pi_{m+1} = \min_{1 \leq j \leq n} \{c_j\} < 0$$

The last inequality follows from the assumption that c is not nonnegative.

Let us assume, then, that we have a π feasible in D, without changing the notation to allow for the possible maneuver above. Now some of the inequality constraints

$$\pi' A_j \leq c_j$$

will be satisfied strictly and some will not. Define the index set J by

$$J = \{j : \pi' A_j = c_j\} \tag{5.4}$$

We know from (5.1) and (5.2) that an x feasible in P is optimal exactly when

$$x_j = 0 \qquad \text{for all } j \notin J \tag{5.5}$$

This amounts to searching for an x that satisfies

$$\begin{aligned} \sum_{j \in J} a_{ij} x_j &= b_i \qquad i = 1, \ldots, m \\ x_j &\geq 0 \qquad j \in J \\ x_j &= 0 \qquad j \notin J \end{aligned} \tag{5.6}$$

This set of equalities uses only the columns of A corresponding to equalities in D, which in turn correspond to the set J; for this reason we call J the set of *admissible columns.* To search for such an x, we invent a new LP, called the *restricted primal* (RP), as follows:

$$\min \xi = \sum_{i=1}^{m} x_i^a$$

$$\begin{aligned} \sum_{j \in J} a_{ij} x_j + x_i^a &= b_i \qquad i = 1, \ldots, m \\ x_j &\geq 0 \qquad j \in J \\ x_j &= 0 \qquad j \notin J \\ x_i^a &\geq 0 \end{aligned} \tag{RP}$$

That is, we introduce artificial variables x_i^a, $i = 1, \ldots, m$, one for each equation in P. The restricted primal RP can then be solved using the ordinary simplex algorithm. If the optimal solution to RP is $\xi_{\text{opt}} = 0$, then we have in fact found a solution to Eq. 5.6 and hence an optimal solution to P. What is of interest next is what happens when $\xi_{\text{opt}} > 0$ in RP.

To answer this question, we need to consider the dual DRP of the restricted primal RP, which is:

$$\max w = \pi' b \tag{5.7}$$

$$\pi' A_j \leq 0 \qquad j \in J \tag{5.8}$$

$$\pi_i \leq 1 \qquad i = 1, \ldots, m \qquad \text{(DRP)} \tag{5.9}$$

$$\pi_i \gtrless 0 \tag{5.10}$$

We denote the optimal solution of DRP, obtained when RP is solved, by $\bar{\pi}$.

We are now in the following position: We have tried to find a feasible x using only admissible columns, but because $\xi_{\text{opt}} > 0$ in RP, we have failed. All we have to show for our effort is the optimum of RP and its corresponding dual in DRP, $\bar{\pi}$. This suggests examining a "corrected" π, given by a linear combination of our original π and $\bar{\pi}$:

$$\pi^* = \pi + \theta\bar{\pi} \tag{5.11}$$

In analogy with the ordinary simplex algorithm, we ask how to choose θ so that π^* stays feasible in D and leads to an improved cost function. Let us first examine the new cost

$$\pi^{*\prime} b = \pi' b + \theta \bar{\pi}' b \tag{5.12}$$

Since RP and DRP are a primal-dual pair, their optimal costs at mutual optimality are equal, so we have

$$\bar{\pi}' b = \xi_{\text{opt}} > 0 \tag{5.13}$$

We should therefore take $\theta > 0$ to increase (and therefore improve) the cost in D.

Next, consider the effect of adding $\theta\bar{\pi}$ to π on feasibility in D. To remain feasible in D, we require

$$\pi^{*\prime} A_j = \pi' A_j + \theta \bar{\pi}' A_j \leq c_j \tag{5.14}$$

When $\bar{\pi}' A_j \leq 0$ this causes no difficulty. In fact, we see immediately that if $\bar{\pi}' A_j \leq 0$ for every j, we can increase θ indefinitely in Eq. 5.11, thereby yielding an indefinitely large cost by Eq. 5.12. This in turn implies that D is unbounded, and hence that the original primal P is infeasible. We are already assured that $\bar{\pi}' A_j \leq 0$ for $j \in J$, since $\bar{\pi}$ is optimal (and therefore feasible) in DRP. Hence, we have established the following fact:

Theorem 5.1 *If $\xi_{\text{opt}} > 0$ in RP and the optimal dual satisfies*

$$\bar{\pi}' A_j \leq 0 \qquad \text{for } j \notin J$$

then P is infeasible.

We therefore need concern ourselves with maintaining feasibility only when

$$\bar{\pi}'A_j > 0 \qquad \text{for some } j \notin J$$

The feasibility criterion becomes in this case

$$\pi^{*\prime}A_j = \pi'A_j + \theta\bar{\pi}'A_j \leq c_j \qquad j \notin J \text{ and } \bar{\pi}'A_j > 0 \tag{5.15}$$

We are now in a situation analogous to that in the ordinary simplex algorithm: We can move θ so much and no more. More precisely, we can state our result in the form of another theorem.

Theorem 5.2 *When $\xi_{\text{opt}} > 0$ in RP and there is a $j \notin J$ with $\bar{\pi}'A_j > 0$, the largest θ that maintains the feasibility of $\pi^* = \pi + \theta\bar{\pi}$ is*

$$\theta_1 = \min_{\substack{j \notin J \\ \text{such that} \\ \bar{\pi}'A_j > 0}} \left[\frac{c_j - \pi'A_j}{\bar{\pi}'A_j}\right] \tag{5.16}$$

The new cost is

$$w^* = \pi'b + \theta_1\bar{\pi}'b = w + \theta_1\bar{\pi}'b > w.$$

When we solve the restricted primal and arrive at an improved solution to D, we redefine the set J and repeat the procedure until either $\xi_{\text{opt}} = 0$ and optimality in P is reached, or until we show by Theorem 5.1 that P is infeasible. A program for the complete primal-dual algorithm is sketched in Fig. 5–2.

```
procedure primal-dual
begin
  infeasible := 'no', opt := 'no';
  let π be feasible in D (comment: possible by (5.3));
  while infeasible = 'no' and opt = 'no' do
  begin set J = {j: π'Aj = cj};
        solve RP by the simplex algorithm;
        if ξopt = 0 then opt := 'yes'
          else if π̄'Aj ≤ 0 for all j ∉ J
               then infeasible := 'yes'
                 else π := π + θ1π̄
                 (comment: Equation 5.16)
  end
end
```

Figure 5–2 Algorithm primal-dual.

5.3 Comments on the Primal-Dual Algorithm

It is an important convenience that at each iteration we can start RP from the optimal solution obtained on the previous iteration. This is so because no variable that is in J and also in the optimal basis of RP at the end of an iteration can leave J at that point. More formally, we have the following.

Theorem 5.3 *Every admissible column in the optimal basis of RP remains admissible at the start of the next iteration.*

Proof If a column A_j is in the optimal basis of RP at the end of an iteration, its relative cost (in RP) is

$$\bar{c}_j = -\bar{\pi}'A_j = 0$$

This implies that

$$\pi^{*\prime}A_j = \pi'A_j + \theta_1\bar{\pi}'A_j = c_j$$

so that j stays in the set J. □

This fact not only allows us to start from our previous feasible solution, but also allows us to use the revised simplex method, since we cannot be embarrassed by having a basic column become inadmissible.

Finally, we consider the finiteness of the primal-dual algorithm. We note first that if the minimum in the calculation of θ_1, Eq. 5.16, occurs for $j = j_0$, then j_0 becomes a new member of the set J, because by (5.15),

$$\pi^{*\prime}A_{j_0} = c_{j_0}$$

Furthermore, the relative cost in RP of the new column j_0 is

$$-\bar{\pi}'A_{j_0} < 0$$

so that at least one pivot takes place when we start the new iteration in RP. If we consider the variables in RP to be all the original columns, admissible or not, plus all the artificial variables, then we move from a bfs to a bfs. In the absence of degeneracy in RP, the cost ξ_{opt} decreases monotonically with each iteration of the primal-dual method, no basis in RP is repeated, and the method is finite. If there are degenerate bfs's in RP, we can guarantee finiteness by using a rule to avoid the repetition of bases, as in any primal simplex algorithm.

We summarize this discussion as follows.

Theorem 5.4 *The primal-dual algorithm of Figure 5–2 correctly solves P in a finite amount of time.*

5.4 The Primal-Dual Method Applied to the Shortest-Path Problem

To illustrate the primal-dual method, we return to the shortest-path problem (SP), as formulated in node-arc form. This will also serve as a forerunner of its effective application to more general network problems.

$$\min c'f$$

$$Af = \begin{bmatrix} +1 \\ 0 \\ \cdot \\ \cdot \\ \cdot \\ 0 \end{bmatrix} \begin{matrix} \leftarrow \text{Row } s \\ \\ \\ \\ \\ \\ \end{matrix} \tag{P}$$

$$f \geq 0$$

where A is the $(m-1) \times n$ node-arc incidence matrix for a directed graph $G = (V, E)$ with m nodes and n arcs, with the row corresponding to the terminal t omitted (it is redundant); $f \in R^n$ is the vector of flow (in this problem with components 0 or 1), and $c \in R^n$ is the cost vector. The dual is

$$\begin{aligned} &\max \pi_s \\ &\pi_i - \pi_j \leq c_{ij} \qquad \text{for all arcs } (i, j) \in E \\ &\pi_i \gtrless 0 \qquad \text{for all } i \\ &\pi_t = 0 \end{aligned} \tag{D}$$

where we must fix $\pi_t = 0$, since its row is omitted from A. The set of admissible arcs is then defined by

$$J = \{\text{arcs } (i, j) : \pi_i - \pi_j = c_{ij}\}$$

and the restricted primal becomes

$$\min \xi = \sum_{i=1}^{m-1} x_i^a$$

$$Af + x^a = \begin{bmatrix} +1 \\ 0 \\ \cdot \\ \cdot \\ \cdot \\ 0 \end{bmatrix} \begin{matrix} \leftarrow \text{Row } s \\ \\ \\ \\ \\ \\ \end{matrix} \tag{RP}$$

$$\begin{aligned} &f_j \geq 0 \qquad \text{for all } j \\ &f_j = 0 \qquad j \notin J \\ &x_i^a \geq 0 \end{aligned}$$

Finally, the dual of the restricted primal is

$$\begin{aligned} &\max w = \pi_s \\ &\pi_i - \pi_j \leq 0 \qquad \text{for all arcs } (i, j) \in J \\ &\pi_i \leq 1 \qquad \text{for all } i \\ &\pi_i \gtrless 0 \end{aligned} \tag{DRP}$$

Now DRP is very easy to solve: since $\pi_s \leq 1$ and we wish to maximize π_s, we try $\pi_s = 1$. If there is no path from π_s to π_t, using only arcs in J, then we can propagate the 1 from s to all nodes reachable by a path from s without violating the $\pi_i - \pi_j \leq 0$ constraints, and an optimal solution to DRP is then

$$\bar{\pi} = \begin{cases} 1 & \text{for all nodes reachable by paths from } s \text{ using arcs in } J \\ 0 & \text{for all nodes from which } t \text{ is reachable using arcs in } J \\ 1 & \text{for all other nodes} \end{cases} \tag{5.17}$$

(Notice that this $\bar{\pi}$ is not unique.) We can then calculate

$$\theta_1 = \min_{\substack{\text{arcs } (i,j) \notin J \\ \text{such that} \\ \bar{\pi}_i - \bar{\pi}_j > 0}} \{c_{ij} - (\pi_i - \pi_j)\}$$

update π and J, and re-solve DRP. If we get to a point where there is a path from s to t using arcs in J, $\pi_s = 0$, and we are optimal, since $\xi_{\min} = w_{\max} = 0$. Any path from s to t using only arcs in J is optimal.

Thus, the primal-dual algorithm *reduces* the shortest-path problem to repeated solution of the simpler problem of finding the set of nodes reachable from a given node.

Example 5.1

Figure 5–3 shows a shortest-path problem somewhat more complicated than our previous example.

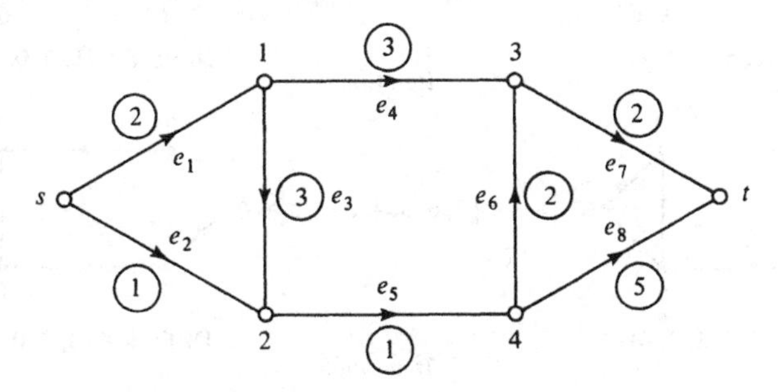

Figure 5–3 An example of the shortest-path problem. The circled numbers are arc weights.

Since the costs are nonnegative, we can begin with $\pi = 0$ in D. Figure 5–4 shows the succession of five iterations of the primal-dual method that finally reaches the optimal $\pi = (6, 5, 5, 2, 4)$ and the corresponding arc set J, which contains the optimal path. Any path from s to t in the final admissible arc set will satisfy complementary slackness and hence be optimal; in this case the optimal path is e_2, e_5, e_6, e_7, with cost $6 = \pi_s$. □

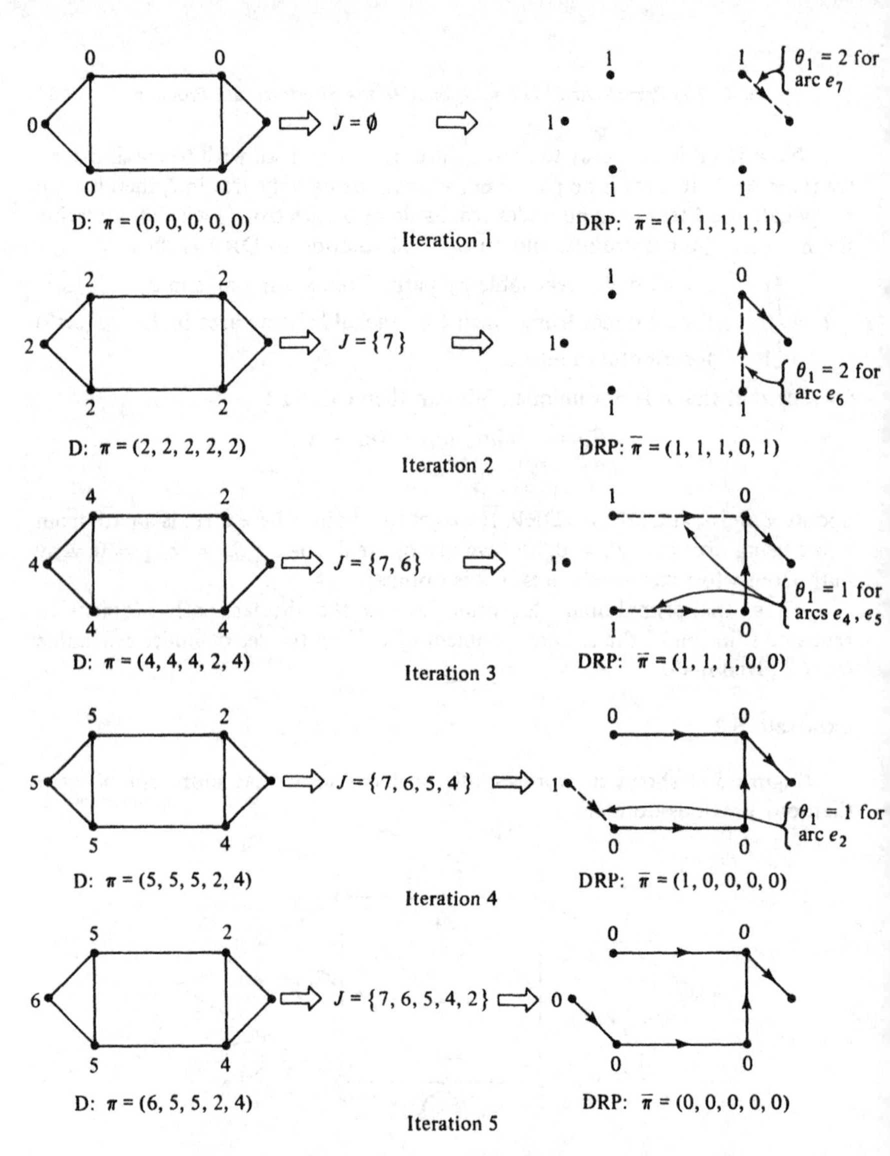

Figure 5–4 Solving a shortest-path problem using the primal-dual algorithm.

We now note some properties of this algorithm that lead to a very simple interpretation of what is happening. First, if we define at any point in the algorithm the set

$$\begin{aligned} W &= \{i \colon t \text{ is reachable from } i \text{ by admissible arcs}\} \\ &= \{i \colon \bar{\pi}_i = 0\} \end{aligned}$$

then the variable π_i remains fixed from the time that i enters W to the conclusion of the algorithm, because the corresponding $\bar{\pi}_i$ will always be zero. Second, every arc that becomes admissible (enters J) stays admissible throughout the algorithm, because once we have

$$\pi_i - \pi_j = c_{ij} \qquad \text{for arc } (i, j) \in E$$

we always change π_i and π_j by the same amount, by Eq. 5.17. Some thought shows that π_i, $i \in W$, is the length of the shortest path from i to t and that the algorithm proceeds by adding to W, at each stage, the nodes in $\bar{W}$ next closest to t. With some streamlining, this algorithm—the primal-dual algorithm applied to the shortest-path problem—is in fact Dijkstra's shortest-path algorithm, which we describe in more detail in Chapter 6 [Di].

Finally, we note that there is a definite bound on the number of steps required by the primal-dual algorithm applied to shortest path. Since the set W grows by at least one node at each stage, there can be no more than $|V|$ stages.

5.5 Comments on Methodology

Let us step back for a moment and take a look at what the primal-dual algorithm accomplishes. We started with the general linear program P and cost vector c; we changed it into the iterative solution of RP, which does not depend on the cost vector c explicitly, but rather through the set J of admissible variables. For the price of the iteration loop in Figure 5–1, we have eliminated the complication of a general cost vector. For want of a better term, we shall say we have "combinatorialized" the cost.

We can also look at matters from the point of view of the dual. In going from D to DRP, we combinatorialize the right-hand side. Schematically, we can write

$$\begin{aligned} &\text{P} \longrightarrow \text{RP combinatorializes cost} \\ &\text{D} \longrightarrow \text{DRP combinatorializes right-hand side} \end{aligned} \tag{5.18}$$

In the shortest-path problem, the right-hand side of P is essentially trivial, and all the *numerical* problem data enters through the cost vector. Therefore, RP and its dual do not depend explicitly on the numerical problem data at all, but only on the admissible set J, and we produce a purely combinatorial subproblem, which we identified in this case as a reachability problem.

The technique of starting with one problem, and iteratively solving subproblems that are "more combinatorial" by applying the primal-dual algorithm is central to combinatorial optimization. It is the basis of almost every efficient algorithm for flow and matching problems.

We next apply the idea to the max-flow problem.

5.6 The Primal-Dual Method Applied to Max-Flow

Our program in this section is as follows: We shall first formulate max-flow in node-arc form. This LP will have an essentially trivial *cost* vector; its numerical input (the capacities) will appear on the *right-hand side*. We shall therefore regard max-flow as a dual problem (remember Eq. 5.18) in order to combinatorialize its data. The subproblems will again be reachability problems, which in this application are problems of finding *augmentation paths.*

In more detail, consider the flow network $N = (s, t, V, E, b)$ with $n = |V|$ nodes and $m = |E|$ arcs, and let the flow in arc (x, y) be denoted by $f(x, y)$. Then an s-t flow of value v is defined by the constraints

$$Af = \begin{cases} +v & \text{Row } s \\ -v & \text{Row } t \\ 0 & \text{other rows} \end{cases} \tag{5.19}$$

$$f \leq b$$

$$f \geq 0$$

where A is the node-arc incidence matrix of the directed graph (V, E), and f and $b \in R^m$ are the flow and capacity vectors, respectively. The problem is to maximize v subject to the constraints in (5.19).

Let us rewrite the problem slightly by using the vector $d \in R^n$ defined by

$$d_i = \begin{cases} -1 & i = s \\ +1 & i = t \\ 0 & \text{otherwise} \end{cases} \tag{5.20}$$

The LP then becomes

$$\begin{aligned} \max\ & v \\ Af + dv &\leq 0 \\ f &\leq b \\ -f &\leq 0 \end{aligned} \qquad \text{(D)} \tag{5.21}$$

(Notice that $Af + dv \leq 0$ implies $Af + dv = 0$, since a deficit in the flow balance at any node implies a surplus at some other.)

We have achieved the goal of writing max-flow as a dual of a standard form LP; we now want to combinatorialize the right-hand side by considering DRP. Comparing D and DRP earlier in this chapter, the rules for doing this become apparent: We replace the right-hand side by 0, cross out the rows not in J, and add constraints $f, v \leq 1$. Thus, DRP becomes

$$\begin{aligned} &\max v \\ &Af + dv \leq 0 &&\text{for all rows} \\ &f \leq 0 &&\text{for rows where } f = b \text{ in } D \\ &-f \leq 0 &&\text{for rows where } f = 0 \text{ in } D \qquad \text{(DRP)} \\ &f \leq 1 \\ &v \leq 1 \end{aligned} \tag{5.22}$$

This problem has the following interpretation. Find a path from s to t (for a flow of value 1) that uses only the following arcs in the following ways: saturated arcs in the backward direction; arcs with zero flow in the forward direction; and other arcs in either direction. As in shortest path, this is a reachability problem.

The next step in the primal-dual algorithm, once such an augmentation path is found, corresponds to augmenting f by as large a flow along the path as possible; that is, until an arc traversed in the backward direction becomes empty (all its flow canceled) or until an arc traversed in the forward direction becomes saturated.

We shall fill in the details of this specialized algorithm for max-flow in the next chapter. It was originally devised by Ford and Fulkerson [FF].

Notice that we solved DRP directly and did not use simplex to solve RP. Theorem 5.4, however, assumes that simplex is used and further assumes that it is implemented in a way that resolves degeneracy to avoid repetition of bases. We therefore have no guarantee of finiteness. If we allow the capacities to be irrational numbers, this is not merely a technical point, because the Ford-Fulkerson algorithm may in fact fail to terminate if no precautions are taken in choosing flow-augmenting paths. We shall deal further with this problem in the next chapter, which will be devoted to a more detailed study of the primal-dual algorithms we have just derived for max-flow and shortest path.

PROBLEMS

1. Prove rigorously that the LP in Eq. 5.21, with inequality constraints on the flow balance equations, is equivalent to the LP with equality constraints.

2. Prove that in the primal-dual shortest-path algorithm, π_i, $i \in W$ is the length of the shortest path from i to t, and that the algorithm adds at each stage the nodes not in W that are next closest to t.

3. Show that the θ_1 calculation in the primal-dual max-flow algorithm corresponds to increasing the flow on the augmentation path until an arc becomes empty or saturated.

4. Where do we use the fact that we chose $b \geq 0$ in P of the general primal-dual algorithm?
5. Consider the primal-dual solution of shortest path. Describe optimal solutions to RP that correspond to the optimal solutions to DRP given by Eq. 5.17. Do so for the suboptimal stages and the final optimal stage of the algorithm.
6. Suppose we allow negative arc weights in the node-arc formulation of shortest path. Prove that the following conditions are equivalent.
   (a) There is a shortest *s-t* walk.
   (b) The dual LP is feasible.
   (c) There is no cycle with negative total cost.
7. Prove that in the primal-dual algorithm the cost of the feasible solution in D increases by a positive amount during each iteration. Explain why this does not imply that the method terminates in a finite number of steps, as it would for simplex.
8. Show that RP in max-flow and shortest path are highly degenerate.

NOTES AND REFERENCES

The primal-dual algorithm for general LP's was first described in

[DFF] DANTZIG, G. B., L. R. FORD, and D. R. FULKERSON, "A Primal-Dual Algorithm for Linear Programs," pp. 171–181 in *Linear Inequalities and Related Systems*, ed. H. W. Kuhn and A. W. Tucker. Princeton, N.J.: Princeton University Press, 1956.

It is presented there as a generalization of

[Ku] KUHN, H. W., "The Hungarian Method for the Assignment Problem," *Naval Research Logistics Quarterly*, 2, nos. 1 and 2 (1955), 83–97

and similar algorithms for more general flow problems, which will be described later in this book.

As mentioned in Sec. 5.4, primal-dual applied to shortest path leads to Dijkstra's algorithm:

[Di] DIJKSTRA, E. W., "A Note on Two Problems in Connexion with Graphs," *Numerische Mathematik*, 1 (1959), 269–71.

Primal-dual applied to max-flow leads to the Ford and Fulkerson algorithm, which is described in

[FF] FORD, L. R., JR., and D. R. FULKERSON, *Flows in Networks*. Princeton, N.J.: Princeton University Press, 1962.

This book, as Dantzig's, remains a standard almost 20 years after its appearance.

6

Primal-Dual Algorithms for Max-Flow and Shortest Path: Ford-Fulkerson and Dijkstra

6.1 The Max-Flow, Min-Cut Theorem

This chapter is devoted to the detailed development of the primal-dual algorithms derived in the last chapter for max-flow and shortest path. This is a point of transition in this book from the study of algorithms for general LP's to more specialized algorithms for certain network problems, all derived, as we have mentioned before, from primal-dual. We are thus headed towards algorithms that are in a sense less numerical and more combinatorial than general simplex algorithms. As we proceed, we shall also use linear programming theory to establish useful facts about a variety of graph-theoretic problems, beginning with the famous max-flow, min-cut theorem of the max-flow problem.

Returning to the node-arc formulation of max-flow in the previous chapter, consider the flow network $N = (s, t, V, E, b)$ with $n = |V|$ nodes and $m = |E|$ arcs; let the flow in arc (x, y) be denoted by $f(x, y)$. Then the max-flow problem is the following LP, which we think of as a dual:

$$\begin{aligned} \max v \\ Af + dv = 0 \\ f \leq b \\ f \geq 0 \end{aligned} \tag{6.1}$$

where $d \in R^n$ is defined, as before, by

$$d_i = \begin{cases} -1 & i = s \\ +1 & i = t \\ 0 & \text{otherwise} \end{cases} \tag{6.2}$$

An important concept in dealing with flow problems in general, and one that played a central role in the invention of the Ford-Fulkerson algorithm, is that of a *cut*.

Definition 6.1

An *s-t cut* is a partition $(W, \bar{W})$ of the nodes of V into sets W and $\bar{W}$ such that $s \in W$ and $t \in \bar{W}$. The *capacity* of an *s-t* cut is

$$C(W, \bar{W}) = \sum_{\substack{(i,j) \in E \\ \text{such that} \\ i \in W, j \in \bar{W}}} b(i, j) \quad \square \tag{6.3}$$

Figure 6–1 illustrates the idea behind the definition of a cut; the capacity of a cut is the sum of the capacities of the "forward" arcs: those which go from nodes in W to nodes in $\bar{W}$. We would expect that the value of an *s-t* flow cannot exceed the capacity of an *s-t* cut, since all the *s-t* flow must pass through the forward arcs of a cut. This result is intimately related to the fact that cuts correspond to feasible solutions to the dual of the max-flow problem, which leads us next to write down the dual of the LP formulation of the max-flow problem in Eq. 6.1.

Assign variables $\pi(x)$ to the first n constraints, which correspond to flow conservation; and assign variables $\gamma(x, y)$ to the next m capacity constraints.

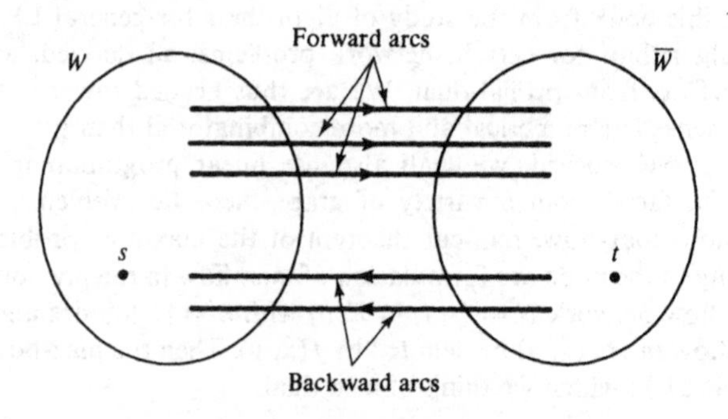

Figure 6–1 A cut in a flow network.

Since the first n constraints are equalities, $\pi(x) \gtrless 0$; and since the next m constraints are inequalities, $\gamma(x, y) \geq 0$. The LP in Eq. 6.1 is exactly in the form of the dual in Def. 3.1, so by symmetry, the primal in that definition is the dual for which we are looking:

$$\begin{aligned} \min \sum_{(x,y) \in E} \gamma(x, y) b(x, y) & \\ \pi(x) - \pi(y) + \gamma(x, y) \geq 0 & \quad \text{for all } (x, y) \in E \\ -\pi(s) + \pi(t) \geq 1 & \\ \pi(x) \gtrless 0 & \\ \gamma(x, y) \geq 0 & \end{aligned} \tag{6.4}$$

The last inequality corresponds to the variable v.

We can now prove Theorem 6.1.

Theorem 6.1 *Every s-t cut determines a feasible solution with cost $C(W, \bar{W})$ to the dual of max-flow as follows:*

$$\begin{aligned} \gamma(x, y) &= \begin{cases} 1 & (x, y) \text{ such that } x \in W, y \in \bar{W} \\ 0 & \text{otherwise} \end{cases} \\ \pi(x) &= \begin{cases} 0 & x \in W \\ 1 & x \in \bar{W} \end{cases} \end{aligned} \tag{6.5}$$

Proof We need to check the inequality constraints in (6.4). There are four cases to consider, since x and y may each be in W or $\bar{W}$. In each case the inequality is easily verified, and the inequality is strict when and only when $x \in \bar{W}$ and $y \in W$. Also, $\pi(s) = 0$ because $s \in W$, and $\pi(t) = 1$ because $t \in \bar{W}$; so

$$-\pi(s) + \pi(t) = 1$$

which verifies the last inequality. Finally, the cost of this dual feasible solution is

$$\sum_{(x,y) \in E} \gamma(x, y) b(x, y) = \sum_{\substack{(x,y) \in E \\ \text{such that} \\ x \in W, y \in \bar{W}}} b(x, y) = C(W, \bar{W}) \quad \square$$

From Theorem 6.1, we get our main result.

Theorem 6.2 (Max-flow, min-cut) *The value v of any s-t flow is no greater than the capacity $C(W, \bar{W})$ of any s-t cut. Furthermore, the value of the maximum flow equals the capacity of the minimum cut, and a flow f and cut $(W, \bar{W})$ are jointly optimal if and only if*

$$\begin{aligned} f(x, y) &= 0 & (x, y) \in E \text{ such that } x \in \bar{W} \text{ and } y \in W \\ f(x, y) &= b(x, y) & (x, y) \in E \text{ such that } x \in W \text{ and } y \in \bar{W} \end{aligned} \tag{6.6}$$

Proof That v is no larger than the capacity of any cut follows directly from the previous theorem. The labeling algorithm in the next section will show that a given maximal flow of value v can always be used to construct a cut with value $C = v$. Equation 6.6 expresses the complementary slackness conditions, which may be seen as follows: If $x \in \bar{W}$ and $y \in W$ the dual inequality

$$\pi(x) - \pi(y) + \gamma(x, y) = 1 - 0 + 0 > 0$$

is strict, as mentioned above, so the corresponding variable $f(x, y)$ must be zero. Similarly, if $x \in W$ and $y \in \bar{W}$, we have $\gamma(x, y) = 1$ from (6.5), so the corresponding primal inequality

$$f(x, y) \leq b(x, y)$$

must be a strict equality. □

Example 6.1

Figure 6–2 shows a max-flow problem with an optimal flow of value 5 and a corresponding min-cut. Notice that every forward arc across the cut is saturated ($f(x, y) = b(x, y)$) and every backward arc is empty ($f(x, y) = 0$), verifying Eq. 6.6, and demonstrating that the flow and cut are optimal. □

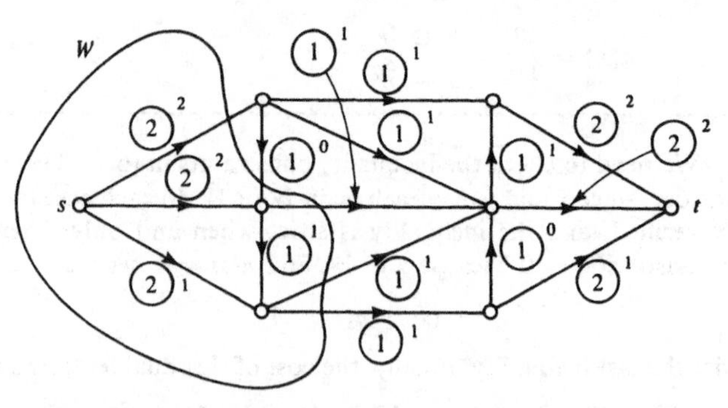

Figure 6–2 A max-flow problem: The arc capacities are circled numbers, a max-flow is given by the uncircled numbers, and a min-cut is indicated by the set W.

6.2 The Ford and Fulkerson Labeling Algorithm

We shall now fill in the details of the primal-dual algorithm for max-flow that we derived at the end of the previous chapter. Recall that the dual of the restricted primal (DRP) was the subproblem of searching for a path from s to t along which flow can be augmented. When the search for such a path fails, the complementary slackness conditions hold, and the primal-dual algorithm is at optimality. The kind of path for which we search is characterized in the following definition.

Definition 6.2

Given a flow network $N = (s, t, V, E, b)$ and a feasible s-t flow f, an *augmentation* (or *augmenting*) *path* P is a path from s to t in the undirected graph resulting from G by ignoring arc directions, with the following properties:

(a) For every arc $(i, j) \in E$ that is traversed by P in the forward direction (called a *forward arc*), we have $f(i, j) < b(i, j)$. That is, forward arcs of P are unsaturated.

(b) For every arc $(j, i) \in E$ that is traversed by P in the reverse direction (called a *backward arc*), we have $f(j, i) > 0$. □

Figure 6–3 shows an augmentation path in the network of Fig. 6–2. The arc (a, e) is a backward arc on P.

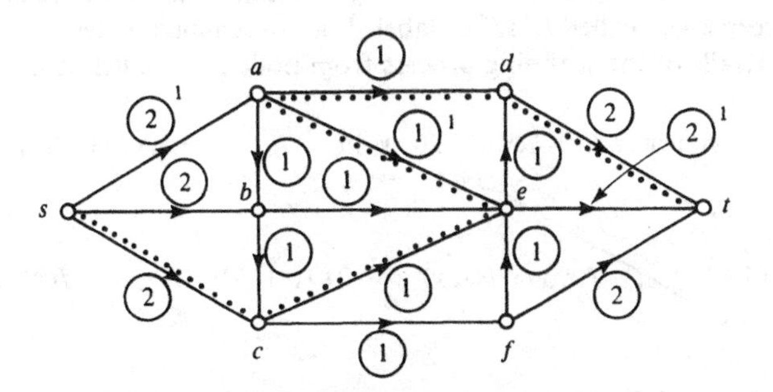

Figure 6–3 The dotted lines indicate an augmentation path. The uncircled numbers show a flow of value 1, and the circled numbers are capacities.

It should be clear that we can increase the flow from s to t while maintaining flow conservation at every node by increasing the flow on every forward arc of P and decreasing it along every backward arc. We can increase the flow until we violate the capacity constraint of a forward arc or empty a backward arc. Thus the maximum amount of flow augmentation possible along P is given by

$$\delta = \min_{\text{arcs of } P} \begin{Bmatrix} b(i, j) - f(i, j) \text{ along a forward arc} \\ f(j, i) \text{ along a backward arc} \end{Bmatrix}$$

In the example of Fig. 6–3, $\delta = 1$; the result of flow augmentation by this amount is shown in Fig. 6–4, where the s-t flow has been increased from a value of 1 to 2.

It remains for us to describe a systematic procedure for finding an augmentation path if one exists. We do this by propagating labels from s until we

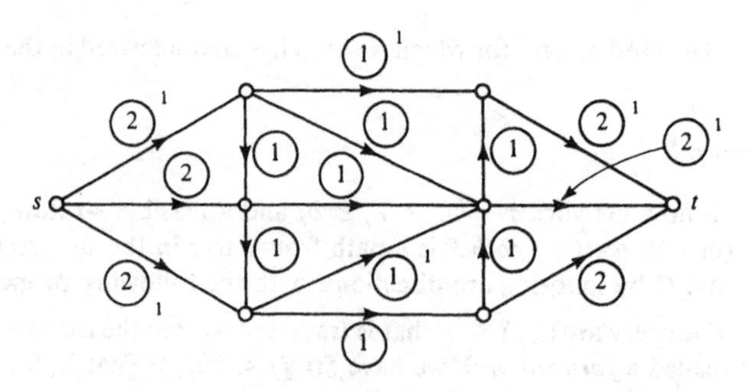

Figure 6–4 The result of flow augmentation in the network of Figure 6–3.

reach t or get stuck. Each node x will have assigned to it a two-part label, *label* $(x) = (L1\,[x], L2\,[x])$. Here $L1\,[x]$ will tell us *from where* x was labeled, and $L2\,[x]$ will tell us the amount of extra flow that can be brought to x from s. The process of labeling outward from a given node x is called *scanning* x, and we shall keep a list, called *LIST*, of labeled but unscanned nodes.

The details of the scanning process from node x are illustrated in Fig. 6–5.

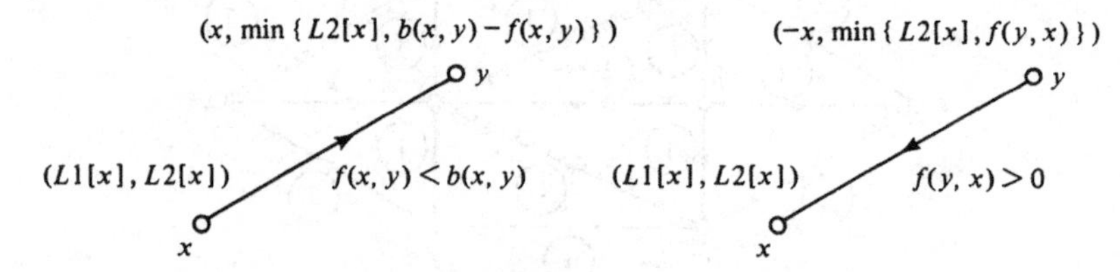

Figure 6–5 The two possible cases of labeling y while scanning x.

There are two cases. If node y is unlabeled and succeeds x, we may label y if $f(x, y) < b(x, y)$, in which case we set

$$L1\,[y]: = x$$

$$L2\,[y]: = \min\,\{L2\,[x], \quad b(x, y) - f(x, y)\}$$

This records that y was labeled from x and that the flow can be augmented by the smaller of $L2\,[x]$ and $b(x, y) - f(x, y)$.

In the case that a node y is unlabeled and precedes x, we may label y from x if $f(y, x) > 0$, in which case we set

$$L1\,[y]: = -x$$

$$L2\,[y]: = \min\,\{L2\,[x], \quad f(y, x)\}$$

Notice that we use a negative sign in $L1$ to tell us that this labeling was across a backward edge.

The algorithm starts by scanning s and adding to *LIST* all nodes labeled from s. The process is then repeated—a node x is selected from *LIST* and scanned, and all nodes labeled from x are added to *LIST*. The process terminates in one of two ways: either t gets labeled, in which case we can reconstruct an augmentation path backwards from t using $L1$; or we empty *LIST*. In the first case we augment the flow along the augmentation path; in the second case we have reached an optimal flow, as we shall see next. A sketch of the algorithm is shown in Fig. 6–6.

```
FORD AND FULKERSON ALGORITHM
Input:  A network N=(s,t,V,A,b)
Output: The maximum flow f of N.
begin
      f := 0; (comment: initialize flow)
again: set all labels to 0, set LIST := {s}, set L2[s] := ∞;
      (comment: initialize for the search for new augmentation path)
      while LIST ≠ ∅ do
      begin
         let x be any node in LIST;
         remove x from LIST;
         scan x;
         if t is labeled then
            begin
               augment flow f along augmentation path;
               go to again
            end
      end
end

procedure scan
   begin
      label forward to all unlabeled nodes adjacent to x by arcs that
      are unsaturated, putting newly labeled nodes on LIST;
      label backward to all unlabeled nodes from which x is adjacent
      by arcs that have positive flows, putting newly labeled nodes
      on LIST;
   end
```

Figure 6–6 The Ford and Fulkerson algorithm for solving the max-flow problem.

We now show (independently of its derivation *via* the primal-dual algorithm) that the algorithm can terminate only at optimality.

Theorem 6.3 *When the Ford and Fulkerson labeling algorithm terminates, it does so at optimal flow.*

Proof At termination of the algorithm, some nodes are labeled and some are not. Call the set of labeled nodes W and the set of unlabeled nodes $\bar{W}$, as illustrated in Fig 6–7. All arcs (x, y) directed from W to $\bar{W}$ must be saturated;

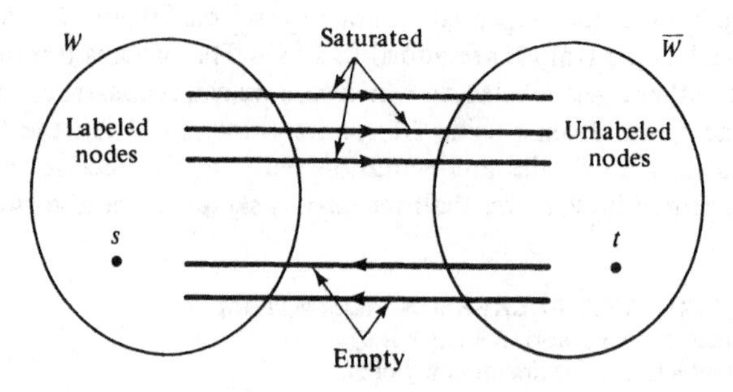

Figure 6–7 The sets W and $\bar{W}$ at termination of the Ford and Fulkerson labeling algorithm.

otherwise y would have been labeled when x was scanned. Similarly, all arcs (y, x) directed from $\bar{W}$ to W must be empty; otherwise y would also have been labeled when x was scanned. Therefore, $(W, \bar{W})$ is a min-cut and the flow must be optimal by Theorem 6.2. □

Example 6.1 (Continued)

Figure 6–8 shows the result of three augmentations starting with the flow of value 2 in Fig. 6–4 and ending with an optimal flow of value 5. The optimal flow is different from the optimal flow in Fig. 6–2 but the min-cut is the same. (We have allowed any intermediate augmentation paths in this example, to illustrate backward labeling. The labeling algorithm might not have found the first augmentation path shown in Fig. 6–8 but might have found some other. However, this does not affect the argument in Theorem 6.3.) □

6.3 The Question of Finiteness of the Labeling Algorithm

We have shown that the Ford and Fulkerson labeling algorithm, when and if it terminates, finds the optimal flow value. But how do we know that the algorithm terminates in a finite number of iterations? The answer may be somewhat surprising, since up to now every algorithm has been shown to be finite. When the capacities b are irrational, the labeling algorithm may not, in fact, terminate.

When the capacities b are integers, it is clear that the algorithm is finite, since each augmentation increases the flow value by at least 1 unit, so if the maximum flow value is v, we can have at most v augmentations. In the same way, if the capacities are rational, we may put them over a common denominator D, scale by D, and use the same argument to conclude that at most Dv augmentations can occur.

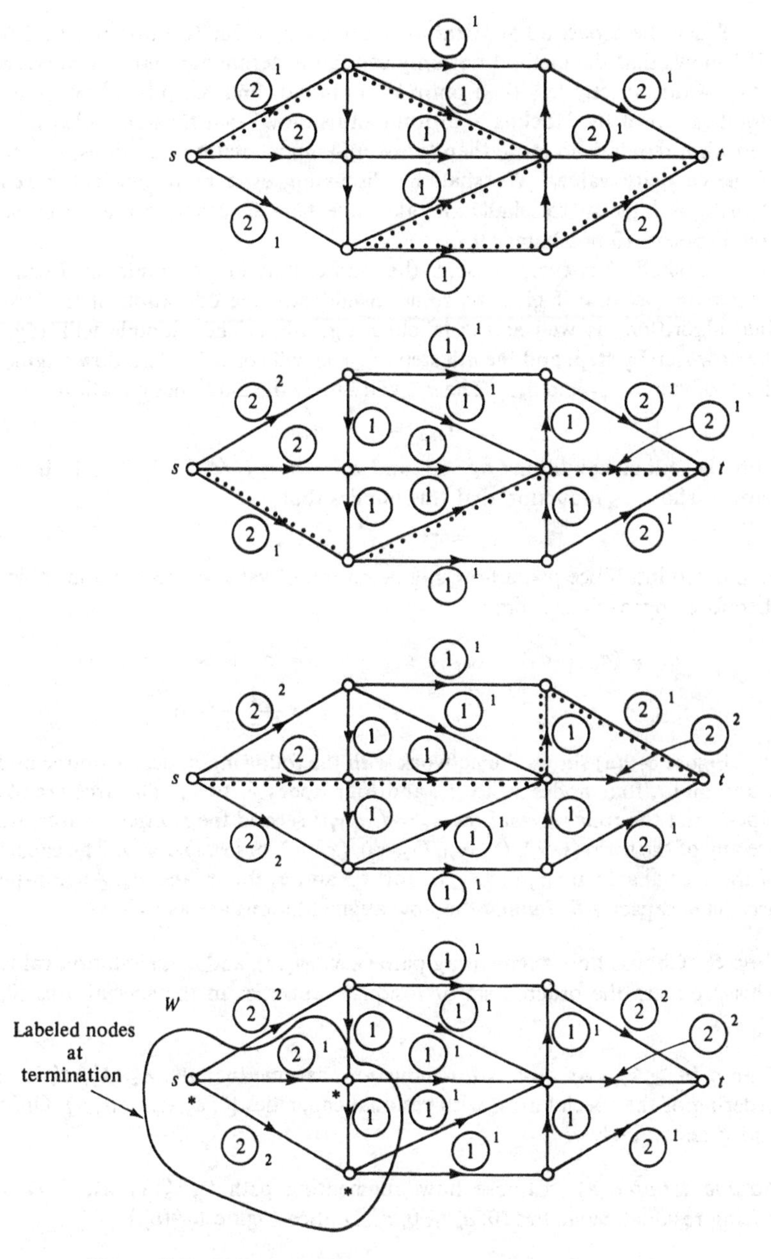

Figure 6–8 Three augmentations from the flow in Figure 6–4, leading to an optimal flow of value 5.

When the capacities are irrational, an example due to Ford and Fulkerson [FF] shows that the method not only can fail to terminate, but can converge to a flow value strictly less than optimal! Edmonds and Karp [EK] introduced a modification of the labeling algorithm and showed that their modified labeling algorithm requires no more than $(n^3 - n)/4$ augmentation iterations, regardless of the capacity values. We shall be discussing even more efficient max-flow algorithms later so we shall not take time here to describe their arguments. (See Problem 10 in Chapter 9.)

We shall, however, present the nonterminating example of Ford and Fulkerson, because it gives us some insight into the operation of the primal-dual algorithm, as well as the labeling algorithm. The example will augment the flow step by step, and the nth step ($n \geq 1$) will consist of two flow augmentations, of value a_{n+1} and a_{n+2}. The a_i's will satisfy the difference equation

$$a_{n+2} = a_n - a_{n+1} \tag{6.6}$$

with the initial conditions $a_0 = 1$ and $a_1 = \sigma = (\sqrt{5} - 1)/2 < 1$. It is not hard to show by induction that this implies that

$$a_i = \sigma^i \qquad i = 0, 1, \ldots \tag{6.7}$$

Step 0 will introduce just a flow augmentation of value a_0, so the total flow will therefore approach as a limit

$$\begin{aligned} a_0 + (a_2 + a_3) + (a_3 + a_4) + \cdots &= a_0 + a_1 + a_2 + \cdots \\ &= \frac{1}{1 - \sigma} = S \end{aligned} \tag{6.8}$$

Figure 6–9(a) shows the network with the following nodes: a source node s, a terminal t, four nodes x_1 to x_4, and four nodes y_1 to y_4. The arcs are of two types: first the four *special* arcs $A_i = (x_i, y_i)$; second the *nonspecial* arcs, which are any of the form (s, x_i), (y_i, y_j), (y_i, x_j), (x_i, y_j), or (y_i, t), $i \neq j$. The capacities of the special arcs are a_0 for A_1, a_1 for A_2, and a_2 for A_3 and A_4. All nonspecial arcs have capacity S. Finally the flow augmentations are as follows.

Step 0 Choose flow augmenting path (s, x_1, y_1, t), and augmentation value a_0. This produces the ordered set of residual capacities in the special arcs $(0, a_1, a_2, a_2)$.

Step n ($n \geq 1$) As a basis for induction, assume that A'_1, A'_2, A'_3, A'_4 is some ordering of the special arcs, with residual capacities $(0, a_n, a_{n+1}, a_{n+1})$. Order x'_i and y'_i accordingly.

Augmentation n(a) Choose flow augmenting path $(s, x'_2, y'_2, x'_3, y'_3, t)$, producing residual capacities $(0, a_{n+2}, 0, a_{n+1})$. (See Figure 6–9(b).)

Augmentation n(b) Choose flow augmenting path $(s, x'_2, y'_2, y'_1, x'_1, y'_3, x'_3, y'_4, t)$, producing residual capacities $(a_{n+2}, 0, a_{n+2}, a_{n+1})$. (See Figure 6–9(c).)

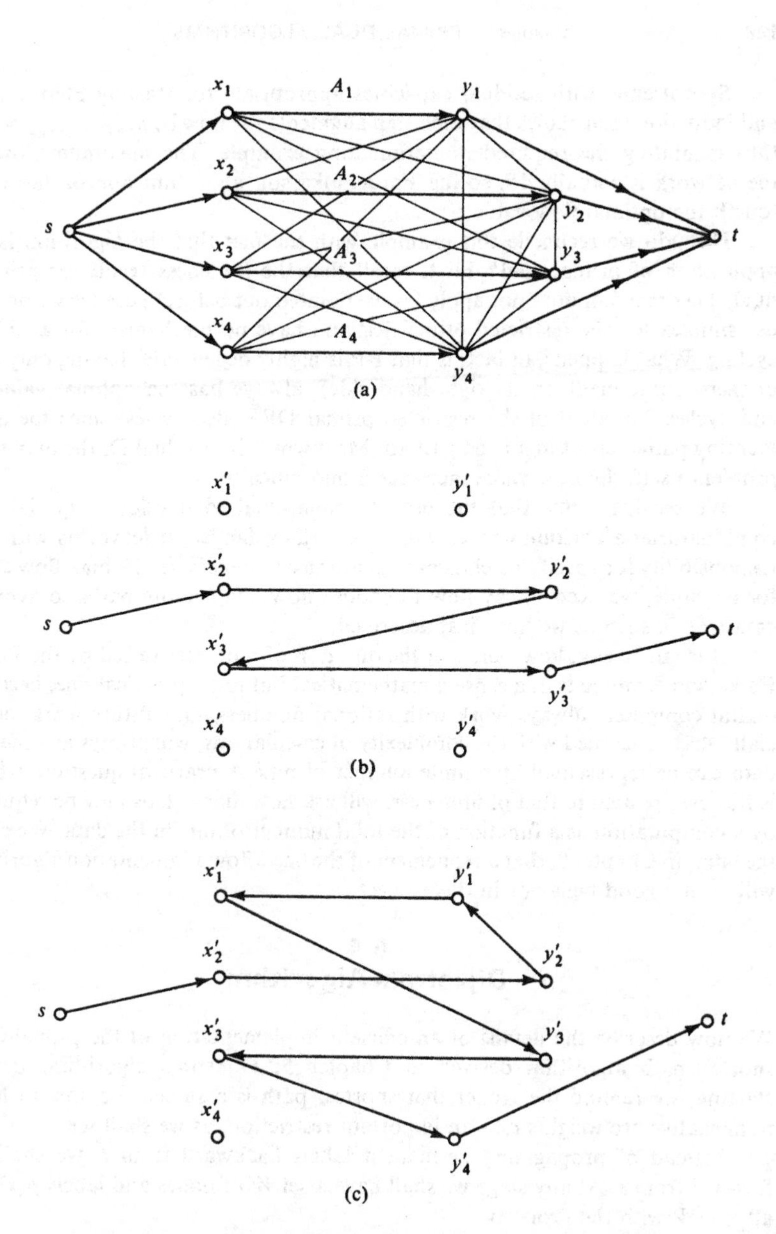

Figure 6–9 The pathological example for the Ford-Fulkerson algorithm. (a) The network. (b) and (c) The first and second parts of Step *n*. Only the arcs on the flow augmenting paths are shown.

Step n ends with residual capacities appropriate for starting Step $n + 1$, and induction then shows that each step augments the flow by $a_{n+1} + a_{n+2} = a_n$, thus exhibiting the required nonterminating example. The maximum flow in the network is actually $4S$, so the Ford-Fulkerson algorithm approaches one-fourth the optimal flow value.

How do we reconcile this example with the fact that the algorithm is an application of primal-dual? First, recall that the finiteness result for primal-dual, Theorem 5.4, does not apply (as we pointed out before) because we do not use simplex for the restricted primal RP and have no mechanism for avoiding cycling. What happens, in fact, is that RP is highly degenerate (having only one nonzero component on its right-hand side), always has the optimal value 1, and cycles. The dual of the restricted primal DRP also cycles, since the augmenting paths repeat in a fixed pattern. Meanwhile, in the dual D, the max-flow problem itself, the flow value increases monotonically.

We see from this that the use of primal-dual to develop a specialized, combinatorial algorithm carries with it a certain danger. It leaves us with the responsibility for specifying choices to guarantee finiteness. In the max-flow case, for example, we need to say how to choose flow augmenting paths to avoid a catastrophe such as we have just described.

It is fair to say, however, that the question of finiteness raised by the Ford-Fulkerson example is in a sense a mathematical but not a practical one, because digital computers always work with rational numbers. Our future work, especially that concerned with the complexity of calculations, will always assume the data can be represented by a finite number of bits. A practical question, which is however related to that of finiteness, will ask how many steps may be required by a computation as a function of the total number of bits in the data. We shall see later, in Chapter 9, that a refinement of the basic flow augmentation algorithm will ensure good behavior in this respect.

6.4 Dijkstra's Algorithm

We now describe the details of an efficient implementation of the primal-dual shortest-path algorithm derived in Chapter 5: Dijkstra's algorithm. Before starting, we remind the reader that shortest path is assumed for now to have nonnegative arc weights c_{ij}—an important restriction, as we shall see.

Instead of propagating permanent labels backward from t, we shall go forward from s. At any stage we shall have a set W of nodes and labels $p(x)$ for all $x \in V$ with the property

$$p(x) = \text{shortest length of any path from } s \text{ to } x, \quad \textit{using only intermediate nodes in } W \tag{6.9}$$

Now consider the node $x \notin W$ with the smallest $p(x)$. The shortest path from s to this x uses only nodes in W as intermediate nodes, for otherwise it could not have the smallest $\rho(x)$ not in W. (We have used the nonnegativity of the weights c_{ij}.) Therefore we can add x to W and update the labels $\rho(y)$ for $y \notin W$ by

$$\rho(y) = \min\{\rho(y), \rho(x) + c_{xy}\} \qquad \text{for all } y \notin W \qquad (6.10)$$

This says that the new $\rho(y)$ for $y \notin W$ is either unaffected by the addition of x to W or that it is given by the shortest distance from s to x through nodes in W, plus the distance directly from x to y. (See Problem 4.) When finally $W = V$, $\rho(x)$ is the shortest distance from s to x with no conditions attached. We start by taking $W = \varnothing$, all $\rho = \infty$, and adding s to W; the algorithm is given in Fig. 6–10. It is assumed that $c_{xy} = \infty$ if arc (x, y) is not present.

```
DIJKSTRA'S ALGORITHM
Input:    A digraph D=(V,A), with costs c_uv≥0 on its arcs; a
          node s ∈ V.
Output:   The shortest distances from s to all v ∈ V in
          the array ρ.
begin     set W := {s}; ρ[s] := 0;
          for all y ∈ V−{s} do ρ[y] := c_sy;
          while  W ≠ V do
          begin  find min{ρ[y] : y ∉ W}, say ρ[x];
                 set W := W ∪ {x};
                 for all y ∈ V−W do
                   ρ[y] := min{ρ[y], ρ[x] + c_xy}
          end
end
```

Figure 6–10 Dijkstra's algorithm.

Figure 6–11 shows the successive stages in applying Dijkstra's algorithm to the shortest-path problem used in Example 1 of Chapter 5. It is easy enough to reconstruct the path in the usual way—by keeping track at each node of where its label comes from via Eq. 6.10. In the example, every time a node is added to W, the corresponding arc is thatched. The result is a tree rooted at s with the shortest paths from s to all other nodes.

The time required by Dijkstra's algorithm can be bounded from above as follows: Each iteration requires a number of steps proportional to the number of nodes not in W, which is at most n. Since there are n iterations (including the initialization), the whole algorithm requires time proportional to at most n^2.

6.5 The Floyd-Warshall Algorithm

We conclude this chapter with a very efficient, simply programmed, and widely used algorithm that finds the shortest paths between all pairs of nodes, all at once. Furthermore, it has the important advantage over Dijkstra's algorithm of working when the arc weights are allowed to be negative and will in fact

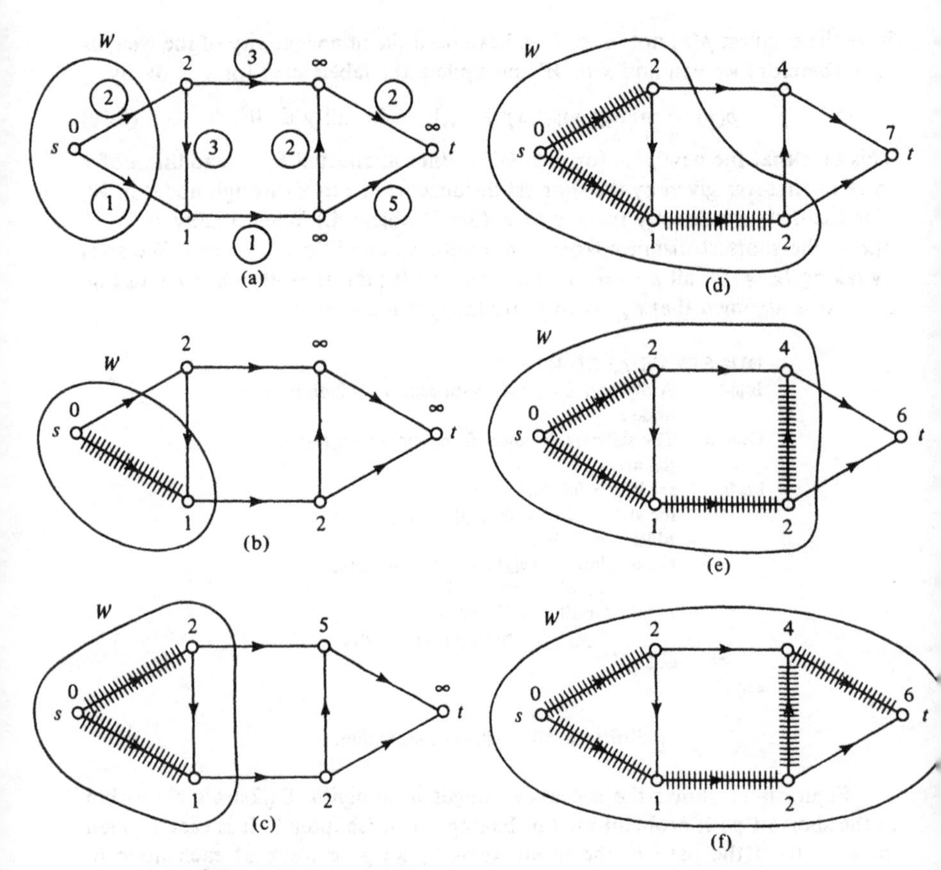

Figure 6–11 Successive stages in the application of Dijkstra's algorithm.

allow us to detect negative-cost cycles. For a change, this algorithm does not appear to be primal-dual.

The method works with an $n \times n$ array of numbers d_{ij}, initially set to the arc weights c_{ij} of the directed graph $G = (V, E)$. For our purposes, we take $c_{ii} = \infty$ for every i. The kernel of the algorithm is the following operation.

Definition 6.4

Given an $n \times n$ distance matrix d_{ij}, a *triangle operation* for a fixed node j is

$$d_{ik} := \min \{d_{ik}, d_{ij} + d_{jk}\} \qquad \text{for all } i, k = 1, \ldots, n \quad \text{but} \quad i, k \neq j$$

Note that we allow $i = k$. □

This operation replaces, for all i and k, the d_{ik} entry with the distance $d_{ij} + d_{jk}$ if the latter is shorter (see Fig. 6–12).

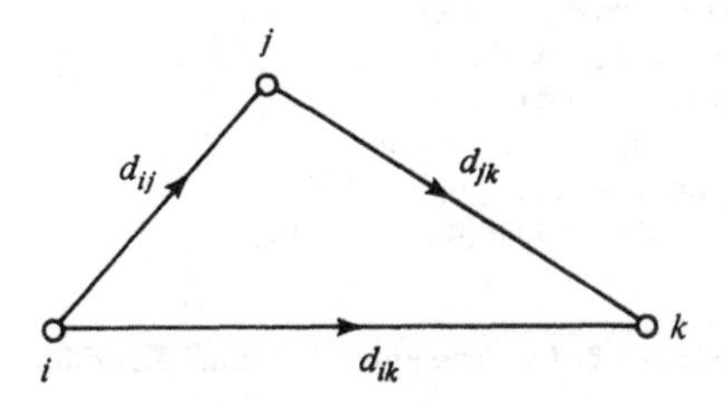

Figure 6–12 The triangle operation for fixed j and all other i and k.

The Floyd-Warshall algorithm is based on the following theorem; the inductive proof that follows is as good a way as any of understanding why the algorithm works.

Theorem 6.4 *If we perform a triangle operation for successive values $j = 1, 2, \ldots, n$, each entry d_{ik} becomes equal to the length of the shortest path from i to k, assuming the weights $c_{ij} \geq 0$.*

Proof We shall show by induction that after the triangle operation for $j = j_0$ is executed, d_{ik} is the length of the shortest path from i to k with intermediate nodes $v \leq j_0$, for all i and k. The basis for $j_0 = 1$ is clear. Assume then that the inductive hypothesis is true for $j = j_0 - 1$ and consider the triangle operation for $j = j_0$:

$$d_{ik} := \min \{d_{ik}, d_{ij_0} + d_{j_0k}\}$$

If the shortest path from i to k with intermediate nodes $v \leq j_0$ does not pass through j_0, d_{ik} will be unchanged by this operation, the first argument in the min-operation will be selected, and d_{ik} will still satisfy the inductive hypothesis. On the other hand, if the shortest path from i to k with intermediate nodes $v \leq j_0$ *does* pass through j_0, d_{ik} will be replaced by $d_{ij_0} + d_{j_0k}$. By the inductive hypothesis, d_{ij_0} and d_{j_0k} are both optimal distances with intermediate vertices $v \leq j_0 - 1$, so $d_{ij_0} + d_{j_0k}$ is optimal with intermediate vertices $v \leq j_0$. This completes the induction. □

The complete algorithm is shown in Fig. 6–13. All the loops are of fixed length, and the algorithm requires a total of $n(n-1)^2$ comparisons.

We can keep track of the shortest paths themselves with another $n \times n$ matrix, say e_{ik}, where we define

e_{ik} = highest-numbered intermediate node on the shortest path from i to k if there is one, and zero otherwise

```
FLOYD-WARSHALL ALGORITHM
Input:  An n×n matrix [c_ij] with nonnegative entries.
Output: An n×n matrix [d_ij], where d_ij is the shortest distance
        from i to j under [c_ij].
begin
  for all i ≠ j do d_ij := c_ij;
  for i = 1, . . . , n do d_ii := ∞;
  for j = 1, . . . , n do
     for i = 1, . . . , n, i ≠ j, do
        for k = 1, . . . , n, k ≠ j, do
           d_ik := min {d_ik, d_ij + d_jk}
end
```

Figure 6–13 The Floyd-Warshall algorithm.

We set $e_{ik} = 0$ initially, and when we do the triangle operations we set

$$e_{ik} := \begin{cases} j & \text{if } d_{ik} > d_{ij} + d_{jk} \\ e_{ik} & \text{otherwise} \end{cases}$$

The shortest path from i to k can then be reconstructed easily from the final e_{ik}-matrix. (See Problem 13.)

Now consider what happens if we allow the arc weights c_{ij} to be negative. First, if there are no cycles of negative length, the shortest paths to which the proof of Theorem 6.4 refers are all well defined, and the algorithm works as with nonnegative weights. If there is a negative-length cycle, it will cause some d_{hh} to become negative during the course of the algorithm. To see this, consider a simple negative-length cycle, and let h be the highest-numbered node on the cycle. Then the proof of Theorem 6.4 shows that the negative-length path from h to h with intermediate nodes $v < h$ will be found by the time $j = h$.

Example 6.2

Figure 6–14 shows a graph with a negative-length cycle, and the successive d_{ij} and e_{ij} matrices in the application of the Floyd-Warshall algorithm are shown in Fig. 6–15. Newly changed entries are circled.

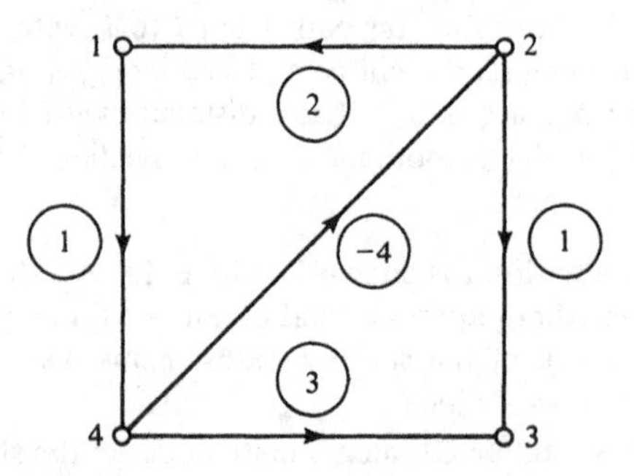

Figure 6–14 A graph with a negative-weight cycle.

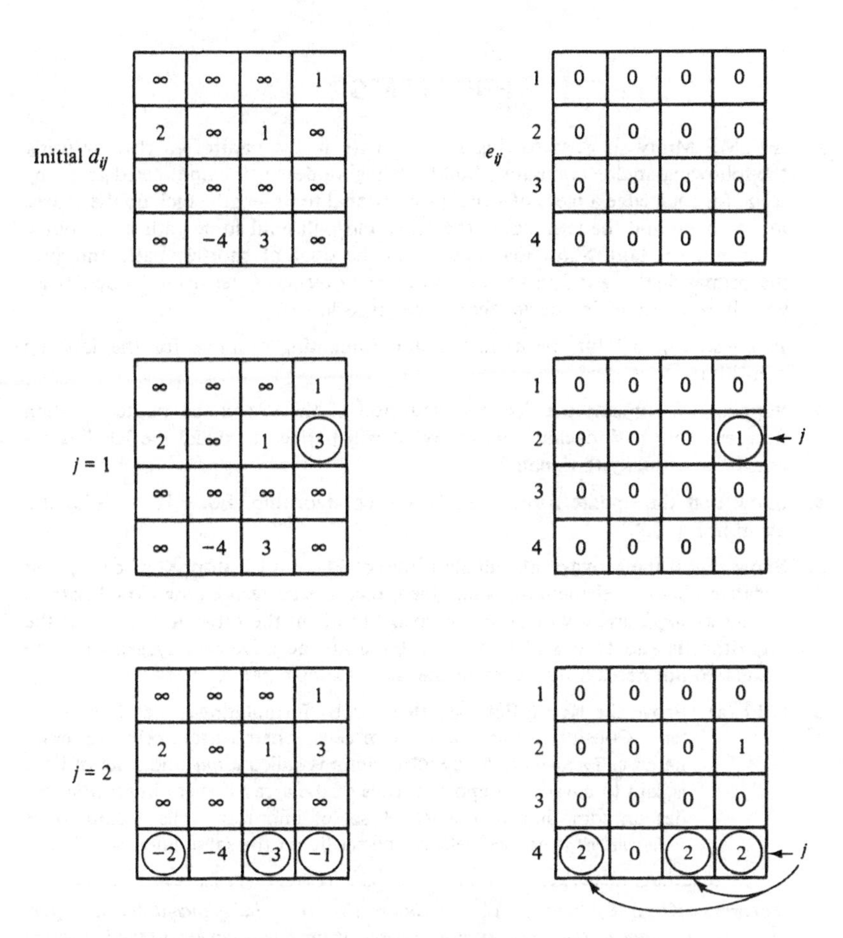

Figure 6–15 Successive distance and e_{ij}-matrices in the application of the Floyd-Warshall algorithm.

We stop when $d_{44} = -1$, resulting from the negative-length cycle 4-2-1-4, which can be found from the last e_{ik}-matrix. □

To summarize the state of affairs, we can solve the shortest-path problem by Dijkstra's algorithm in an n-node graph with nonnegative weights in time proportional to n^2. When the weights are allowed to be negative, we can use the Floyd-Warshall algorithm to solve shortest path where there are no negative-weight cycles or else to find a negative-weight cycle, in time proportional to n^3. Dijkstra's algorithm can also be extended to the negative-weight case, but it then also takes time proportional to n^3 [Ne, BL].

PROBLEMS

1. [FF, Mi] Minty suggests solving shortest path in the undirected case by using the following analog computer: Build a string model for the undirected graph by using for each edge a piece of string proportional to its length. Pick up the source in one hand and the terminal in the other and pull until some path from source to terminal is taut. Show that this solves the dual of shortest path. Interpret the primal-dual algorithm for this problem in terms of the string model. *Can you think of an analog device that solves max-flow?

2. Establish Eq. 6.7 for the a_i in the nonterminating example for the labeling algorithm.

3. Suppose we truncate the decimal expansion of the a_i's in the nonterminating flow example to k decimal places. What will happen to make the labeling algorithm eventually terminate?

4. Show that the update formula in Dijkstra's algorithm, Eq. 6.10, satisfies the claim in Eq. 6.9.

5. Show that if the Floyd-Warshall algorithm of Fig. 6–13 is stopped when the first negative diagonal element d_{ii} is obtained, then the corresponding closed path is in fact a *simple* cycle with negative cost. Show, on the other hand, that if the algorithm is run to completion when there are negative-cost cycles, then the results do not necessarily represent the cost of simple paths.

6. [FF] (a) Prove the König-Egerváry theorem by formulating a max-flow problem: Consider an $m \times n$ array of cells, where certain cells are designated as *important*. A row or column is called a *line* and a set of lines is said to *cover* the important cells of the array if every important cell lies on some line of the set. A set of important cells is said to be *independent* if no two cells of the set lie on the same line.

Theorem (König-Egerváry) *The maximum size of an independent set of important cells is equal to the smallest size of a set of lines that covers all the important cells.*

(b) Explain how to obtain an independent set of maximum size and a minimum line cover from the solution to the max-flow problem.

*(c) Find a bound on the number of steps this algorithm takes in terms of m, n, and p, the number of important cells.

7. Show how the Ford-Fulkerson algorithm can be applied to the case where capacities are put on the *nodes* of a flow network.

8. Suppose a directed graph G represents a communication network. The maximum number of node-disjoint paths from node s to t is called the *s-t connectivity*. The *s-t vulnerability* is the minimum number of nodes (besides s, t) whose removal disconnects s from t. Prove that the s-t connectivity equals the s-t vulnerability. (This is a version of Menger's theorem [FF].)

*9. Suppose that for each arc in a flow network, we have a *lower* bound on the flow as well as an upper bound; that is, for each arc (i, j), we demand $l_{ij} \leq f_{ij} \leq b_{ij}$.

   (a) Show how to apply the Ford-Fulkerson algorithm to find if there is a feasible s-t flow of value v. (*Hint:* Start with an infeasible flow and work towards feasibility.)

   (b) Show how to find a feasible flow of minimum value.

   (c) Show how to find a feasible flow of maximum value.

10. [FF] Suppose there are n men, n women, and m marriage brokers. Each broker has a list of some of the men and women as clients and can arrange marriages between any pairs of men and women on that list. In addition, we restrict the number of marriages that broker i can arrange to a maximum of b_i. All marriages are heterosexual, and all men and women are monogomous. Translate the problem of finding a solution with the most marriages into one of finding the maximum flow in a capacitated flow network.

11. Consider a flow network, and suppose we want to find paths between all pairs of nodes with maximum capacity in the sense that the smallest capacity arc on a path is as large as possible. Modify the Floyd-Warshall algorithm so that it solves this problem in $O(n^3)$ time for an n-node network.

12. Why are we justified in excluding the cases $i = j$ and $k = j$ in the Floyd-Warshall algorithm, even when negative weights are allowed? Will the algorithm work properly if we admit these cases?

13. Give the details of reconstructing shortest paths in the Floyd-Warshall algorithm.

NOTES AND REFERENCES

The nonterminating example of the labeling algorithm is from

[FF] FORD, L. R., JR., and D. R. FULKERSON, *Flows in Networks*, pp. 21–22. Princeton, N.J.: Princeton University Press, 1962.

Edmonds and Karp suggest modifying the labeling algorithm so that each flow augmentation is taken along an augmentation path with the fewest possible arcs at each stage. They then establish the bound of $(n^3 - n)/4$ on the number of flow augmentations (see also Problem 10 of Chapter 9).

[EK] EDMONDS, J., and R. M. KARP, "Theoretical Improvement in Algorithmic Efficiency for Network Flow Problems," *J. ACM*, 19, no. 2 (April 1972), 248–64.

Reference to Dijkstra's paper was given in Chapter 5. Dijkstra's algorithm is extended to the case where negative arc weights are allowed in

[Ne] NEMHAUSER, G. L., "A Generalized Permanent Label Setting algorithm for the Shortest Path between Specified Nodes," *J. Math. Analysis and Appl.*, 38 (1972), 328–34.

[BL] BAZARAA, M. S., and R. W. LANGLEY, "A Dual Shortest Path Algorithm," *J. SIAM*, 26, no. 3 (May 1974), 496–501.

The idea here is first to find a feasible solution π to the dual problem and then to apply the standard algorithm to the problem with modified weights $c_{ij} - \pi_i + \pi_j \geq 0$. [BL] obtain a time bound proportional to $|V|^3$ for such a procedure.

The Floyd-Warshall algorithm is from

[Fl] FLOYD, R. W., "Algorithm 97: Shortest Path," *Comm. ACM*, 5, no. 6 (1962), 345.

[WA] Warshall, S., "A Theorem on Boolean Matrices," *J. ACM*, 9, no. 1 (1962), 11–12.

The idea of the algorithm is described in the very general setting of closed semirings, and attributed essentially to Kleene, in

[AHU] AHO, A. V., J. E. HOPCROFT, and J. D. ULLMAN, *The Design and Analysis of Computer Algorithms*. Reading, Mass.: Addison-Wesley Publishing Co., Inc., 1974.

The string model for shortest path in Problem 1 is from [FF, p. 131] and

[Mi] MINTY, G. J., "A Comment on the Shortest Route Problem," *OR*, 5, no. 5 (October 1957), 724.

For more about the shortest-path problem and its solution by dynamic programming, see Chapter 18.

7

Primal-Dual Algorithms for Min-Cost Flow

7.1 The Min-Cost Flow Problem

We shall now study an important and general class of network problems, those with both nontrivial cost *and* capacity constraints. Our approach will be to apply and specialize the primal-dual algorithm, as we did for max-flow and shortest path, and there are two ways of doing this: We can consider our original problem as the *dual* D, combinatorialize the capacities, and arrive at min-cost subproblems; or we can consider our original problem as the *primal* P and combinatorialize the cost. We start with the first option, which maintains a feasible solution to our original problem D at all times and does not deal explicitly with the primal P or its restriction.

The problem we want to consider can be defined as follows.

Definition 7.1

Let $N = (s, t, V, E, b)$ be a flow network with underlying directed graph $G = (V, E)$, a weighting on the arcs $c_{ij} \in R^+$ for every arc $(i, j) \in E$, and a flow value $v_0 \in R^+$. The min-cost flow problem is to find a feasible *s-t* flow of value v_0 that has minimum cost. In the form of an LP:

$$\begin{aligned} &\min c'f \\ &Af = -v_0 d \quad \text{every node} \\ &f \leq b \quad \text{every arc} \\ &f \geq 0 \quad \text{every arc} \end{aligned} \tag{7.1}$$

where as usual A is the node-arc incidence matrix and

$$d_i = \begin{cases} -1 & i = s \\ +1 & i = t \\ 0 & \text{otherwise} \end{cases} \tag{7.2}$$ □

In contrast with max-flow, min-cost flow asks for a flow of *fixed* value that is cheapest among all such flows.

7.2 Combinatorializing the Capacities— Algorithm Cycle

We proceed by writing min-cost flow as the dual of a primal in standard form:

$$\begin{aligned} &\max -c'f \\ &Af \leq -v_0 d \\ &f \leq b \quad \text{for every arc} \\ &-f \leq 0 \quad \text{for every arc} \end{aligned} \tag{D}$$

We have replaced the equalities in the flow conservation equations with inequalities, using the same justification as in the max-flow problem. For any feasible flow, every one of these inequalities will be satisfied with equality. There is no difficulty in finding an initial flow of value v_0 in D, using max-flow, for example.

The dual of the restricted primal, DRP, can now be written by inspection, as we did for max-flow:

$$\begin{aligned} &\max -c'f \\ &Af = 0 \\ &f \leq 0 \quad \text{for saturated arcs} \\ &f \geq 0 \quad \text{for empty arcs} \\ &f \geq -1 \quad \text{for all arcs (since } -c \leq 0) \end{aligned} \tag{DRP}$$

We have here replaced $Af \leq 0$ by $Af = 0$ to get back to a clear indication of flow conservation at every node. A feasible flow satisfying $Af = 0$ has a special meaning for us, which we now distinguish by a special name.

Definition 7.2

A feasible flow f that satisfies $Af = 0$ is called a *circulation*. Its cost is $c'f$. □

The optimal solution to DRP is a circulation of a special kind: It must have no negative flow on an empty arc, no positive flow on a saturated arc, and no value less than -1 on any arc. It is now convenient to define a new weighted capacitated network in which these constraints are embodied.

Definition 7.3

Given a feasible flow f in weighted flow network N, define the *incremental* weighted flow network $N'(f)$ as follows: N' will have the same node set as N. For each arc $e = (i, j)$ in N with flow v, capacity d, and cost c, put two arcs in N': one arc (i, j) with capacity $d - v \geq 0$ and cost c; and another arc (j, i) with capacity $v \geq 0$ and cost $-c$. Omit all arcs with zero capacity. (See Fig. 7–1.) □

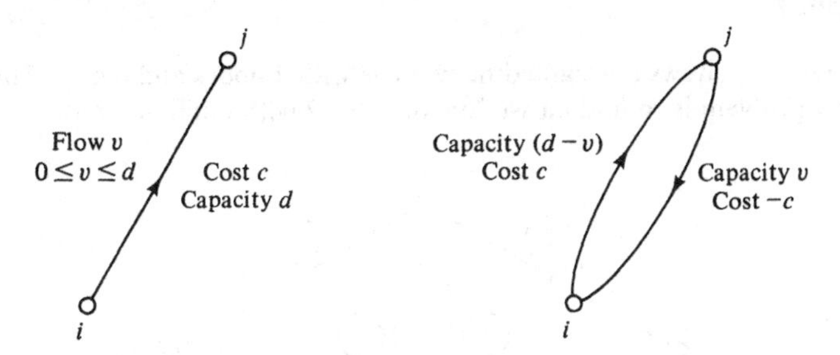

Figure 7–1 An arc in a weighted flow network and the corresponding arcs in the incremental network $N'(f)$.

From this definition it follows that a path from s to t in $N'(f)$ with weight x determines an s-t augmentation path in N; and that an increase in the value of the s-t flow f in N along this path by one unit results in an increase in the total cost of the flow by x units. Likewise, a circulation $\bar{f}$ in $N'(f)$ of cost x determines a new s-t flow $f + \bar{f}$ in N of the same value, with a cost increase of x.

We can now state the optimality condition for DRP in the following form.

Theorem 7.1 *An s-t flow f in a network is an optimal min-cost flow if and only if there are no cycles in $N'(f)$ with negative cost.*

Proof The primal-dual algorithm tells us that the flow f is optimal if and only if the optimal solution to DRP has cost zero, which is equivalent to there being no negative-cost circulations in $N'(f)$. The incremental network $N'(f)$ has a negative-cost circulation if and only if it has a negative-cost cycle. □

The primal-dual algorithm can now be implemented by starting with any flow f of value v_0 and searching for a negative-cost cycle using either the Floyd-Warshall algorithm or an extension of Dijkstra's algorithm to find negative-weight cycles. The loop in the primal-dual algorithm is closed by adding as much of the cyclic flow $\bar{f}$ on such a cycle as possible to the original f. The final algorithm *cycle,* due to Klein [Kl], is shown in Figure 7–2.

```
procedure cycle
begin
  use the max-flow algorithm to find a flow of value v0;
  while there is a negative-cost cycle C in N' do
        augment flow on C until N' no longer contains C
end
```

Figure 7–2 The primal-dual algorithm cycle for min-cost flow.

Example 7.1

Figure 7–3 shows a weighted network N with 4 nodes and 6 arcs. The min-cost flow problem is to find an s-t flow of value 2 with minimum cost.

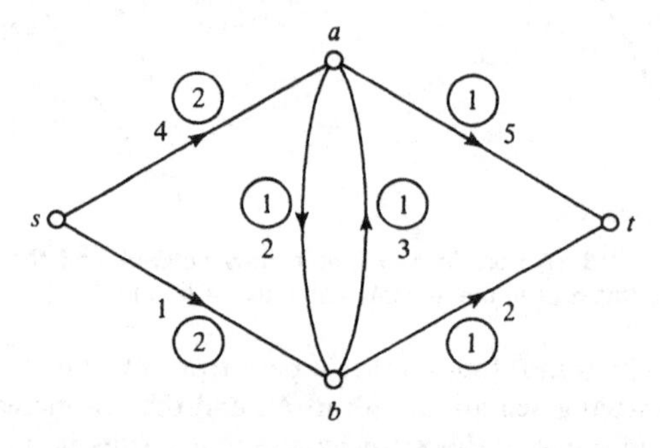

Figure 7–3 A weighted flow network illustrating min-cost flow. The circled numbers are capacities and the uncircled numbers are costs.

Suppose there is a flow of 1 unit along the path (s, a, t), and 1 unit along (s, a, b, t), with a total cost of 17. The incremental network N' is shown in Fig. 7–4. Note that the negative-cost cycle (s, b, a, s) is a circulation $\bar{f}$ of cost -5. We can add 1 unit of the circulation $\bar{f}$ to our original f, producing a flow of 1 unit along (s, a, t) and 1 along (s, b, t), with a total cost of $17 - 5 = 12$. The new incremental network, shown in Fig. 7–5, has no negative-cost cycles, and therefore the flow is now optimal. □

We shall call algorithm cycle a *problem-feasible algorithm,* because we always have a feasible solution to the original problem. This is analogous to calling the simplex algorithm *primal-feasible,* but that term would be hopelessly confusing here, since we can consider our original problem either a primal or a dual.

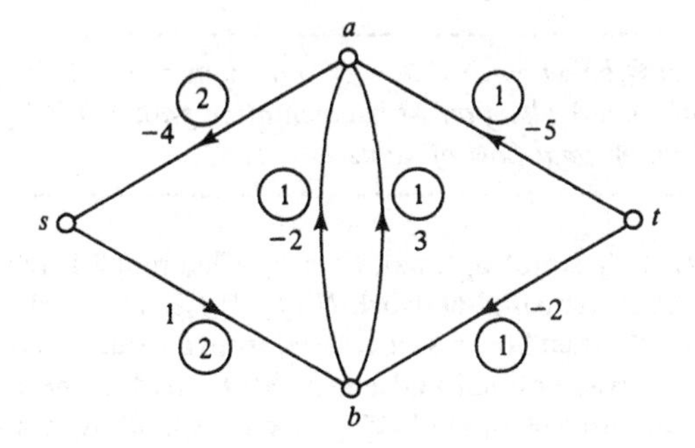

Figure 7–4 The first incremental network N' in Example 7.1.

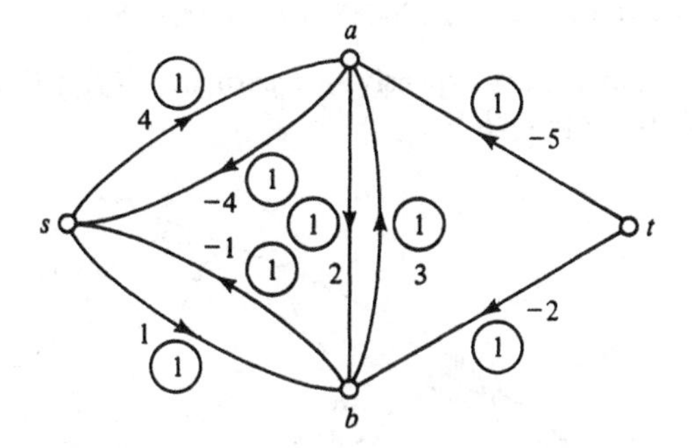

Figure 7–5 The second incremental network N' in Example 7.1.

7.3 Combinatorializing the Cost— Algorithm Buildup

The second option in applying primal-dual to min-cost flow is to combinatorialize the cost, considering the problem as a primal P. This gets complicated, since we have to deal with the corresponding dual D—to find the admissible set, for example. This is done quite explicitly by Ford and Fulkerson [FF], where they let the value v of the flow be a variable. They maximize a cost function

$$pv - c'f \tag{7.3}$$

for increasing values of the number p, thereby obtaining a sequence of flows of increasing value, each of minimum cost. The net effect of all this, however, is to arrive at an algorithm that avoids the dual D after all. The key result can be summarized in the following theorem.

Theorem 7.2 *Let f_1 be an optimal flow of value v in an instance of min-cost flow. Let f_2 be a flow of value 1 along an s-t augmentation path P in $N'(f_1)$ of least cost. Then $f_1 + f_2$ is an optimal flow of value $v + 1$.*

Proof If $f_1 + f_2$ is not optimal, then by Theorem 7.1, there is a negative-cost cycle C in the incremental network $N'(f_1 + f_2)$. This cycle appeared in the incremental network when flow was increased from a value of v to $v + 1$, because the flow of value v was optimal and hence $N'(f_1)$ had no negative-cost cycles. Therefore C has an arc $e = (i, j)$ of cost $-c$ corresponding to an arc (j, i) on P, as shown in Fig. 7–6. Replace the arc (j, i) on P by $C - \{e\}$. The effect on the cost of the path P is to increase it by

$$-c + (\text{cost of remainder of } C) = \text{cost of } C < 0$$

Therefore P was not a least expensive s-t path in $N'(f_1)$; this contradiction proves that $f_1 + f_2$ is optimal. □

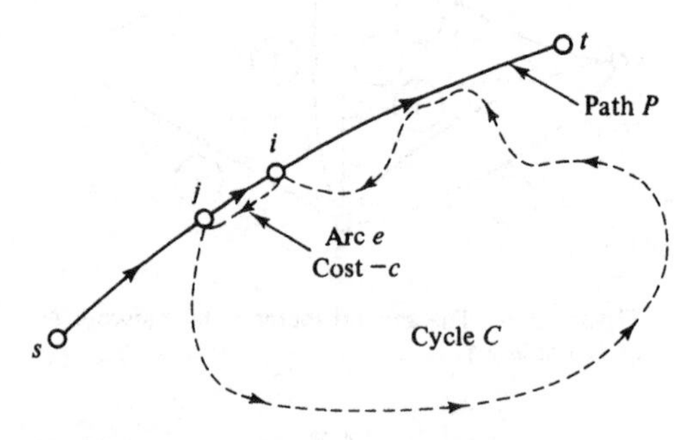

Figure 7–6 The construction in the proof of Theorem 7.2: The negative-cost cycle C leads to a lower cost path P.

By Theorem 7.2, we can build up an optimal flow step by step, by adding flows along augmentation paths of least cost in N'. At each stage we find a least-cost augmentation path P in N', and then augment the flow along P until the flow reaches the value v, or until P is no longer the least-cost augmentation path because one of its arcs disappears due to the saturation or emptying of the corresponding arc in N. Notice that some arcs in N' can have negative cost and the shortest-path algorithm to find P must deal with this case. However, because we always have, at any stage, an optimal flow of some value $f < v$, there are never negative-cost cycles in N'. The algorithm is outlined in Fig. 7–7.

In contrast to cycle, buildup does not produce a feasible flow of value v_0 until it terminates, so we call it a *problem-infeasible* algorithm.

```
procedure buildup
begin
  while flow f<v0 do
    begin find a shortest path P from s to t in N';
          augment the flow along P until it
          reaches v0 or until P is no longer
          a least-cost augmentation path
    end
end
```

Figure 7–7 Algorithm buildup for min-cost flow.

Example 7.2

If we begin with a zero flow in the network of Fig. 7–3, the least-cost s-t path is (s, b, t), and augmentation along this path produces a flow of value 1 and cost 3. A least-cost s-t path in the resulting incremental network N' is (s, a, t), producing the same optimal flow of value 2 and cost 12 as did algorithm cycle in Example 7.1. □

7.4 An Explicit Primal-Dual Algorithm for the Hitchcock Problem—Algorithm Alphabeta

We shall now take up a very famous problem which is a special case of min-cost flow. It was originally formulated around 1941 by several people, including Hitchcock [Hi], and bears his name. It is motivated by the following situation: We have m sources of some commodity, each with a supply of a_i units, $i = 1, \ldots, m$, and n terminals, each of which has a demand of b_j units, $j = 1, \ldots, n$. Furthermore, we know the unit cost c_{ij} of sending the commodity from source i to terminal j. How do we satisfy the demands at minimum cost? As an LP, we have the following.

Definition 7.4

Given $m, n \in Z^+$; source supplies $a_i \in R^+$, $i = 1, \ldots, m$; terminal demands $b_j \in R^+, j = 1, \ldots, n$; and $c_{ij} \in R^+, i = 1, \ldots, m$ and $j = 1, \ldots, n$; an instance of the *Hitchcock problem* is the following LP with variables f_{ij}:

$$\min \sum_{i,j} c_{ij} f_{ij}$$

$$\sum_{j=1}^{n} f_{ij} = a_i \qquad i = 1, \ldots, m \qquad (7.4)$$

$$\sum_{i=1}^{m} f_{ij} = b_j \qquad j = 1, \ldots, n \qquad (7.5) \qquad \text{(P)}$$

$$f_{ij} \geq 0$$

where

$$\sum_{i=1}^{m} a_i = \sum_{j=1}^{n} b_j \quad \square$$

We note that we could have formulated the problem with the inequalities

$$\sum_{j=1}^{n} f_{ij} \leq a_i \qquad i = 1, \ldots, m \tag{7.6}$$

$$\sum_{i=1}^{m} f_{ij} \geq b_j \qquad j = 1, \ldots, n \tag{7.7}$$

expressing the facts that a_i is available as a supply and b_j is a demand that must be met. The equalities in (7.4) and (7.5) can be used without loss of generality, however, since we can always introduce a fictitious $(n + 1)$st terminal with demand

$$b_{n+1} = \sum_{i=1}^{m} a_i - \sum_{j=1}^{n} b_j \tag{7.8}$$

and costs

$$c_{i,n+1} = 0 \qquad i = 1, \ldots, m$$

This extra terminal uses up the excess supply in Eq. 7.8 (assumed nonnegative) at no cost. (Notice that to work, this depends on the assumption $c_{ij} \geq 0$. See Problem 3.)

When all the a_i and b_j are 1, the Hitchcock problem is called the *assignment problem*, which will reappear later in this book. In fact, when we apply primal-dual to Hitchcock, we are in the position of reversing history, since Kuhn's "Hungarian method" [Ku] for the assignment problem was the precursor to the general primal-dual algorithm.

Our plan is to combinatorialize the cost in Hitchcock and to examine the dual D explicitly, rather than bypass it as we did in the previous section. Assign variables α_i and β_j to Equalities 7.4 and 7.5, respectively, yielding the dual

$$\max w = \sum_{i=1}^{m} a_i \alpha_i + \sum_{j=1}^{n} b_j \beta_j$$

$$\alpha_i + \beta_j \leq c_{ij} \qquad \text{for all } i = 1, \ldots, m \text{ and } j = 1, \ldots, n \tag{D}$$

$$\alpha_i, \beta_j \gtrless 0$$

An initial feasible solution to the dual can be written immediately.

$$\begin{aligned} \alpha_i &= 0 \\ \beta_j &= \min_{1 \leq i \leq m} \{c_{ij}\} \end{aligned} \tag{7.9}$$

We next define the *admissible set* IJ of indices of variables in the restricted primal by the pairs (i, j) for which equality is achieved in D.

$$IJ = \{(i, j) \colon \alpha_i + \beta_j = c_{ij}\}$$

Next, the restricted primal RP is defined by

$$\min \xi = \sum_{i=1}^{m+n} x_i^a$$

$$\begin{aligned}
\sum_j f_{ij} + x_i^a &= a_i & i &= 1, \ldots, m \\
\sum_i f_{ij} + x_{m+j}^a &= b_j & j &= 1, \ldots, n \\
x_i^a &\geq 0 & i &= 1, \ldots, m+n \\
f_{ij} &\geq 0 & (i, j) &\in IJ \\
f_{ij} &= 0 & (i, j) &\notin IJ
\end{aligned} \tag{RP}$$

where we have called the artificial variables x_i^a, $i = 1, \ldots, m + n$.

The cost in RP can be written

$$\xi = \sum_{i=1}^{m} a_i + \sum_{j=1}^{n} b_j - 2 \sum_{(i,j) \in IJ} f_{ij} \tag{7.10}$$

Minimizing ξ is therefore equivalent to maximizing the total flow on admissible arcs. We may then rewrite RP without the artificial variables but with inequality constraints as

$$\max \sum_{(i,j) \in IJ} f_{ij}$$

$$\begin{aligned}
\sum_j f_{ij} &\leq a_i & i &= 1, \ldots, m \\
\sum_i f_{ij} &\leq b_j & j &= 1, \ldots, n \\
f_{ij} &\geq 0 & (i, j) &\in IJ \\
f_{ij} &= 0 & (i, j) &\notin IJ
\end{aligned} \tag{RP'}$$

This is the max-flow problem shown in Fig. 7–8: we create a super-source s

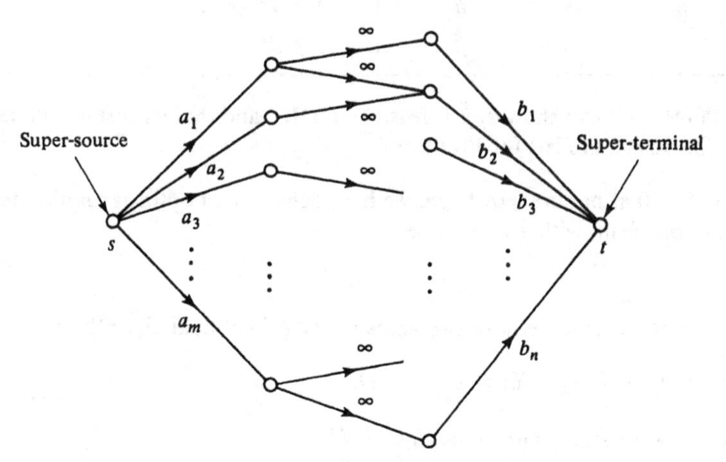

Figure 7–8 A max-flow problem equivalent to the restricted primal of a Hitchcock problem; the infinite capacity arcs correspond to admissible indices in the dual.

and a super-terminal t; from s to each of m sources we install an arc with capacity a_i; from each of n terminals we install an arc with capacity b_j. From a source to a terminal we install arc (i, j) with infinite capacity exactly when $(i, j) \in IJ$. This ensures that the variable f_{ij} can be greater than zero only when the index pair (i, j) is admissible.

The primal-dual algorithm proceeds by improving the dual feasible solution α, β with the optimal dual, say $\bar{\alpha}, \bar{\beta}$, of the restricted primal. We saw in Chapter 6 that the set of labeled nodes at termination of the Ford-Fulkerson algorithm determines an optimal s-t cut and hence also an optimal solution. We need some terminology for discussing the application of the labeling algorithm to Hitchcock.

Definition 7.5

When we achieve optimality after application of the labeling algorithm to RP, we say we are at *nonbreakthrough*. At nonbreakthrough, let

$$I^* = \{i\text{: source } i \text{ is labeled}\} \tag{7.11}$$
$$J^* = \{j\text{: terminal } j \text{ is labeled}\} \quad \square$$

We can now set down an optimal dual to the restricted primal.

Lemma 7.1 *At nonbreakthrough in the solution of RP', an optimal solution to the dual of RP is given by*

$$\begin{aligned} \bar{\alpha}_i &= 1 & i &\in I^* \\ \bar{\alpha}_i &= -1 & i &\notin I^* \\ \bar{\beta}_j &= -1 & j &\in J^* \\ \bar{\beta}_j &= 1 & j &\notin J^* \end{aligned} \tag{7.12}$$

Proof We can show $\bar{\alpha}, \bar{\beta}$ is feasible in DRP and that its cost is optimal for RP. The details are left for Problem 4. $\square$

If $\xi = 0$ at nonbreakthrough, we have achieved an optimal solution to our original problem, with a flow value

$$\sum_{(i,j)\in IJ} f_{ij} = \sum_i a_i = \sum_j b_j$$

If $\xi > 0$, recall that we have two cases in the primal-dual algorithm:

Case 1: $\bar{\alpha}_i + \bar{\beta}_j \leq 0$ for all $(i, j) \notin IJ$.

Case 2: $\bar{\alpha}_i + \bar{\beta}_j > 0$ for some $(i, j) \notin IJ$.

Case 1 implies that the primal was infeasible and hence must be impossible, because our formulation of P always has a feasible solution. Therefore Case 2 holds and we calculate

$$\theta_1 = \min_{\substack{i,j \\ \text{such that} \\ \bar{\alpha}_i+\bar{\beta}_j>0 \\ \text{and } (i,j)\notin IJ}} \left[\frac{c_{ij} - \alpha_i - \beta_j}{\bar{\alpha}_i + \bar{\beta}_j}\right] = \min_{\substack{i \in I^* \\ j \notin J^*}} \left[\frac{c_{ij} - \alpha_i - \beta_j}{2}\right] \tag{7.13}$$

The last equation follows since $\bar{\alpha}_i + \bar{\beta}_j$ can exceed zero only when $i \in I^*$ and $j \notin J^*$, in which case it equals 2; in this case we are assured that $(i, j) \notin IJ$, for otherwise j would have been labeled. The new dual solution π is then obtained by

$$\alpha_i^* = \begin{cases} \alpha_i + \theta_1 & i \in I^* \\ \alpha_i - \theta_1 & i \notin I^* \end{cases}$$
$$\beta_j^* = \begin{cases} \beta_j - \theta_1 & j \in J^* \\ \beta_j + \theta_1 & j \notin J^* \end{cases} \tag{7.14}$$

As it turns out, no arc that carries flow can become inadmissible, just as no basic column can become inadmissible in the general primal-dual algorithm. (See Problem 5.) This enables us to continue labeling in the Ford-Fulkerson algorithm from the flow pattern that was optimal at the previous nonbreakthrough. This completely determines the algorithm, for eventually we reach the maximum flow $\sum a_i = \sum b_j$. Figure 7–9 shows an outline of the entire algorithm, which we shall call alphabeta.

```
procedure alphabeta
begin
   choose α, β feasible in D;
   while flow is not maximum do
   begin
      solve the max-flow problem RP using only admissible arcs;
      find labeled rows and columns at non-breakthrough, say I* and J*;
      calculate θ1 and update α, β
      (comment: Equations 7.13 and 7.14)
   end
end
```

Figure 7–9 The primal-dual procedure alphabeta for Hitchcock.

We end this section with a general comment about where the primal-dual methodology has led us. Combinatorializing the cost in Hitchcock yields max-flow as a subproblem. Max-flow is solved by combinatorializing the capacities, yielding the completely combinatorial subproblem of finding a flow-augmenting path. We have thus used the primal-dual idea in a nested fashion to replace the two data vectors, cost and capacities, by two nested loops around a very simple combinatorial problem, as illustrated in Figure 7–10.

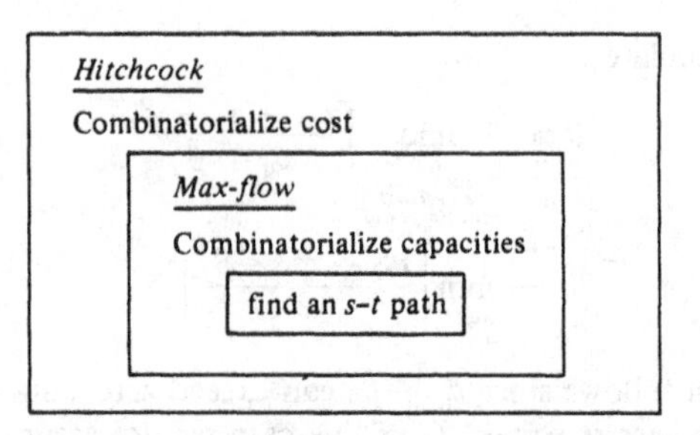

Figure 7–10 A schematic representation of the alphabeta algorithm as two nested loops.

If we use a version of Dijkstra that handles negative-cost arcs to solve the subproblems in cycle or buildup, then the same interpretation applies to those algorithms.

7.5 A Transformation of Min-Cost Flow to Hitchcock

That Hitchcock is a special case of min-cost flow is clear from the construction in Figure 7–8 with all arcs admissible. We simply supply all the sources from a super-source and collect flow from all the terminals at a super-terminal. What is perhaps surprising, however, is that min-cost flow is a special case of Hitchcock! We mean by this that given an instance of min-cost flow, we can construct an instance of Hitchcock that has the same solution. Later on in this book, this idea will play an important role in our understanding of intractable problems, and we shall use the terminology that "min-cost flow *transforms* to Hitchcock." The transformation described next is due to Wagner [Wa, FF].

Given an instance of min-cost flow, we shall construct an instance of Hitchcock according to the correspondence below:

Min-Cost Flow	Hitchcock
arc (i, j)	source ij
node i	terminal i
cost c_{ij}	arc (ij, j) with cost c_{ij} and infinite capacity
——	arc (ij, i) with cost zero and infinite capacity
capacity b_{ij}	supply b_{ij} to source ij

To specify the demands, we first need the notation

$$b_{iV} = \sum_{\substack{\text{all } j \text{ such that} \\ (i, j) \in E}} b_{ij} \tag{7.15}$$

That is, b_{iV} is the total capacity *out* of node i. The demand at terminal i is then

$$\begin{array}{ll} b_{iV} - v_0 & i = s \\ b_{iV} + v_0 & i = t \\ b_{iV} & i \neq s, t \end{array} \tag{7.16}$$

The construction is illustrated in Figure 7–11.

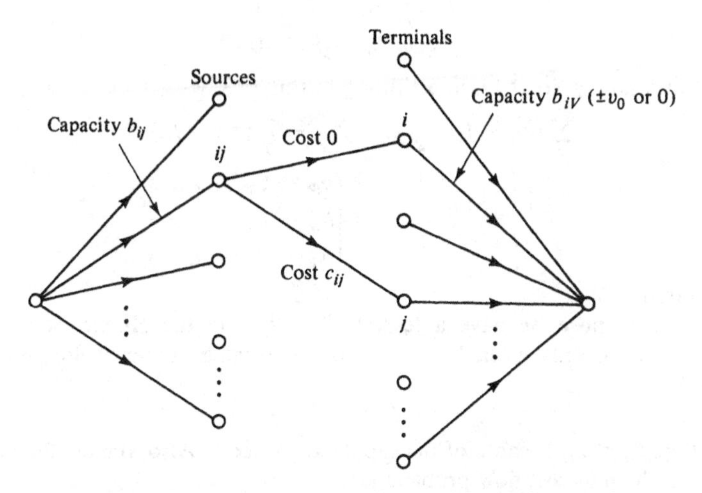

Figure 7–11 The Hitchcock problem constructed from a min-cost flow problem.

The Hitchcock problem is to find a flow $f_{ij,k}$ such that

$$f_{ij,j} + f_{ij,i} = b_{ij} \tag{7.17}$$

(the supply at source ij is used completely);

$$\sum_j (f_{ij,i} + f_{ji,i}) = \begin{cases} b_{iV} - v_0 & i = s \\ b_{iV} + v_0 & i = t \\ b_{iV} & i \neq s, t \end{cases} \tag{7.18}$$

(the demand into node i is filled); and

$$f_{ij,k} \geq 0 \qquad \text{for all } i, j, k \tag{7.19}$$

to

$$\min \sum_{\substack{\text{all} \\ \text{sources} \\ ij}} f_{ij,j} c_{ij,j} \tag{7.20}$$

Note that the total supply equals the total demand in Hitchcock. We now prove the desired result.

Lemma 7.2 *The original instance of min-cost flow and the constructed instance of Hitchcock are equivalent in the sense that a feasible flow in either corresponds to a feasible flow in the other with the same cost.*

Proof First, let f_{ij} be a feasible flow in the min-cost flow problem. Then, in the Hitchcock problem, if we take

$$f_{ij,j} = f_{ij} \qquad \geq 0 \tag{7.21}$$

$$f_{ij,i} = b_{ij} - f_{ij} \geq 0 \tag{7.22}$$

these flows satisfy Eq. 7.17. Substituting in Eq. 7.18, we get

$$\sum_j (b_{ij} - f_{ij} + f_{ji}) = b_{iV} + \sum_j (f_{ji} - f_{ij})$$

$$= \begin{cases} b_{iV} - v_0 & i = s \\ b_{iV} + v_0 & i = t \\ b_{iV} & i \neq s, t \end{cases} \tag{7.23}$$

as required.

Next, suppose we have a feasible flow $f_{ij,k}$ in the Hitchcock problem. This is nonzero only when $k = i$ or j. In the original min-cost flow problem, define

$$f_{ij} = f_{ij,j} \tag{7.24}$$

Then $0 \leq f_{ij} \leq b_{ij}$ because of the supply at source ij. Also, the net flow out of node i in the min-cost flow problem is

$$\sum_j f_{ij,j} - \sum_j f_{ji,i} = \sum_j (b_{ij} - f_{ij,i}) - \sum_j f_{ji,i}$$

$$= b_{iV} - \sum_j (f_{ij,i} + f_{ji,i})$$

$$= \begin{cases} v_0 & i = s \\ -v_0 & i = t \\ 0 & i \neq s, t \end{cases} \tag{7.25}$$

where we used Eqs. 7.17 and 7.18, so that all the constraints are satisfied.

Finally, it is easy to see that the costs of flows which correspond by Eqs. 7.21 and 7.24 are equal in their respective problems. □

7.6 Conclusion

We have used the primal-dual idea to develop a variety of algorithms for path and flow problems, and we shall exploit it further for matching problems. In the case of shortest path, it was easy to see that Dijkstra's algorithm took no more than $|V|^2$ steps. But even for the next simplest primal-dual algorithm, Ford-Fulkerson for max-flow, there is no easy way to obtain an upper bound on the required time; that must wait for Chapter 9. We next turn our attention to more general problems of defining and predicting the time complexity of algorithms.

PROBLEMS

1. Show how alphabeta can be modified simply so that all the numbers involved are integers.
2. Prove that the transformation of min-cost flow to Hitchcock is efficient in the following sense: the total number of elementary steps required for the construction of the instance of Hitchcock is bounded by a polynomial function of the number of bits required to represent the input to the original min-cost flow problem.
3. Why does the restriction of Hitchcock to one with equality constraints rely on the costs being nonnegative? Investigate the dependence of cycle, buildup, and alphabeta on this assumption.
4. Prove Lemma 7.1, giving an optimal solution to DRP in alphabeta. Is this a unique solution?
5. Prove directly from alphabeta that only an empty arc can become inadmissible. Does this result follow from Theorem 5.3, which states the analogous fact for basic columns in RP? Construct an example in which an arc does in fact become inadmissible.
6. Prove that a negative-cost circulation contains a negative-cost cycle.
7. Solve the following Hitchcock problem, using cycle, buildup, and alphabeta:

a \ b	1	3	2	2	2
3	3	2	3	1	2
2	1	5	4	5	2
1	4	4	3	2	1
4	5	1	3	5	2

   The matrix entries are the c_{ij}.

8. [Zal] Show how to transform a min-cost circulation problem with lower and upper bounds on arc flows and an initial flow that violates some upper and lower bounds to an ordinary min-cost flow problem. (*Hint:* Devise an incremental network with zero flow and thus only violated lower bounds. Change flow variables and introduce a fictitious source and terminal.) How can one accomplish this transformation so that all costs are nonnegative?
9. (The Caterer Problem [FF]) A caterer needs to supply r_i napkins on N successive days, $i = 1, \ldots, N$. The caterer can buy new napkins at p cents each, or launder

them at a fast laundry that takes m days and costs f cents a napkin, or launder them at a slow laundry that takes $n > m$ days and costs $s < f$ cents a napkin. At the end of each day the caterer must decide how many dirty napkins to send to the fast laundry, how many to the slow laundry, and how many to hold over. The requirement r_i on any day is met from those clean napkins available from both laundries, plus any new napkins that must be purchased.

(a) Formulate this problem as a min-cost flow problem. (*Hint:* Use N "supply" vertices and N "demand" vertices, plus some others.) Explain the arcs in your solution.

(b) Solve the problem with $N = 3$, $r_1 = 3$, $r_2 = 2$, $r_3 = 4$, $m = 1$ (napkins sent to the fast laundry are ready the next day), $n = 2$, $p = 10$, $f = 6$, and $s = 3$, using any method. *Prove*, however, that your answer is optimal. What is its cost?

10. (The Min-Cost Multicommodity Flow Problem [To]) A flow network is given which has q source-terminal pairs (s_i, t_i), instead of just one. There is to be a flow of value r_k, $k = 1, \ldots, q$, between s_k and t_k, thought of as a flow of commodity k. Each arc (i, j) has capacity b_{ij}, which is a limit on the total flow of all commodities on the directed arc (i, j); and a cost c_{ij} per unit total flow on that arc. All flows are positive on directed arcs. The problem is to find a feasible flow of minimum cost.

(a) Formulate this problem in node-arc form. How many rows and columns are there? If the Dantzig-Wolfe decomposition method is used, what are the subproblems?

(b) Formulate the problem in arc-chain form. How many rows and columns are there? If the revised simplex algorithm is used, what is the column-selection problem?

(c) Show that the formulations in (a) and (b) are equivalent.

11. (Capacitated Spanning Tree Problem [GJ]) We are given a complete undirected graph $G = (V, E)$, a symmetric distance matrix $[d_{ij}]$, an integer "traffic-generation rate" A_i for each node $i \in V$, integer capacity B, and a distinguished node s, called the *central site*. We want to find a spanning tree with minimum cost and total traffic no more than B on any edge, assuming traffic A_i is routed from every node to s. Show how to solve the special case where all A_i are zero or 1, and $B = 1$, by using a min-cost flow algorithm.

NOTES AND REFERENCES

The problem-feasible algorithm cycle for min-cost flow is due to

[Kl] KLEIN, M., "A Primal Method for Minimal Cost Flows," *Management Science*, 14, no. 3 (November 1967), 205–20.

Ford and Fulkerson attribute Theorem 7.2, from which algorithm buildup follows, to W. S. Jewell, R. G. Busacker and P. J. Gowen, and M. Iri, all in technical reports from 1958 to 1961.

[FF] FORD, L. R., JR., and D. R. FULKERSON, *Flows in Networks*. Princeton, N.J.: Princeton University Press, 1962.

When the primal-dual algorithm is used to combinatorialize cost in the min-cost circulation problem with lower bounds as well as upper bounds on arc flows, the *out-of-kilter method* results:

[Fu] FULKERSON, D. R., "An Out-of-Kilter Method for Minimal Cost Flow Problems," *J. SIAM*, 9, no. 1 (1961), 18–27.

Both [FF] and the following book by E. L. Lawler have full treatments of the algorithm.

[La] LAWLER, E. L., *Combinatorial Optimization: Networks and Matroids*. Holt, Rinehart & Winston, New York: 1976.

This min-cost circulation problem can be reduced to the min-cost flow problem and solved by cycle or buildup, as suggested by the recent work of N. Zadeh. (See Problem 8.)

[Za1] ZADEH, N., "A Simple Alternative to the Out-of-Kilter Algorithm," Technical Report No. 35, Dept. of Operations Research, Stanford University, May 31, 1979.

The following report shows many interrelations between simplex, dual-simplex, and out-of-kilter for min-cost flow.

[Za2] ZADEH, N., "Near-Equivalence of Network Flow Algorithms," Technical Report No. 26, Dept. of Operations Research, Stanford University, December 1, 1979.

The question of the time requirements of min-cost flow algorithms is problematical. From the theoretical point of view, Edmonds and Karp showed that there exists a polynomial algorithm for min-cost flow; that is, one that requires no more steps than a polynomial function of the number of bits in the problem description. (This idea will be made more precise in the next chapter.)

[EK] EDMONDS, J., and R. M. KARP, "Theoretical Improvements in Algorithmic Efficiency for Network Flow Problems," *J. ACM*, 19, no. 2 (April 1972), 248–64.

They modify a version of buildup by using a "scaling" technique, which can be described very roughly as follows: First, a zeroth-order approximation to the original problem is solved by approximating the capacities by 1-bit numbers, 0 or 1. The resultant flow is multiplied by 2, and a better approximation to the original problem is solved, one with 2-bit numbers for capacities, and so on. It will be shown in the next chapter that *any* LP can be solved in polynomial time, so the theoretical content of Edmonds and Karp's scaling method is subsumed by this more general result.

From a practical point of view, neither the scaling method nor the general polynomial algorithm for LP (the ellipsoid method) is a serious candidate for the most efficient min-cost flow algorithm, at least at this time. We must rely on empirical results for the relative efficiencies of various implementations of cycle, buildup, simplex, dual simplex, out-of-kilter, and even alphabeta applied to a transformation of Hitchcock. See, for example,

[BGK] BARR, R. S., F. GLOVER, and D. KLINGMAN, "An Improved Version of the Out-of-Kilter Method and a Comparative Study of Computer Codes," *Math. Prog.*, 7, no. 1 (August 1974), 60–86.

[GKK] GLOVER, F., D. KARNEY, and D. KLINGMAN, "Implementation and Computational Comparisons of Primal, Dual, and Primal Dual Computer Codes for Minimum Cost Network Flow Problems," *Networks*, 4, no. 3 (1974), 191–212.

[Mu] MULVEY, J. V., "Testing of a Large-Scale Network Optimization Program," *Math. Prog.*, 15, no. 3 (November 1978), 291–314.

The problem of finding a version of cycle, buildup, or alphabeta that has a polynomial time bound on the number of iterations that is independent of the costs and capacities is an important open problem at the time of this writing. ([EK] states this question explicitly.) We have already accomplished this goal for Dijkstra and shall do likewise for Ford-Fulkerson. That the algorithms need to be restricted in their choices to ensure good behavior is shown by the pathological examples of N. Zadeh.

[Za3] ZADEH, N., "A Bad Network Problem for the Simplex Method and Other Minimum Cost Flow Algorithms," *Math. Prog.*, 5, no. 3 (December 1973), 255–66.

Zadeh gives examples here of modified Hitchcock problems that require an exponential number of iterations when certain versions of cycle, buildup, and alphabeta are applied. In the following reference, Zadeh gives an example of min-cost flow for which a version of cycle can take an arbitrarily large number of iterations, in analogy with the Ford and Fulkerson nonterminating example for the labeling algorithm.

[Za4] ZADEH, N., "More Pathological Examples for Network Flow Problems," *Math. Prog.*, 5, no. 2 (October 1973), 217–24.

Ford and Fulkerson attribute formulation of the Hitchcock problem to several people, including, of course, Hitchcock

[Hi] HITCHCOCK, F. L., "The Distribution of a Product from Several Sources to Numerous Localities," *J. Math. Phys.*, 20, no. 2 (April 1941), 224–30.

and T. C. Koopmans, A. N. Tolstoi, L. Kantorovitch, and M. K. Gavurin, all around the same time.

When all the supplies and demands in Hitchcock are unity, we have the *assignment problem.* Alphabeta, often called simply the primal-dual method for Hitchcock, is a generalization of Kuhn's Hungarian method, named for its dependence on a result of J. Egerváry (see Problem 6 of Chapter 6).

[Ku] KUHN, H. W., "The Hungarian Method for the Assignment Problem," *Naval Research Logistics Quarterly*, 2, nos. 1 and 2 (1955), 83–97.

As mentioned in the text, alphabeta was in turn the precursor to the general primal-dual algorithm for LP's.

The transformation from min-cost flow to Hitchcock is attributed in [FF] to

[Wa] WAGNER, H. M., "On a Class of Capacitated Transportation Problems," *Management Science*, 5 (1959), 304–18.

It is a remarkable harbinger of things to come (see Problem 2 and Chapter 15). Ford and Fulkerson dismiss the transformation as a practical way to solve min-cost flow, but that is not a completely obvious conclusion.

Problem 10 is from

[To] TOMLIN, J. A., "Minimum-Cost Multicommodity Network Flows," *OR*, 14, no. 1 (February 1966), 45–51.

The special case of the capacitated spanning tree problem of Problem 11 is mentioned in

[GJ] GAREY, M. R., and D. S. JOHNSON, *Computers and Intractability: A Guide to the Theory of NP-Completeness*. San Francisco, California: W. H. Freeman & Company, Publishers, 1979.

For more on the problem, which is in general very hard, see

[VSC] VAN SICKLE, L., and K. M. CHANDY, "Computational Complexity of Network Design Algorithms," *Information Processing* 77, ed. B. Gilchrist. North Holland Publishing Co., 1977.

[Pa] PAPADIMITRIOU, C. H., "The Complexity of the Capacitated Tree Problem," *Networks*, 8 (1978), 217–30.

8

Algorithms and Complexity

8.1 Computability

The wide applicability of the simplex algorithm and its variants to problems with hundreds and thousands of variables and constraints would be impossible without the existence of the fast digital computers of today. The same is true for many commonly used techniques for solving numerical problems, simulating physical or social processes and manipulating information: These methods cannot be realistically applied by hand, except to instances too small in size to be of any practical significance. It is now widely accepted that the limits of the human computer have been surpassed by the needs of today's science and technology. Are there limits to the potential of electronic computers?

Clearly, ill-defined, nonmathematical tasks such as "solve the energy problem" or "outsmart humans" cannot be performed by computers. Computers can only carry out *algorithms*; that is, precise and universally understood sequences of instructions that solve any instances of rigorously defined computational problems. The methods taught in elementary schools for performing arithmetic operations on decimal integers are typical algorithms. They are precise methods that can be applied to any integers—no matter how large—and they are correct, in the sense that they are guaranteed to terminate with the right answer. They are so dryly and literally written that we can trust their execution to machines. It should not be a surprise that this intuitive concept of an algorithm can be defined rigorously. The corresponding mathematical object is called a *Turing machine*, after the British mathematician Alan M. Turing, who invented it in 1936 (see Sec. 15.5).

Are there well-defined mathematical problems for which there is no algorithm? By brilliant arguments, Turing showed that such *undecidable* problems do exist. A typical one is the so-called *halting problem*: Given a computer program with its input, will it ever halt? Turing proved that there is no algorithm that solves correctly all instances of this problem. It is possible to find some heuristic ways to detect some infinite loop patterns by examining the program and the input, but there will always be subtleties that escape our analysis. Of course, we may simply run the program and report success if we reach an **end** statement. Unfortunately, this scheme is not an algorithm, because it is not guaranteed *itself* to halt!

8.2 Time Bounds

If mathematical formalisms like Turing machines led the mathematicians of the 1930s to the study of undecidable problems, the digital computers of today present us with different challenges.

All of the computational problems discussed in this book are of the decidable kind. That is, in principle there is an algorithm that would correctly solve any instance of the problem. This is not always considered satisfactory, however, because excessive time requirements may render an algorithm completely useless.

Example 8.1

The traveling salesman problem (TSP, see Example 1.1) is certainly a decidable problem, since an instance of the TSP can be solved by finding the best among a finite set of tours. Thus a computer could solve any instance of the TSP by systematically examining and evaluating all tours and then choosing the shortest one.

The number of tours of n cities is $(n - 1)!/2$. So the implementation of the above algorithm in a computer would require about $n!$ steps (elementary instructions). The solution by this algorithm of a modestly sized instance of the TSP—for instance, finding the best tour of the state capitals of the United States—would thus require many billions of years, even under the most optimistic assumptions about the speed of computers in the future (50! has about 65 decimal digits). □

Example 8.2

The minimum spanning tree problem (MST, see Example 1.2) can also be solved by exhaustively examining all spanning trees and choosing the best. Since there are n^{n-2} spanning trees with n nodes (for a proof see [Ev]), the time requirements of the exhaustive approach are again unbearable. However, there is a much better algorithm for this problem. In Chapter 12 we develop an algorithm for the MST that requires a number of elementary steps proportional to n^2, when applied to n points. This is a very practical algorithm. Using this algorithm, the MST of the U.S. state capitals can be computed in a few seconds by most computers. □

The most widely accepted performance measure for an algorithm is the time it spends before producing the final answer. This amount of time may vary vastly from one computer to another because of differences in speed and instruction repertoire. In the analysis of algorithms in this book, we express the time requirements of algorithms in terms of the number of elementary steps—arithmetic operations, comparisons, branching instructions, and so on—required for the execution of the algorithm on a hypothetical computer. We assume, that is, *that all these kinds of operations require unit time.*

The number of steps required by an algorithm is not the same for all inputs. In the simplex algorithm, for instance, the number of elementary steps required to solve the $m \times n$ LP

$$\max c'x, \qquad \text{subject to } Ax \leq b, \quad x \geq 0$$

may vary considerably with the parameters A, b, and c, even if their dimensions are kept constant. In an extreme case, if $c \leq 0$ the starting feasible solution is optimal and no pivoting is required, whereas, for different choices of the parameters, a significant number of iterations may be required to reach the optimum.

To smooth such sharp contrasts in the behavior of an algorithm from one input to another, we consider *all* inputs of a given size *n together*, and we define the complexity of the algorithm for that input size to be the *worst-case* behavior of the algorithm on any of these inputs. Then the complexity of an algorithm is a function of the size of the input, such as $10n^3$, 2^n, and $n \log n$.†

In studying the complexity of an algorithm, we are often interested only in the behavior of the algorithm when supplied with very large inputs, because it is these inputs that are going to determine the limits of the applicability of the algorithm. Differences such as that between an algorithm of complexity $10n^3$ and one of $9n^3$ can be made irrelevant by a technological breakthrough that induces a tenfold increase in the speed of computers. On the other hand, slower growing terms (like the $5n$ term in the bound $n \log n + 5n$) will eventually be

†When no base is given explicitly, all logarithms are understood to be to the base 2.

overwhelmed by faster growing terms for large enough n, (in our example, for $n \gg 1000$). We are interested, therefore, in the *rate of growth* of the complexity of the algorithm. To deal with rates of growth of functions the following formalism is helpful.

Definition 8.1

Let $f(n)$, $g(n)$ be functions from the positive integers to the positive reals.

(a) We write $f(n) = O(g(n))$ if there exists a constant $c > 0$ such that, for large enough n, $f(n) \leq cg(n)$.

(b) We write $f(n) = \Omega(g(n))$ if there exists a constant $c > 0$ such that, for large enough n, $f(n) \geq cg(n)$.

(c) We write $f(n) = \Theta(g(n))$ if there exist constants $c, c' > 0$ such that, for large enough n, $cg(n) \leq f(n) \leq c'g(n)$.

We may write $f(n) \asymp g(n)$ instead of $f(n) = \Theta(g(n))$. It is easy to see that $\asymp$ is an equivalence relation. The equivalence class of $f(n)$ in this equivalence relation (that is, the set of all functions $g(n)$ such that $f(n) = \Theta(g(n))$) is called the *rate of growth* of $f(n)$. □

Using this notation, the rate of growth of the complexity of an algorithm may be bounded from above by phrases like "takes time $O(n^3)$."

8.3 The Size of an Instance

We measure the complexity of an algorithm as a function of the size of the input of the algorithm. But what is the size of the input? The input in combinatorial optimization problems is a combinatorial object: a graph, a set of integers (possibly arranged in vectors and matrices), a family of finite sets, and so on. To submit this input for solution by a computer, we must somehow *encode* it, or *represent* it, as a sequence of symbols over some fixed alphabet such as bits, typewriter symbols, or ASCII characters. We shall not define precisely how we encode combinatorial objects into sequences of symbols. These encodings may be done in any of a number of obvious and simple ways, some of which we illustrate in the examples below. Moreover, it will almost always turn out that all of these encodings are essentially equivalent for our purposes. On the rare occasions that this issue may present us with some subtle difficulty, we shall explicitly point out the problem and the cure.

Once we have decided that the input of an algorithm is represented as a sequence (or *string*) of symbols, we define the *size* of the input to be the *length of this sequence*, that is, the number of symbols in it.

Example 8.3

In many problems—for example, in the problem of testing whether an integer is prime—an instance is simply an integer. There are many economical ways of representing integers; the most common ones are arithmetic systems to some fixed base, such as decimal and binary. In these systems, the number of symbols required in order to represent an integer n is $\lceil \log_B n \rceil$,† where $B \geq 2$ is the *base*. We see that, no matter what base is used, the size of the representation of n is $\Theta(\log n)$—recall that

$$\log_B n = \frac{\log n}{\log B}$$

and $\log B$ is a constant, once we have fixed B. □

Example 8.4

What is the size of a linear program? As before, we assume that entries of A, b and c are integers. Thus the size of a linear program would be the number of symbols required in order to write A, b, and c. Since this can be done by listing the elements of the matrices in binary (or decimal), using appropriate delimiters to stand for the horizontal and vertical lines of the tableau, the size of an $m \times n$ LP is $\Theta(mn + \lceil \log |P| \rceil)$, where P is the product of all nonzero coefficients. □

Example 8.5

In many interesting problems the input is just a graph. What is the size of a graph?

A graph may be represented in many ways. For example we can associate with any graph $G = (V, E)$ its $|V| \times |V|$ *adjacency matrix* $A_G = [a_{ij}]$, such that $a_{ij} = 1$ if $[v_i, v_j] \in E$, and $a_{ij} = 0$ otherwise. Nevertheless, this may not be the most economical representation of a graph. A graph (V, E) may have up to $\binom{|V|}{2} = \Theta(|V|^2)$ edges. However, many graphs are *sparse* in that the number of their edges is far less than $\binom{|V|}{2}$. For example, we may have a graph with 100 nodes and 500 edges. Representing this graph by its adjacency matrix would require 10,000 digits for recording all entries. Simply listing the edges one by one would be more economical.

One useful way of representing a graph is by its *adjacency lists*. For each node $v \in V$, we record the set $A(v) \subseteq V$ of nodes adjacent to it (see Fig. 8–1).

The size of this representation depends on the sum of the lengths of the lists. Since each edge adds 2 to this total length (one for the list of one endpoint plus one for the other), it follows that we have a total of $2|E|$ elements to write down.

†By $\lceil x \rceil$ we denote the smallest integer q such that $q \geq x$, and by $\lfloor x \rfloor$, the largest integer q such that $q \leq x$.

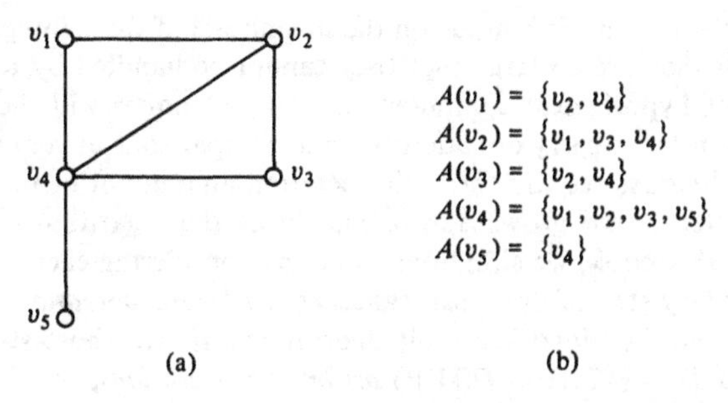

Figure 8–1

There is another factor, however, that affects the total length of the representation. Even for graphs of moderate size, we cannot have a different letter for each node: Recall that our alphabet must be of a fixed finite size. We shall then have to use subscripts to distinguish among vertices. Since we have $|V|$ vertices, we need approximately $\Theta(\log |V|)$ bits—or decimal places—to do this. It follows that $\Theta(|E| \log |V|)$ symbols are required in order to represent the graph $G = (V, E)$.

Why, then, is it that in practice we say that a graph (V, E) can be encoded in $O(|E|)$ space? The reason is that today's computers usually treat all integers in their range—typically from zero to 2^{31}—the same. They assign the same space, a *word*, to both 3 and 7^{10}. Since we may be almost sure that graphs with more than a trillion nodes will never come up in an application, $O(|E|)$ such words are adequate for storing the adjacency lists of a graph using indices in this range to identify the vertices. Hence $|E|$ is a reasonable approximation to the size of a graph, and analyzing the complexity of graph algorithms using $|E|$ as a parameter is an acceptable practice.

Nevertheless, sometimes we use *both* $|V|$ and $|E|$ as parameters for characterizing the complexity of an algorithm. This is a convention with roots in people's tendency to consider $|V|$ as the prime measure of size of a graph, probably because in most applications of graph theory V is the point of departure for the construction of G. Naturally, $|V|$ and $|E|$ satisfy $|E| \leq |V|(|V| - 1)/2$, and we can assume that $|E| \geq |V|/2$ (for instance, if our graph has no isolated points). However, $|E|$ may vary greatly within these bounds to make G *dense* ($|E| = \Theta(|V|^2)$) or *sparse* ($|E|$ is far less than its maximum value). Hence, an $O(|E|^3)$ algorithm may be preferable to an $O(|V|^3 |E|)$ algorithm when the graph is sparse, whereas the opposite choice is justified for dense graphs. □

Some combinatorial problems, such as the TSP, the shortest-path problem, and the MST, have an input consisting, at least in part, of integers. Algorithms for these problems usually involve operations such as addition and comparison of integers. In the Floyd-Warshall algorithm for the shortest-path problem, for example, most steps consist of pairwise comparisons and additions of integers.

Since we have no explicit bounds on the magnitude of these integers, it may be the case that they are so large that they cannot be handled by the finite word length of our hypothetical computer. Special techniques will then have to be employed in order to carry out additions and comparisons of very large integers, and such techniques require for each operation an amount of time (number of elementary steps) that grows approximately as the logarithm of the integers involved. In this book, we simplify matters by considering each such operation as an elementary step of unit cost, called an *arithmetic operation*. For example, we shall say that the Floyd-Warshall algorithm solves the shortest-path problem in a digraph $D = (V, A)$ in $O(|V|^3)$ *arithmetic operations*, or in $O(|V|^3)$ *time*, for short. What we really mean by this is that the number of elementary operations is bounded by $O(|V|^3 \log M)$, where M is the largest integer appearing in this instance. We can adopt this convention without changing the essence of our results, because the algorithms in which we shall be interested involve operations on integers that are not significantly larger than those in the input.†

A completely different (and more subtle) situation arises when the total *number* of arithmetic operations depends on the magnitude of the integers involved. We shall see in the next section that the labeling algorithm for the max-flow problem described in Chapter 6 presents such behavior.

8.4 Analysis of Algorithms

Deriving tight upper bounds for the time requirements of an algorithm is not always straightforward; it can be as ingenious and artful as the design of the algorithm.

Example 8.6

The Floyd-Warshall algorithm for the shortest-path problem for a digraph $D = (V, A)$ has a time complexity of $O(|V|^3)$ arithmetic operations. In this case the analysis is especially easy, and in fact the time requirements of the algorithm do not vary considerably with the digraph and the distances appearing in the input. The key observation is that the algorithm consists essentially of three "nested loops," each of which is executed about $|V|$ times; the innermost "triangle" operation can be implemented with just two arithmetic operations. □

Example 8.7

What is the complexity of the labeling algorithm for solving the maximum flow problem (Chapter 6) for a network $N = (s, t, V, A, b)$? We observe that the algorithm has an initialization step, which can be carried out in time proportional to $|V|$, and many iteration steps. Each iteration step involves the

†In the simplex method all numbers that we handle are rationals, with both numerator and denominator bounded in absolute value by the size of the instance (recall the proof of Lemma 2.1).

scanning and labeling of vertices. In order to bound the complexity of each iteration, we observe that in the scanning process each arc (v, u) of N can be visited at most twice—once for the scanning of v and once for u. Thus the labeling process requires a number of arithmetic steps of the order $O(|A|)$. On the other hand, following the labels backwards can be done in $O(p)$ steps, where p is the length of (number of arcs in) the augmenting path discovered. We note that there can be no repetitition of nodes in the augmenting path; hence we have $p \leq |V|$. It follows that each iteration of the algorithm takes time $O(|V| + |A|) = O(|A|)$.

Consequently, the overall algorithm has complexity $O(S \cdot |A|)$, where S is the number of iterations involved. Because the capacities of all the arcs are integral, we notice, by an easy induction on the number of iterations, that the flow remains integral at all stages. It follows that the increments also have positive integer values, and so at each iteration the flow is augmented by at least one. Thus, if v is the value of the maximum flow, we have $S \leq v$. There is something definitely wrong with this estimate, however: we bound the complexity of solving a problem in terms of its solution! What we really need is an a priori estimate of the complexity, expressed in terms of the input. In order to obtain this, we just need to observe that $v \leq \sum_{(x,y) \in A} b(x, y)$. We thus conclude that the labeling algorithm has complexity

$$O\left(\left(\sum_{(x,y) \in A} b(x, y)\right) \cdot |A|\right)$$

We shall see in the next chapter that this worst-case bound is actually achievable. □

Example 8.8

The simplex algorithm, like the labeling algorithm of Example 8.7, involves an initialization phase together with a number of iterations. If the dimensions of the matrix A are $m \times n$, the initialization requires $O(nm)$ arithmetic operations. Similarly, each iteration can be viewed roughly as a matrix-vector multiplication, and hence it can also be done in $O(mn)$ arithmetic operations.

Because cycling is avoided, the simplex method can, at worst, visit all basic feasible solutions, and there are at most $\binom{m+n}{m}$ basic feasible solutions. In fact, in Sec. 8.6, we show that indeed there are LP's that may cause simplex to perform many (though not exactly as many as $\binom{m+n}{n}$) iterations. □

8.5 Polynomial-Time Algorithms

When should we consider a computational problem satisfactorily solved? The answer obviously lies in the performance of the known algorithms for this problem. If there is an algorithm for this problem that is not too time-consuming—

the prime criterion we use here—then the problem may be considered solved, and not otherwise. In fact, as we have pointed out, it is the rate of growth of the best known time bound that is going to determine the practical utility of the algorithm. What rates of growth, then, should we consider as acceptable solutions to computational problems?

Today there is general agreement among computer scientists that an algorithm is a practically useful solution to a computational problem only if its complexity grows *polynomially* with respect to the size of the input. For example, algorithms of complexity $O(n)$ or $O(n^3)$ are acceptable in this school of thought. (Note that the rate of growth of a polynomial is completely specified by its degree.) Naturally, algorithms for which the asymptotic complexity is not a polynomial itself but is *bounded* by a polynomial, also qualify. Examples are $n^{2.5}$ and $n \log n$.

To understand the significance of polynomially bounded algorithms as a class, let us consider the remaining algorithms, those that violate all polynomial bounds—for large enough instances, that is. We usually refer to these as *exponential* algorithms, because 2^n is the paradigm of nonpolynomial rates of growth. Other examples of exponential rates of growth are k^n (any fixed $k > 1$), $n!$, 2^{n^2}, n^n and $n^{\log n}$†. It is obvious that, when the size of the input grows, any polynomial algorithm will eventually become more efficient than any exponential one (see Table 8.1). Another positive feature of polynomial algorithms is that,

TABLE 8.1
The growth of polynomial and exponential functions.

Function	Approximate Values		
n	10	100	1000
$n \log n$	33	664	9966
n^3	1000	1,000,000	10^9
$10^6 n^8$	10^{14}	10^{22}	10^{30}
2^n	1024	1.27×10^{30}	1.05×10^{301}
$n^{\log n}$	2099	1.93×10^{13}	7.89×10^{29}
$n!$	3,628,800	10^{158}	4×10^{2567}

in a sense, they take better advantage of technological advances. For example, each time a technological breakthrough increases the speed of computers tenfold, the size of the largest instance solved by a polynomial algorithm in an hour, for instance, will be multiplied by a constant between 1 and 10. In contrast, an exponential algorithm will experience only an *additive* increase in the size of the instance it can solve in a fixed amount of time (Table 8.2). Finally, we may

†Rates of growth such as $n^{\log n}$ that are faster than any polynomial but slower than 2^{n^ϵ} for all $\epsilon > 0$ are sometimes called *subexponential*.

TABLE 8.2
Polynomial-time algorithms take better advantage of technology.

Function	Size of Instance Solved in One Day	Size of Instance Solved in One Day in a Computer 10 Times Faster
n	10^{12}	10^{13}
$n \log n$	0.948×10^{11}	0.87×10^{12}
n^2	10^6	3.16×10^6
n^3	10^4	2.15×10^4
$10^8 n^4$	10	18
2^n	40	43
10^n	12	13
$n^{\log n}$	79	95
$n!$	14	15

comment that polynomial algorithms have nice "closure" properties: Polynomial algorithms may be combined to solve special cases of the same problem; a polynomial algorithm may invoke another polynomial algorithm as a "subroutine," and the resulting algorithm will still be polynomial.

When differentiating between polynomial and exponential algorithms, special care must be taken when the time bound involves the numerical input of the problem. For the labeling algorithm, for example, the time bound derived is

$$O((\sum_{(x,y)\in A} b(x, y)) \cdot |A|)$$

(Example 8.7). The asymptotic growth of this function looks at first glance perfectly polynomial, because there is no obvious exponentiation or equivalent operation. Nevertheless, it is not a polynomially bounded function in terms of the size of the input, and this is because it is expressed in terms of the numerical input of the instance. To demonstrate this, suppose that in a network all arcs have capacity $2^{|A|}$. We could encode this instance of the max-flow problem by listing $|A|$ integers, each $|A|$ bits long; thus the size of this instance is $O(|A|^2)$. However, the bound now becomes $O(2^{|A|} \cdot |A|^2)$, not a polynomial function of the size of the instance. In Chapter 16 we shall introduce the term *pseudopolynomial* for algorithms with time bounds such as this.

So would an n^{80} algorithm be a practical solution to a problem? Probably not. The time required to solve instances of size 3 is already astronomical, and an exponential algorithm may perform better for all reasonable inputs. The thesis that polynomial-time algorithms are "good" seems to weaken when pushed to extremes. Experience, however, comes to its support. For most problems, once *any* polynomial-time algorithm is discovered, the degree of the polynomial quickly undergoes a series of decrements as various researchers improve on the idea. Usually, the final rate of growth is $O(n^3)$ or better. In contrast, exponential algorithms are usually as time-consuming in practice as they are in theory, and they are quickly abandoned once a polynomial-time algorithm for the same problem is discovered.

These empirical rules are by no means universally accepted, and in some cases they seem to fail completely. In fact, in the next two sections we examine the background of an on-going controversy concerning the simplex algorithm and a recently discovered polynomial-time algorithm for LP. This controversy is the most serious challenge today to the thesis that *polynomial-time* is a synonym of *practical.*

8.6 Simplex Is Not a Polynomial-Time Algorithm

The most prominent algorithm discussed in previous chapters is simplex. It is therefore natural to try to apply to it the mathematical criterion of "goodness" of algorithms proposed in the previous section and ask if simplex is a polynomial-time algorithm. In this section we shall present a simple argument, due to Klee and Minty [KM], which establishes that the simplex algorithm in *not* polynomial-time.

The simplex algorithm, as shown in Figure 2–3, has an incompletely specified step: the choice of j. Worst-case analysis of such an algorithm means finding (a) the most unfavorable instance, and (b) the most unfortunate sequence of choices at the incompletely specified steps. So in order to show that simplex is not polynomial, it would suffice to exhibit a family of instances on which it *may* use an exponential number of pivots. This can be done by exhibiting an exponential—in the size of the LP—sequence of basic feasible solutions x_1, x_2, $\ldots, x_k$ such that x_i and x_{i+1} are adjacent and satisfy $c'x_{i+1} < c'x_i$ for $i = 1, \ldots, k-1$.

At this point, it is useful to assist the argument by some geometric intuition. We know (Theorem 2.2) that the bfs's of an LP are the vertices of the corresponding polytope. Furthermore, by Theorem 2.10, adjacent nondegenerate bfs's of the LP correspond to vertices that are joined by an *edge* (1-dimensional face) of the polytope (in Figure 8–2(a), for example, (0, 0, 1) and (0, 1, 1) are adjacent). Finally, suppose that we have oriented the polytope so that the direction of decreasing costs is upwards. Thus we wish to find a sequence of exponentially many vertices that are one adjacent to the next, each higher than the previous. But first, we have to start with a polytope that has exponentially many vertices to begin with. An example of such a polytope is the *cube*:

$$0 \leq x_j \leq 1 \qquad j = 1, 2, 3$$

(see Figure 8–2(a)).

The 3-dimensional cube has 6 faces—many games are based on this fundamental fact—and 8 vertices. In general, it is easy to see that the *d-dimensional cube* (or *d-hypercube*) defined by the inequalities

$$0 \leq x_j \leq 1 \qquad j = 1, 2, \ldots, d$$

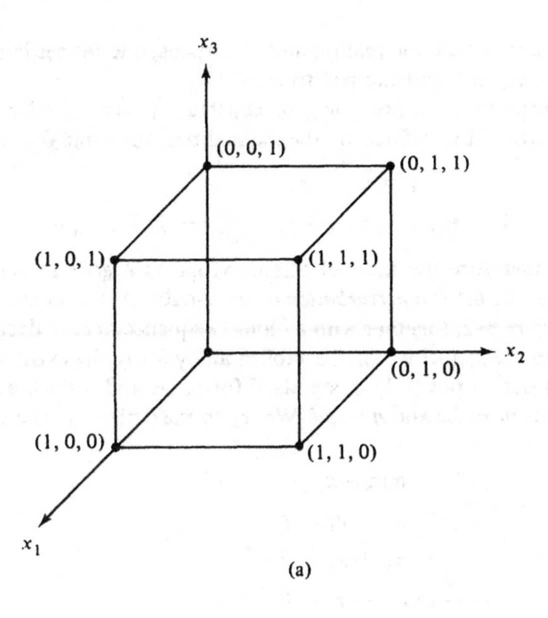
x_3
(0, 0, 1)
(0, 1, 1)
(1, 0, 1)
(1, 1, 1)
x_2
(0, 1, 0)
(1, 0, 0)
(1, 1, 0)
x_1

(a)

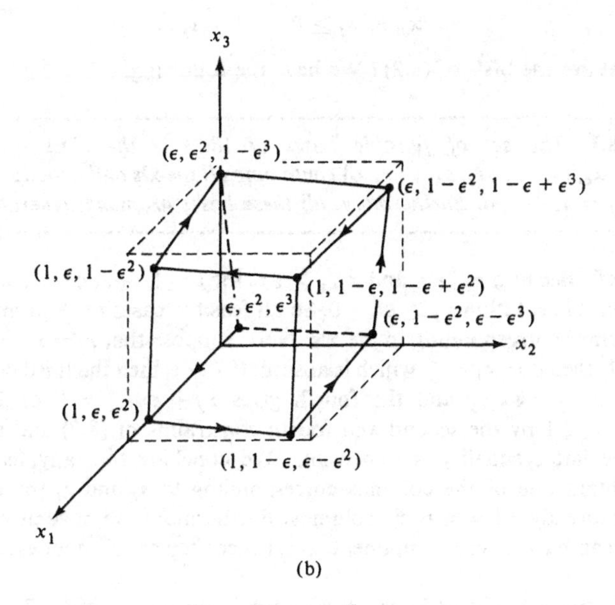
x_3
$(\epsilon, \epsilon^2, 1-\epsilon^3)$
$(\epsilon, 1-\epsilon^2, 1-\epsilon+\epsilon^3)$
$(1, \epsilon, 1-\epsilon^2)$
$(1, 1-\epsilon, 1-\epsilon+\epsilon^2)$
$\epsilon, \epsilon^2, \epsilon^3$
$(\epsilon, 1-\epsilon^2, \epsilon-\epsilon^3)$
x_2
$(1, \epsilon, \epsilon^2)$
$(1, 1-\epsilon, \epsilon-\epsilon^2)$
x_1

(b)

Figure 8–2

has $2d$ faces, one for each inequality, and 2^d vertices, one for setting each subset of $\{x_1, x_2, \ldots, x_d\}$ to 1 and the rest to zero.

The polytope that we are going to construct is very similar to the cube (see Figure 8–2(b)). It is defined by the inequalities, for some $0 < \epsilon < \frac{1}{2}$,

$$\begin{aligned} 1 \geq x_1 \geq \epsilon \\ 1 - \epsilon x_{j-1} \geq x_j \geq \epsilon x_{j-1} \qquad j = 2, 3, \ldots, d \end{aligned} \tag{8.1}$$

The d-cube is therefore the limit of this polytope as ϵ goes to zero; to put it otherwise, Polytope 8.1 is a *perturbation* of the d-cube. A 3-dimensional example is shown in Figure 8–2, together with a "long" sequence of cost-decreasing adjacent vertices. In the sequel we shall establish analytically the existence of such a sequence. In order to put (8.1) in standard form, we add d slack and d surplus variables. Hence $m = 2d$ and $n = 3d$. We try to maximize x_d. The complete LP is shown below:

$$\begin{aligned} \min\ -x_d \\ x_1 - r_1 = \epsilon \\ x_1 + s_1 = 1 \\ x_j - \epsilon x_{j-1} - r_j = 0 \\ x_j + \epsilon x_{j-1} + s_j = 1 \qquad j = 2, 3, \ldots, d \\ x_j, r_j, s_j \geq 0 \qquad j = 1, \ldots, d \end{aligned} \tag{8.2}$$

What are the bfs's of (8.2)? We have the following.

Lemma 8.1 *The set of feasible bases of (8.2) is the class of subsets of* $\{x_1, \ldots, x_d, r_1, \ldots, r_d, s_1, \ldots, s_d\}$ *containing all the x's and exactly one of* s_j, r_j *for each* $j = 1, \ldots, d$. *Furthermore, all these bases are nondegenerate.*

Proof Because $x_1 \geq \epsilon$ and $x_{j+1} \geq \epsilon x_j$ for $j = 1, \ldots, d-1$, we have that in any feasible solution $x_j \geq \epsilon^j > 0$. So all feasible bases of (8.2) must contain all d columns corresponding to the x's. Next, suppose that $r_j = s_j = 0$ for some j. If $j = 1$, then $\epsilon = x_1 = 1$, which is absurd. If $j > 1$, then the third constraint of (8.2) gives $x_j = \epsilon x_{j-1}$ and the fourth gives $x_j + \epsilon x_{j-1} = 1$, or $2\epsilon x_{j-1} = 1$. Now, $x_{j-1} \leq 1$ by the second and fourth constraints of (8.2) and also $\epsilon < \frac{1}{2}$; hence the latter equality is impossible. We conclude that any feasible basis must contain one of the columns corresponding to s_j and r_j for each j. But these are already $2d = m$ basic columns. Furthermore, we have proved that no such bfs can have a zero component, and hence they are all nondegenerate. □

We write a bfs of (8.2) as x^S, where S is the subset of $\{1, 2, \ldots, d\}$ that corresponds to nonzero r's in x^S. The value of x_j in x^S will be denoted by x_j^S. We need the following lemmas.

Lemma 8.2 *Suppose that $d \in S$ but $d \notin S'$; then $x_d^S > x_d^{S'}$. Furthermore, if $S' = S - \{d\}$, $x_d^{S'} = 1 - x_d^S$.*

Proof Since $d \in S$, $s_d = 0$ and the fourth constraint in (8.2) gives $x_d^S = 1 - \epsilon x_{d-1}^S$. Now $x_{d-1}^S \leq 1$ and $\epsilon < \frac{1}{2}$; hence $x_d^S > \frac{1}{2}$. On the other hand, because $d \notin S'$, we have $r_d = 0$ and the third constraint in (8.2) gives $x_d^{S'} = \epsilon x_{d-1}^{S'} < \frac{1}{2}$. Consequently $x_d^{S'} < x_d^S$.

To show the second part, notice that if $S = S' \cup \{d\}$ then $x_{d-1}^{S'} = x_{d-1}^S$. So $x_d^{S'} = \epsilon x_{d-1}^{S'} = 1 - (1 - \epsilon x_{d-1}^S) = 1 - x_d^S$. □

Lemma 8.3 *Let the subsets of $\{1, 2, \ldots, d\}$ be enumerated in such a way that $x_d^{S_1} \leq x_d^{S_2} \leq \cdots \leq x_d^{S_{2^d}}$. Then the inequalities are strict, and the bfs's x^{S_j} and $x^{S_{j+1}}$ are adjacent for $j = 1, 2, \ldots, 2^d - 1$.*

Proof We use induction on d. It certainly holds for $d = 1$: we have two bfs's, namely $(x_1, r_1, s_1) = (\epsilon, 0, 1 - \epsilon)$ and $(1, 1 - \epsilon, 0)$. The bfs's have unequal x_1's and they are certainly adjacent. For the induction step, suppose that the lemma is established for the d-cube, and let $S_1, \ldots, S_{2^d}$ be the appropriate enumeration. Now $S_1, \ldots, S_{2^d}$ are also subsets of $\{1, 2, \ldots, d + 1\}$, and in fact $x_{d+1}^{S_j} = \epsilon x_d^{S_j}$. Therefore, by the induction hypothesis, $x_{d+1}^{S_1} < x_{d+1}^{S_2} < \cdots < x_{d+1}^{S_{2^d}}$. Now consider the remaining subsets of $\{1, \ldots, d + 1\}$, namely, $S'_j = S_j \cup \{d + 1\}, j = 1, \ldots, 2^d$. We have, by Lemma 8.2, that $x_{d+1}^{S'_j} > x_{d+1}^{S_{2^d}}$ and $x_{d+1}^{S'_j} = 1 - x_{d+1}^{S_j}$. Hence

$$x_{d+1}^{S_1} < \cdots < x_{d+1}^{S_{2^d}} < x_{d+1}^{S'_{2^d}} < \cdots < x_{d+1}^{S'_1}$$

By the induction hypothesis x^{S_j} and $x^{S_{j+1}}$ are adjacent, and so are $x^{S'_j}$ and $x^{S'_{j+1}}$. Also, $x^{S_{2^d}}$ and $x^{S'_{2^d}}$ are adjacent as well, since the latter basis results from the former by adding r_{d+1} and omitting s_{d+1}. The lemma follows. □

We can now prove the main result of this section.

Theorem 8.1 *For every $d > 1$ there is an LP with $2d$ equations, $3d$ variables, and integer coefficients with absolute value bounded by 4, such that simplex may take $2^d - 1$ iterations to find the optimum.*

Proof Take $\epsilon = \frac{1}{4}$ and multiply all equations of (8.2) by 4, so that all coefficients are integers. Since the objective of (8.2) is to maximize x_d, the exponentially long chain of adjacent bfs's whose existence is established by Lemma 8.3 has decreasing costs. The theorem follows. □

Results similar to Theorem 8.1 are known for almost all variations of simplex, including several heuristic pivoting rules, the primal-dual simplex of Chapter 5, and others; see the problems and references at the end of the chapter.

Until recently whether there can be *any* polynomial-time algorithm for LP was a most perplexing question. There was conflicting evidence about the answer. On the one hand, LP was certainly one of the problems (together with the TSP and many others; see Chapter 15) which seemed to defy all reasonable attempts at the development of a polynomial-time algorithm. On the other hand, LP had two positive features that made it completely different from the other classical problems in that class. First, LP has a strong *duality theory*, which is conspicuously lacking for all the other hard combinatorial problems (see Section 16.1). And secondly, LP has an algorithm, the simplex method, which—although exponential in its worst case—certainly works empirically on instances of seemingly unlimited size.

In the next section, we examine a startling recent development that resolved this conundrum.

8.7 The Ellipsoid Algorithm

In the spring of 1979 the Soviet mathematician L.G. Khachian published a proof that a certain algorithm for LP is polynomial, thus resolving a long-standing open question. Khachian's result is based on work of other Soviet mathematicians on nonlinear programming (see references) and is drastically different from most previous approaches to LP in that it almost completely disregards the combinatorial nature of the problem.

In the following subsections we introduce and discuss this algorithm. First we have to establish some background results, at times very interesting in themselves.

8.7.1 LP, LI, and LSI

Formally, *linear programming* (LP) (in standard form) is the following computational problem:

> Given an integer $m \times n$ matrix A, m-vector b and n-vector c, either
> (a) Find a rational n-vector x such that $x \geq 0$, $Ax = b$, and $c'x$ is minimized subject to these conditions, or
> (b) Report that there is no n-vector x such that $x \geq 0$ and $Ax = b$, or
> (c) Report that the set $\{c'x\colon Ax = b, x \geq 0\}$ has no lower bound.

Consider now the problem of *linear inequalities* (LI), defined as follows:

> Given an integer $m \times n$ matrix A and m-vector b, is there an n-vector x such that $Ax \leq b$?

For convenience, we shall assume that $m \geq n$ in LI and LSI to follow, although this is not really restrictive.

It turns out that LI is almost as hard as LP, at least as far as the existence of polynomial-time algorithms is concerned. To establish this, we need some preliminaries.

We start by introducing a very common programming technique called *binary search*. Suppose that we wish to determine an (unknown) integer x between 1 and B by asking questions of the form "Is $x > a$?" for some a of our choice. We can do this by first asking whether x is in the upper or lower half of the interval $[1, B]$, then asking whether x is in the upper or lower half of the new (smaller by a factor of 2) interval, and so on, until the interval in which we are certain x lies contains exactly one integer: x. This will happen after $\lceil \log B \rceil$ such questions—$\lceil \log B \rceil$ can be alternatively defined as the number of times that B has to be divided by 2 to obtain a number at most equal to 1. Figure 8–3 illustrates binary search for $B = 32$ and $x = 11$. $\lceil \log 32 \rceil = 5$ questions are enough.

Question	*Answer*	*Interval of possible values of x after the question*
		1____________32
Is $x > 16$?	NO	1______16
Is $x > 8$?	YES	9____16
Is $x > 12$?	NO	9__12
Is $x > 10$?	YES	11_12
Is $x > 11$?	NO	$11 = x$.

Figure 8–3 A sequence of five questions of the form *Is x > a?* for different values of *a*, which results in the determination of $x = 11$.

We summarize this discussion for future reference as follows.

Lemma 8.4 *An integer x between* 1 *and B can be determined by* $\lceil \log B \rceil$ *questions of the form "Is* $x > a$*?".*

Consider an $m \times n$ LP in the standard form:

$$\begin{aligned} \min\ & c'x \\ Ax &= b \\ x &\geq 0 \end{aligned} \tag{8.3}$$

Its *size* is

$$L = mn + \lceil \log |P| \rceil, \tag{8.4}$$

where P is the product of the nonzero (integer) coefficients appearing in A, b, and c (recall Example 8.4). We now state the following version of Lemma 2.1.

Lemma 8.5 *The bfs's of (8.3) are n-vectors of rational numbers, both the absolute value and the denominators of which are bounded by* 2^L.

Proof Similar to Lemma 2.1. □

Lemma 8.6 *Suppose that two bfs's x_1, x_2 of (8.3) satisfy $K2^{-2L} < c'x_1, c'x_2 \leq (K+1)2^{-2L}$ for some integer K. Then $c'x_1 = c'x_2$.*

Proof Suppose that $c'x_1 \neq c'x_2$. Then by Lemma 8.5, $c'x_1$ and $c'x_2$ are distinct rational numbers with denominators at most 2^L; hence $|c'x_1 - c'x_2| \geq 2^{-2L}$, a contradiction. □

We can now prove this theorem.

Theorem 8.2 *There is a polynomial-time algorithm for LP iff there is a polynomial-time algorithm for LI.*

Proof For the *only if* direction, we can answer any instance of LI by determining whether the corresponding LP (with slack variables added and unrestricted variables removed; see Sec. 2.1) is feasible. Hence a polynomial-time algorithm for LP would imply a polynomial-time algorithm for LI. Conversely, suppose that we have an algorithm $\mathcal{A}$ that solves LI in polynomial time. We shall describe a polynomial-time algorithm for LP that uses $\mathcal{A}$. Suppose that the input consists of the LP in (8.3).

1. Our algorithm first determines whether the LP is feasible by invoking $\mathcal{A}$ once, with input the inequalities $Ax \geq b$, $Ax \leq b$, and $x \geq 0$. If $\mathcal{A}$ answers no, we report infeasibility and halt.
2. Next, we check for feasibility the inequalities $Ax \geq b$, $Ax \leq b$, $x \geq 0$, $c'x \leq -2^{2L} - 1$. Since $c'x$ is bounded from below by -2^{2L} if bounded at all, whenever $\mathcal{A}$ determines that these inequalities are satisfiable, we report unboundedness and halt.
3. Otherwise, we know that the problem has an optimal bfs, $\hat{x}$. We first determine an integer $-2^{4L} \leq K \leq 2^{4L}$ such that $K2^{-2L} < c'\hat{x} \leq (K+1)2^{-2L}$. We can do this by *binary search* (Lemma 8.4) in $4L + 1$ invocations of $\mathcal{A}$ with the inequalities
$$Ax \geq b, \qquad Ax \leq b, \qquad x \geq 0, \qquad 2^{2L}c'x \leq a$$
for various values of a. Since $\mathcal{A}$ is polynomial-time, K can thus be determined in polynomial time.
4. Finally, we determine the basis that corresponds to $\hat{x}$. For $k = 1, \ldots, n$, we check whether the inequalities $Ax \leq b$, $Ax \geq b$, $K \leq 2^{2L}c'x \leq K + 1$; $x \geq 0$, $x_k \leq 0$, and $x_j \leq 0$ for $j \in S(k)$ are all satisfiable. Here $S(k)$ is the set of indices less than k for which the answer was yes. It should be obvious that any m columns not in $S(n+1)$ can be chosen to be the basis of $\hat{x}$, and that there will be at least m columns not in $S(n+1)$. Having thus determined the basis (within degeneracy), we

can find $\hat{x}$ efficiently by simply inverting it—inversion of an $n \times n$ integer matrix M can be carried out in polynomial time (in the size of M) by Gaussian elimination [Ga]. □

By Theorem 8.2, the seemingly restrictive problem LI captures the complexity of LP. To put it otherwise, the complexity of Phase I of the ordinary simplex method (which has the purpose just to find a feasible point; see Sec. 2.8) is as great as the complexity of the whole problem!

We now introduce yet another related problem, that of *linear strict inequalities* (LSI):

Given an $m \times n$ integer matrix A and m-vector b, is there an n-vector x such that $Ax < b$?

Not surprisingly, LSI is no easier than LI.

Lemma 8.7 *The system of linear inequalities*

$$a_i'x \leq b_i, \qquad i = 1, \ldots, m \tag{8.5}$$

has a solution iff the system of linear strict *inequalities*

$$a_i'x < b_i + \epsilon, \qquad i = 1, \ldots, m \tag{8.6}$$

has a solution, where $\epsilon = 2^{-2L}$.

Proof If (8.5) has a solution, then this solution also satisfies (8.6). For the opposite direction, suppose that (8.6) has a solution. From the assumed solution of (8.6), we shall construct a solution $\hat{x}$ of (8.5). The construction may be considered a tricky version of that of Theorem 2.1, where we showed how to construct a bfs starting from an arbitrary feasible solution.

Let x_0 be a solution of (8.6). Consider the set of row vectors $I = \{a_i : b_i \leq a_i'x_0 < b_i + \epsilon\}$. We may assume that, for all j, $a_j = \sum_{a_i \in I} \beta_{ji} a_i$ for some numbers β_{ji}. Because, if some a_j were independent of the a_i's in I, then the system

$$a_i'z = 0, \qquad a_i \in I$$

$$a_j'z = 1$$

would have a solution, z_0. By taking $x_1 = x_0 + \lambda z_0$ for sufficiently small λ, we can create another solution x_1 of (8.6), which however has one more vector in the set I; after at most m steps, this must stop.

Therefore, for all j, $a_j = \sum_{a_i \in I'} \beta_{ji} a_i$ for some linearly independent subset I' of I. By Cramer's rule, the β_{ji}'s are all quotients of determinants of absolute value less than 2^L: $\beta_{ji} = D_{ji}/|D|$. Consider the solution $\hat{x}$ to the equations $a_i'x = b_i$, $a_i \in I'$. We have, for each j:

$$
\begin{aligned}
|D|(a'_j\hat{x} - b_j) &= \sum_{a_i \in I'} D_{ji} a'_i \hat{x} - |D| b_j \\
&= \sum_{a_i \in I'} D_{ji} b_i - |D| b_j \quad \text{(by the definition of } \hat{x}) \\
&= -\sum_{a_i \in I'} D_{ji}(a'_i x_0 - b_i) + |D|(a'_j x_0 - b_j) \\
&\qquad \text{(By adding and subtracting } |D| a'_j x_0) \\
&< \epsilon\Big(\sum_{a_i \in I'} |D_{ji}| + |D|\Big) \quad (\text{Because } |a'_i x_0 - b_i| < \epsilon \text{ for } i \in I' \\
&\qquad \text{and } a'_j x_0 - b_j < \epsilon \text{ for all } j) \\
&< 2^{-2L}(m+1)2^L < 1
\end{aligned}
$$

Therefore, for all j, $|D|(a'_j\hat{x} - b_j) < 1$. Moreover, the denominators of all components of $\hat{x}$ divide $|D|$, by the definition of D, and hence $|D|(a'_j\hat{x} - b_j)$ is an integer. We conclude that $a'_j\hat{x} - b_j \leq 0$ for all j, and therefore $\hat{x}$ is the desired solution to (8.5). □

Corollary *If there is a polynomial-time algorithm for LSI, then there is a polynomial-time algorithm for LI.*

Proof Given a set of linear inequalities as in (8.5), we may check equivalently whether the system

$$2^{2L} a'_i x < 2^{2L} b_i + 1, \qquad i = 1, \ldots, m \tag{8.7}$$

is satisfiable. The LP (8.7) has size at most twice the square of that of (8.5). □

In Subsec. 8.7.3, we present a polynomial-time algorithm for LSI.

8.7.2 Affine Transformations and Ellipsoids

In this subsection we introduce some standard concepts from linear algebra and certain facts concerning them (Lemmas 8.8, 8.9, and 8.10) without proofs.

Let Q be an $n \times n$ *nonsingular* matrix, and t an n-vector. The transformation $T: R^n \longrightarrow R^n$ defined as $T(x) = t + Q \cdot x$ for each $x \in R^n$ is called an *affine transformation*. Since Q is nonsingular, T is a uniquely invertible transformation. The inverse of T is an affine transformation itself.

The *unit sphere* is the set

$$S_n = \{x \in R^n : x'x \leq 1\}$$

If T is an affine transformation, then $T(S_n)$ is called an *ellipsoid*. Alternatively, $T(S_n) = \{y \in R^n : (y - t)'B^{-1}(y - t) \leq 1\}$, where $B = QQ^T$. A matrix such as B—the product of a nonsingular matrix by its transpose—is *positive definite*; that is, $x'Bx > 0$ for all nonzero $x \in R^n$.

Affine transformations preserve set inclusion.

Lemma 8.8 *If $S \subseteq S' \subseteq R^n$, then $T(S) \subseteq T(S')$.*

Example 8.9

In 2-dimensional space, let $t = (2, 3)'$, and $Q = \begin{bmatrix} 2 & 0 \\ 0 & 1 \end{bmatrix}$. Then $B^{-1} = \begin{bmatrix} \frac{1}{4} & 0 \\ 0 & 1 \end{bmatrix}$ and the ellipsoid $T(S_2)$ is as shown in Fig. 8–4. In general, the axes of the ellipsoid may not be parallel to the coordinate axes, but they will always be orthogonal to one another. The center of the ellipsoid is always t. □

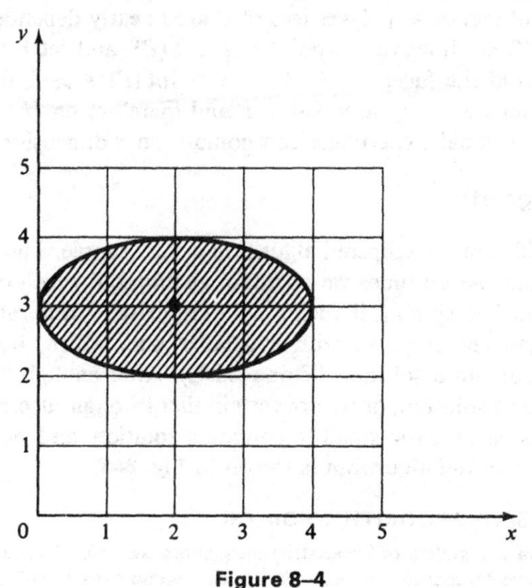

Figure 8–4

Lemma 8.9 *Suppose that a subset S of R^n has volume V. Then T(S) has volume* $V \cdot |\det(Q)|$.

For instance, the ellipsoid of Figure 8–4 has volume 2π.

Consider an affine transformation R with $t = 0$, and a matrix (also denoted by R) with the following special property: $RR^T = I$. Such transformations are called *rotations*—it is readily seen that they map the unit sphere to itself.

Lemma 8.10 *Let $a \in R^n$ be a vector of length $\|a\|$. There is a rotation R such that* $Ra = (\|a\|, 0, \ldots, 0)$.

Finally, we need a fact from the theory of convex polytopes. Consider a convex polytope P in R^n. We know that P may be written as

$$P = \{x \in R^n\colon Ax \leq b\}$$

for some $m > n$, $m \times n$ matrix A and m-vector b. Let the *interior* of P be defined as follows:

$$\text{Int}(P) = \{x \in R^n: Ax < b\}$$

Lemma 8.11 *If Int $(P) \neq \emptyset$, then there exist $n + 1$ linearly independent vertices of P.*

Proof If all sets of $n + 1$ vertices of P are linearly dependent, P lies on a hyperplane H. Take, however, a point $x \in$ Int (P), and let ϵ be the smallest distance of x from the facets of P. Since $x \in$ Int (P), $\epsilon > 0$, the sphere with center x and radius ϵ lies totally within P and therefore on H. This is absurd: no $(n - 1)$-dimensional hyperplane can contain an n-dimensional sphere. □

8.7.3 The Algorithm

The main idea of the ellipsoid algorithm is very simple. The algorithm proceeds in iterations. At all times we maintain an ellipsoid which contains a solution to the given LSI system, if such a solution exists. An iteration consists of replacing the current ellipsoid with a smaller one, which, however, is also guaranteed to contain a solution (if one exists). After enough iterations either we must discover a solution, or we are certain that through successive shrinkings the ellipsoid has become too small to contain a solution, and we report that no solution exists. The full algorithm is shown in Fig. 8–5.

THE ELLIPSOID ALGORITHM FOR LSI.

Input: An $m \times n$ system of linear strict inequalities $Ax < b$, of size L

Output: an n-vector x such that $Ax < b$, if such a vector exists; "no" otherwise.

1: (Initialize) Set $j := 0$, $t_0 := 0$, $B_0 := n^2 2^{2L} \cdot I$

(**Comment:** j counts the number of iterations so far. The current ellipsoid is $E_j = \{x: (x - t_j)' B_j^{-1} (x - t_j) \leq 1\}$).

2: (Test) **If** t_j is a solution to $Ax < b$ **then return** t_j;

If $j > K = 16n(n + 1)L$ **then return** "no";

3: (Iteration) Choose any inequality in $Ax < b$ that is violated by t_j; say $a' t_j \geq b$.

Set

$$t_{j+1} := t_j - \frac{1}{n+1} \frac{B_j a}{\sqrt{a' B_j a}};$$

$$B_{j+1} := \frac{n^2}{n^2 - 1}\left[B_j - \frac{2}{n+1} \frac{(B_j a)(B_j a)'}{a' B_j a}\right];$$

$$j := j + 1;$$

go to 2

Figure 8–5

Example 8.10

Suppose that for some j we have $t_j = (0, 0)'$, $B_j = \begin{bmatrix} 9 & 0 \\ 0 & 4 \end{bmatrix}$, and that one of the inequalities reads $x + y < -1$. The situation is depicted in Fig. 8–6(a). Remember that we somehow know that the solution of the LSI instance, if it

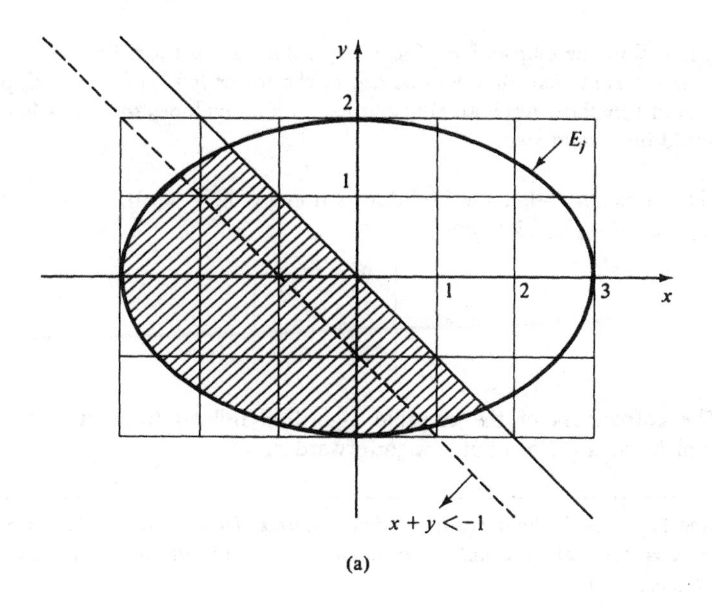

(a)

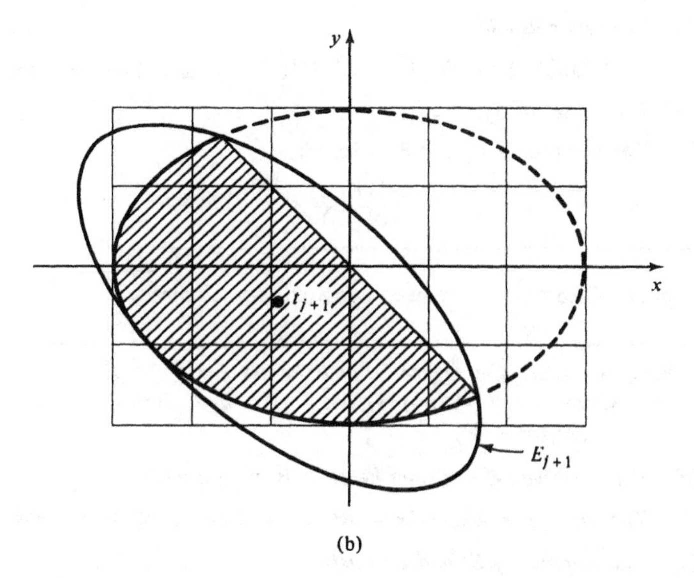

(b)

Figure 8–6 (a) The ellipsoid at the jth iteration. (b) The ellipsoid at the next iteration.

exists, is within the ellipsoid E_j. Since any solution must satisfy $x + y < -1$, we know for sure that all solutions are in the lower left half of the ellipsoid. If we could somehow draw an ellipsoid E_{j+1} which includes the lower left half, this would be a valid step.

That is *exactly* what the iteration is doing (Fig. 8–6(b)). We obtain $t_{j+1} = (-3/\sqrt{13}, -4/(3\sqrt{13}))'$, and

$$B_{j+1} = \begin{bmatrix} \frac{84}{13} & \frac{-32}{13} \\ \frac{-32}{13} & \frac{496}{117} \end{bmatrix}. \quad \square$$

The correctness of the ellipsoid algorithm follows from the next theorem, which has a tedious but straightforward proof.

Theorem 8.3 *Let B_j be a positive definite matrix, let $t_j \in R^n$, and let a be any nonzero n-vector. Let B_{j+1} and t_{j+1} be as in Step 3 of the ellipsoid algorithm. Then the following hold.*

(a) *B_{j+1} is positive-definite (or, equivalently, $E_{j+1} = \{x \in R^n: (x - t_{j+1})'B_{j+1}^{-1}(x - t_{j+1}) \leq 1\}$ is an ellipsoid).*

(b) *The semiellipsoid*

$$\tfrac{1}{2}E_j[a] = \{x \in R^n: (x - t_j)'B_j^{-1}(x - t_j) \leq 1, \quad a'(x - t_j) \leq 0\}$$

is a subset of E_{j+1}.

(c) *The volumes of E_j and E_{j+1} satisfy*

$$\frac{\text{vol}(E_{j+1})}{\text{vol}(E_j)} < 2^{-1/2(n+1)}$$

To prove Theorem 8.3, we need two auxiliary lemmas.

Lemma 8.12 *Consider the sphere S_n and the set $E = \{x \in R^n: (x - t)'B^{-1}(x - t) \leq 1\}$, where $t = (-1/(n+1), 0, \ldots, 0)'$ and $B = \text{diag}(n^2/(n+1)^2, n^2/(n^2-1), \ldots, n^2/(n^2-1))$.*

(a) *B is positive-definite (and hence E is an ellipsoid).*

(b) *The hemisphere $\tfrac{1}{2}S_n = \{x \in R^n: x'x \leq 1 \text{ and } x_1 \leq 0\}$ is a subset of E.*

(c) *The volumes of S_n and E satisfy*

$$\frac{\text{vol}(E)}{\text{vol}(S_n)} < 2^{-1/2(n+1)}$$

Proof

(a) $B = QQ^T$, where

$$Q = \text{diag}\left(\frac{n}{n+1}, \frac{n}{\sqrt{n^2-1}}, \ldots, \frac{n}{\sqrt{n^2-1}}\right)$$

(b) Suppose that $x \in \frac{1}{2}S_n$. Then

$$\begin{aligned}(x-t)'B^{-1}(x-t) &= \frac{(n+1)^2}{n^2}\left(x_1 + \frac{1}{n+1}\right)^2 + \frac{n^2-1}{n^2}\sum_{i=2}^{n} x_i^2 \\ &= \frac{n^2-1}{n^2}x'x + \frac{2n+2}{n^2}x_1^2 + \frac{2n+2}{n^2}x_1 + \frac{1}{n^2} \\ &\leq 1 + \frac{2n+2}{n^2}(x_1^2 + x_1)\end{aligned}$$

(Because $x \in S_n$)

$$\leq 1$$

(Because $x \in \frac{1}{2}S_n$)

(c) vol (E)/vol $(S_n) = \det Q$, by Lemma 8.9, where Q is as in Part (a). Because Q is diagonal,

$$\det Q = \frac{n}{n+1}\left(\frac{n^2}{n^2-1}\right)^{(n-1)/2}$$

Because for all $x > 0$, $1 + x \leq e^x$, $1 - x \leq e^{-x}$

$$\frac{n}{n+1} = 1 - \frac{1}{n+1} \leq e^{-1/(n+1)}$$

$$\frac{n^2}{n^2-1} = 1 + \frac{1}{n^2-1} \leq e^{1/(n^2-1)}$$

Thus

$$\det Q < \exp\left(\frac{n-1}{2(n^2-1)} - \frac{1}{n+1}\right) < 2^{-1/2(n+1)} \qquad \square$$

Lemma 8.13 *Let B_j be a positive-definite matrix, $t_j \in R^n$, and let a be any nonzero n-vector. Let t_{j+1} and B_{j+1} be obtained as in Step 3 of the ellipsoid algorithm. Let $\frac{1}{2}S_n$ and E be as in previous lemma. Then there exists an affine transformation T such that*

(*a*) $T(S_n) = \{x \in R^n: (x - t_j)'B_j^{-1}(x - t_j) \leq 1\}$;

(*b*) $T(E) = \{x \in R^n: (x - t_{j+1})'B_{j+1}^{-1}(x - t_{j+1}) \leq 1\}$;

(*c*) $T(\frac{1}{2}S_n) = \{x \in R^n: (x - t_j)'B_j^{-1}(x - t_j) \leq 1, a'(x - t_j) \leq 0\}$.

Proof By hypothesis, B_j is positive-definite, and hence $B_j = QQ^T$ for some nonsingular matrix Q. Also, by Lemma 8.10, there exists a rotation R^T such that $R^TQ^Ta = (\|Q^Ta\|, 0, \ldots 0)'$. The transformation T is defined thus: $T(x) = t_j + QRx$. We shall check the three conditions.

(a)

$$\begin{aligned} T(S_n) &= \{T(x): x'x \leq 1\} \\ &= \{x: (T^{-1}(x))'T^{-1}(x) \leq 1\} \\ &= \{x: (x - t_j)'(Q^{-1})^T RR^T Q^{-1}(x - t_j) \leq 1\} \\ &= \{x: (x - t_j)'B_j^{-1}(x - t_j) \leq 1\} \end{aligned}$$

(b) First notice that

$$\begin{aligned} B_{j+1} &= \frac{n^2}{n^2 - 1}\left[B_j - \frac{2}{n+1}\frac{B_j aa' B_j^T}{a'B_j a}\right] \\ &= \frac{n^2}{n^2 - 1}\left[B_j - \frac{2}{n+1}\frac{QRR^TQ^T aa' QRR^TQ^T}{a'QRR^TQ^T a}\right] \\ &= \frac{n^2}{n^2 - 1}\left[B_j - \frac{2}{n+1}\frac{QR\,\text{diag}\,(\|Q^Ta\|^2, 0, \ldots, 0)R^TQ^T}{\|Q^Ta\|^2}\right] \\ &\qquad \text{(Because } R^TQ^Ta = (\|Q^Ta\|, 0, \ldots, 0)'\text{)} \\ &= \frac{n^2}{n^2 - 1}\left[B_j - \frac{2}{n+1}QR\,\text{diag}\,(1, 0, \ldots, 0)R^TQ^T\right] \\ &= \frac{n^2}{n^2 - 1}QR\,\text{diag}\left(\frac{n-1}{n+1}, 1, \ldots, 1\right)R^TQ^T \\ &= QRBR^TQ^T \end{aligned}$$

where B is as in Lemma 8.12.

Also,

$$\begin{aligned} (x - t_{j+1}) &= \left(x - t_j + \frac{QRR^TQ^Ta}{(n+1)\sqrt{a'QRR^TQ^Ta}}\right) \\ &= \left(x - t_j + \frac{QR(\|Q^Ta\|, 0, \ldots, 0)'}{(n+1)\|Q^Ta\|}\right) \\ &= QR(T^{-1}(x) - t) \end{aligned}$$

Therefore

$$\begin{aligned} T(E) &= \{T(x): (x - t)'B^{-1}(x - t) \leq 1\} \\ &= \{x: (T^{-1}(x) - t)'B^{-1}(T^{-1}(x) - t) \leq 1\} \\ &= \{x: (x - t_{j+1})'(Q^{-1})^T RB^{-1}R^TQ^{-1}(x - t_{j+1}) \leq 1\} \\ &= \{x: (x - t_{j+1})'B_{j+1}^{-1}(x - t_{j+1}) \leq 1\} \end{aligned}$$

(c) The condition in (c) now follows easily from the condition in (a) and Lemma 8.8, by observing that

$$T(\{x \in R^n: x_1 \leq 0\}) = \{x \in R^n: a'(x - t_j) \leq 0\} \qquad \square$$

Proof of Theorem 8.3

(a) By (b) of Lemma 8.13, $T(E) = E_{j+1}$; also, by (a) of Lemma 8.12, $E = T'(S_n)$ for some affine transformation T'. Hence $E_{j+1} = T \cdot T'(S_n)$ is an ellipsoid (the composition of two affine transformations is also an affine transformation).

(b) $E_j[a] = T(\frac{1}{2}S_n)$—by (c) of Lemma 8.13—and $\frac{1}{2}S_n \subseteq E$. Hence $E_j[a] \subseteq T(E) = E_{j+1}$.

(c) By Lemma 8.9 and by (c) of Lemma 8.12

$$\frac{\text{vol}(E_{j+1})}{\text{vol}(E_j)} = \frac{\text{vol}(T(E))}{\text{vol}(T(S_n))}$$
$$= \frac{\det(QR)\,\text{vol}(E)}{\det(QR)\,\text{vol}(S_n)} < 2^{-1/2(n+1)}$$

□

We can also prove this lemma.

Lemma 8.14 *If an LSI system of size L has a solution, then the set of solutions within the sphere $\|x\| \leq n2^L$ has volume at least $2^{-(n+2)L}$.*

Proof If $Ax < b$ has a solution, we know that

$$Ax < b$$
$$x_i < 2^L \qquad i = 1, \ldots, n$$

has a solution. Hence the polytope

$$Ax \leq b$$
$$x_i \leq 2^L \qquad i = 1, \ldots, n$$

has an interior point. By Lemma 8.11, it has $n + 1$ linearly independent vertices $\{v_0, v_1, \ldots, v_n\}$. All interior points of the convex hull of these vertices are solutions of $Ax < b$ within the sphere $\|x\| < n2^L$. The volume of this convex hull is

$$\frac{1}{n!}\left|\det\begin{pmatrix} 1 & 1 & \cdots & 1 \\ v_0 & v_1 & \cdots & v_n \end{pmatrix}\right| \neq 0$$

Each v_i can be written as u_i/D_i, where u_i is an integer vector and D_i is a determinant of absolute value at most 2^L. Thus the volume of the convex hull is at least $\left(n! \prod_{i=0}^{n} |D_i|\right)^{-1} > 2^{-(n+2)L}$. □

Theorem 8.4 *The ellipsoid algorithm correctly decides whether a system of LSI's has a solution.*

Proof If the algorithm returns a point t_j at Step 2, then trivially this point is a solution. Suppose now that the algorithm returns "no", and yet the system has a solution. By Lemma 8.14, there is a set S of solutions within the sphere $E_0 = \{x \in R^n \colon x'B_0^{-1}x \leq 1\}$ with volume at least $2^{-(n+2)L}$. By Theorem 8.3(b), S continues to be a subset of E_j for $j = 0, \ldots, K$. However, by Theorem 8.3(c), $\text{vol}(E_K) < \text{vol}(E_0) \cdot 2^{-K/2(n+1)} < (2 \times n^2 2^{2L})^n \cdot 2^{-8nL} < 2^{-(n+2)L}$ and so S cannot be contained in E_K; this is a contradiction. □

8.7.4 Arithmetic Precision

There is one last issue that we have to settle before we can proclaim the ellipsoid algorithm to be a polynomial-time algorithm for the problem of linear strict inequalities and therefore, by Theorem 8.2 and Lemma 8.7, for LP. Unlike all other computations discussed in this book, the ellipsoid algorithm cannot be carried out by integer or rational arithmetic. This is immediately manifested by the square root in Figure 8–5. Consequently, in any computer implementation we shall have to *approximate* all intermediate results by rationals. A simple scheme for doing so is the following: Any real x is represented by a binary integer with up to P bits, multiplied by some power of 2 (possibly negative); the resulting rational closest to x is denoted by $\hat{x}$. For example, let $P = 4$. If $x = 37$, then $\hat{x} = 9 \times 2^2 = 36$; if $x = 3.156$, then $\hat{x} = 13 \times 2^{-2} = 3.25$. We say that our computation is carried out *with precision P*. Suppose that our precision P is fixed. It is easy to see that for all real numbers x, $|x - \hat{x}|/|x| \leq 2^{-P}$; we can write this as follows:

$$\hat{x} = x(1 + \theta 2^{-P}) \qquad \text{for some } -1 < \theta < 1$$

or simply

$$\hat{x} = x(1 + \theta 2^{-P}) \tag{8.9}$$

omitting the explicit statement of the bounds for θ. We shall use the notation in (8.9) very often in the sequel. When we write it with a vector or matrix, we allow θ to be chosen differently for each component.

We shall carry out the ellipsoid algorithm with precision P. An arithmetic operation will not in general return the precise result r (sum, product, square root, and so on) but will instead return $\hat{r}$, the approximation of r with precision P. Performing such operations with arguments that are real numbers approximated with precision P involves two steps: handling (that is, adding, subtracting, or comparing) the two exponents of 2; and performing the operation on two P-bit integers. For example, the addition $9 \times 2^2 + 13 \times 2^{-2}$ is carried out as follows: We first compare and subtract the two exponents and discover that the second is smaller by 4. We find the integer closest to $13/2^4$, namely, 1. We add 1 to 9 and obtain the answer $\hat{r} = 10 \times 2^2$. We check whether $10 \geq 16$—that is, whether our P-bit integer has "overflowed"—and, since it has not, we report $10 \times 2^2 = 40$ as our result.

The first step, comparing, adding, or subtracting exponents, is easy to do. Any number appearing in the ellipsoid algorithm has an exponent that is at most *exponential* in L. (This is because we have a polynomial (total) number of operations, and at each operation the exponent at most doubles.) So, if we also keep the exponents in binary, the first step of any arithmetic operation can be performed in time polynomial in L. The second step, performing arithmetic operations on binary integers with P bits, can be carried out in $O(P^2)$ elementary bit operations by the well-known methods taught in elementary schools. The question then becomes how large P has to be so that the ellipsoid algorithm is still correct. If P can be bounded by a polynomial in L, the size of the instance, then we can finally claim that the ellipsoid algorithm is polynomial.

Let us recall the ellipsoid algorithm. It constructs a sequence of ellipsoids $E_0 = (t_0, B_0), E_1 = (t_1, B_1), \ldots$, for $j = 0, 1, \ldots$. The quantities t_{j+1} and B_{j+1} are calculated from t_j and B_j by the rules

$$t_{j+1} = F(t_j, B_j)$$
$$B_{j+1} = G(t_j, B_j)$$

where the transformations F and G are given explicitly in Figure 8–5. The *modified* ellipsoid algorithm for precision P also creates a sequence of ellipsoids, $E'_0 = E_0, \ldots, E'_j = (t'_j, B'_j \cdot (1 + \delta)^j), \ldots$ and so on, where $\delta > 0$ is to be determined, and the t'_{j+1} and B'_{j+1} are calculated as follows:

$$t'_{j+1} = \hat{F}(t'_j, B'_j)$$
$$B'_{j+1} = \hat{G}(t'_j, B'_j)$$

The hats over F and G indicate that the computation is carried out with all intermediate results approximated with precision P. Notice that the t'_{j+1} and B'_{j+1} are calculated as in the original ellipsoid algorithm, the only difference being that in our definition of E'_j we include an extra factor of $(1 + \delta)$ at each step, and thus our estimate of volume will be multiplied by $(1 + \delta)^n$ per step. Intuitively, this extra factor will guarantee that, if E'_j contains the set S of solutions of $Ax < b$, then so does E'_{j+1}, and Theorem 8.4 is still valid, despite the round-off error introduced by finite-precision arithmetic. Notice that the $(1 + \delta)$ factors are not actually taken into account in the computation; they are simply used in the definition of the ellipsoids E'_j that are employed in our argument. We can do this because the formula for the calculation of B_{j+1} from B_j (Figure 8-5) is *homogeneous* in B_j, and thus multiplying all entries of B_j by a scalar has the effect of multiplying all entries of B_{j+1} by the same scalar.

We are going to argue that, with appropriate choices of P and δ, the semiellipsoid

$$\tfrac{1}{2}E'_j = \{x : x \in E'_j, \quad a'(x - t'_j) \leq 0\}$$

is contained in E'_{j+1}. It suffices to show that the ellipsoid

$$H_j = (F(t'_j, B'_j), G(t'_j, B'_j)(1 + \delta)^j)$$

is contained in E'_{j+1}, since, by Theorem 8.3, $\frac{1}{2}E'_j \subseteq H_j$. In fact, it suffices to show that all points on the *boundary* of H_j are also within E'_{j+1}: that is

$$(x - F(t'_j, B'_j))'G(t'_j, B'_j)^{-1}(x - F(t'_j, B'_j)) = (1 + \delta)^j \tag{8.10}$$

implies

$$x \in E'_{j+1} \tag{8.11}$$

Now, (8.11) is equivalent to

$$(x - \hat{F}(t'_j, B'_j))'\hat{G}(t'_j, B'_j)^{-1}(x - \hat{F}(t'_j, B'_j)) \leq (1 + \delta)^{j+1} \tag{8.12}$$

Let r be any number appearing in the ellipsoid algorithm (r could be an entry of B_j or t_j for some j, or some intermediate result). It is not hard to prove that $|r| \leq M = 2^{cKn}$ for some $c > 0$, where K is the number of iterations (see Problem 16; the same result holds for the modified ellipsoid algorithm with precision P, which we are analyzing). Thus any arithmetic operation carried out with precision P on such numbers produces a result with absolute error at most $M \cdot 2^{-P}$. Any sequence of p such operations produces a *cumulative* error of at most $pM \cdot 2^{-P}$. Because the computations of F' and G' involve $O(n^3)$ arithmetic operations, we have (omitting the arguments B'_j and t'_j of F, $\hat{F}$, G, and $\hat{G}$ for simplicity)

$$\hat{F} = F + \theta M 2^{cn-P}, \text{ for some } c > 0$$

(We use c in different equations to denote positive constants, not necessarily related to each other.) Notice that here θ is a vector. It follows that

$$x - \hat{F} = x - F + \theta M 2^{cn-P} \tag{8.13}$$

Let us assume that the LSI instance that we are examining has a solution—otherwise the ellipsoid algorithm is certainly correct. It is not hard to show (Problem 15) that, for every x on the boundary of the ellipsoid H_j we have $|x - F| \geq 2^{-2nL}$. Thus (8.13) yields

$$x - \hat{F} = (x - F)(1 + \theta M 2^{cnL-P}),$$

and hence, for any x satisfying (8.10) we have

$$(x - \hat{F})'\hat{G}^{-1}(x - \hat{F}) = (x - F)'\hat{G}^{-1}(x - F)(1 + \theta M 2^{cnL-P})^2 \tag{8.14}$$

Consider now an entry $\hat{g}$ of $\hat{G}^{-1}$. It is the quotient of the determinant $\hat{D}$ of a minor of $\hat{G}$, divided by the determinant of $\hat{G}$. Because an $n \times n$ determinant is the sum of $n!$ products, and each factor of each product has a maximum possible error of $M2^{cn-P}$, as in (8.13), it follows that

$$\det(\hat{G}) = \det(G) + \theta M^n 2^{cnL-P}$$

By Theorem 8.4, and since the jth iteration was not the last, $\det(B'_j) \geq 2^{-(n+2)L}(1 + \delta)^{-jn}$ and hence $\det(G) \geq 2^{-(n+2)L-1}(1 + \delta)^{-jn}$. Thus,

$$\det(\hat{G}) = \det(G) \cdot (1 + \theta M^n 2^{cnL-P})(1 + \delta)^{Kn} \tag{8.15}$$

Similarly, if D is the subdeterminant of G corresponding to $\hat{D}$

$$\hat{D} = D + \theta M^n 2^{cnL-P}$$

and if g is the entry of G^{-1} corresponding to $\hat{g}$

$$\hat{g} = g + \theta M^{n+1} 2^{cnL-P}(1 + \delta)^{Kn}$$

Hence (8.14) becomes

$$\begin{aligned}(x - \hat{F})'\hat{G}^{-1}(x - \hat{F}) &= (x - F)'G^{-1}(x - F)(1 + \theta M 2^{cnL-P})^2 \\ &\quad + \theta M^{n+3} 2^{cnL-P}(1 + \delta)^{Kn} \\ &= (1 + \delta)^j(1 + \theta M^{n+3} 2^{cnL-P}(1 + \delta)^{Kn})\end{aligned}$$

If we choose $K = 32n(n + 1)L$, $\delta = 1/Kn$, and $P = cKn^2$ for large enough c we obtain Inequality 8.12, and thus the half-ellipsoid $\frac{1}{2}E'_j$ is indeed contained in E'_{j+1}.

It remains to show that the volumes of the ellipsoids E'_j decrease geometrically. By part (c) of Theorem 8.3, Equation 8.15, and by the definition of E'_{j+1}, we have

$$\frac{\text{vol}\,(E'_{j+1})}{\text{vol}\,(E'_j)} < 2^{-1/2(n+1)} \cdot (1+\delta)^n \cdot (1 + M^n 2^{cL-P}(1+\delta)^{Kn})$$

Again, the choice of $P = cKn^2$ for large enough c yields

$$\frac{\text{vol}\,(E'_{j+1})}{\text{vol}\,(E'_j)} < 2^{-1/4(n+1)}$$

(Notice that P is polynomially bounded in L.) By the same argument as in Theorem 8.4, after $K = 32n(n+1)L$ iterations the modified ellipsoid algorithm will either converge to a solution, or will confidently report that no solution exists. We have finally shown the following.

Theorem 8.5 *There is a polynomial-time algorithm for LP.*

PROBLEMS

1. Describe the well-known method for subtracting decimal integers as an algorithm. You may use the style in which algorithms are presented in this book. The input to your algorithm is n, the number of digits in the two given integers, and two n-element arrays A and B containing the decimal digits of the integers.
2. It can be proved that there is no algorithm for any of the three problems below. Examine each of these problems and convince yourself at least that some obvious approaches to solving them do not work.
   (a) *The halting problem* described in Section 8.1. (If you are ambitious, try to argue why there can be no algorithm that solves this one. *Hint:* Suppose there were . . .)
   (b) In *Post's correspondence problem*, you are given a dictionary between two languages; that is, a finite set of pairs of words. Each pair contains one word from Language 1 and one from Language 2. You are asked whether there is a *phrase* (sequence of words, possibly with repetitions) that has the same meaning in both languages. (Assume that words in these languages are juxtaposed with no intermediate blanks.)

   For example, the dictionary could be

Word Number	Language 1	Language 2
1	cab	ba
2	add	ad
3	dad	adddad
4	bad	db

The answer to this instance is yes, because the phrase consisting of words 2, 4, 2 and 3 is addbadadddad in both languages.

(c) You are given a finite set of *tiles* of different shapes, and you are asked whether you can tile the whole plane by using an infinite supply of each shape. For example, the shapes could be as follows.

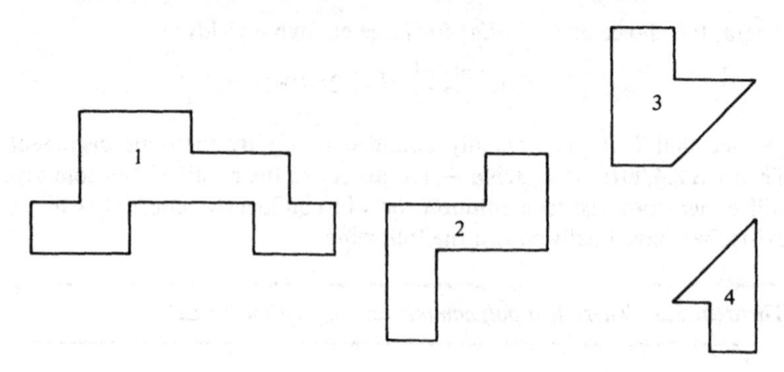

The answer to this instance is yes, because of the following tiling.

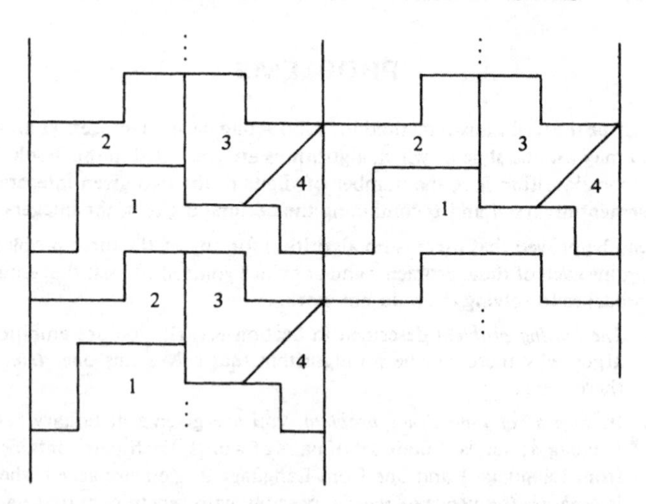

3. Give a detailed description of an algorithm which, given an $n \times n$ distance matrix, systematically generates all tours of n cities and chooses the cheapest (recall Example 8.1).

4. Give algorithms for solving each of the problems below. In each problem fix a representation of the input, and give an upper bound on the time required by your algorithm as a function of the size of the input.

(a) Given a graph $G = (V, E)$, is there a circuit $[u, v, w, z, u]$ of G such that $[v, z], [u, w] \notin E$?

(b) Given $n > 2$ lines on the plane $\{a_i x + b_i y = c_i : i = 1, \ldots, n\}$, with a_i, b_i, c_i integers for all i, is there a point lying on all of them?

(c) Given an integer p, is it a prime?

*(d) Given integers x, y, and z, is there an integer $n > 0$ such that $x^n + y^n = z^n$?

(e) Given a graph $G = (V, E)$ and $s, t \in V$, is there a path from s to t in G?

(f) Given a graph $G = (V, E)$ and $s, t \in V$, how many paths from s to t are there in G?

*(g) Given a position in chess, is it a forced win by White? (*Question:* How large can the instances of this one be? Is the notion of *rate of growth* of the complexity of an algorithm meaningful in this problem?)

(h) Given a position of *n-dimensional tic-tac-toe*, the game played on a

$$\underbrace{3 \times 3 \times \ldots \times 3}_{n \text{ times}}$$

board, is it a forced win by $\times$?

*(i) A graph can be used to represent the map of a system of *tunnels* in which a *fugitive* is hidden. Given a graph G and an integer $k > 0$, are k searchers enough for arresting the fugitive, even if the latter is assumed to be of infinite cunning and luck?

(j) Solve Post's correspondence problem (Part b of Problem 2), only with *no repetitions of words allowed* in a phrase.

(*Warning:* Your algorithms for some of these problems may have to be *very* inefficient.)

5. A procedure that invokes itself is called *recursive*. A frequent instance of recursion is the following: We wish to solve an instance of a problem that has size n, say. If n is small, then the instance is easy to solve. Otherwise, we may know how to solve the instance once we have the solutions to one or more *smaller* instances. We obtain the solutions to these smaller instances by invoking the same procedure. For example, suppose that we are given n and wish to compute $n!$. If $n = 0$, then naturally $n! = 1$. Otherwise, suppose that we have already computed $(n - 1)!$, the solution to the smaller instance. Then we could obtain $n!$ by multiplying $(n - 1)!$ by n. This suggests the following recursive algorithm.

```
procedure factorial(n)
if n = 0 then return 1
   else return n · factorial(n−1)
```

(a) Simulate step-by-step the execution of factorial(3).

(b) Give a recursive program that solves the Tower of Hanoi problem (Part b of Problem 1 in Chapter 1).

(c) We may use recursion in order to *sort* an array of 2^n distinct integers (that is, rearrange them so that they are in nondecreasing order). The idea is the following: It is quite easy to merge two arrays that are already sorted into a single sorted array. So, if the first and second halves of an array are sorted, it easy to sort the array itself. We can use recursion to sort the two halves.

*(d) Based on the algorithm in (c), show that an array of N elements can be sorted in $O(N \log N)$ time.

6. Prove the following.
   (a) $(\log n)^k = O(n^\epsilon)$ for all integers k and all $\epsilon > 0$.
   (b) $2^{n+1} \asymp 2^n$.
   (c) $2^{2^{n+1}} \not\asymp 2^{2^n}$.
   (d) $(\log n)^{\log n} = \Omega(n^k)$ for all integers k.

7. Let f and g be defined as follows.

$$f(n) = \begin{cases} n^2 & \text{if } n \text{ is prime} \\ n^3 & \text{otherwise} \end{cases}$$

$$g(n) = \begin{cases} n^2 & \text{if } n \text{ is odd} \\ n^3 & \text{otherwise} \end{cases}$$

   Which of the following are true?
   (a) $f(n), g(n) = \Omega(n^2)$
   (b) $f(n) = O(g(n))$
   (c) $g(n) = O(f(n))$

8. Suppose that in the simplex algorithm of Figure 2–3, we always choose the column j for which $\bar{c}_j < 0$ is smallest. Show that with this rule, simplex would solve the LP in (8.2) in one pivot step.

9. Suppose that in the simplex algorithm of Figure 2–3, we always choose the column j which, after pivoting, yields the largest decrease in the cost. Show that with this rule, simplex would solve the LP of (8.2) in one pivot step.

*10. Suppose that in the simplex algorithm of Figure 2–3, we choose one of the columns j for which $\bar{c}_j < 0$ *at random*, with equal probability that each such column is selected. Show that with this rule, simplex would solve the LP of (8.2) in $O(n)$ *expected* number of pivot steps.

11. Prove Lemma 8.8.

12. Prove Lemma 8.10.

*13. An iteration of the ellipsoid algorithm computes a new ellipsoid E_{j+1} which includes the semiellipsoid

$$\tfrac{1}{2}E_j[a] = E_j \cap \{x \in R^n : a'(x - t_j) \leq 0\}$$

   Show that the following formulas compute a new ellipsoid E'_{j+1} that includes the set

$$\tfrac{1}{2}E'_j[a] = E_j \cap \{x \in R^n : a'(x - t_j) \leq b\}$$

   (This set is, in the example of Figure 8–6(a), the smaller region cut from E_j by the broken line.)

$$t_{j+1} := t_j - \frac{1+n\Delta}{n+1}\frac{B_j a}{\sqrt{a'B_j a}}$$

$$B_{j+1} := (1-\Delta^2)\frac{n^2}{n^2-1}\left[B_j - \frac{1+n\Delta}{n+1}\frac{2}{1+\Delta}\frac{(B_j a)(B_j a)'}{a'B_j a}\right]$$

where $\Delta = (a'x - b)/\sqrt{a'B_j a}$. Argue that this modified iteration is correct in the sense that Theorem 8.4 is still valid, and that it may accelerate considerably the convergence of the ellipsoid method.

14. (A *product form* of the ellipsoid method) Recall the following equation from the proof of Lemma 8.13:

$$B_{j+1} = QRBR^TQ^T$$

We can rewrite this equation as

$$Q_{j+1} = Q_j R_j G,$$

where

(i) For each j, $Q_j Q_j^T = B_j$;

(ii) $G = \text{diag}\left(\frac{n}{n+1}, \frac{n}{\sqrt{n^2-1}}, \ldots, \frac{n}{\sqrt{n^2-1}}\right)$;

(iii) R_j is the rotation, depending on Q_j, which transforms $Q_j a$ to $(\|Q_j a\|, 0, \ldots, 0)$.

Based on this equation, prove that Q_j may be written as $Q_0 \prod_{i=1}^{j-1} F_i$ and give an explicit formula for the factor F_j in terms of a and Q_j.

15. (a) Show that if an ellipsoid $E = \{x: (x-t)'B^{-1}(x-t) \leq 1\}$ contains a sphere of radius r, then it also contains the sphere $\{x: (x-t)'(x-t) \leq r^2\}$.

(b) Consider an n-dimensional simplex in R^n with vertices $v_0, \ldots, v_n$. Let

$$v_c = \frac{1}{n+1}\sum_{j=0}^{n} v_j$$

be the center of gravity of the simplex, and suppose that the v_j's have rational coefficients with denominators bounded by 2^L. Show that there is a sphere with center v_c and radius $r = 2^{-2nL}$ totally within the simplex.

(c) From (a) and (b) conclude that, if an LSI system has a solution and the ellipsoid E is guaranteed to contain all solutions within the sphere $\|x\| \leq n2^L$, then any point on the boundary of E has a distance from the center of E at least 2^{-2nL}.

16. (a) Show that, if $[r_{ij}]$ is a rotation, then

$$\sum_{j=1}^{n} r_{ij}^2 = 1$$

(b) Based on Problem 14 and (a) above, show that all entries r of t_j and B_j in the ellipsoid algorithm satisfy

$$|r| \leq 2^{(2j+3)\log n + 2L}$$

NOTES AND REFERENCES

Turing machines as a formalism for algorithms were introduced in

[Tu] TURING, A. M., "On Computable Numbers, with an Application to the Entscheidungsproblem," *Proc. London Math. Soc., Ser. 2,* 42 (1936), 230–65. *Corrigendum,* ibid, 43 (1937), 544–46.

Other, equivalent, models of algorithms are introduced in the following references.

[Po] POST, E. L., "Formal Reductions of the General Combinatorial Decision Problem," *Amer. J. Math.,* 65 (1943).

[Ma] MARKOV, A. A., "Theory of Algorithms," Akad. Nauk. SSR Mate'm. Inst. Trudy, 42, Moscow 1954 (in Russian).

[ER] ELGOT, C. C., and A. ROBINSON, "Random Access Stored Program Machines," *J. ACM,* 11, no. 4 (1964), 365–99.

It is remarkable that a polynomial-time algorithm in any of these diverse models can be translated into a polynomial-time algorithm in any other model. See, for example, Chapter 1 of

[AHU] AHO, A. V., J. E. HOPCROFT, and J. D. ULLMAN, *The Design and Analysis of Computer Algorithms.* Reading, Mass.: Addison-Wesley, 1974.

The dichotomy between polynomial and exponential algorithms can be traced, interestingly enough, to John von Neumann, the father of the modern computer.

[vN] VON NEUMANN, J., "A Certain Zero-Sum Two-Person Game Equivalent to the Optimal Assignment Problem," in *Contributions to the Theory of Games II,* ed. H. W. Kuhn and A. W. Tucker. Princeton, N.J.: Princeton Univ. Press, 1953.

More explicit mention of polynomial-time computation was first made in the following articles:

[Co] COBHAM, A., "The Intrinsic Computational Difficulty of Functions," pp. 24–30 in *Proc. 1964 Int. Congress for Logic Methodology and Phil. of Science,* ed. Y. Bar-Hillel. Amsterdam: North Holland, 1964.

[Ed] EDMONDS, J., "Paths, Trees, and Flowers," *Canad. J. Math.,* 17 (1965), 449–67.

The results of Sec. 8.6 are from

[KM] KLEE, V., and G. J. MINTY, "How Good is the Simplex Algorithm?" pp. 159–175 in *Inequalities III,* ed. O. Shisha. New York: Academic Press, Inc., 1972.

Pathological examples with exponentially many pivots exist for the versions of simplex in which we choose the most negative column (Problem 8), the column with the greatest gain in the cost (Problem 9) and other variants

[Je] JEROSLOW, R. J., "The Simplex Algorithm with the Pivot Rule of Maximizing Criterion Improvement," *Discrete Math.,* 4 (1973), 367–78.

as well as for the primal-dual method and its variants:

[Za1] ZADEH, N., "A Bad Network Problem for the Simplex Method and Other Minimum Cost Flow Algorithms," *Math. Prog.*, 5 (1973), 255–66.

[Za2] ———, "More Pathological Examples for Network Flow Problems," *Math. Prog.*, 5 (1973), 217–24.

The progenitor of the ellipsoid algorithm was a more general method due to Shor:

[Sh1] SHOR, N. Z., "Utilization of the Operation of Space Dilatation in the Minimization of Convex Functions," *Kibernetika*, 6 (1970), 6–12. Translated in *Cybernetics*, 6, 7–15.

The ellipsoid algorithm for convex—not necessarily linear—constraints was described in

[Sh2] SHOR, N. Z., "Cut-off Method with Space Extension in Convex Programming Problems," *Kibernetika*, 13, no. 1 (1977), 94–95. Translated in *Cybernetics*, 13, 94–96.

[JN] JUDIN, D. B., and A. S. NEMIROVSKII, "Informational Complexity and Effective Methods for the Solution of Convex Extremal Problems," *Ekonomika; Matematicheskie Metody*, 12, no. 2 (1976), 357–69 (in Russian).

The important consequences of these results for linear programming were first pointed out in

[Kh] KHACHIAN, L. G., "A Polynomial Algorithm for Linear Programming," *Doklady Akad. Nauk USSR*, 244, no. 5 (1979), 1093–96. Translated in *Soviet Math. Doklady*, 20, 191–94.

Our exposition of the ellipsoid algorithm was influenced by

[AS] ASPVALL, B., and R. E. STONE, "Khachiyan's Linear Programming Algorithm," *Journal of Algorithms*, 1, no. 1 (1980).

Despite the great theoretical value of the ellipsoid algorithm, it is not clear at all that this algorithm can be practically useful. The most obvious among many obstacles is the large precision apparently required. On the other hand, there is no conclusive evidence that the ellipsoid algorithm (and all of its possible improvements) must be inherently impractical. Some of these issues are discussed in

[Da] DANTZIG, G. B., "Comments on Khachian's Algorithm for Linear Programming," Technical Report SOL 79-22, Dept. of Operations Research, Stanford Univ., 1979.

[GT] GOLDFARB, D., and M. J. TODD, "Modifications and Implementation of the Schor-Khachian Algorithm for Linear Programming," Dept.of Computer Science, Cornell Univ., 1980.

Some consequences of the ellipsoid algorithm for complexity issues in combinatorial optimization are examined in

[KP] KARP, R. M., and C. H. PAPADIMITRIOU, "On Linear Characterizations of Combinatorial Optimization Problems," *Proc. 21st Annual Symposium on Foundations of Computer Science*, IEEE (1980), 1–9.

[GLS] GRÖTSCHEL, M., L. LOVÁSZ, and A. SCHRIJVER, "The Ellipsoid Method and its Consequences in Combinatorial Optimization," Report 80-151-OR, Univ. of Bonn, 1980.

In the next chapter we develop a polynomial time algorithm for a special case of LP, namely the max-flow problem of Chapter 6. Unlike the ellipsoid algorithm, however, this algorithm involves a number of arithmetic operations which is bounded by a polynomial in the *number of integers* in the input (not their total length L). Let us call such an algorithm *strongly polynomial* if it is also a polynomial algorithm in our ordinary sense (that is, if the integers involved in the arithmetic operations are not exponentially long). An important open question in the wake of the ellipsoid algorithm, is whether there is a strongly polynomial algorithm for LP. An example of such an algorithm would be simplex with a pivoting rule that guarantees termination within a polynomial (in m and n) number of pivots. The results in Section 8.6 and References [Je], [Za1], and [Za2] do not exclude the possibility that such a pivoting rule exists. It was recently pointed out by N. Zadeh that an attractive pivoting rule for which no exponential counterexample is known is the following: "Among all columns with $\bar{c}_j < 0$ choose the one that has entered the basis so far the fewest times." In fact, there is no known strongly polynomial algorithm even for an immediate generalization of max-flow, the min-cost flow problem of Chapter 7 (see the Notes and References of Chapter 7). In Chapter 16 we shall discuss *pseudopolynomial* algorithms, a concept which is exactly the opposite of strongly polynomial algorithms.

Proofs that the problems defined in Problem 2 are undecidable can be found in Chapter 6 of

[LP] LEWIS, H. R., and C. H. PAPADIMITRIOU, *Elements of the Theory of Computation*. Englewood Cliffs, N.J.: Prentice-Hall Publishing Co., Inc., 1981.

That the procedure for inverting a matrix known as *Gaussian elimination* is a polynomial-time algorithm when applied to integer matrices (proof of Theorem 8.2) follows from the fact that all intermediate results are rational numbers with numerators and denominators equal to subdeterminants of the original matrix; see

[Ga] GANTMACHER, F. R., *Matrix Theory*, vol. 1. Chelsea, 1959, Chap. 2.

For a proof that there are n^{n-2} spanning trees of n nodes (Example 2), see Chapter 2 of

[Ev] EVEN, S., *Graph Algorithms*. Potomac, Maryland: Computer Science Press, 1979.

9

Efficient Algorithms for the Max-Flow Problem

In the previous chapter, we introduced a formalism for describing algorithmic efficiency whereby the performance of any algorithm can be evaluated in a uniform mathematical setting. By using these concepts, we noticed that the simplex algorithm for linear programming, although a very clever and practically useful algorithm, is not "good" according to the rigid criteria of this school. In this chapter, we examine from the same viewpoint an important special case of the linear programming problem, namely, the max-flow problem discussed in Chapter 6. We shall show that, like simplex, the labeling algorithm that we have developed for this problem may require in the worst case an exponential amount of time. Fortunately, there is an efficient algorithm for the max-flow problem, which is, in fact, a rather simple modification of the labeling algorithm. Besides solving the max-flow problem, this algorithm can be adapted for the efficient solution of some other interesting combinatorial optimization problems.

We first examine a fundamental graph algorithm called *search*. Different variants of search are at the heart of many of the graph algorithms explained in this and the coming chapters, as well as the rudimentary labeling algorithm of Chapter 6.

9.1
Graph Search

A graph $G = (V, E)$ is represented by its *adjacency lists* $A(v)$, $v \in V$ (see Example 8.5). As always, we assume that $|E| \geq \frac{1}{2}|V|$—assuming, for example, that G has no isolated nodes.

Consider the algorithm search shown in Fig. 9–1. It is based on the following idea: we start with a vertex v_1 and "mark" it. Then we repeat this with the vertices adjacent to v_1, and the vertices adjacent to these, and so on.

```
ALGORITHM SEARCH
Input:  A graph G, represented by its adjacency lists; a node v1.
Output: The same graph with the nodes reachable by paths from
        the node v1 "marked".
begin
 ┌──────────┐
 │ Q:= {v1}; │
 └──────────┘
  while Q ≠ ∅ do
  begin
    let v be any element of Q;
    remove v from Q;
  ┌──────────┐
  │ mark v;  │
  └──────────┘
    for all v' ∈ A(v) do
  ┌──────────────────────────────────────┐
  │ if v' is not marked then add v' to Q; │
  └──────────────────────────────────────┘
  end
end
```

Figure 9–1 The algorithm search.

In the set Q we maintain a list of all vertices that are available for marking—that is, are adjacent to a previously marked node and not yet marked. The process terminates when Q becomes empty.

Theorem 9.1 *The algorithm search of Fig. 9–1 marks all nodes of G connected to v_1 in $O(|E|)$ time.*

Proof Suppose that a node v is connected to v_1 by a path. It can be shown by induction on the length of this path that v will be marked. On the other hand, suppose that there is no path from v_1 to v. It can be shown, also by easy induction on the number of executions of the **while** loop of search, that v will not be marked.

For the proof of the time bound, the complexity of search has three components.

1. Initialization; this takes constant time.
2. Maintaining the set Q; there are at most $2|V|$ additions and removals of elements of Q. Each can be done by two or three elementary operations (see Figs. 9–2 and 9–4); an $O(|V|)$ bound follows.

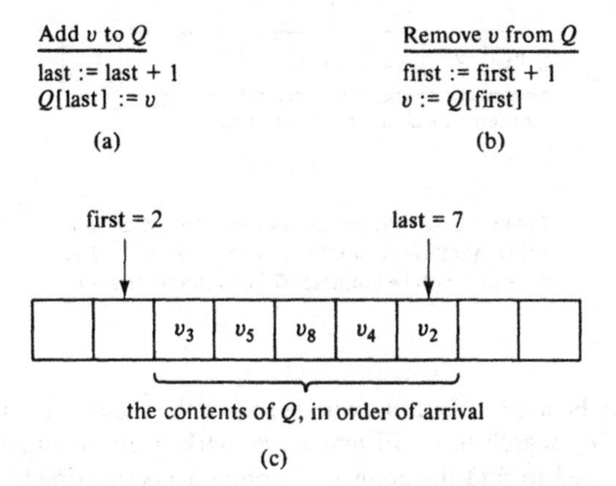

Figure 9–2 Programs that implement Q as a queue. The variables *last* and *first* are initialized to zero and Q is an array of $|V|$ entries. Q is empty iff *first* = *last*.

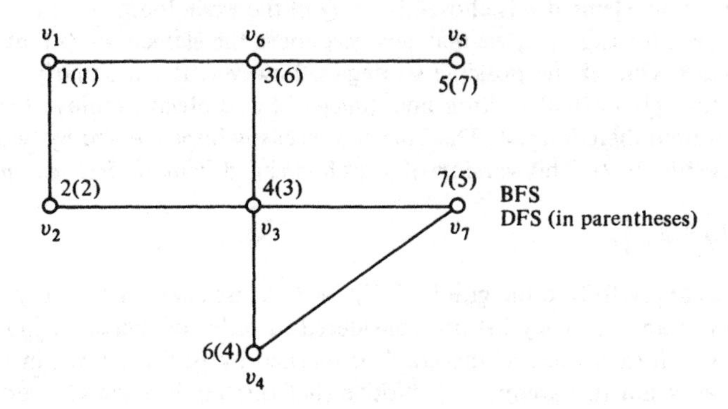

Figure 9–3

3. Searching the adjacency lists; we do constant work for each element of each adjacency list. Since the sum of the lengths of the adjacency lists is $2|E|$, the time required is $O(|E|)$.

The overall $O(|E|)$ bound follows. □

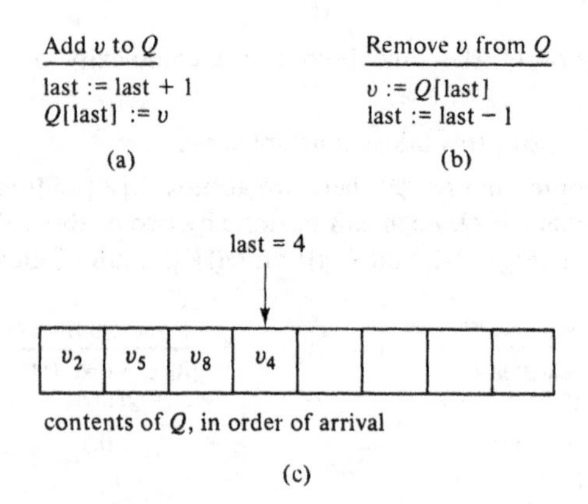
Add v to Q
last := last + 1
Q[last] := v
(a)
Remove v from Q
v := Q[last]
last := last − 1
(b)
last = 4
v_2
v_5
v_8
v_4
contents of Q, in order of arrival
(c)

Figure 9–4 Programs that implement Q as a LIFO queue (or *stack*). Initially, *last* = 0. Q is an array with $|V|$ entries. Q is empty iff *last* = 0.

Example 9.1

Search can be used to find out whether a graph is *connected*: G is connected iff, after applying search to it, all nodes are marked. In an equally direct way, search can be used to find the connected components (maximal connected subgraphs) of G (Problem 1). □

Search, as shown in Fig. 9–1, is not completely specified. We must define exactly how the element v is chosen from Q in the **while** loop. Several rules can be employed; for example, we may always choose the element of Q that has the lowest index. One of the possible strategies, however, is particularly intuitive: Imagine that Q is a real waiting line, (*queue !*), and always remove the vertex that has waited there longest. This rule can be easily implemented by two simple programs (Fig. 9–2). This version of search is called *breadth-first search* (BFS).

Example 9.2

Let us apply BFS to the graph of Fig. 9–3. As we shall customarily assume, the nodes in an adjacency list are considered in order of increased index. The order in which each node of the graph is marked by BFS is shown in Fig. 9–3 (the numbers not in parentheses). Notice that this kind of search amounts to visiting the nodes in order of increasing length of the shortest path from v_1—from which comes the name *breadth-first search*. (Proof: Problem 3.) □

Depth-first search (DFS) is another version of search in which Q is maintained in a last-in, first-out (LIFO) manner (Fig. 9–4). This algorithm performs a deep probe, creating a path as long as possible, and returns to assume a new probe only when no new nodes can be reached from the tip of the path.

Example 9.2 (Continued)

The order of visit of the nodes of the graph in Fig. 9–3 by DFS is shown in parentheses. ☐

The search algorithm can also be applied to *digraphs*. We can also represent a digraph $D = (V, A)$ by its adjacency lists: $A(v)$ is the set of all nodes $v' \in V$ such that $(v, v') \in A$. Note that from this viewpoint graphs are just a special case of digraphs, namely, those that are *symmetric* in that $u \in A(v)$ iff $v \in A(u)$. The algorithm search (and its special cases DFS and BFS) can thus be applied to digraphs without any changes.

Example 9.3

In Fig. 9–5 we show a digraph $D = (V, A)$ and its adjacency lists. The order in which BFS marks the nodes of D is shown. The numbers in parentheses correspond to DFS. ☐

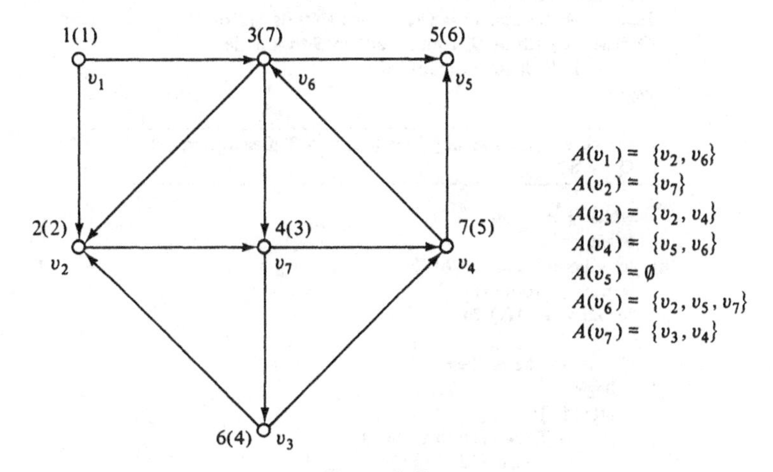

Figure 9–5

Because of the asymmetry inherent in digraphs, the notion of connectedness is more subtle than the corresponding notion in graphs, and a connectedness test requires some nontrivial use of search (see Problems 4 and 5).

What is really most important about the algorithm search of Figure 9–1 is that it is not just one algorithm, but a *template* for a whole *class* of algorithms. By varying the pieces of program in the "boxes" of Fig. 9–1, we can create numerous algorithms that do different things to the same graph, following the same general pattern of search. For an easy example, suppose in Fig. 9–1 that instead of *mark* v we insert the code *count*:=*count*+*1*, *order*[v]:=*count* wherever *mark* is used, and initialize *count* to zero in the first box. We thus obtain an algorithm that records the order in which search visits the nodes (recall Figs. 9–3 and 9–5). For another example, it is not hard to see that the Ford-Fulkerson labeling algorithm for max-flow, discussed in Chapter 6, is in fact a more sophisticated variant of search. A related example, which we examine in detail next, is the task of finding paths in digraphs.

Suppose that we are given a digraph $D = (V, A)$ and two sets of nodes $S, T \subseteq V$, the *sources* and *targets*, respectively. We wish to find a path in D leading from any node of S to any node of T. Not surprisingly, we can solve this problem by an easy modification of search. The only differences are that now Q is initialized to S instead of $\{v_1\}$, and for each v' we consider, we do slightly more work than just marking it. In particular, we set *label* $[v']$—where *label* is an array with $|V|$ entries—equal to the predecessor v of v', which caused v' to be considered. These labels are going to be most helpful in actually recovering the path from S to T. The vertices in S have label 0, by convention. The whole algorithm is shown in Fig. 9–6.

```
ALGORITHM FINDPATH
Input:  A digraph D = (V,A); two subsets S,T of V.
Output: A path in D from a node in S to a node
        in T, if such a path exists.
begin
  ┌─────────────────────────────────────────────────────────┐
  │ for all v ∈ S do label [v] := 0, if v ∈ T then return (v); │
  │ Q := S;                                                   │
  └─────────────────────────────────────────────────────────┘
  while Q ≠ Ø do
  begin
  let v be any element of Q;
  remove v from Q;
  for all v' ∈ A(v) do
  ┌────────────────────────────────┐
  │ if v' is unlabeled then          │
  │   begin                          │
  │   label [v']:=v;                 │
  │   if v' ∈ T then return path(v') │
  │             else add v' to Q     │
  │   end                            │
  └────────────────────────────────┘
  end
  return "no S−T path in D"
end

  procedure path(v)
(comment: it returns a sequence of nodes from some s ∈ S to v)
  if label[v] = 0 then return (v)
    else return path(label [v]) || (v).
(comment: path is recursive, see Problem 5 in Chapter 8;
|| stands for concatenation—that is, juxtaposition—of paths)
```

Figure 9–6 The algorithm findpath.

Example 9.4

Suppose that we wish to find a path from S to T in the digraph D of Fig. 9–7(a). The labels of the nodes of D at the termination are shown in Fig. 9–7(b). We implement search as BFS: As a result, we shall find the *shortest* path from S to T; that is, the one with the fewest arcs. It is finally found that $v_9 \in T$. Then path is called with argument v_9. Path(v_9) is successively evaluated as

$$\begin{aligned}\text{path}(v_9) &= \text{path}(v_8) \,\|\, (v_9)\\ &= \text{path}(v_6) \,\|\, (v_8, v_9)\end{aligned}$$

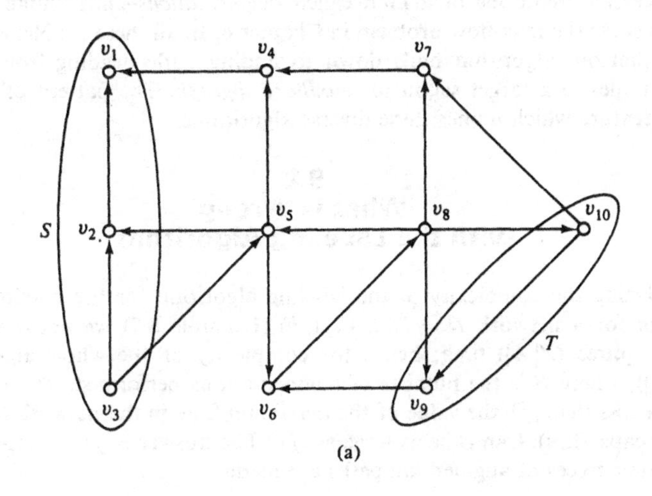

(a)

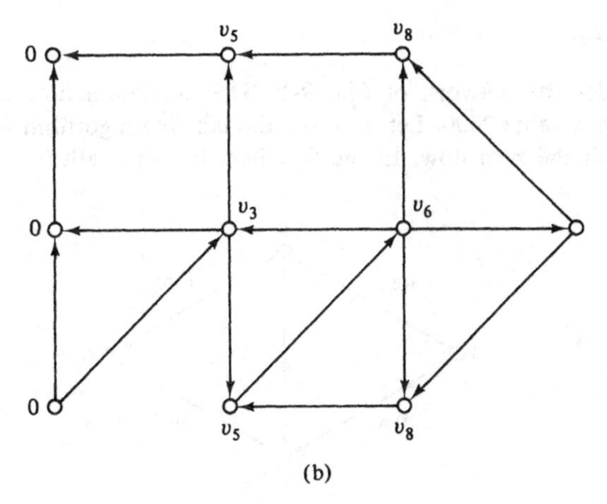

(b)

Figure 9–7

$$= \text{path}(v_5) \,||\, (v_6, v_8, v_9)$$
$$= \text{path}(v_3) \,||\, (v_5, v_6, v_8, v_9)$$
$$= (v_3, v_5, v_6, v_8, v_9)$$

which is indeed the shortest S-T path. □

Finding paths in digraphs in this manner has surprisingly many applications in solving several much more complex combinatorial optimization problems. In this and the next three chapters, we are going to develop efficient algorithms for certain combinatorial problems that are susceptible to solution by successive iterations of a kind called *augmentations*—not unlike the ones used to solve the max-flow problem in Chapter 6. In all these problems we shall notice that our algorithm boils down to finding paths leading from a set of source nodes to a target set in an *auxiliary digraph*. The pattern of search is then a feature which unifies these diverse algorithms.

9.2 What Is Wrong with the Labeling Algorithm

In analyzing the complexity of the labeling algorithm for the maximum flow problem for a network $N = (s, t, V, A, b)$ (Example 8.7) we noted that each stage requires $O(|A|)$ time; hence the complexity of the whole algorithm is $O(S|A|)$, where S is the number of augmentations performed. The number S must be less than $|\hat{f}|$, the value of the maximum flow in the network (assuming integer capacities). Can S be as large as $|\hat{f}|$? The answer is yes, if repeated unfortunate choices of augmenting paths are made.

Example 9.5

Consider the network of Fig. 9–8. The maximum flow in this network obviously has value 2000. Let us apply the labeling algorithm to this network, starting with the zero flow. In the first iteration the path (s, u, v, t) is a legal

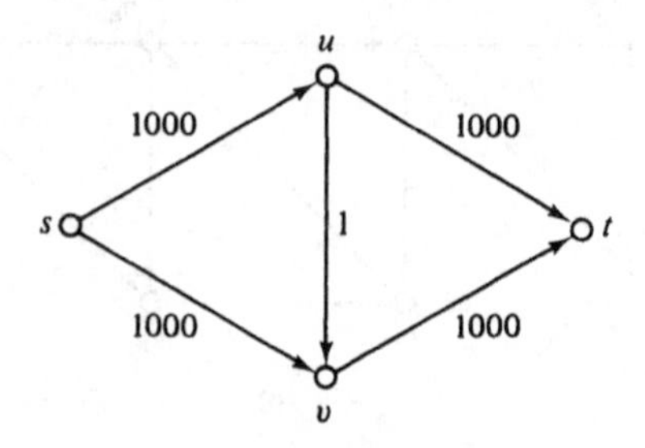

Figure 9–8

augmenting path, which achieves a flow of value 1. In the second iteration we may choose to augment the flow along the path (s, v, u, t); since (v, u) is by now a reverse arc, this path is also a proper choice, according to the labeling algorithm. The value of the flow is now 2. It is easy to see that augmenting paths may alternate between (s, u, v, t) and (s, v, u, t), adding 1 to the value of the flow in each iteration. Thus, we shall go through 2000 iterations before we reach the optimal flow. If, instead of 1000, the capacities of the four arcs are taken to be *any* large integer M, the number of iterations will grow as fast as $2M$. Thus the labeling algorithm, in this worst case, requires an exponential number of steps (recall the discussion in Sec. 8.5). □

Fortunately, we can avoid this exponential complexity by being a little more careful about our choice of the augmenting path. The choices we made in Example 9.5 are unfortunate in that the augmenting paths are longer than necessary. That is, if we had picked the (s, v, t) augmenting path instead of the path (s, u, v, t), which is one arc longer, this would have resulted in a larger gain in the flow and would have reduced substantially the total number of iterations required. To turn the above observation into a proposal for improving the labeling algorithm, let us find in each iteration the augmenting path relative to the existing flow that is *as short as possible.* To do this does not require extra complexity per iteration; it can be achieved, as we shall see in Sec. 9.3, simply by performing the labeling process in a breadth-first manner.

This is not to say that the improvement of the flow along short paths is always greater than the improvement along longer paths; for example, if we interchange the capacities of the arcs (s, v) and (u, v) in Fig. 9–8, the augmentation along (s, u, v, t) now would result in better gains than that along (s, v, t). However, by *systematically* choosing the shortest possible augmenting path, we avoid pathologies like the one pointed out in Example 9.5, namely, *repeatedly* choosing small augmentations many times. In fact, if we always choose the shortest possible augmenting path relative to the given flow, we can prove that we can have no more than $|V| \cdot |A|$ stages. This follows, roughly speaking, from the fact that each arc (v, u) can be used under this rule at most $|V|$ times as the bottleneck in the augmenting path (see Problem 9). We do not examine these arguments in detail now since, in Sec. 9.4, we shall obtain a stronger result.

The following example demonstrates another aspect of the "wastefulness" of the original version of the labeling algorithm.

Example 9.6

Let us apply the labeling algorithm to the network of Fig. 9–9. If we carry out the first stage we may discover the augmenting path (s, v, z, t) and augment the flow. The labels at the end of the first stage are shown in Fig. 9–9. In fact, this augmenting path is as short as possible, and thus we have even incorporated our previous suggestion in the algorithm.

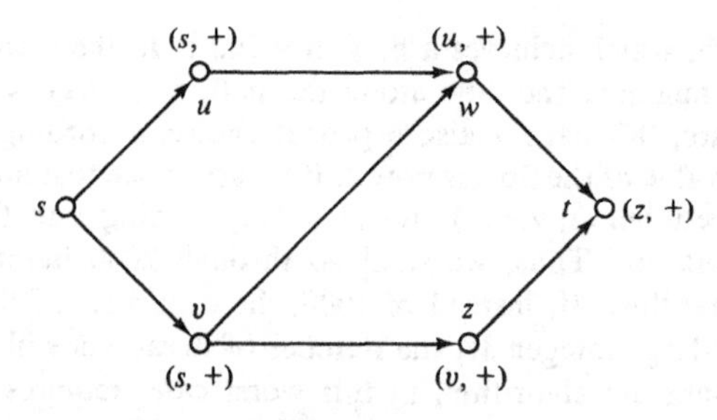

Figure 9–9

At this point, the labeling algorithm would require us to erase all labels and proceed to the next stage. In the next stage we would discover the augmenting path (s, u, w, t). The interesting fact, however, is that in order to discover this path, we used almost exclusively labels that were available in the previous stage. That is, *this path could be found by using the information available during the previous stage.* Erasing the labels is therefore a loss of potentially useful information, and it is thus advisable not to proceed to the next iteration unless we are certain that no profitable use of the labeling can be made. □

In the improved labeling algorithm that we are going to describe in Sec. 9.4, we combine these two ideas—namely, augmenting along shortest paths and augmenting more than once in each iteration—to design an efficient algorithm for maximum flow.

9.3 Network Labeling and Digraph Search

To implement the improvements of the labeling algorithm discussed in the previous section without substantially increasing the complexity of each stage, we shall first look at the process of labeling from a somewhat different viewpoint.

Suppose that we wish to apply the labeling routine to a network $N = (s, t, V, A, b)$ with initial flow zero. We need not examine capacities and flows in this case; it is a priori certain that all arcs in A are forward, and that there are no backward arcs. Consequently, our task of labeling the network in order to discover an augmenting path is very much the same as applying findpath of Fig. 9–6 to the digraph (V, A) with $S = \{s\}$ and $T = \{t\}$. Once an augmenting path is discovered, we augment the flow along it and continue.

How about the subsequent stages? Can they also be thought of as digraph searches? The additional elements that are present when the flow is nonzero are the backward arcs, certain missing (that is, saturated) forward arcs, and some modifications of the capacities. It turns out that labeling a network N

with respect to a flow f is equivalent to applying findpath to a network $N(f)$ defined as follows.†

Definition 9.1

Given a network $N = (s, t, V, A, b)$ and a feasible flow f of N, we define the network $N(f) = (s, t, V, A(f), ac)$, where $A(f)$ consists of the following arcs:

(a) If $(u, v) \in A$ and $f(u, v) < b(u, v)$, then $(u, v) \in A(f)$ and $ac(u, v) = b(u, v) - f(u, v)$.

(b) If $(u, v) \in A$ and $f(u, v) > 0$, then $(v, u) \in A(f)$ and $ac(v, u) = f(u, v)$.

We call $ac(u, v)$ the *augmenting capacity* of $(u, v) \in A(f)$. □

If A contains both arcs (v, u) and (u, v), then $N(f)$ may have multiple copies of these arcs. We can avoid this in many ways. For example, we can replace any such arc (u, v) in A by the two arcs (u, w) and (w, v) of the same capacity, where w is a new node. We shall henceforth assume that $N(f)$ has no multiple arcs.

We shall now show an interesting property of $N(f)$. Take any s-t cut $(W, \bar{W})$ of $N(f)$. The value of this cut is, naturally, the sum of the augmenting capacities of all arcs of $N(f)$ going from W to $\bar{W}$. However, such an arc (u, v) may be either a *forward* arc (Case a of the definition of $N(f)$) or a *backward* arc (Case b). In the first case, $ac(u, v) = b(u, v) - f(u, v)$; in the second case, $ac(u, v) = f(v, u)$. Thus the value of $(W, \bar{W})$ in $N(f)$ is the value of $(W, \bar{W})$ in N minus the forward flow across the cut, plus the reverse flow across the cut. Note that the two last terms are exactly $-|f|$, where by $|f|$ we denote $\sum_{(s,u) \in A} f(s, u)$, the value of the flow f. We conclude that if a cut in N has value C, the same cut in $N(f)$ has value $C - |f|$. Thus, if the *minimum* cut in N has value $|\hat{f}|$, then the minimum cut in $N(f)$ has value $|\hat{f}| - |f|$. However, in both networks, the value of the minimum cut equals the value of the maximum flow, by the max-flow min-cut theorem. Thus, we have shown the following proposition.

Proposition 1 If $|\hat{f}|$ is the value of the maximum flow in N, then the value of the maximum flow in $N(f)$ is $|\hat{f}| - |f|$.

Example 9.7

Let us consider the network N shown in Fig. 9–10(a) with the indicated flow f. The network $N(f)$ associated with N and f is shown in Fig. 9–10(b); the augmenting capacities of the arcs of $N(f)$ are also shown. If we were to

†Recall the incremental network in Chapter 7.

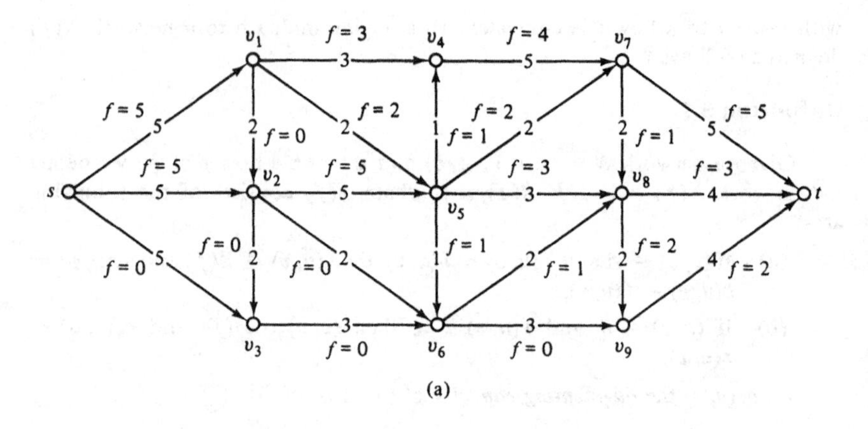

(a)

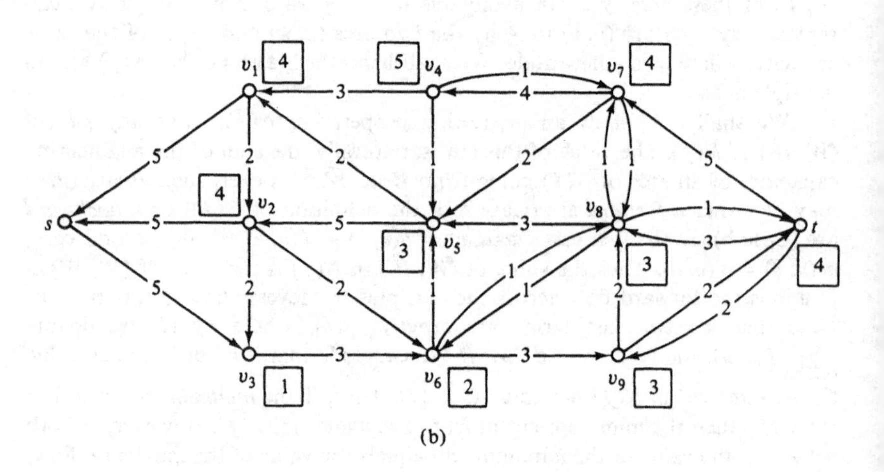

(b)

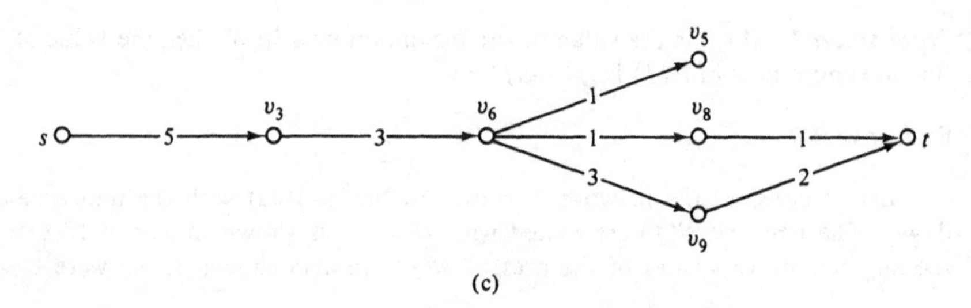

(c)

Figure 9–10

apply the labeling routine to the network N with respect to the flow f, we would have to propagate labels through forward and backward arcs, starting from s. Because $N(f)$ contains exactly all these pertinent forward and backward arcs, the labeling routine with respect to f amounts simply to applying search to $N(f)$ starting from s. □

This viewpoint is extremely helpful when it comes to materializing our idea of augmenting along the shortest possible path. This is so because shortest augmenting paths in N with respect to f correspond to shortest paths (that is, with the least number of arcs) from s to t in $N(f)$. We know how to implement search to discover shortest paths: We simply use the technique of breadth-first search—that is, we process the nodes to be scanned as in a queue. Breadth-first search partitions the vertices of $N(f)$ into disjoint *layers* (layer 0, layer 1, . . .) according to their distance (that is, number of arcs in the shortest path) from s. For example, in Fig. 9–10(b), we show the layer of each vertex of $N(f)$. Here t is at layer 4.

Keeping in mind that we are interested only in *shortest* s-t paths in $N(f)$, we can perform some further simplifications. First, we note that the shortest s-t path cannot pass through a vertex in a higher layer than that of t. In our example, the shortest path from s to v_4 in $N(f)$ is of length 5, and hence no shortest s-t path can pass through v_4; we can thus discard v_4 and all arcs incident on it without losing any useful information for this stage. In fact, we can similarly discard any vertices other than t in the *same* layer as t (like v_1 and v_7 in our example) because they cannot be on any shortest s-t path either. There are even more parts of $N(f)$ that we do not need consider. Any shortest path from s to t will start from a layer-0 node (that is, s), then visit a layer-1 node, a layer-2 node, and so on. In other words, a shortest path can contain only arcs that go from layer j to layer $j + 1$, for some j. So we can also discard without any loss of relevant information any arc that goes from a layer to a lower layer (like (v_8, v_6) in Fig. 9–10(b)) or any arc that joins two nodes of the same layer (like (v_8, v_5)). The resulting subnetwork $AN(f)$ of $N(f)$, called the *auxiliary network with respect to* f, is shown in Fig. 9–10(c).

The auxiliary network $AN(f)$ has a special structure best captured by the following definition.

Definition 9.2

A *layered network* $L = (s, t, U, A, b)$ is a network with vertex set U equal to the disjoint union of the sets $U_0, \ldots, U_d$, such that $U_0 = \{s\}$, $U_d = \{t\}$, and $A \subseteq \bigcup_{j=1}^{d} (U_{j-1} \times U_j)$. □

Thus the auxiliary network $AN(f)$ is a layered network. Notice that $AN(f)$ is easy to construct. We can create it while carrying out the breadth-first search on $N(f)$ by keeping only the arcs that lead us to new nodes and only the nodes that are at lower levels than t. Hence creating the auxiliary network can be done in $O(|A(f)|) = O(|A|)$ time.

Using the auxiliary network, we can very easily find the shortest augmenting path with respect to the current flow. Furthermore, materializing our second idea for improving the labeling algorithm, namely, performing as many augmentations at the same stage as possible, is now just one step further.

Definition 9.3

Let $N = (s, t, V, A, b)$ be a layered network. An augmenting path in N with respect to some flow g is called *forward* if it uses no backward arcs. A flow g of N is called *maximal* (not *maximum*) if there is no forward augmenting path in N with respect to g. □

All maximum flows are therefore maximal; a maximal flow, g, however, may not be maximum, as shown in Fig. 9–11.

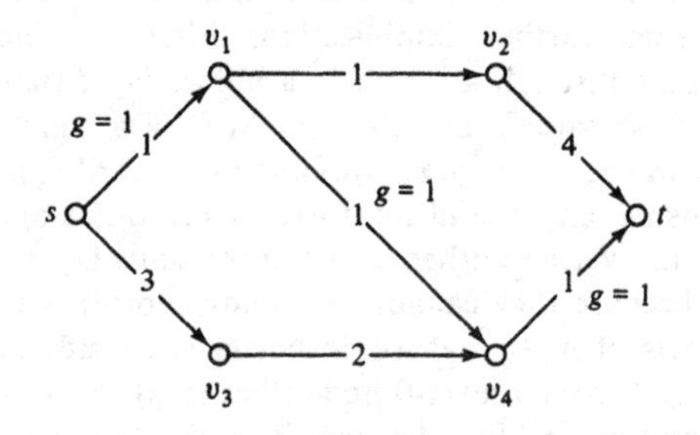

Figure 9–11

Finding a maximal flow in the auxiliary network with respect to the current flow corresponds to finding as many shortest augmenting paths at a stage as possible. This concept is the heart of our max-flow algorithm, described in the next section.

9.4 An $O(|V|^3)$ Max-Flow Algorithm

Our max-flow algorithm—like many other combinatorial optimization algorithms described in the forthcoming chapters—operates in *stages*. At each stage we construct the network $N(f)$, where f is the current flow, and from it we find the auxiliary network $AN(f)$. We then find a *maximal* flow g in $AN(f)$, add g to f, and repeat. Adding g to f entails adding $g(u, v)$ to $f(u, v)$ for all forward arcs (u, v) in $AN(f)$, and *subtracting* $g(u, v)$ from $f(v, u)$ for all backward arcs (u, v) in $AN(f)$. We end when s and t are disconnected in $N(f)$, a sign that f is optimal.

The main part of our algorithm is a clever way of finding a *maximal* flow g in a layered network $AN(f)$. Our method does not augment along single paths; this time we try to *push* flow through nodes along *many* paths at the same time. Let us illustrate this by way of an example. Consider the layered network of Fig. 9–12. Let us define the *throughput* of a node v to be the maximum amount of flow that can be pushed through v; that is, *throughput*$[v]$ is the sum of the capacities of incoming arcs to v (∞ if $v = s$), or the sum of the capacities of outgoing arcs (∞ if $v = t$), whichever is smaller. In Fig. 9–12(a), therefore, *throughput*$[v_1] = 3$ and *throughput*$[v_3] = 4$. If we try to push 4 units of flow through v_3, we first observe that they have to be sent to v_6. Next, if we try to push the 4 units of flow that now come into v_6, we notice that we cannot, and so we have to get involved in a time-wasting backtracking. Let us examine, however, the real reasons for this. The vertex that we started with, v_3, has a rather large throughput. So it was quite likely that we could get stuck at a vertex of lower throughput—v_6 in our example. However, this could never happen had we started from the vertex with the *lowest* throughput. Let us therefore try a node with minimum throughput, such as v_1. We distribute the 3 units of flow between the arcs (v_1, v_3) and (v_1, v_4) up to their capacity. The flow into v_3 now has to go to (v_3, v_6). At this point we could just continue pushing this unit of flow all the way to t, in a depth-first manner, but we do not. Instead, we first wait to accumulate all the flow that has to go through v_6 and process it once and for all. This means that we never process a node unless all nodes in previous layers have been processed, or, equivalently, that we go from node to node in a breadth-first manner.

We now see in detail how to process a node. Let us take v_4, which is next. Two units of flow must be pushed out of it. We do this in an organized manner: We examine all edges one by one and fill them to their full capacity, or up to the point that all the required units have been pushed. So, at v_4 we might first examine (v_4, v_6) and saturate it. It would also be legitimate to examine (v_4, v_7) first and push all the flow of 2 units through it; what would be contrary to our strategy would be to unjustifiably distribute 1 unit to each of (v_4, v_8) and (v_4, v_7). Thus we push a total of 3 units from v_6 to v_9—it was a good thing that we waited and did not have to process v_6 twice—and finally 3 units from v_9 to t. Notice that throughout this step we need not worry about whether the nodes processed will have an adequate throughput. This is because at all layers, the total amount of entering flow, distributed among the nodes of the layer, is equal to *throughput*$[v_1]$. Since v_1 was the node with the smallest throughput, no node will ever have to handle an amount of flow larger than its throughput, and hence no backtracking is necessary.

In order to complete this into a legal flow, we have to bring some flow from s to v_1. However, this is completely symmetric to what we have done. It can be carried out by traversing arcs backwards, starting from v_1, processing the nodes layer by layer in the same breadth-first way and the incoming arcs in the same organized manner, effectively *pulling* flow. In our example, in fact, this is very

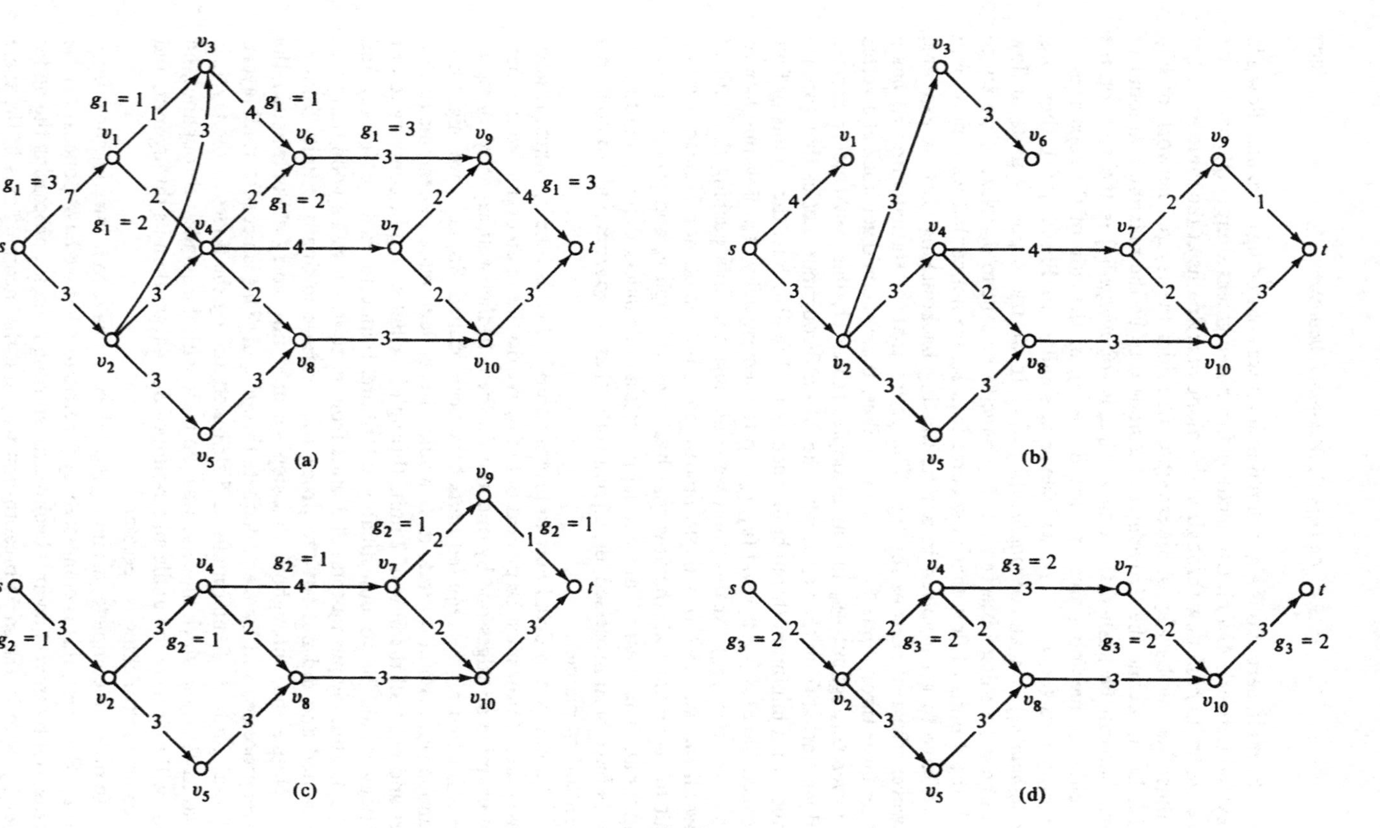

(a)
s v_1 v_2 v_3 v_4 v_5 v_6 v_7 v_8 v_9 v_{10} t
$g_1 = 3$
$g_1 = 1$
$g_1 = 2$
$g_1 = 1$
$g_1 = 2$
$g_1 = 3$
$g_1 = 3$
(b)
s v_1 v_2 v_3 v_4 v_5 v_6 v_7 v_8 v_9 v_{10} t
(c)
s v_2 v_4 v_5 v_7 v_8 v_9 v_{10} t
$g_2 = 1$
$g_2 = 1$
$g_2 = 1$
$g_2 = 1$
$g_2 = 1$
(d)
s v_2 v_4 v_5 v_7 v_8 v_{10} t
$g_3 = 2$
$g_3 = 2$
$g_3 = 2$
$g_3 = 2$
$g_3 = 2$

Figure 9–12

easy to do: Just pull three units of flow from s to v_1. The resulting flow g_1 is depicted in Fig. 9–12(a).

Here g_1 is not a maximal flow, and therefore we need to continue the same routine. We modify the capacities to reflect the achievement of g_1, and we obtain the layered network of Fig. 9–12(b)—we omit arcs with zero capacities. If we try to find the node with the smallest throughput, we observe that v_1 and v_6 have throughput zero. Therefore we can remove these nodes—and all adjacent arcs—from the network. Finally, we arrive at the network of Fig. 9–12(c).

Now v_9 has the smallest throughput, namely, 1. We therefore find a flow g_2 using the same method—in fact, this flow happens to be simply an augmenting path. After updating capacities and deleting useless arcs and nodes, we arrive at Fig. 9–12(d). Now v_4 has the smallest throughput, and the resulting flow g_3—again a path—is shown in Fig. 9–12(d). After updating capacities and deleting useless nodes and arcs, we observe that the node s is deleted. As we shall see, this means that the flow $g = g_1 + g_2 + g_3$ already obtained is maximal, and the stage is over. The whole algorithm is shown in Fig. 9–13. To show its correctness and time bound, we need some preliminary results.

```
MAX-FLOW ALGORITHM
Input:  A network N=(s,t,V,A,b).
Output:  The maximum flow f of N.
begin
  f:=0, done:="no"; (comment: initialize)
  while done = "no" do
  begin (comment: a new stage)
    g:=0;
    construct the auxiliary network AN(f)=(s,t,U,B,ac);
    if t is not reachable from s in AN(f) then done:="yes"
      else repeat
          begin
          while there is a node v with throughput[v]=0 do
          if v=s or t then go to incr
            else delete v and all incident arcs from AN(f);
          let v be the node in AN(f) for which
          throughput[v] is the smallest (comment: nonzero);
          push(v, throughput [v]);
          pull(v, throughput [v])
          end
incr: f:=f+g
   end
end

procedure push(y,h)
(comment: it increases the flow g by h units pushed from y to t in
a "systematic" way)
(comment: the procedure pull is completely analogous)
```

Figure 9–13 The max-flow algorithm.

```
begin
  Q:={y}
  for all u∈U−{y} do req[u] :=0;
  req[y] :=h
  (comment: req[u] denotes how many units of
            flow must be pushed out of u)

    while Q ≠ ∅ do
    begin
      let v be an element of Q;
      remove v from Q;
      (comment: Q must be a queue)
      for all v' such that (v,v')∈B and until req[v]=0 do

    begin
    m:=min (ac[v,v'], req [v]);
    ac [v,v'] :=ac[v,v']−m;
    if ac[v,v']=0 then remove (v,v') from B;
    req [v] :=req [v]−m;
    req [v'] :=req [v']+m;
    add v' to Q;
    g[v,v']:=g[v,v']+m
    end

    end
end
```

Figure 9–13 *(continued)*

Lemma 9.1 *An arc a of AN(f) is removed from B at some stage only if there is no forward augmenting path with respect to flow g in AN(f) that passes through a.*

Proof If an arc a is deleted at a stage, it may either be that $g(a) = ac(a)$, or that $a = (v, u)$ and either v or u have throughput zero. Suppose that $g(a) = ac(a)$. This means that a may appear in an augmenting path in $AN(f)$ with respect to g only as a backward arc. Hence the desired result holds. Let us now consider the case when v has throughput zero—the case of u is identical. Then v has zero input, and hence there is no arc that can be used in a forward path entering v. Hence $a = (v, u)$ cannot be used in *any* forward path. □

Lemma 9.2 *At the end of each stage, g is a maximal flow in AN(f).*

Proof If an arc is useless as far as forward augmenting paths with respect to g are concerned, it will continue to be so throughout the stage, since the only changes are removals of arcs and decreases in the capacities. Therefore, at

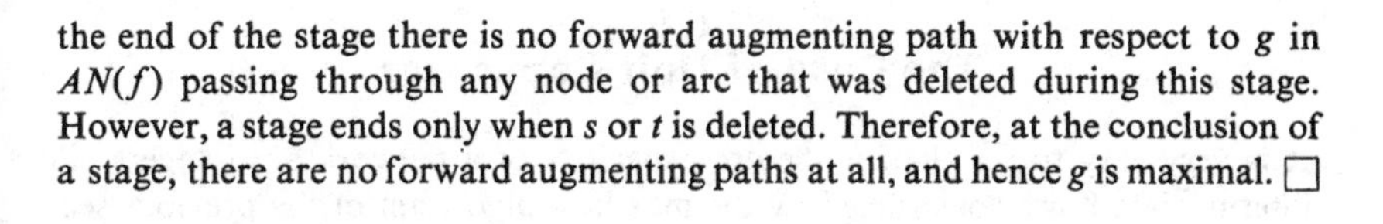

the end of the stage there is no forward augmenting path with respect to g in $AN(f)$ passing through any node or arc that was deleted during this stage. However, a stage ends only when s or t is deleted. Therefore, at the conclusion of a stage, there are no forward augmenting paths at all, and hence g is maximal. □

Lemma 9.3 *The s-t distance in $AN(f + g)$ at some stage is strictly greater than the s-t distance in $AN(f)$ at the previous stage.*

Proof It is easy to see that the auxiliary network $AN(f + g)$ is the same as the auxiliary network of $AN(f)$ with respect to g. In $AN(f)$, however, there are no forward augmenting paths with respect to the flow g, because g is maximal by Lemma 9.2. Hence all augmenting paths have length greater than the s-t distance in $AN(f)$. We conclude that the s-t distance in $AN(f + g)$ is greater than the s-t distance in $AN(f)$. □

Theorem 9.2 *The algorithm of Fig. 9–13 correctly solves the max-flow problem for a network $N = (s, t, V, A, b)$ in $O(|V|^3)$ arithmetic operations.*

Proof In the last stage, s and t are disconnected, and so the maximum flow in $N(f)$ is zero. However, by Proposition 1, this maximum flow has value $|\hat{f}| - |f|$, where $|\hat{f}|$ is the value of maximum flow in N. We conclude that $|f| = |\hat{f}|$, and hence the optimum is attained at the conclusion of the algorithm.

For the time bound, we first note that we have $O(|V|)$ stages. This is because, by Lemma 9.3, the s-t distance in $AN(f)$ is increasing from one stage to another. Therefore, since it cannot surpass $|V|$, we have at most $|V|$ stages. At each stage, the time requirements are dominated by the total number of times T that the different arcs are processed (second box of the procedure push, Fig. 9–13). We write $T = T_s + T_p$ to denote that a step of processing an arc at a stage can be either *saturating* (that is, the arc is filled to its capacity) or *partial.* Once an arc suffers a saturating step, it is deleted from the arc set; thus an arc can have at most one such step, and hence $T_s = O(|A|)$. However, we may have many partial steps for the same arc. The observation here is that we have at most $|V|$ executions of the push and pull procedures at each stage, because each such execution results in the deletion of the node v with the lowest throughput where this execution was started. Furthermore, in each execution of the push and pull procedures, we have at most $|V|$ partial steps (one for each node) because of our systematic manner of processing the nodes. Therefore, we have at each stage at most $|V|^2$ partial steps, and $T = T_s + T_p = O(|A|) + O(|V|^2) = O(|V|^2)$. So we have a total of $O(|V|^3)$ arithmetic operations for all stages, and the theorem is proved. □

9.5
The Case of Unit Capacities

It is very easy to see that, if the arc capacities of a network are integers, all intermediate flows constructed by the max-flow algorithm of the previous section (as well as by any "reasonable" max-flow algorithm) are also integers. Thus if a network has *unit* capacities, the flow values will always be zero or 1, and all auxiliary networks of our max-flow algorithm will also have unit augmenting capacities. The case of unit capacities is particularly interesting because, as is shown in the problems and in the next chapter, it has many applications to certain important purely combinatorial problems, such as bipartite matching. Furthermore, it can be shown that our max-flow algorithm performs especially well in the unit capacity case.

We start by analyzing the complexity of each stage of our algorithm when applied to networks with unit arc capacities.

Lemma 9.4 *A stage of the algorithm of Fig. 9–13 applied to a network $N = (s, t, V, A)$ with unit arc capacities can be carried out in $O(|A|)$ time.*

Proof One need only observe that when all augmenting capacities are 1, all arcs that are processed once become saturated immediately. □

Thus, by taking into account the unit capacities, we have been able to improve our upper bound on the complexity of each stage of the max-flow algorithm, at least for sparse networks. It turns out that the other factor in the complexity of the algorithm—the number of stages required—can also be improved upon. We first need the following lemma.

Lemma 9.5 *In a network N with unit arc capacities, the distance l between s and t cannot be greater than $2|V|/\sqrt{|\hat{f}|}$, where $|\hat{f}|$ is the value of the maximum flow.*

Proof Let V_i denote the subset of V with distance from s equal to i. We note that the set of arcs going from V_i to V_{i+1}, $i < l$, is a genuine s-t cut in N, and hence its value, that is, the total number of arcs, should be $|\hat{f}|$ or more. However, the maximum possible number of arcs going from $|V_i|$ to $|V_{i+1}|$ is $|V_i| \cdot |V_{i+1}|$. Thus $|V_i| \cdot |V_{i+1}| \geq |\hat{f}|$, $i = 1, \ldots, l-1$; hence at least one of $|V_i|, |V_{i+1}|$ is no smaller than $\sqrt{|\hat{f}|}$. Thus at least every other level $V_1, \ldots, V_l$ must contain $\sqrt{|\hat{f}|}$ or more vertices; it follows that $l/2 \cdot \sqrt{|\hat{f}|} \leq |V|$ and $l \leq 2|V|/\sqrt{|\hat{f}|}$. □

Theorem 9.3 *For networks with unit arc capacities, the algorithm of Fig. 9–13 requires at most $O(|V|^{2/3} \cdot |A|)$ time.*

Proof By Lemma 9.4, it suffices to show that S, the number of stages required, is $O(|V|^{2/3})$. We first note that $S \leq |\hat{f}|$ because at every stage we increment the value of the flow by at least 1. So, if $|\hat{f}| \leq |V|^{2/3}$, the result follows immediately. Now, if $|\hat{f}| > |V|^{2/3}$, consider the stage after which the value of the flow exceeds $|\hat{f}| - |V|^{2/3}$ for the first time; let g be the flow at the beginning of this stage. The remaining flow value is $|V|^{2/3}$ or less, and hence at most $|V|^{2/3}$ stages will follow. Because the distance l from s to t in $N(f)$ increases monotonically from stage to stage, the number of stages so far cannot be greater than the value of l in $N(g)$. However, by Lemma 9.5 applied to $N(g)$, $l \leq 2|V|/\sqrt{|\hat{f}| - |g|}$, since the value of the maximum flow in $N(g)$ is $|\hat{f}| - |g|$. Because g was taken so that $|\hat{f}| - |g| \geq |V|^{2/3}$, the number of stages so far does not exceed $2|V|^{2/3}$. Hence the *total* number of stages S is at most $3|V|^{2/3} + 1$, and the theorem follows. □

When the networks with unit arc capacities have a somewhat more special structure, our algorithm can be shown to perform even better. The special case that we shall attack now requires not only that the network have unit arc capacities, but also that each vertex have either indegree 1 or zero, or outdegree 1 or zero. We shall refer to such networks as *simple*. For example, the network shown in Fig. 9–14 is simple.

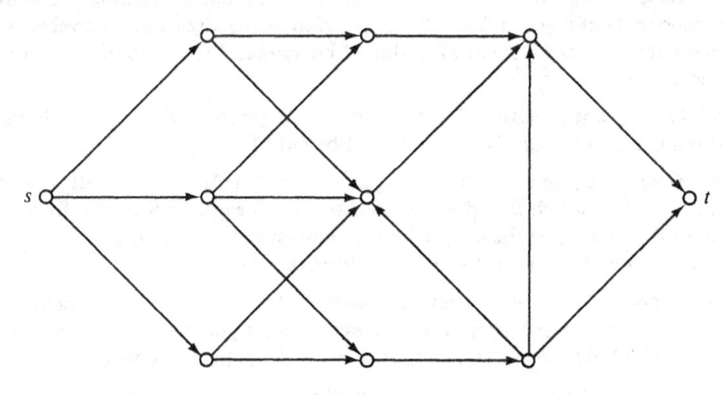

Figure 9–14

Lemma 9.6 *In a simple network N, the distance l between s and t cannot exceed $|V|/|\hat{f}|$, where $|\hat{f}|$ is the value of the maximum flow.*

Proof Let $V_1, V_2, \ldots, V_l$ stand for the sets of vertices of distance 1, 2, . . . , l respectively, from s, as in the proof of Lemma 9.5. We shall argue that $|V_i| \geq |\hat{f}|$. All of the maximum flow $|\hat{f}|$ must go through V_i. However, because the network is simple, each vertex of V_i can handle only one unit of this flow, because either its indegree or its outdegree is 1. Hence $|V_i| \geq |\hat{f}|$ and $|V| \geq \sum_{i=1}^{l} |V_i| \geq l|\hat{f}|$. It follows that $l \leq |V|/|\hat{f}|$. □

Theorem 9.4 *For simple networks, the algorithm of Fig. 9–13 requires at most* $O(|V|^{1/2}|A|)$ *time.*

Proof As in Theorem 9.3, it suffices to show that $S = O(|V|^{1/2})$. Again, if $|\hat{f}| \leq |V|^{1/2}$, the result follows directly. Otherwise, consider the stage after which the value of the flow exceeds $|\hat{f}| - |V|^{1/2}$ for the first time; let g be the value of the flow at the beginning of this stage. In $N(g)$ the value of the optimal flow is $|V|^{1/2}$ or more, and therefore—by Lemma 9.6—$l \leq |V|^{1/2}$ (notice that if N is a simple network and f a 0-1 flow, then $N(f)$ is also simple; see Problem 10). Thus there have been at most $|V|^{1/2}$ stages so far, and there are at most $|V|^{1/2}$ remaining stages. Hence $S = O(|V|^{1/2})$, and the theorem follows. □

PROBLEMS

1. Describe an algorithm that uses a variant of search for finding the connected components of a graph G. At the conclusion of the algorithm, the vertices of the first connected component of G should be marked "1," those of the second "2," and so on.

2. Write a variant of search that tests whether a graph $G = (V, E)$ is a forest. The running time of your algorithm should be $O(|V|)$.

3. Let $G = (V, E)$ be a graph, and $v_1, u, w \in V$. Let BFS (u), and BFS (w) denote the orders in which u and w are marked in a breadth-first search from v_1, and let $d(u)$ and $d(w)$ be the lengths of the shortest paths from v_1 to u and w, respectively. Show that BFS (u) $\leq$ BFS (w) implies $d(u) \leq d(w)$.

4. A digraph $D = (V, A)$ is *strongly connected* if for every $v, u \in V$ there is a path from v to u and one from u to v. Describe an algorithm that, given a digraph $D = (V, A)$, decides in $O(|A|)$ time whether D is strongly connected.

*5. Give an $O(|A|)$ algorithm that finds the *strongly connected components* of a digraph $D = (V, A)$.

6. Give an $O(|E|)$ algorithm for deciding whether a graph $G = (V, E)$ is bipartite.

7. A graph $G = (V, E)$ is *biconnected* if for any two edges $e_1, e_2 \in E$, there is a circuit of G containing both e_1 and e_2. Give an $O(|E|)$ algorithm for determining whether a given graph $G = (V, E)$ is biconnected.

8. Rewrite depth-first search as a *recursive algorithm* (recall Problem 5 in Chapter 8) without using the array Q.

9. Let $N = (s, t, V, A, b)$ be a network. Before the jth stage of the Ford-Fulkerson algorithm, we have a flow denoted by f_{j-1}. At the jth stage we find an augmenting path p_j in $N(f_{j-1})$ and augment the flow along it by δ_j, where δ_j is the minimum of the augmenting capacities of the arcs in p_j. We call an arc of p_j that has augmenting capacity δ_j a *j-bottleneck*.

   (a) Show that if (u, v) is a j-bottleneck *and* a j'-bottleneck for some $j' > j$, then (v, u) must appear on p_i for some $j < i < j'$.

   Suppose now that at each stage j, p_j is the *shortest* s-t path in $N(f_{j-1})$.

   (b) Show that the distance (that is, length of the shortest path in $N(f_j)$) from s to any $v \in V$ cannot decrease from one stage to the next.

   (c) Suppose that (u, v) is a j-bottleneck *and* a j'-bottleneck for some $j' > j$. Show that the distance from s to t must increase from Stage j to Stage j'.

   (d) Conclude that, under the shortest augmenting path rule, the Ford-Fulkerson algorithm terminates after $O(|V| \cdot |A|)$ augmentations.

10. Show that if N is a simple network and f is a 0-1 flow of N, then $N(f)$ is also simple.

11. Show that there is a polynomial-time algorithm for the following versions of the max-flow problem by *reducing* each version to the original form of the max-flow problem.

   (a) The network has many sources and sinks.

   (b) The network is undirected.

   (c) The nodes, as well as the arcs, have capacities.

   (d) The network is undirected and the nodes have capacities.

   *(e) There are both upper and lower bounds on the value of the flow through each arc.

12. (a) Show that the *s-t connectivity* of a digraph $D = (V, A)$ (see Problem 8 of Chapter 6) can be computed in $O(|V|^{1/2} \cdot |A|)$ time.

   (b) Repeat (a) for an undirected graph (same definition of s-t connectivity).

13. The *connectivity* of a graph $G = (V, E)$ is the minimum s-t connectivity, taken over all pairs $s, t \in V$.

   (a) If $c(G)$ is the connectivity of G, show that $c(G) \leq 2|E|/|V|$.

   *(b) Show that the connectivity of G can be computed by calculating at most $c(G) \cdot |V|$ s-t connectivities.

   (c) Conclude from (a) and (b) that $c(G)$ can be computed in $O(|V|^{1/2} \cdot |E|^2)$ time.

14. (a) Construct networks $N = (s, t, V, A, b)$, for infinitely many values of $|V|$, for which the algorithm of Fig. 9–13 runs in $\Omega(|V|^3)$ time.

   (b) Construct networks N with unit capacities, for infinitely many values of $|V|$, for which the algorithm of Fig. 9–13 runs in $\Omega(|V|^{2/3} \cdot |A|)$ time.

(c) Construct simple networks N with unit capacities, for infinitely many values of $|V|$, for which the algorithm of Fig. 9–13 runs in $\Omega(|V|^{1/2} \cdot |A|)$ time.

NOTES AND REFERENCES

The idea of search in graphs is very natural and thus old; see, for example,

[Be] BERGE, C., *Theory of Graphs and its Applications.* New York: John Wiley & Sons, Inc., 1958.

For more sophisticated applications of depth-first search, see

[Ta1] TARJAN, R. E., "Depth-first Search and Linear Graph Algorithms," *J. SIAM Comp.*, 1, no. 2 (1972), 146–60.

[Ta2] ———, "Finding Dominators in Directed Graphs," *J. SIAM Comp.*, 3 (1974), 62–89.

[HT1] HOPCROFT, J. E., and R. E. TARJAN, "Dividing a Graph into Triconnected Components," *J. SIAM Comp.*, 2 (1973), 135–58.

[HT2] HOPCROFT, J. E., and R. E. TARJAN, "Efficient Planarity Testing," *J. ACM*, 21 (1974), 549–58.

Problems 4, 5, and 7 are from [Ta1].

The first polynomial-time algorithm for max-flow was the modification of the Ford-Fulkerson labeling algorithm suggested in Problem 9 and is due to

[EK] EDMONDS, J., and R. M. KARP, "Theoretical Improvements in Algorithmic Efficiency for Network Flow Problems," *J. ACM*, 19, no. 2 (1972), 248–64.

Independently, E. A. Dinits discovered a faster algorithm by exploiting the idea of augmenting along many paths simultaneously.

[Di] DINITS, E. A., "Algorithm for Solution of a Problem of Maximal Flow in a Network with Power Estimation," *Soviet Math. Dokl.*, 11 (1970), 1277–80.

(The improbable last two words in the title are most likely a bad translation for "complexity analysis.") Several improved algorithms followed:

[Ka] KARZANOV, A. V., "Determining the Maximal Flow in a Network with the Method of Preflows," *Soviet Math. Dokl.*, 15 (1974), 434–37.

[Ch] CHERKASKI, B. V., "Algorithm of Construction of Maximal Flow in Networks with Complexity $O(|V|^2\sqrt{|E|})$ Operations," *Math. Methods of Solutions of Economic Problems*, 7 (1977), 117–25 (in Russian).

[Ga] GALIL, Z., "A New Algorithm for the Maximal Flow Problem," *Proc. 19th Symp. on Foundations of Computer Science*, IEEE (October 1978), 231–45.

[GN] GALIL, Z., and A. NAAMAD, "Network Flow and Generalized Path Compression," *Proc. 11th Annual ACM Symp. on Theory of Computing*, ACM (May 1979), 13–26.

[Sh] SHILOACH, Y., "An $O(n \cdot I \log^2 I)$ Maximum-Flow Algorithm," Tech. Report STAN-CS-78-802, Computer Science Dept., Stanford University, 1978.

The fastest known algorithm for *sparse* networks is due to D. Sleator and R. E. Tarjan; see

[Sl] SLEATOR, D. D., (unpublished Ph.D. dissertation, Stanford University, 1980).

The running times of these algorithms are tabulated below.

Reference	Complexity
[EK]	$O(\|V\|\|A\|^2)$
[Di]	$O(\|V\|^2\|A\|)$
[Ka], [MKM]	$O(\|V\|^3)$
[Ch]	$O(\|V\|^2\|A\|^{1/2})$
[Ga]	$O(\|V\|^{5/3}\|A\|^{2/3})$
[GN], [Sh]	$O(\|V\|\|A\|\log^2\|V\|)$
[Sl]	$O(\|V\|\|A\|\log\|V\|)$

For dense networks, $O(|V|^3)$ is the best bound known. The algorithm of Fig. 9–13 is that of [Ka], as simplified considerably in

[MKM] MALHOTRA, V. M., M. P. KUMAR, and S. N. MAHESHWARI, "An $O(|V|^3)$ Algorithm for Finding Maximum Flows in Networks," *Inf. Proc. Letters*, 7, no. 6 (October 1978), 277–78.

In the case that the network is planar, faster algorithms exist; see

[IS] ITAI, A., and Y. SHILOACH, "Maximum Flow in Planar Networks," *J. SIAM Comp.*, 8, no. 2 (1979), 135–50.

Theorems 9.3 and 9.4, and Problems 12 and 13 are from

[ET] EVEN, S., and R. E. TARJAN, "Network Flow and Testing Graph Connectivity," *J. SIAM Comp.*, 4, no. 4 (1975), 507–18.

10

Algorithms for Matching

A *matching* in a graph is a set of edges, no two of which share a node. Given a graph, it is a well-known problem to find a matching that has as many edges as possible; in another version we are also given weights for the edges, and the challenge is to find the matching that has the largest total weight. Both problems have attracted the interest of researchers during the last two decades. They can be stated simply, have intuitive appeal, and have many applications. Furthermore, there are some very interesting algorithms that solve these problems in an efficient manner. These algorithms can be considered as further instances of the concept of *augmentation*, familiar to us from the solutions to the max-flow problem discussed in Chapters 6 and 9. However, in the case of matching, detecting and performing augmentations efficiently can become extremely subtle. In both the unweighted and the weighted versions of the matching problem, all these issues are simplified considerably when the underlying graph is *bipartite*. In this chapter and the next, we present and analyze algorithms for the problems of matching and weighted matching. In both problems we handle the bipartite case first, because this case, being considerably simpler in structure, helps illustrate the basic ideas involved in solving the general problem.

10.1
The Matching Problem

A *matching* M of a graph $G = (V, E)$ is a subset of the edges with the property that no two edges of M share the same node. In the graph of Fig. 10–1, for example, $M_1 = \{[v_2, v_3], [v_4, v_5], [v_6, v_8], [v_7, v_{10}]\}$ (shown as heavy lines in Fig. 10–1) and $M_2 = \{[v_1, v_2], [v_3, v_5], [v_4, v_7], [v_6, v_8], [v_9, v_{10}]\}$ are matchings. In fact, M_2 is a *maximum* matching, since a matching of G obviously can never have more than $|V|/2 = 5$ edges. Given a graph $G = (V, E)$, the *matching problem* is to find a maximum matching M of G. When the cardinality of a matching is $\lfloor |V|/2 \rfloor$, the largest possible in a graph with $|V|$ nodes, we say the matching is *complete*, or *perfect*.

Let us consider a graph G together with a fixed matching M of G. Edges in M are called *matched* edges; the other edges are *free*. If $[v, u]$ is a matched edge, then u is the *mate* of v. Nodes that are not incident upon any matched edge are called *exposed*; the remaining nodes are *matched*. A path $p = [u_1, u_2, \ldots u_k]$ is called *alternating* if the edges $[u_1, u_2], [u_3, u_4], \ldots, [u_{2j-1}, u_{2j}], \ldots$ are free, whereas $[u_2, u_3], [u_4, u_5], \ldots, [u_{2j}, u_{2j+1}], \ldots$ are matched. Vertices that lie on an alternating path starting with an exposed vertex and have odd rank on this path are called *outer*. Those with even rank are called *inner*. Referring to the graph G and matching M_1 of Figure 1, we see that $p_1 = [v_1, v_2, v_3, v_5, v_4, v_8]$ and $p_2 = [v_1, v_4, v_5, v_6, v_8, v_7, v_{10}, v_9]$ are alternating paths. The existence of these alternating paths establishes that the vertices v_1, v_3, v_4 and v_5, v_8, v_{10} are outer, because each occupies an odd position in one of these paths, and v_1 is exposed. These are not all the outer vertices of G with respect to M_1: v_9 and v_6 are also outer, because $p_3 = [v_9, v_8, v_6]$ is also an alternating path. An alternating path $p = [u_1, u_2, \ldots, u_k]$ is called *augmenting* if both u_1 and u_k are exposed vertices. In Fig. 10–1, $p_2 = [v_1, v_4, v_5, v_6, v_8, v_7, v_{10}, v_9]$ is an augmenting path. The significance of augmenting paths for the matching problem is due to the following fact.

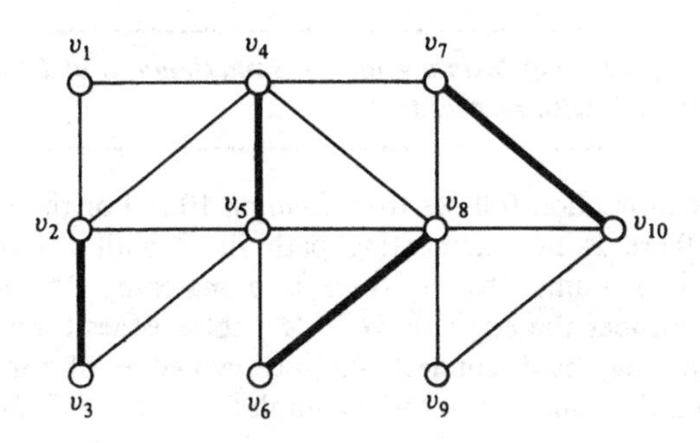

Figure 10–1

Lemma 10.1 *Let P be the set of edges on an augmenting path $p = [u_1, u_2, \ldots, u_{2k}]$ in a graph G with respect to the matching M. Then $M' = M \oplus P$† is a matching of cardinality $|M| + 1$.*

Proof Let us first show that $M \oplus P$ is a matching; that is, no two edges in $M \oplus P$ share the same node of G. Suppose that it is the case that two edges e, e' in $M \oplus P$ are incident upon the same node. Since $M \oplus P = (M - P) \cup (P - M)$ we have three cases:

1. $e, e' \in M - P$
2. $e, e' \in P - M$
3. $e \in M - P, \quad e' \in P - M$

In Case 1, we have two edges in M sharing the same node, a contradiction. In Case 2, we note that the edges in $P - M$ are the edges of the form $[u_{2j-1}, u_{2j}]$ and hence two of them cannot be incident upon the same node. For Case 3, suppose that an edge $e' = [u_{2j-1}, u_{2j}]$ in $P - M$ has a node in common with another edge $e \in M - P$. Assume, with no loss of generality, that this node is u_{2j}. But u_{2j} is a node of the edge $e'' = [u_{2j}, u_{2j+1}] \in M$, and hence two edges e'' and e of M have a common node, a contradiction. It follows that M' is a matching. Now P contains $2k - 1$ edges; k of them are free ($[u_1, u_2], [u_3, u_4], \ldots, [u_{2k-1}, u_{2k}]$) and $k - 1$ belong to M. Hence, $M' = M \oplus P$ has $|M| + 1$ edges. □

For example, the augmenting path $p_2 = [v_1, v_4, v_5, v_6, v_8, v_7, v_{10}, v_9]$ with respect to the matching M_1 in Fig. 10–1 can be used to augment M_1 to $M_3 = M_1 \oplus P_2 = \{[v_1, v_4], [v_2, v_3], [v_5, v_6], [v_7, v_8], [v_9, v_{10}]\}$. Because M_3 is maximum (it has $|V|/2 = 5$ edges), it makes no sense to look for augmenting paths in G with respect to M_3; there can be no augmenting paths with respect to a maximum matching, since such a path could be used, by Lemma 10.1, to augment the matching. It turns out that the converse is true as well.

Theorem 10.1 *A matching M in a graph G is maximum if and only if there is no augmenting path in G with respect to M.*

Proof One direction follows from Lemma 10.1. For the other direction, suppose that there is no augmenting path in G with respect to M, and yet M is not maximum. That is, there is a matching M' of G such that $|M'| > |M|$. Consider the edges in $M \oplus M'$; these edges form a subgraph of G (this subgraph may be disconnected). Since two edges of a matching cannot be incident upon the same vertex, the subgraph $G' = (V, M \oplus M')$ has a special

†If S, T are sets, then $S \oplus T$ denotes the *symmetric difference* of S and T, defined as $S \oplus T = (S - T) \cup (T - S)$.

structure—all vertices have degree 2 or less. If the degree of a vertex is 2, one of the edges is in M and the other in M'. Thus all connected components of G' will be either paths or circuits of even length. In all circuits, we have the same number of edges in M as in M'. Because $|M'| > |M|$, it must be the case that in one of the paths we have more edges from M' than from M, and hence this path is an augmenting path. However, this contradicts our assumption that there are no augmenting paths in G with respect to M, and the theorem is proved. □

Theorem 10.1 is a characterization of maximum matchings, not unlike the characterization of maximum flows in terms of augmenting paths (Theorem 6.3). One is thus tempted to devise the analog of the max-flow algorithm for matching: Start with any matching (for example, the empty one), and repeatedly discover augmenting paths. Indeed, all known algorithms for matching are based on exactly this idea. Nevertheless, the details of such algorithms are quite involved. A case where these ideas are applicable in a more or less straightforward manner is the matching problem in bipartite graphs. The analogy to the max-flow problem is clearer there as well.

10.2 A Bipartite Matching Algorithm

Because the matching problem for bipartite graphs is a special case of the matching problem discussed in the previous section (and consequently Theorem 10.1 applies), we shall solve it by repeatedly discovering an augmenting path p with respect to the current matching M and augmenting the current matching to $M \oplus P$.

How can we organize the search for augmenting paths with respect to a matching M of a bipartite graph $B = (V, U, E)$ in a systematic and efficient way? Consider the graph B and matching M of Fig. 10–2(a). Naturally, a search for augmenting paths must start by constructing alternating paths from the exposed vertices. Because an augmenting path must have one endpoint in V and the other in U, it is no loss of generality to start growing alternating paths only from exposed vertices of V (v_2 in our example). In fact, we may search for all possible alternating paths from v_2 simultaneously in a breadth-first manner. Starting from v_2, we may search for alternating paths by considering all vertices adjacent to v_2, namely u_2 and u_6 (see Fig. 10–2(b)). Since v_2 is exposed, all the adjacent corresponding edges are free. By the definition of an alternating path, we now have to look for matched edges emanating from u_2 and u_6. This is a straightforward step, since all nodes have at most one mate. Naturally, if either u_2 or u_6 were exposed, our task would be over: We would have discovered an augmenting path. However, this is not the case, and thus we add the nodes v_3 and v_5 to our set of alternating paths. By our construction, v_3 and v_5 are *outer* vertices. We continue growing alternating paths from v_3 and v_5. Note that since the node

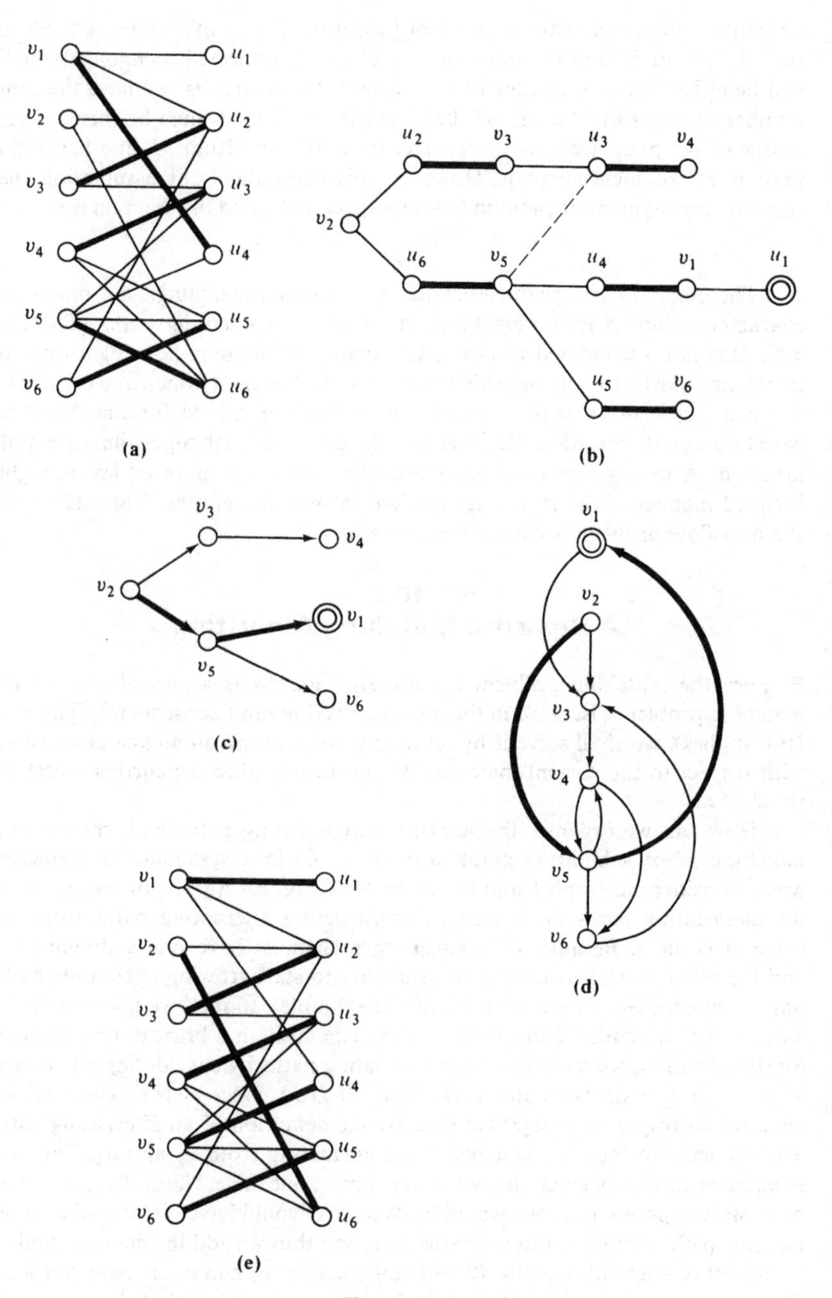

v_1
u_1
v_2
u_2
v_3
u_3
v_4
u_4
v_5
u_5
v_6
u_6
(a)
u_2
v_3
u_3
v_4
v_2
u_6
v_5
u_4
v_1
u_1
u_5
v_6
(b)
v_3
v_4
v_2
v_1
v_5
v_6
(c)
v_1
v_2
v_3
v_4
v_5
v_6
(d)
v_1
u_1
v_2
u_2
v_3
u_3
v_4
u_4
v_5
u_5
v_6
u_6
(e)

Figure 10–2

u_3 is reached from v_3 before the scanning of v_5, we omit the edge $[v_5, u_3]$ (the dotted line in Fig. 10–2(b)). Obviously, by so doing we are missing only redundant augmenting paths. We discover the new outer vertices v_4, v_1, and v_6 (see Fig. 10–2(b)) and finally we notice that the outer vertex v_1 is adjacent to the exposed vertex u_1. We have thus discovered an augmenting path, and we may augment M by using it (see Fig. 10–2(e)).

The process of searching for augmenting paths is thus quite reminiscent of breadth-first search. However, if we look back at the construction of the alternating paths of Fig. 10–2(b), we may notice that this breadth-first search has a special structure. The searches from odd-numbered levels (v_2 is at the zeroth level) are rather trivial, because the next vertex is always the mate of the current vertex. We can thus simplify this search technique by *ignoring* odd-numbered levels and going directly from outer vertices to new outer vertices; in our example, the desired search would then proceed as shown in Fig. 10–2(c). Obviously, this corresponds to searching a digraph (V, A), where $(v_1, v_2) \in A$ if and only if v_2 can be the next outer vertex after v_1 in an augmenting path; that is, v_1 is adjacent to the mate of v_2. The node set of this *auxiliary* digraph is V, since obviously only vertices of V may become outer vertices in alternating paths starting from V. The auxiliary graph corresponding to our example of Fig. 10–2(a) is shown in Fig. 10–2(d). The search for augmenting paths shown in Fig. 10–2(c) can easily be seen to be a breadth-first search of the auxiliary graph starting from the node v_2.

Our algorithm uses two arrays, *mate* and *exposed*, besides the array *label* used for search. The array *mate* has $|V| + |U|$ entries and is our way of representing the current matching. For $v \in V$, *exposed*$[v]$ is a node of U that is exposed and is adjacent to v; if no such node exists, *exposed*$[v] = 0$. Obviously, if a node $v \in V$ with *exposed*$[v] \neq 0$ is encountered in our search, we have discovered an augmenting path. The procedure augment is recursive (recall the procedure path of Fig. 9–6). In our example of Figure 10–2, we first call augment(v_1); *exposed*$[v_1] = u_1$, and so augment changes the matching of v_1 from u_4 to u_1 and then calls augment(v_5). The matching of v_5 is changed from u_6 to u_4, and then augment(v_2) is called. Now, *label*$[v_2] = 0$, and thus augment matches v_2 to u_6 and stops. The full algorithm is shown in Fig. 10–3.

Theorem 10.2 *The algorithm of Fig. 10–3 correctly solves the matching problem for a bipartite graph* $B = (V, U, E)$ *in* $O(\min(|V|, |U|) \cdot |E|)$ *time.*

Proof The algorithm terminates if there is no path in the auxiliary graph from an exposed node of V to a target vertex. By the construction of the auxiliary graph, this means that there is no augmenting path in B with respect to the current matching, and hence—by Theorem 10.1—the current matching is optimal.

For the time bound, we note that a matching in B can have no more than

```
BIPARTITE MATCHING ALGORITHM
Input:  A bipartite graph B=(V,U,E).
Output: The maximum matching of B, represented by the array mate.
begin
   for all v ∈ V∪U do mate[v]:=0; (comment: initialize)
stage: begin
   for all v ∈ V do exposed[v]:=0;
   A:=∅; (comment: begin construction of the auxiliary graph (V,A))
   for all [v,u] ∈ E do
   if mate[u]=0 then exposed[v]:=u else
      if mate[u]≠v then A:=A ∪ (v, mate[u]);

   ┌-----------------------------------------------------------------┐
   ¦ Q:=∅;                                                           ¦
   ¦ for all v ∈ V do if mate[v]=0 then Q:=Q∪{v}, label[v]:=0;       ¦
   └-----------------------------------------------------------------┘

   while Q≠∅ do
      begin
      let v be a node in Q;
      remove v from Q;
      ┌-----------------------------------------------------┐
      ¦ if exposed[v]≠0 then augment(v), go to stage;       ¦
      ¦ else                                                ¦
      ¦   for all unlabeled v' such that (v,v') ∈ A do      ¦
      ¦   label[v']:=v, Q:=Q∪{v'};                          ¦
      └-----------------------------------------------------┘

      end
   end
end

procedure augment(v)
   if label[v]=0 then mate[v]:=exposed[v],
                           mate[exposed[v]]:=v;
   else begin
         exposed[label[v]]:=mate[v];
         mate[v]:=exposed[v];
         mate[exposed[v]]:=v;
         augment(label[v])
         end
```

Figure 10–3 The bipartite matching algorithm.

min ($|V|, |U|$) edges. Since each augmentation increases the cardinality of the matching by 1, we can have at most min ($|V|, |U|$) stages. It remains to show that the complexity of each stage is $O(|E|)$. The construction of the auxiliary graph and the calculation of the array *exposed* requires $O(|E|)$ time, because $|E|$ edges have to be considered. Looking for the desired directed path in the auxiliary graph is done in $O(|A|) = O(|E|)$ time; finally, augmenting the matching requires $O(|V|)$ time. The theorem follows. ☐

10.3 Bipartite Matching and Network Flow

It turns out that we can improve upon the $O(\min(|V|, |U|) \cdot |E|)$ algorithm of the previous section by relating the matching problem for bipartite graphs to the max-flow problem for networks. In particular, we are going to *reduce* the bipartite matching problem to the max-flow problem for simple networks. By this, we mean that we shall show how to solve the bipartite matching problem efficiently by making use of *any* algorithm that solves the max-flow problem efficiently.

Given any bipartite graph $B = (V, U, E)$, we define the following network with unit arc capacities—$N(B) = (s, t, W, A)$, where s and t are two new nodes; W is the union of $\{s, t\}$, V, and U; and A is a set of arcs consisting of three categories:

1. The arcs (s, v) for all $v \in V$;
2. The arcs (u, t) for all $u \in U$;
3. The arcs (v, u) for all $v \in V$ and $u \in U$ such that $[v, u] \in E$.

The network $N(B)$ for the bipartite graph B of Fig. 10–4(a) is shown in Fig. 10–4(b). Let us first notice that for any bipartite graph B, $N(B)$ is a *simple* network. This is so because all nodes in V have indegree 1 in $N(B)$ and all nodes in U have outdegree 1.

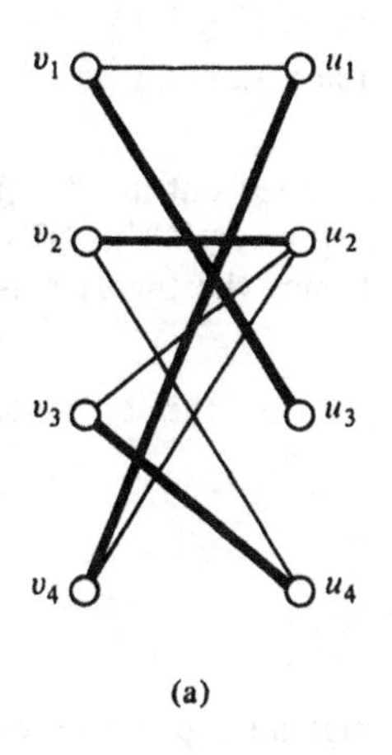

(a)

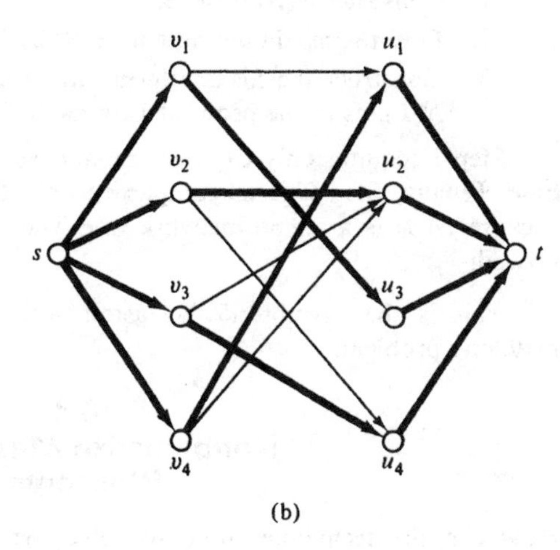

(b)

Figure 10–4

Lemma 10.2 *The cardinality of the maximum matching in a bipartite graph B equals the value of the maximum s-t flow in $N(B)$.*

Proof Given any matching M of B, we can construct a feasible flow in $N(B)$ as follows: For each matched node v of V—that is, each node v for which $[v, u] \in M$ for some $u \in U$—we have a unit flow through (s, v). For each matched node u of U, we have a unit flow through (u, t). Finally, for each edge $[v, u] \in M$, we add a unit flow through (v, u) (see Fig. 10–4). Obviously, the resulting flow is feasible and its value equals $|M|$. Conversely, given a maximum flow f in $N(B)$, we know that there is an equivalent flow with integral (0-1 in our case) flow values (for example, the one discovered by the algorithm of Figure 9–13). Starting from such a flow, we construct a matching M containing all edges $[v, u] \in E$ for which $f(v, u) = 1$. By the construction of $N(B)$, this is bound to be a legitimate matching, because if two edges in E share a node v, then one of the corresponding arcs of $N(B)$ must have zero flow; otherwise, the capacity of the arc (s, v)—or (v, t) if $v \in U$—will be violated. □

Notice that the proof of Lemma 10.2 is constructive; it allows us to find the maximum matching of B given an integral maximum flow in $N(B)$ in $O(|V|)$ time.

Theorem 10.3 *We can solve the matching problem for bipartite graphs in $O(|V|^{1/2} \cdot |E|)$ time.*

Proof The algorithm is the following:

1. Construct $N(B)$ from B.
2. Find the maximum flow in $N(B)$ by the algorithm of Fig. 9–13.
3. Construct the maximum matching in B from the maximum flow in $N(B)$, as in the proof of Lemma 10.2.

Step 1 requires only $O(|E|)$ time, and Step 3 can be carried out in $O(|V|)$ time. Finally, Step 2 is of complexity only $O(|V|^{1/2} \cdot |A|) = O(|V|^{1/2} \cdot |E|)$, because $N(B)$ is a simple network (see Theorem 9.4). Hence the theorem is proved. □

This is the asymptotically fastest algorithm known for the bipartite matching problem.

10.4 Nonbipartite Matching: Blossoms

How can the techniques used to solve the bipartite matching problem be extended to nonbipartite graphs? The reduction to max-flow of the previous section does not seem to carry over (for a reduction of nonbipartite matching to an *extension* of the max-flow problem, which, unfortunately, is not any easier to solve than the matching problem with which we started, see Problems 6 and 7). Theorem 10.1, however, holds for general graphs, and hence the approach taken in Sec. 10.2 (start with the empty matching and repeatedly discover augmenting paths and augment along these) can be extended to the nonbipartite matching problem. Unfortunately, the lack of the bipartite structure makes the task of finding augmenting paths far more difficult.

Let us examine how our auxiliary digraph technique must be modified to work for general graphs. First, in the bipartite case, we knew that vertices of U could not be outer in any alternating path starting from V. Consequently, our auxiliary graph involved only vertices from V. In the general nonbipartite case there is no such restriction, and hence the auxiliary graph must contain all vertices. The arcs of the auxiliary digraph still represent pertinent information: An arc (v, u) means that u may be the next outer vertex in an alternating path on which v is outer. Hence augmenting paths correspond to paths on the auxiliary graph from an exposed vertex to a target node (that is, a node v with $exposed[v] \neq 0$), exactly as they did in the bipartite case.

What makes the general matching problem considerably more involved is that *the converse correspondence may not hold.* That is, there may be paths in the auxiliary graph leading from exposed nodes to target nodes that do not correspond to augmenting paths relative to the current matching. Let us see this by way of an example. Consider the graph G of Fig. 10–5(a) and the matching M shown. Here M is not maximum, because there is an augmenting path with respect to M in G, namely, $p = [v_1, v_2, v_3, v_4, v_5, v_6, v_7, v_8, v_9, v_{10}]$. Naturally, p corresponds to a path $p' = (v_1, v_3, v_5, v_7, v_9)$ of the auxiliary digraph relative to M, shown in Fig. 10–5(b). Nevertheless, there are paths in the auxiliary digraph that *do not* correspond to legitimate augmenting paths. For example, consider the path $q' = (v_1, v_8, v_6, v_5, v_7, v_9)$. There is no augmenting path in G with the sequence of vertices of q' as outer vertices. If we try the correspondence $(u_1, u_2, \ldots, u_k) \leftrightarrow [u_1, mate[u_2], u_2, \ldots, mate[u_k], u_k, exposed[u_k]]$—by which we can recover the augmenting path in the bipartite case—we get the *walk* $q = [v_1, v_9, v_8, v_7, v_6, v_4, v_5, v_6, v_7, v_8, v_9, v_{10}]$. Naturally q is not an augmenting path; in fact, it is not a path at all. Nodes and edges are repeated in it, such as the matched edges $[v_8, v_9]$ and $[v_6, v_7]$. These repetitions are not reflected in q' because each time different endpoints of the edges are used as "outer." Also, $M \oplus Q$ (where Q is the set of edges in the walk q) is not a matching.

Since our auxiliary digraph technique worked successfully for bipartite graphs and odd circuits are the only feature that bipartite graphs lack (recall Proposition 1.1) we are led to suspect that it is the odd circuits that cause this malfunctioning. And this is exactly what is happening. Until it meets the odd circuit $c = [v_6, v_4, v_5, v_6]$ (Fig. 10–5), q is a perfectly legitimate alternating path. At this point q traverses this circuit and starts "folding" on itself, with the disastrous consequences that we observed above.

Not all odd circuits can cause such undesirable behavior. For example, the

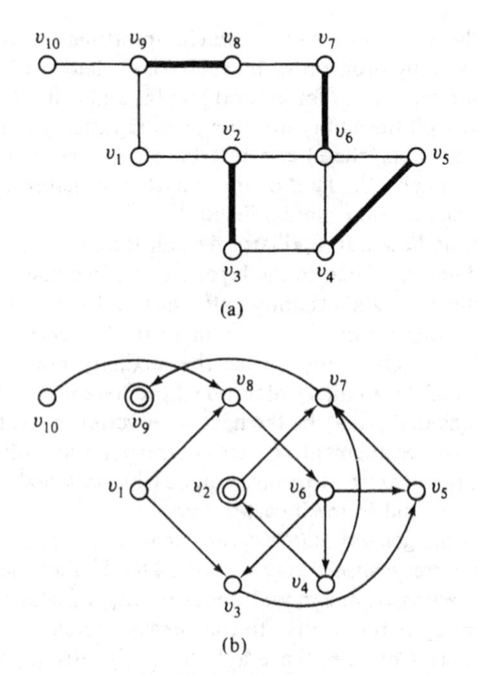

Figure 10–5

circuit of nine nodes in the graph shown in Fig. 10–6(a) does not fool the auxiliary digraph algorithm. This is because, intuitively, this circuit is too "sparse" in matched edges, and hence the augmenting path cannot traverse it and fold. The suspect odd circuits are those that are as dense in matched edges as possible; that is, circuits with $2k + 1$ nodes that have k matched edges (Fig. 10–6(b)). Such circuits are called *blossoms* and play an important role in the theory of matching. Notice that, like most of the matching terminology, a blossom is defined only in the context of some fixed matching.

We have thus located the cause of the failure of the auxiliary digraph technique: It is the presence of blossoms. If a graph has no blossoms with respect to the current matching, then it is effectively bipartite (even though it may have nonblossom odd circuits like the graph of Fig. 10–6(a)) and the search for augmenting paths can be done as in the bipartite case. Consequently, we have to modify our algorithm so that (1) it can sense the existence of blossoms, and (2) it can go on finding valid augmenting paths, even in the presence of blossoms. To keep our algorithm simple, we shall search for augmenting paths from one exposed vertex at a time, as opposed to all exposed vertices simultaneously, as we did in our bipartite matching algorithm.

To detect the existence of blossoms is not particularly hard. A blossom

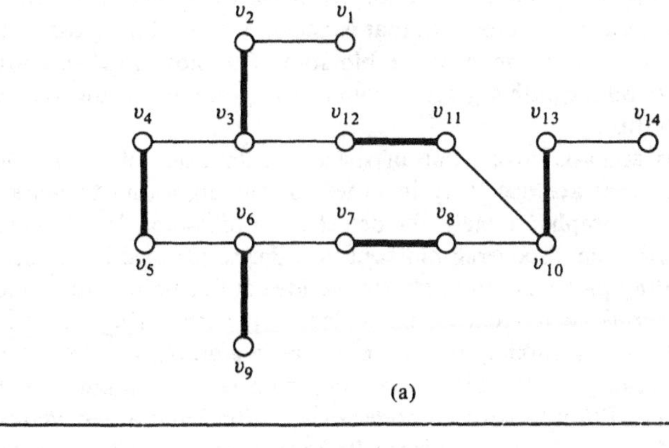

(a)

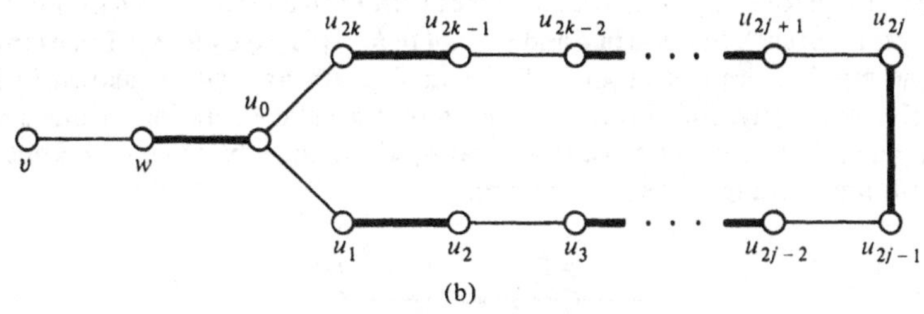

(b)

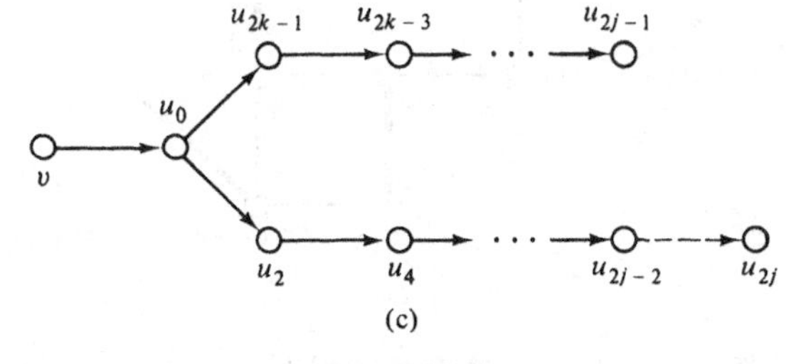

(c)

Figure 10–6

becomes relevant to our algorithm when it is first discovered; that is, when all of its vertices are either outer or mates of outer vertices. This situation is depicted in Fig. 10–6(c) for the blossom shown in Fig. 10–6(b). At this point, we first notice something abnormal. The next vertex in our search, u_{2j} (see the dotted arc in Fig. 10–6(c)), is the mate of vertex u_{2j-1}, which is already outer. This means that our search attempts to use the matched edge $[u_{2j}, u_{2j-1}]$ twice, and the existence of a blossom is thus established. To actually *find* the blossom is not much harder: We backtrack from both u_{2j} and u_{2j-1} to find the two paths leading from the exposed node v to these nodes. We find the *latest* outer vertex that these paths have in common. This node, u_0 in our example of Figs. 10–6(b) and 10–6(c), is the *basis* of the blossom; that is, the only node of the blossom that is not matched with another node of the blossom. The blossom then consists of all outer vertices on the path segments from u_0 to u_{2j} and u_{2j-1}, together with the mates of all but u_0.

The previous analysis covers one of the two extensions of the auxiliary digraph algorithm that are necessary in order for the algorithm to work for general, nonbipartite graphs: namely, the detection of blossoms. It still remains to see how to handle the discovered blossom in order to proceed in our search for valid augmenting paths. A strikingly simple idea at this point is to get rid of the blossom by *shrinking* it—that is, by replacing it with a single node. To formalize the notion of shrinking, if b is a blossom in a graph $G = (V, E)$ with respect to the matching M, the graph resulting from G by shrinking b is $G/b = (V/b, E/b)$, where V/b is V with all nodes of b omitted and a new vertex v_b added to replace b; E/b is E with all edges with both endpoints in b omitted and all edges $[v, u]$ such that u is in b and v is not in b, replaced by $[v, v_b]$. For example, for the graph G and blossom b shown in Fig. 10–7(a), G/b is shown in Fig. 10–7(b). Naturally, for shrinking to represent a valid operation in our search for augmenting paths, we must show that by shrinking a blossom we do not add or omit augmenting paths in our graph.

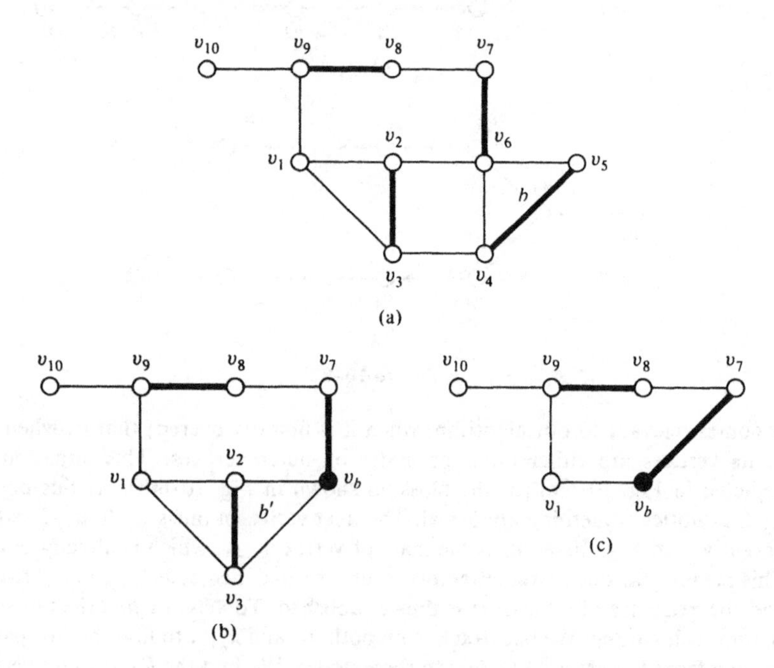

Figure 10–7

Lemma 10.3 *Suppose that while searching for an augmenting path from an exposed node u in a graph G with respect to a matching M, we discovered a blossom b. Then there is an alternating path from u to any node of b ending with a matched edge.*

Proof Since b was discovered from u, there is an alternating path u to the basis of b, as shown in Figure 10–8. Then a vertex v of b is reachable from u by two alternating paths (see Fig. 10–8). One of these must end with a matched edge. □

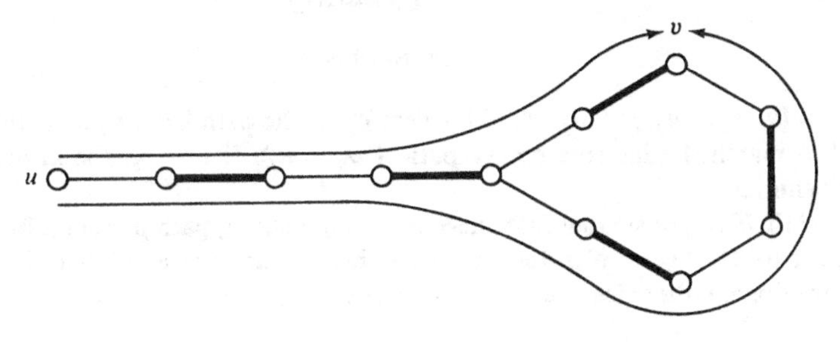

Figure 10–8

Suppose that G is a graph, M a matching, and b a blossom of G with respect to M. Then M/b is the matching that results from M if we omit all matched edges of b and change any matched edge of the form $[v, u]$ with u a node of b to $[v, v_b]$.

Theorem 10.4 *Suppose that, while searching for an augmenting path from a node u of a graph G with respect to a matching M, we discover a blossom b. Then there is an augmenting path from u in G with respect to M iff there is one from u (v_b if u is the basis of b) in G/b with respect to M/b.*

Proof *If* Suppose that there is an augmenting path p from u in G/b with respect to M/b. If p does not pass through v_b, then p itself is an augmenting path in G. If p passes through v_b, we may write $p = [u, p', w, v_b, w', p'']$†, where p and p'' are paths. Either $[w, v_b]$ or $[w', v_b]$ is matched. Suppose that $[w, v_b]$ is matched (the other case is similar). Then $[w, u_0] \in M$ and $[w', u_j] \in E - M$ for the basis u_0 and another node u_j of b (Figure 10–9). The augmenting path from u in G

†In our notation for paths, we use expressions such as $[u, p, w, p', v]$, where u, w, and v are nodes, to denote (in the above example) a path p with endpoints u and w, followed by a path p' with endpoints w and v. Also, $[p, q]$ is the concatenation of two paths, with no endpoints designated. If p is a path, p^R denotes its *reverse*.

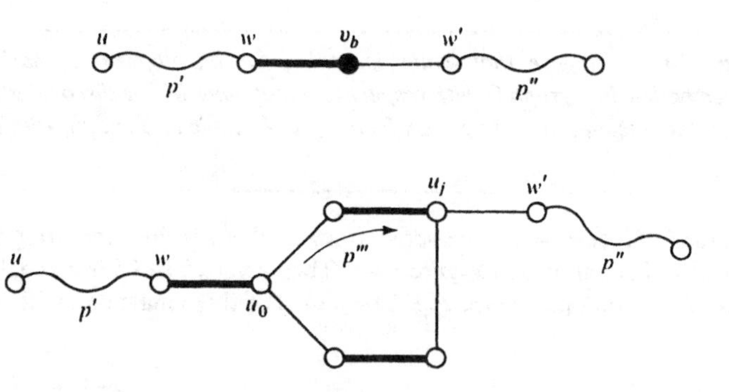

Figure 10–9

is thus $[u, p', w, u_0, p''', u_j, w', p'']$, where p''' is the path from u_0 to u_j that ends with a matched edge (the empty path if $u_0 = u_j$). If $u = v_b$, the argument is very similar.

Only if Suppose now that there is an augmenting path p from u in G with respect to M. Again, if p does not pass through any vertex of the blossom b, we are done. Otherwise, we have two cases.

Case 1 Suppose that p enters b at the basis u_0 with a matched edge (see Fig. 10–10(a)). Let u_j be the last node of b in the path. If $u_j = u_0$, we are done, because p is also augmenting in G/b, with u_0 replaced by v_b. Otherwise, $p = [u, p', u_0, p'', u_j, w, p''']$ and $[u_j, w] \notin M$. Then $[u, p', v_b, w, p''']$ is an augmenting path in G/b.

Case 2 Suppose that p enters b with a free edge. We distinguish between two subcases.

Subcase (a) Suppose that p leaves b through its basis by a matched edge. This case is similar to Case 1.

Subcase (b) In this situation, p leaves b through a free edge. Thus $p = [u, p', u_i, p'', u_j, w, p''']$ (see Fig. 10–10(b)). By Lemma 10.3, there is an alternating path q, from u to the basis u_0 of b, ending with a matched edge. We must consider three possibilities.

1. First suppose that p''' and q do not intersect. Then $[u, q, v_b, w, p''']$ is an augmenting path.
2. Suppose now that p''' and q intersect (see Fig. 10–10(c)). Thus $q = [u, q', x, q'', u_0]$ and $p''' = [p^{(4)}, x, p^{(5)}]$ where $p^{(5)}$ does not have a node in common with q. Then $[u, p', v_b, q''^R, x, p^{(5)}]$ is an augmenting path, where we assumed that q'' and p' have no nodes in common.
3. This is the same as (2), except that p' and q'' intersect (see Figs. 10–10(d) and 10–10(e)). Let y be the latest node of q'' which is either on p' or

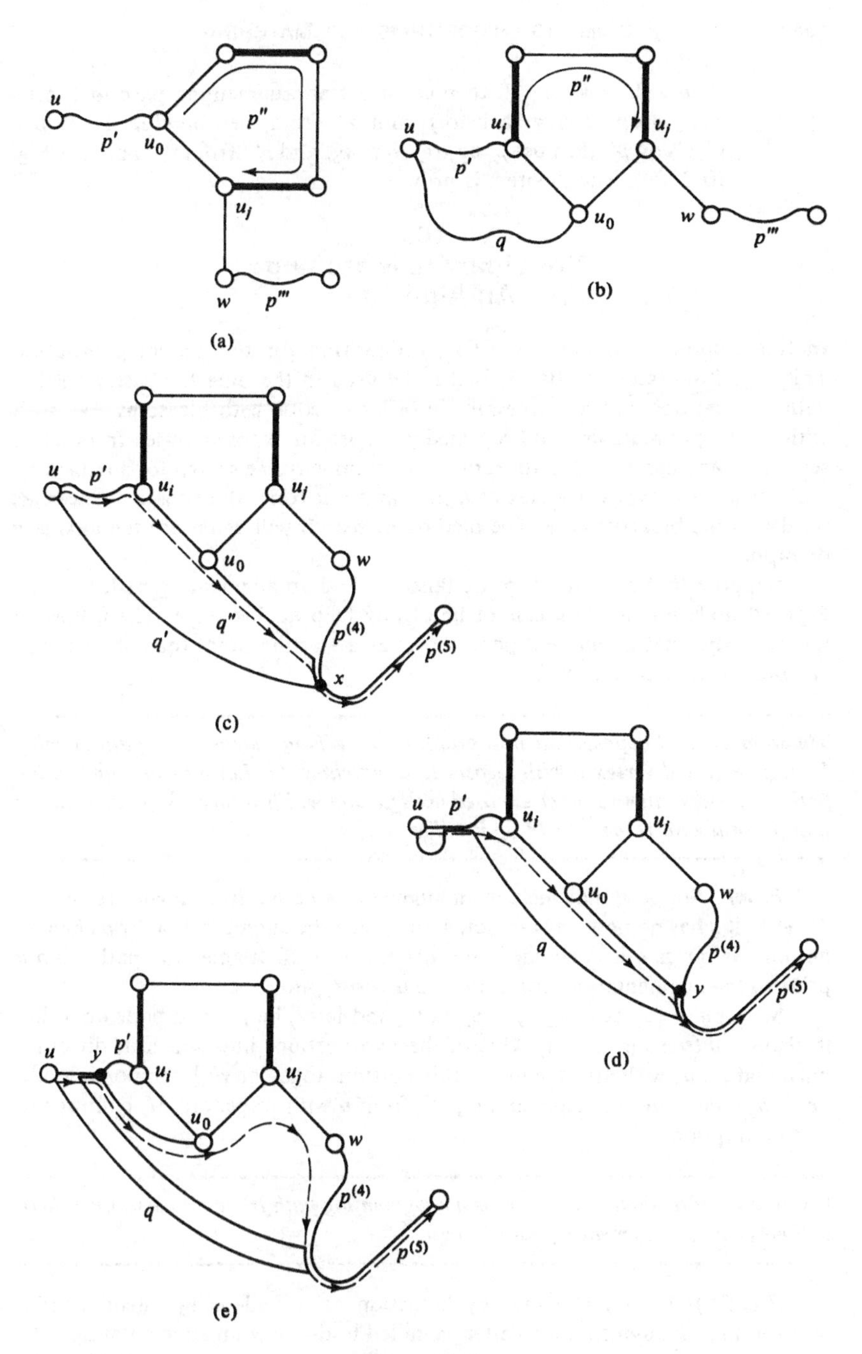
u
p'
u_0
p''
u_j
w
p'''
(a)
p''
u_i
u_j
u
p'
u_0
q
w
p'''
(b)
u
p'
u_i
u_j
u_0
w
q''
q'
p^(4)
p^(5)
x
(c)
u
p'
u_i
u_j
u_0
w
q
p^(4)
y
p^(5)
(d)
u
y
p'
u_i
u_j
u_0
w
p^(4)
q
p^(5)
(e)

Figure 10–10

on p'''. If y is on p''', then construct an augmenting path in G/b by using p up to v_b, q back to y, and p''' from then on (Fig. 10–10(d)). If y is on p', then use p' up to y, q to v_b, and p''' from then on (see Fig. 10–10(e)). The theorem is proved. □

10.5 Nonbipartite Matching: An Algorithm

In this section we present an $O(|V|^4)$ algorithm for the matching problem. This algorithm (see Fig. 10–13) is the one used in the bipartite case, together with the features that are necessary in order to cope with blossoms. We start with the empty matching and repeatedly choose an exposed vertex in order to search for an augmenting path from it. For simplicity, we search for augmenting paths from one exposed vertex at a time and not from all exposed vertices, as we did in the bipartite case. The method of search will again use the auxiliary digraph.

Suppose that at some stage we failed to find an augmenting path from an exposed node u. Clearly u cannot be of any help at this stage. The following theorem says that u will be a poor choice as a starting point for our search in *all* subsequent stages as well.

Theorem 10.5 *Suppose that in a graph G there is no augmenting path starting from an exposed vertex u with respect to a matching M. Let p be an augmenting path with endpoints two other exposed nodes v and w. Then there is no augmenting path from u with respect to $M \oplus P$ either.*

Proof Suppose that there is an augmenting path q from u with respect to $M \oplus P$. If q has no node in common with p, then the augmentation by p changes nothing as far as q is concerned, and hence q was an augmenting path from u prior to the augmentation, contrary to our assumption.

So, let $q = [u_1 \equiv u, u_2, \ldots, u_k \equiv u']$ and let u_j be the first node on q that is also on p (see Fig. 10–11). One of the two portions into which u_j divides p must end on u_j with an edge in M; this portion, together with the portion of q up to u_j constitutes an augmenting path from u with respect to M, contrary to our assumption. □

Corollary *If at some stage there is no augmenting path from a node u, then there will never be an augmenting path from u.*

Proof It is easy to show by induction on k and using Theorem 10.5 that for all k, k augmentations after we failed to discover an augmentating path from u there will still be no augmenting path from u. □

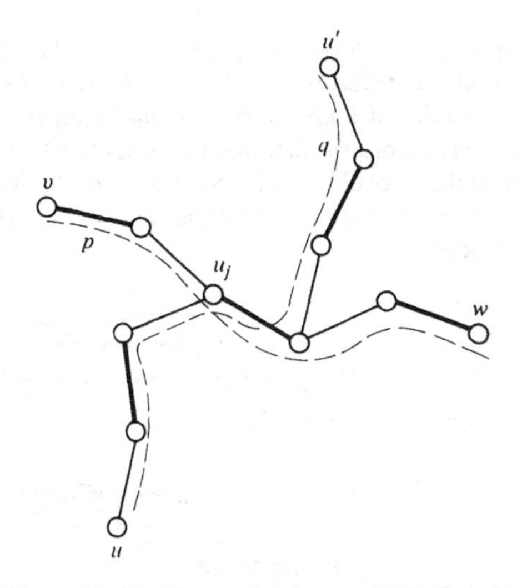

Figure 10–11 The proof of theorem 10.5. Heavy lines are edges in $M \oplus P$.

Therefore, once our search has failed to find an augmenting path from a node u, u will never again be considered by the algorithm as a potential starting point of an augmenting path. In our algorithm we make sure that this is the case by maintaining an array *considered*, initially zero. If a node u is used as the starting point of a search, we set *considered* $[u] = 1$ to signify that we shall never consider u again.

As in the bipartite case, the auxiliary digraph (V, A) is our basic aid for performing our search. Also, the array *exposed* is again used to identify target nodes—that is, nodes that are adjacent to exposed nodes other than the one we are considering. In order to be able to detect blossoms, we use an array *seen* with $|V|$ entries, all initially zero, where *seen*$[v] = 1$ means that v is the mate of an outer vertex; if v is made outer sometime in this search, both endpoints of the edge $[v, mate[v]]$ are outer, and hence a blossom has been discovered. Once the first blossom is discovered, it is shrunk. According to Theorem 10.4, this is a legitimate operation, because it preserves the existence of augmenting paths from the node being considered. Thus we give this current blossom a name b—the next available integer, say, starting from $|V| + 1$, since the first $|V|$ integers are set aside as subscripts of the vertices of V—and the current graph becomes, from G, G/b (Fig. 10–7(b)). Subsequent to the discovery of b, we may discover in our search another blossom b' (see Fig. 10–7(b)). We shrink this one too, to obtain $G/b/b'$ (Fig. 10–7(c)) as our current graph, and so on.

After every such discovery of a blossom b, we must record some useful information (effectively *shrink b*). This is done by the procedure blossom used in Figure 10–13 (not shown explicitly). Blossom does the following.

1. It finds all nodes of b by backtracing the *label* array until the first common node in these two paths—the *basis* of b—is met (Figure 10–12). The nodes of b are exactly the nodes on these two paths and their mates. We record the fact that v belongs to b by setting *blossom*$[v] = b$—notice that v could be a blossom node itself. We also record for future reference the basis of b and the exact cyclic order in which the nodes of b occur.

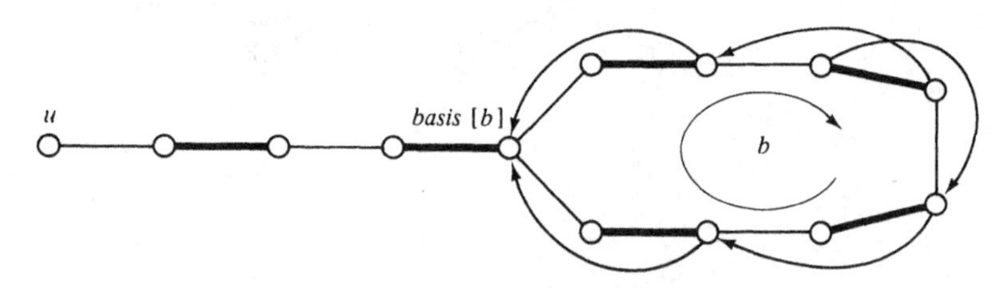

Figure 10–12

After we have identified all nodes of b, we must check whether there is a node u_j among them such that *exposed*$[u_j] \neq 0$. If so, we must augment the matching from v_b, as explained in a few paragraphs.

2. We replace any instance of a node x of b in the auxiliary graph A, the queue Q and the array *label* by the new node v_b. This has the effect of "shrinking" the nodes of b into one node, v_b.

The procedure of searching, discovering, and shrinking blossoms continues until an augmenting path is found in the current graph from u—or from v_b, where v_b is the most recently shrunk blossom that contains u—or until no such path or further blossoms can be found by searching the auxiliary graph.

If we find an augmenting path p in the current graph, we must construct from it an augmenting path in the original graph G. This may be nontrivial, since p can contain several blossom nodes and hence it cannot be used for augmentation as it is. Consider, for example, the augmenting path $p = [v_{10}, v_9, v_8, v_7, v_{b'}, v_1]$ in Fig. 10–7(c). Recovering the augmenting path in the original graph from p can be done by repeated applications of the construction in the proof of Theorem 10.4 (*if* part), which establishes the existence of this path. We shall illustrate the simple principle involved by the example of Fig. 10–7. In $p = [v_{10}, v_9, v_8, v_7, v_{b'}, v_1]$ we have one blossom node, $v_{b'}$. We find the node next to $v_{b'}$ in the path that is connected to $v_{b'}$ by a free edge—v_1 in our example. We find the node of b' which is adjacent—in the graph before the shrinking of b'—to v_1. This will not always be simple, because, instead of v_1, we might have another blossom node. In the analysis of the efficiency of the algorithm (see the proof of Theorem 10.6) we shall show exactly how we do this efficiently. Once we find one such vertex of b'—v_2 or v_3; let us say v_2—we find the unique path from the basis of b' to v_2 that ends with a matched edge, $[v_b, v_3, v_2]$ in our example. We then replace $v_{b'}$ in p by this path to obtain $p' = [v_{10}, v_9, v_8, v_7, v_b, v_3, v_2, v_1]$.

We must now repeat the same process to replace v_b in p'. This time v_3 is the node that is connected to v_b by a free edge. We find that this edge corresponds to the edge $[v_4, v_3]$ in G. We find the unique path within b from the basis v_6 to v_4 that ends with a matched edge, namely, $[v_6, v_5, v_4]$. Finally, we replace v_b in p' by this path, to obtain the final augmenting path $p'' = [v_{10}, v_9, v_8, v_7, v_6, v_5, v_4, v_3, v_2, v_1]$. The procedure augment, not shown explicitly in Fig. 10–13, implements the above ideas. The whole algorithm is shown in Fig. 10–13.

```
NONBIPARTITE MATCHING ALGORITHM
Input:  A graph G=(V,E).
Output: A maximum matching of G in terms of the array mate.
begin
   for all v ∈ V do mate[v]:=0, considered[v]:=0;
stage: while there is a u ∈ V with considered[u]=0 and mate[u] = 0 do
   begin
      considered[u]:=1, A:=∅;
      for all v ∈ V do exposed [v]:=0
      (comment: construct the auxiliary digraph)
      for all [v,w] ∈ E do (comment: repeat for both [v,w] and [w,v])
      if mate[w]=0 and w≠u then exposed[v]:=w else
         if mate[w]≠v, 0 then A:=A∪{(v,mate [w])};
      for all v ∈ V do seen[v]:=0;
      Q:={u}; label [u]:=0; if exposed[u]≠0 then augment(u), go to stage;
      (comment: initialize for search)

      while Q≠∅ do
      begin
         let v be a node in Q; (comment: Q is a queue)
         remove v from Q;
         for all unlabeled nodes w ∈ V such that (v,w) ∈ A do

         begin
            Q:=Q∪{w}, label [w]:=v;
            seen[mate[w]]:=1;
            if exposed[w]≠0 then augment(w), go to stage;
            if seen[w] = 1 then blossom(w)
         end

      end
   end
end
```

Figure 10–13 The nonbipartite matching algorithm.

Theorem 10.6 *The algorithm of Fig.* 10–13 *correctly finds a maximum matching in a graph* $G = (V, E)$ *in* $O(|V|^4)$ *time.*

Proof The algorithm halts when unsuccessful attempts have been made to find augmenting paths from all exposed nodes of G. By Theorem 10.4, we may conclude that, at the time that each such attempt was made, there was no augmenting path from this node in the original graph G. Hence, by Theorem 10.5, there is no augmenting path in G and consequently, by Theorem 10.1, the current matching is optimal.

For the time bound, since the loop labeled *stage* is executed at most $|V|$ times (once for each vertex), it suffices to show that each stage can be done in $O(|V|^3)$ time. The construction of the auxiliary digraph can be carried out in $O(|V|^2)$ time, and so can the calculation of the array *exposed*. At each search we are going to have at most $|V|$ executions of the procedure blossom, because each such execution decreases the number of vertices in the current graph by at least two. Let us now show that blossom can be executed in $O(|V|^2)$ time. Backtracing the labels and finding the latest common vertex can certainly be done in $O(|V|^2)$ time. Updating the auxiliary digraph can be done in $O(|A|) = O(|E|) = O(|V|^2)$ time; also the array *label* and Q can be updated in $O(|V|)$ time. Hence all of the procedure blossom can be done in $O(|V|^2)$ time.

The procedure augment is executed at most once during each search. In each execution we may have to make up to $O(|V|)$ blossom expansions (recall the discussion of augment preceding the theorem). Each expansion amounts essentially to finding the node u_j of the blossom that is connected to w in the original graph. This can be done by examining all points that are eventually shrunk into b—and w, if w is also a blossom—and finding a combination that is an edge in E. The first part can be done in $O(|V|^2)$ time by starting with each node v and finding *blossom*[v], then *blossom*[*blossom*[v]], and so on, at most $O(|V|)$ times, until we hit b, or w, respectively. Also, the second part can be done in $O(|E|)$ time by checking for each edge $[v, u] \in E$ whether v is eventually shrunk into b and u is eventually shrunk into w. The rest of augment can be done in $O(|V|)$ time for each blossom expansion; hence we conclude that the augmentation following each search can be done in $O(|V|^3)$ time. The theorem follows. □

Example 10.1

Let us apply the algorithm of Fig. 10–13 to the graph shown on the top of page 239.

The first seven stages are trivial. Augmenting paths consisting of single edges are discovered immediately and thus the following edges become matched: $[v_1, v_2]$, $[v_3, v_4]$, $[v_5, v_6]$, $[v_7, v_8]$, $[v_9, v_{10}]$, $[v_{11}, v_{12}]$, $[v_{13}, v_{14}]$. The resulting matching is shown on the middle of page 239.

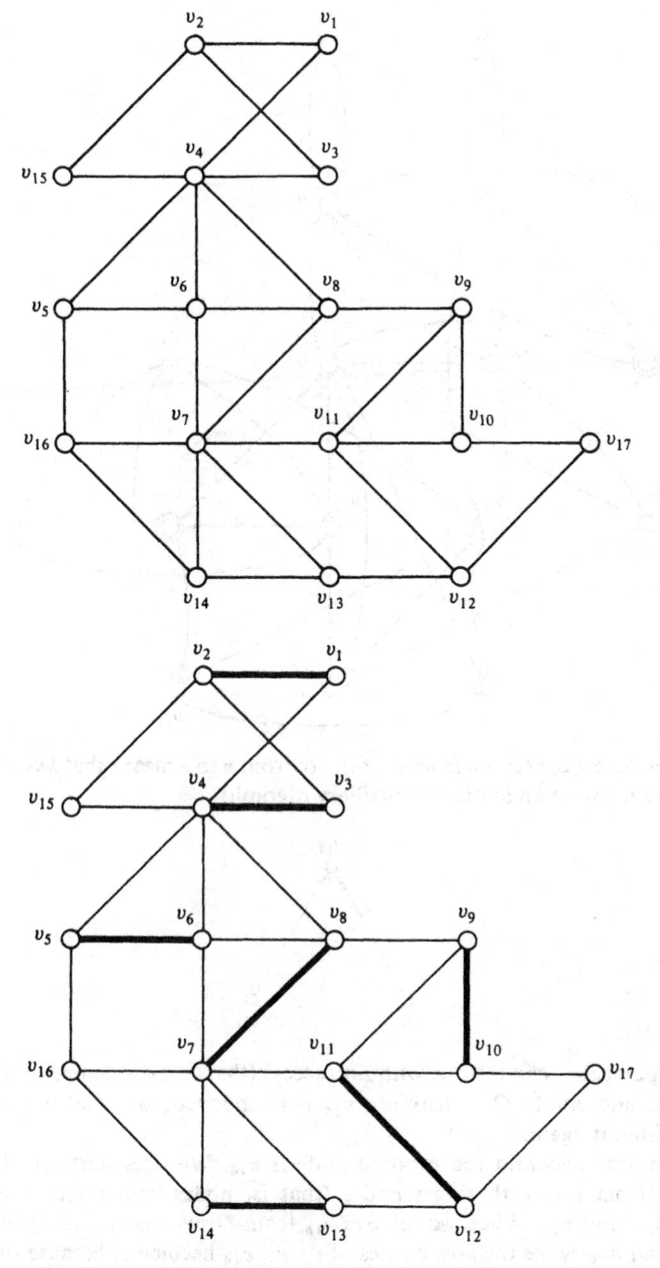

In the next stage, we pick an exposed and unconsidered node, v_{15}, and construct the auxiliary digraph with respect to the current matching. The nodes v with $exposed[v] \neq 0$ are v_5, v_7, v_{10}, v_{12}, and v_{14}.

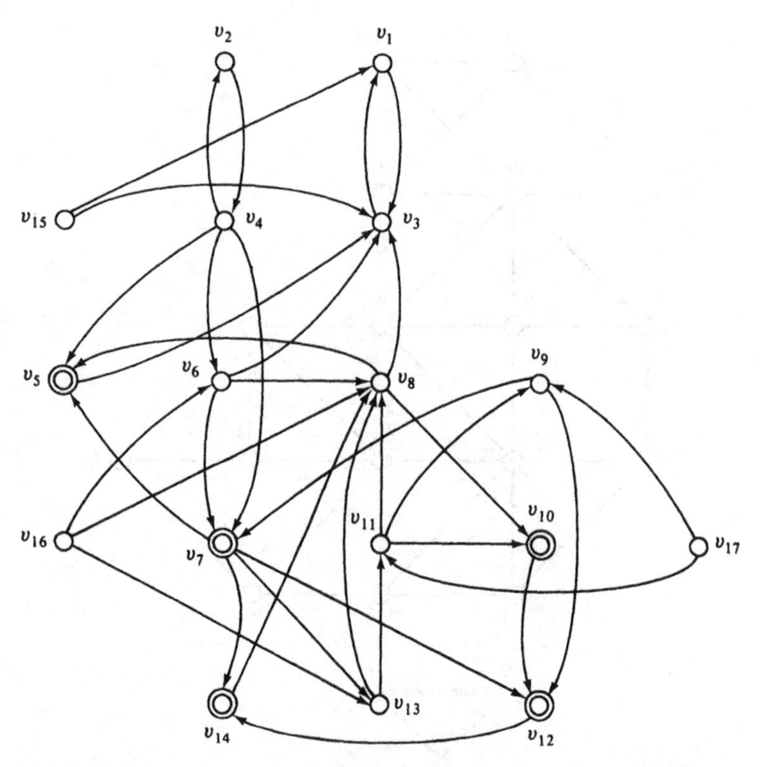

We show below the search from v_{15} (an arc from v to u means that *label* $[v] = u$; it is the *reverse* of an arc in the auxiliary digraph).

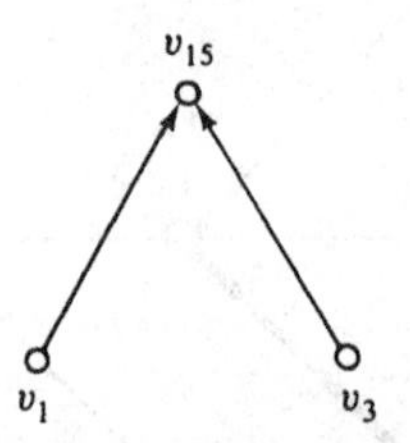

At this point the search stops with no success (that is, no augmenting path, no blossom, and empty Q). Therefore v_{15} is abandoned; we shall never start a search from it again.

The next unconsidered exposed node is v_{16}. We thus start searching the digraph from v_{16}, with target nodes (that is, nodes with nonzero *exposed*) v_2, v_4, v_{10}, and v_{12}. When we remove v_{16} from Q, we insert into Q the nodes v_6, v_8, and v_{13}, while the *seen* entries of v_5, v_7, v_{14} become 1, because the mates of these nodes have been labeled. We next remove v_6 from Q. This has the effect of adding v_3 to Q with *seen*$[v_4] = 1$. Then we examine v_7. The situation is as shown on the top of next page.

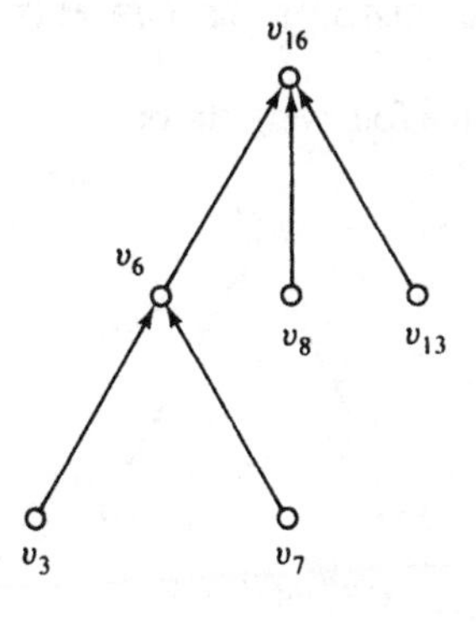

$$Q = \{v_8, v_{13}, v_3, v_7\}$$

At this point we notice that *seen*$[v_7] = 1$; hence v_7 is the mate of an already outer vertex—v_8—and a blossom has been discovered. To find all vertices of the blossom, we trace the array *label* back from both v_7 and its mate, v_8. The first common node in these backwards paths is v_{16}, and so v_{16} is the basis of the blossom. All nodes encountered in this backtracing, and their mates, comprise the nodes of the blossom. We give to this blossom a new index, v_{18}, we replace all entries of Q and *label* that happen to be nodes of the blossom by v_{18}, and we resume the search. The modified auxiliary graph is shown below.

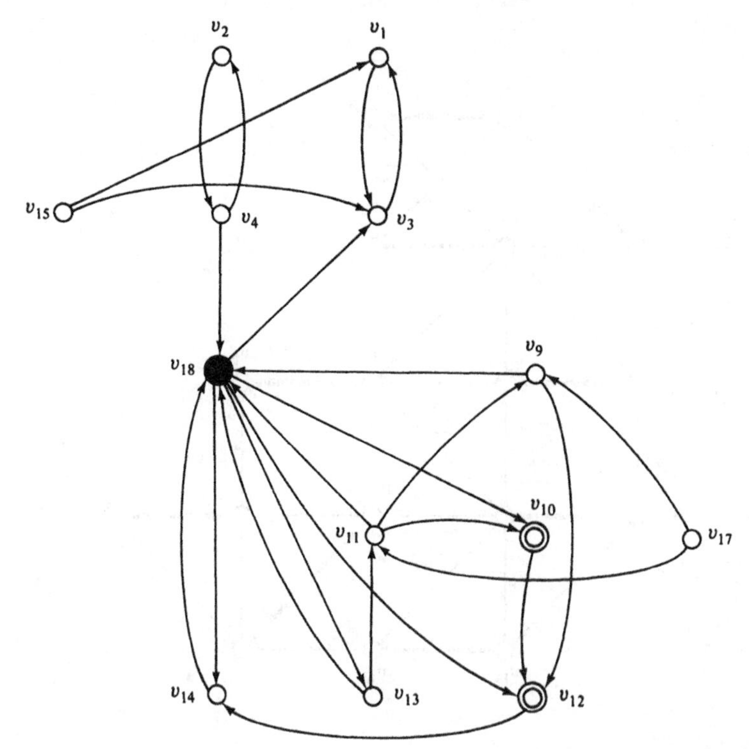

Our search continues from the following state:

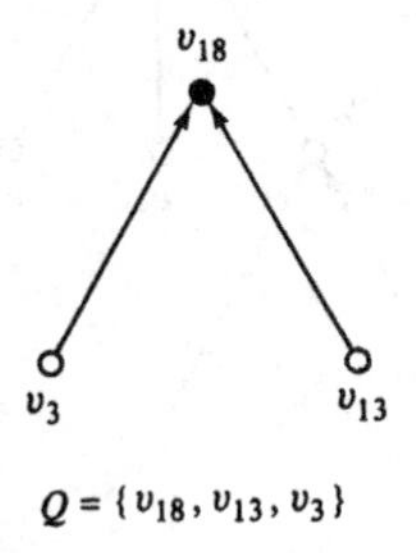

$Q = \{v_{18}, v_{13}, v_3\}$

We next remove v_{18} from the queue, and add v_{10}, a vertex with *exposed*$[v_{10}] = v_{17} \neq 0$.

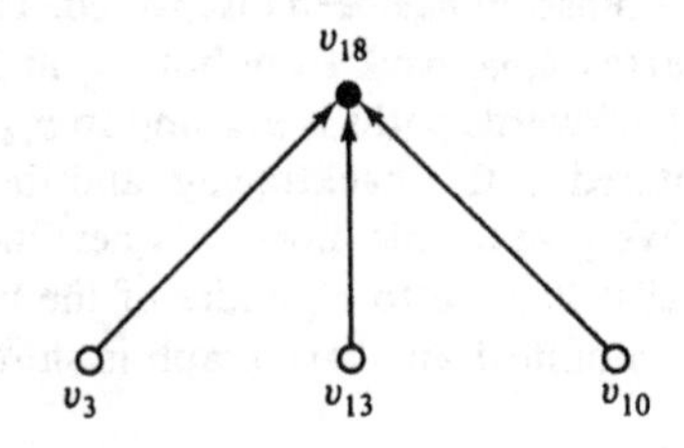

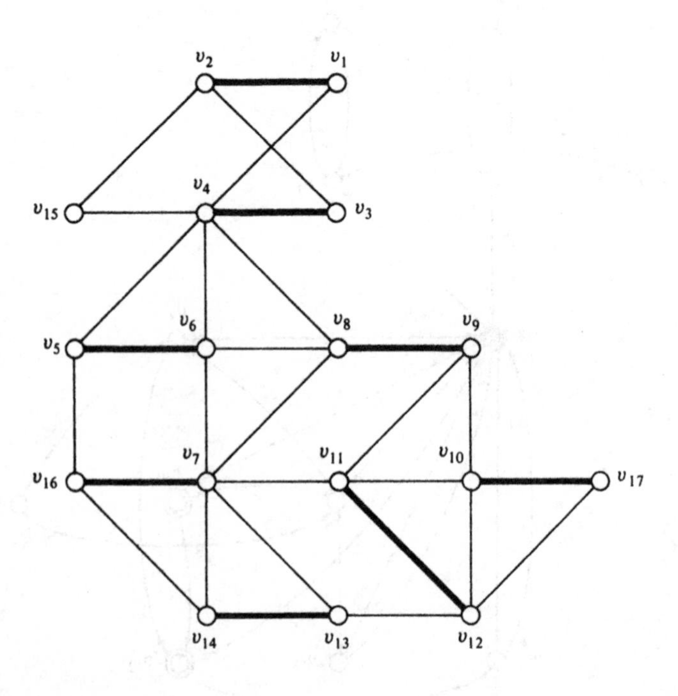

The augmenting path in the current graph is $p' = [v_{18}, mate[v_{10}], v_{10}, exposed[v_{10}]] = [v_{18}, v_9, v_{10}, v_{17}]$. To replace p' by a path in the *original* graph, we examine all edges incident upon v_9 and discover that $[v_{18}, v_9]$ in fact corresponds to the original edge $[v_8, v_9]$; furthermore, the path from the basis of v_{18} to v_8 that ends with a matched edge is $[v_{16}, v_7, v_8]$. Thus we replace v_{18} in p' by $[v_{16}, v_7, v_8]$ and obtain the final augmenting path $p = [v_{16}, v_7, v_8, v_9, v_{10}, v_{17}]$, whereby the current matching is augmented, as shown on page 242.

Next we discover that there is no unmatched and unconsidered vertex of G, and hence the current matching is optimal. □

PROBLEMS

1. Show that in a bipartite graph, the cardinality of the maximum matching equals the cardinality of the smallest set of nodes that covers all edges (that is, every edge is incident upon a node in the set).

2. Show that a bipartite graph $B = (V, U, E)$ with $|V| = |U| = n$ has a matching of cardinality n iff for all $S \subseteq V$ we have $|\{u \in U: [v, u] \in E \text{ for some } v \in S\}| \geq |S|$.

3. An *edge cover* of a graph $G = (V, E)$ is a subset C of E such that
$$V = \bigcup_{[u,v]\in C} \{u, v\}.$$
Suppose that G has no isolated points. Show that the cardinality of the minimum edge cover C of G equals the cardinality of the maximum matching M of G, increased by $|V| - 2|M|$. Give an efficient algorithm for finding the minimum edge cover of a graph.

*4. If $G = (V, E)$ is a graph with $V = \{v_1, v_2, \ldots, v_n\}$, let us define the *Tutte matrix* of G, $T(G)$, to be the $n \times n$ matrix defined as follows:
$$T(G)_{ij} = \begin{cases} x_{ij} & \text{if } [v_i, v_j] \in E \text{ and } i > j \\ -x_{ji} & \text{if } [v_i, v_j] \in E \text{ and } i < j \\ 0 & \text{otherwise} \end{cases}$$
where the x_{ij}'s are indeterminates (that is, variables). Show that G has a complete matching iff $\det(T(G)) \not\equiv 0$.

5. Let $\mathcal{J}$ be a set of *jobs* to be executed, on two processors, with all jobs requiring the same amount of time (say, 1). Suppose that there is a directed acyclic graph $P = (\mathcal{J}, A)$ such that, if $(J_1, J_2) \in A$, then J_1 must be executed before J_2. The *Two-Processor Scheduling Problem* is: Given P, find an *optimal schedule*; that is, a function $S: \mathcal{J} \longrightarrow \{1, 2, \ldots, T\}$ such that
    1. for all $i \leq T$, $|\{J \in \mathcal{J}: S(J) = i\}| \leq 2$;
    2. if $(J_1, J_2) \in A$ then $S(J_1) < S(J_2)$;
    3. T is as small as possible.

(a) Let $G_P = (\mathcal{J}, E_P)$, where $[J_1, J_2] \in E_P$ iff there is no path from J_1 to J_2, nor one from J_2 to J_1, in P. Suppose that M is the maximum matching of G_P. Show that the smallest T achievable must obey

$$T \geq |\mathcal{J}| - |M|$$

(recall Problem 3).

*(b) Show that there is always a schedule with $T = |\mathcal{J}| - |M|$. Give an efficient algorithm for finding the optimal schedule (compare with Theorem 15.5).

6. In a directed graph, every arc has a *head* and a *tail*. A *bidirected graph* (V, A) is a set of nodes and a set of arcs, except that now arcs are allowed to have either a head and a tail, or two heads, or two tails. A *bidirected network* $N = (s, t, V, A, b)$ is a network with (V, A) a bidirected graph. Also, s is no arc's head and t is no arc's tail. A *flow* f in N is a function from A to Z^+ such that for all $a \in A, f(a) \leq b(a)$ and for each $v \in V - \{s, t\}$ we have

$$\sum_{\substack{v \text{ is a} \\ \text{tail of} \\ a}} f(a) = \sum_{\substack{v \text{ is a} \\ \text{head of} \\ a}} f(a)$$

The *value* of f is

$$\sum_{\substack{s \text{ is a} \\ \text{tail of} \\ a}} f(a)$$

Show that we can efficiently *reduce* the nonbipartite matching problem to a bidirected max-flow problem with unit capacities. (Compare with Sec. 10.3.)

***7.** Show that any bidirected flow instance with unit capacities can be reduced efficiently to an instance of the matching problem.

8. Show by an example that if $G = (V, E)$ is a graph, M a matching of G, and b a blossom (a circuit with $2m + 1$ edges, m of which are in M), then the following may fail to be true: There is an augmenting path in G with respect to M iff there is one in G/b with respect to M/b. Compare with Theorem 10.4.

9. Apply the algorithm of Fig. 10–13 to augment the matching shown below.

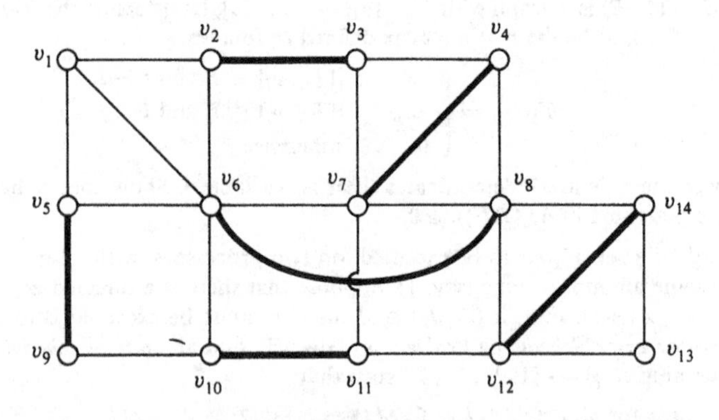

10. The purpose of this problem is to show that the nonbipartite matching algorithm of Fig. 10–13 can be implemented to run in $O(|V^3|)$ time, by designing the procedures blossom and augment appropriately.

(a) Suppose that for each edge e of the current graph which did not exist in the previous current graph, we keep the edge $prev[e]$ of the previous current graph to which e corresponds. Show that, using this device, each execution of augment takes $O(|V|^2)$ time.

*(b) Show how to implement blossom (which now also has to update the array $prev$) so that the total time spent in calls to blossom is $O(|V|^2)$ per stage.

11. The *directed matching problem* is the following: Given a digraph $D = (V, A)$, find a subdigraph (V, M) such that the indegree and the outdegree of each node of V in (V, M) is 1. Give an efficient algorithm for the directed matching problem by reducing it to bipartite matching.

12. The *b-matching problem* is the following: Given a bipartite graph $B = (V, U, E)$ and a function $b: V \cup U \longrightarrow Z^+$, is there a subset $M \subseteq E$ such that for each $v \in V \cup U$, v is incident upon b edges in M?

(a) Give an $O((\sum_{v \in V \cup U} b(v))^{2.5})$ algorithm for b-matching.

(b) Give an $O(|V|^3)$ algorithm for b-matching.

***13.** Give a polynomial-time algorithm for b-matching in nonbipartite graphs. (*Hint:* Construct a new graph that has $b(v)$ new nodes for each node v and two new nodes for each edge; find a complete matching in this graph.)

14. The *bottleneck matching problem* is the following: Given a graph $G = (V, E)$ and a *weight* $w: E \longrightarrow Z^+$, find a complete matching M such that $\max_{e \in M} w(e)$ is as small as possible. Give a polynomial-time algorithm for bottleneck matching.

***15.** Give a polynomial-time algorithm for the following problem: Given a graph $G = (V, E)$, a partition of V into two sets A and B, and two integers a and b, find a matching M such that at least a nodes in A and at least b nodes in B are incident upon edges in M.

16. The *bin-packing* problem is the following. We are given n integers $\{c_1, \ldots, c_n\}$ and another integer B—the *bin capacity*—and we are asked to find a partition of these integers into *bins* such that in each bin the sum of all integers in the bin does not exceed B, and there are *as few bins as possible*. Show that, if the c_j's satisfy $c_j > B/3$, then bin-packing can be formulated as a matching problem. Then solve it in a much easier way.

NOTES AND REFERENCES

Algorithms for bipartite matching have been known for some time; see, for example,

[Ha] HALL, M., JR., "An Algorithm for Distinct Representatives," *American Math. Monthly*, 63 (1956), 716–17.

[FF] FORD, L. R., and D. R. FULKERSON, *Flows in Networks*. Princeton, N.J.: Princeton Univ. Press, 1962.

Theorem 10.1 (the augmenting path theorem) was proved independently by both

[Be] BERGE, C., "Two Theorems in Graph Theory," *Proc. National Acad. of Science*, 43 (1957), 842–44.

[NR] NORMAN, R. Z., and M. O. RABIN, "An Algorithm for a Minimum Cover of a Graph," *Proc. American Math. Society*, 10 (1959), 315–19.

It is interesting that for the longest time it was believed that this theorem *immediately* suggests an efficient algorithm for nonbipartite matching. The intricacy of the problem and its elegant solution were first pointed out by Jack Edmonds.

[Ed] EDMONDS, J., "Paths, Trees and Flowers," *Canad. J. Math*, 17 (1965), 449–67.

For the bipartite case, the $O(|V|^{1/2}|E|)$ algorithm of Sec. 10.3 is due to

[HK] HOPCROFT, J. E., and R. M. KARP, "A $n^{5/2}$ Algorithm for Maximum Matching in Bipartite Graphs," *J. SIAM Comp.*, 2 (1973), 225–31.

The fact that this algorithm is a special case of the max-flow algorithm applied to a simple network was first pointed out by

[ET] EVEN, S., and R. E. TARJAN, "Network Flow and Testing Graph Connectivity," *J. SIAM Comp.*, 4, no. 4 (1975), 507–12.

The fastest algorithm known for nonbipartite matching is described in

[MV] MICALI, S., and V. V. VAZIRANI, "An $O(\sqrt{|V|} \cdot |E|)$ Algorithm for Finding Maximum Matching in General Graphs," *Proc. Twenty-first Annual Symposium on the Foundations of Computer Science*, Long Beach, California: IEEE (1980), 17–27.

An earlier algorithm, less efficient for sparse graphs, is explained in

[EK] EVEN, S., and O. KARIV, "An $O(n^{2.5})$ Algorithm for Maximum Matching in General Graphs," *Proc. Sixteenth Annual Symp. on Foundations of Computer Science*, Berkeley, California: IEEE (1975), 100–12.

[Ka] KARIV, O., *An $O(n^{2.5})$ Algorithm for Maximum Matching in General Graphs*, (unpublished Ph.D. Thesis, Technion, Haifa, Israel, 1977).

Problem 3 is from [NR]. Problem 4 is from

[Tu] TUTTE, W. T., "The Factors of Graphs," *Canad. J. Math.*, 4 (1952), 314–28.

Problem 5 is from

[FKN] FUJII, M., T. KASAMI, and K. NINAMIYA, "Optimal Sequencing of Two Equivalent Processors," *J. SIAM*, 17 (1969), 784–89,

and Problem 15 from

[PY] PAPADIMITRIOU, C. H., and M. YANNAKAKIS, "The Complexity of Restricted Spanning Tree Problems," pp. 460–70 in *Automata, Languages, and Programming*, ed. H. A. Maurer. Berlin: Springer-Verlag, 1979.

11

Weighted Matching

11.1 Introduction

A much more involved version of the matching problem is the one in which we are given, besides the graph $G = (V, E)$, a number $w_{ij} \geq 0$ for each edge $[v_i, v_j] \in E$, called the *weight* of $[v_i, v_j]$. We are supposed to find a matching of G with the largest possible sum of weights. Clearly, the unweighted matching problem that we have attacked in the previous chapter (sometimes called the *cardinality* matching problem) is a special case of the weighted matching problem: Just let $w_{ij} = 1$ for all $[v_i, v_j] \in E$.

It is not hard to see that in the weighted matching problem, we can do away with the graph G by adopting the convention that the underlying graph is always complete, and letting the weights of those edges that were missing in G be equal to zero. Furthermore, we can always assume that we have an even number of nodes—otherwise, add a new node with edges of weight zero incident upon it. Similarly, if we are dealing with the bipartite case of the problem, we may assume that the graph is a complete bipartite graph with two sets of nodes that are equal in size. Also, we may observe at this point that the optimal solutions will always be complete matchings (since $w_{ij} \geq 0$), and hence we can alternatively formulate these problems as *minimization* problems by simply considering the *costs* $c_{ij} = W - w_{ij}$, where W is larger than all the w_{ij}'s. We shall make these assumptions throughout this chapter.

As in the cardinality matching problem, the weighted matching problem is considerably easier in its bipartite case. The bipartite weighted matching problem is also known as the *assignment problem*, because it can be applied in principle in order to calculate the best assignment of tasks to workers, assuming that we know the value w_{ij} produced by the ith worker at the jth task. The assignment problem can be described by a concise linear program, which is a special case of the Hitchcock problem (see Sec. 7.4), and can therefore be solved by the primal-dual method—called, in this particular instance, the *Hungarian method.* The Hungarian method solves the weighted matching problem for a complete bipartite graph with $2 \cdot |V|$ nodes in $O(|V|^3)$ arithmetic operations.

If we try the same line of attack in the nonbipartite case, we see that the linear programming formulation fails to describe exactly the weighted matching problem. By this we mean that the linear program has basic feasible solutions that do not correspond to matchings. Not surprisingly, this pathology again has its roots in the presence of *blossoms*. We must therefore detect these blossoms and discover augmenting paths in spite of them. However, in the case of weighted matching, the graph-theoretic part of our task is easier, both conceptually and algorithmically. The hard part is that of perfecting the linear programming formulation so it describes the problem exactly and thence develop a primal-dual algorithm. The way this is done is a beautiful application of duality theory. The complexity of the weighted matching algorithm for general graphs that we develop in Section 11.3 is $O(|V|^4)$; in Problem 14 we outline the refinements necessary to bring the complexity down to $O(|V|^3)$.

11.2 The Hungarian Method for the Assignment Problem

We shall develop a simple algebraic formulation for the weighted bipartite matching problem. We let x_{ij} be a set of variables for $i = 1, \ldots, n$ and $j = 1, \ldots, n$, where n is the number of nodes in the node sets of the complete bipartite graph $B = (V, U, E)$. Here $x_{ij} = 1$ means that the edge $[v_i, u_j]$ is included in the matching, whereas $x_{ij} = 0$ means that it is not. Naturally, for a set of such values to represent a complete matching, we must have

$$\begin{aligned} \sum_{j=1}^{n} x_{ij} &= 1 \qquad i = 1, \ldots, n \\ \sum_{i=1}^{n} x_{ij} &= 1 \qquad j = 1, \ldots, n \qquad \text{(A)} \\ x_{ij} &\geq 0 \end{aligned}$$

Our goal is to minimize $\sum_{i,j} c_{ij}x_{ij}$. We have thus developed a linear program (in fact, a special case of the Hitchcock problem of Chapter 7, with $m = n$ and with all a's and b's equal to 1) which, in some sense, describes the assignment problem. Unfortunately, there is no way to use linear inequalities to write our implicit assumption that the x_{ij}'s must attain only the values zero and 1. Thus our linear program might very well have a *fractional* optimal solution, which would not correspond to any feasible matching. An example is shown in Fig. 11–1.

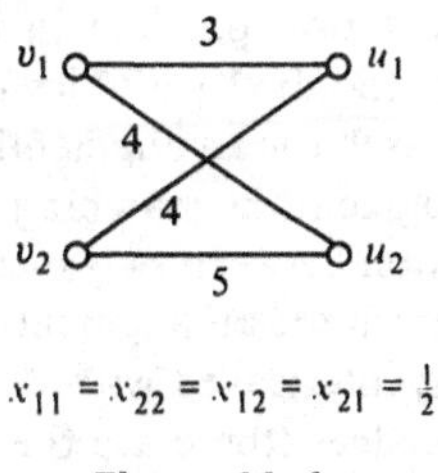

Figure 11–1

Fortunately, as in a similar situation concerning the max-flow problem (Sec. 9.5), some mysterious, friendly power is at work. Although such fractional solutions obviously exist, they are *never* basic feasible solutions of the linear program. Consequently, if we wish to solve this linear program with simplex—any variation thereof—we may be certain that the final optimal solution will be nonfractional, that is, one that corresponds to a real matching.

In Chapter 13 we are going to develop a theory of *total unimodularity* in order to understand this phenomenon completely. For the time being, we shall derive a constructive proof that an optimal solution to (A) that corresponds to a "real" complete matching always exists. Our constructive proof simply entails applying the primal-dual method to this linear program by combinatorializing its cost, exactly as we did with the more general Hitchcock problem in Chapter 7.

Lemma 11.1 *The alphabeta algorithm of Fig. 7–9 applied to the program* $\min \sum_{i,j=1}^{n} c_{ij}x_{ij}$ *subject to constraints (A) yields the optimal solution to the assignment problem.*

Proof It certainly produces the optimal solution to the linear program. It remains to establish that the x_{ij}'s of the optimal solution produced are integral, that is, 0-1. This, however, follows from the fact that the optimal solution found by the alphabeta algorithm is the solution of a max-flow problem with integer—in fact, 0-1—capacities (recall Fig. 7–8). ☐

We shall modify the alphabeta algorithm somewhat in order to apply it to the assignment problem. As a result, it will acquire the appearance of a matching algorithm, reminiscent of those developed in the previous chapter. Moreover, its asymptotic complexity will be reduced substantially. Recall that the algorithm solves an ever-changing restricted primal (RP) problem, which is in our case a max-flow problem. In particular, it is easy to see that RP is an unweighted bipartite matching problem for the bipartite graph consisting of the currently admissible edges (that is, those that have $c_{ij} = \alpha_i + \beta_j$ for the dual variables α_i and β_j; recall Sec. 5.1), which we shall solve essentially by applying the algorithm of Fig. 10–3. While searching for an augmenting path, we may encounter a *nonbreakthrough*—no augmentation possible in the present set of admissible edges. Then we shall change the dual variables α_i and β_j to make new edges admissible and resume the search for an augmenting path. Once an augmenting path is found, we use it to augment the matching and start all over.

Let us call the computation between two successive augmentations a *stage*. Thus a stage consists of a search for an augmenting path in the bipartite graph made up of admissible edges, interleaved with dual variable modifications that change the set of admissible edges (these are the calls of the procedure modify in Figure 11–2). To modify the dual variables, we need to calculate (recall Eq. 7.13)

$$\theta_1 = \tfrac{1}{2} \min_{ij} (c_{ij} - \alpha_i - \beta_j)$$

where the minimum is taken over all labeled nodes $v_i \in V$ and unlabeled nodes $u_j \in U$ (we are thinking of a complete bipartite graph with node sets $V = \{v_1, \ldots, v_n\}$ and $U = \{u_1, \ldots, u_n\}$). To recompute this quantity every time that we modify dual variables by comparing the n^2 candidates could be very costly. We solve this problem by keeping and updating† the two arrays *slack*$[u_j]$ and *nhbor*$[u_j]$ for $j = 1, \ldots, n$, where *slack*$[u_j]$ is the minimum of $(c_{ij} - \alpha_i - \beta_j)$ over all labeled vertices v_i, and *nhbor*$[u_j]$ is the particular labeled vertex v_i with which *slack*$[u_j]$ is achieved. Thus *slack*$[u_j] = 0$ means that $[$*nhbor*$[u_j], u_j]$ is an admissible edge. To compute θ_1, we therefore choose the *smallest nonzero slack*, and this takes only $O(n)$ time. The algorithm is shown in Fig. 11–2, and the analysis of its performance is summarized as follows.

Theorem 11.1 *The Hungarian Method of Fig. 11–2 correctly solves the assignment problem for a complete bipartite graph with 2n nodes in $O(n^3)$ arithmetic operations.*

Proof It can easily be verified that the algorithm of Fig. 11–2 is an implementation of the alphabeta algorithm applied to the assignment problem; hence, its correctness follows by Lemma 11.1.

For the time bound, first notice that we can have $n + 1$ stages, since each stage, except for the last, augments the matching by one edge. To analyze each

†The reader should not miss the similarity between this data structure maneuver and the corresponding one in Dijkstra's algorithm, in Chapter 6.

THE HUNGARIAN METHOD

Input: An $n \times n$ matrix $[c_{ij}]$ of nonnegative integers.

Output: An optimal complete matching (given in terms of the array *mate*) of the complete bipartite graph $B = (V,U,E)$ with $|V| = |U| = n$ under the cost c_{ij}.

```
begin
for all vi ∈ V do mate[vi] :=0, αi :=0;
for all uj ∈ U do mate[uj] :=0, βj :=min_i{cij};
(comment: initialize)
for s:=1, . . . , n do (comment: repeat for n stages)
  begin
  A:=∅;
  for all v ∈ V do exposed[v]:=0, label[v]:=0;
  for all u ∈ U do slack[u]:=∞;
  for all vi,uj with vi ∈ V, uj ∈ U, and αi+βj=cij do
    if mate[uj]=0 then exposed [vi]:=uj
      else if vi ≠ mate [uj] then A:=A∪{(vi, mate [uj])};
  (comment: construct the auxiliary graph)

 ------------------------------------------------------------------
 | Q:=∅;
 | for all vi ∈ V do
 |   if mate [vi]=0 then
 |     begin
 |     if exposed [vi]≠0 then augment(vi), go to endstage;
 |     Q:=Q∪{vi};
 |     label[vi] :=0;
 |     for all uk ∈ U do
 |     if 0<cik−αi−βk<slack[uk] then slack[uk] :=cik−αi−βk,nhbor[uk] :=vi;
 |     end;
 ------------------------------------------------------------------

search:  while Q≠∅ do
           begin
           let vi be any node in Q;
           remove vi from Q;
           for all unlabeled vj ∈ V with (vi,vj) ∈ A do

          ---------------------------------------------------------
          | begin
          | label[vj] :=vi;
          | Q:=Q∪{vj};
          | if exposed[vj] ≠ 0 then augment(vj), go to endstage;
          | for all uk ∈ U do
          | if 0<cjk−αj−βk<slack[uk] then slack[uk]:=cjk−αj−βk, nhbor[uk] :=vj;
          | end
          ---------------------------------------------------------

           end;
           modify;
           go to search
endstage:  end
         end
```

Figure 11–2 The Hungarian method.

```
procedure modify
(comment: it calculates θ1, updates the α's and β's, and activates new
          nodes to continue the search)
begin
θ1:=½ min {slack[u]>0};
      u∈U
for all vi∈V do
  if vi is labeled then αi:=αi+θ1 else αi:=αi−θ1;
for all uj∈U do
  if slack[uj]=0 then βj:=βj−θ1 else βj:=βj+θ1;
for all u∈U with slack[u]>0 do
  begin
  slack[u] := slack[u]−2θ1;
  if slack[u]=0 then (comment: new admissible edge)
    if mate[u]=0 then exposed[nhbor[u]]:=u, augment(nhbor[u]), go to endstage;
    else (comment: mate[u]≠0)
    label[mate[u]]:=nhbor[u],Q:=Q∪{mate[u]}, A:=A∪{(nhbor[u], mate[u])};
  end
end
```

Figure 11–2 *(continued)*

stage, let us examine the complexities of search and dual variable modification separately. The search can be done in $O(n^2)$ operations, because no vertex is ever enqueued for the second time in the same stage, and the removal of a vertex from Q costs $O(n)$ operations. For each dual variable modification, on the other hand, either a new vertex $v \in V$ is labeled, or we have an augmentation and the end of the stage. Thus we can have up to n dual variable modifications at a stage. Each modification takes $O(n)$ time. Consequently, each stage requires $O(n^2)$ operations, and the bound is proved. □

Example 11.1 (The matrix form of the Hungarian method)

One way of looking at the assignment problem and the Hungarian method is in terms of a matrix. The costs c_{ij} are the entries of an $n \times n$ matrix, and the elements of V and U are identified with the rows and columns of the matrix, respectively. Hence the following is an instance of the assignment problem to be solved by the Hungarian method.

α \ β	3	2	1	1	2
0	7	**2**	**1**	9	4
0	9	6	9	5	5
0	**3**	8	3	**1**	8
0	7	9	4	2	**2**
0	8	4	7	4	8

The integers in the margin of the rows and columns are the initial values of α_i and β_j, respectively. Boldface numbers denote admissible edges. Therefore our first three searches will be successful immediately, and the edges $[v_1, u_2]$, $[v_3, u_1]$, and $[v_4, u_5]$ are matched. In our next search, however, the situation is as depicted below.

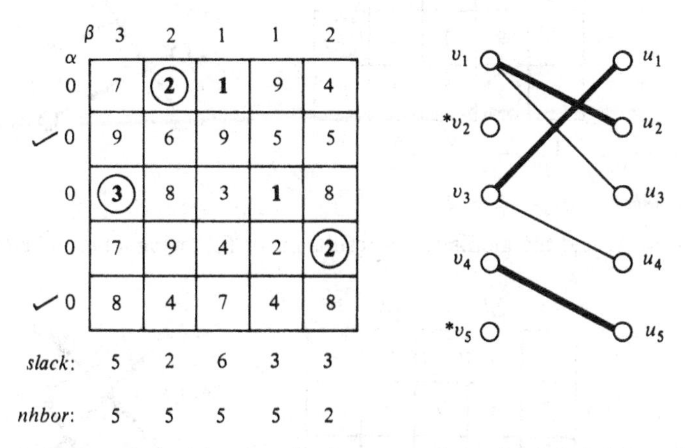

	α \ β	3	2	1	1	2
	0	7	**2**	**1**	9	4
✓	0	9	6	9	5	5
	0	**3**	8	3	**1**	8
	0	7	9	4	2	**2**
✓	0	8	4	7	4	8
	slack:	5	2	6	3	3
	nhbor:	5	5	5	5	2

The graph of the admissible edges is shown on the right (labeled v-nodes are marked by asterisks). On the left, matched edges are circled, and labeled rows (v-nodes) are checked. The arrays $slack[u_j]$ and $nhbor[u_j]$ are shown under the corresponding columns.

Clearly, the situation calls for a modification of dual variables. The smallest nonzero slack is 2, so $\theta_1 = 1$. The results are shown below.

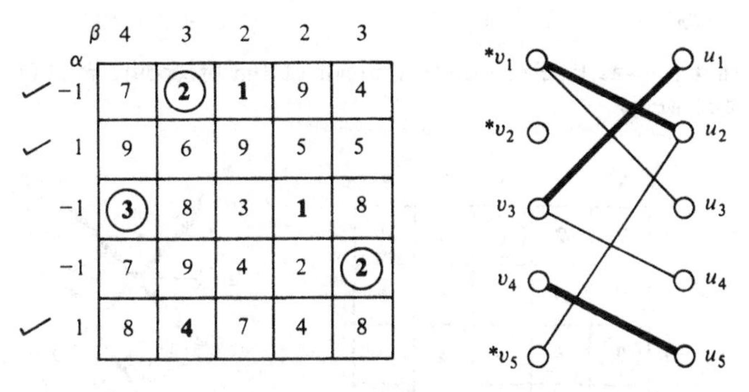

	α \ β	4	3	2	2	3
✓	−1	7	**2**	**1**	9	4
✓	1	9	6	9	5	5
	−1	**3**	8	3	**1**	8
	−1	7	9	4	2	**2**
✓	1	8	**4**	7	4	8

The edge $[v_5, u_2]$ is added to the admissible graph; no edges are deleted. Resuming the search, we notice that v_1, a node with nonzero *exposed* entry, is labeled. The matching is augmented, and the search starts again, to stop as shown on the top of page 254.

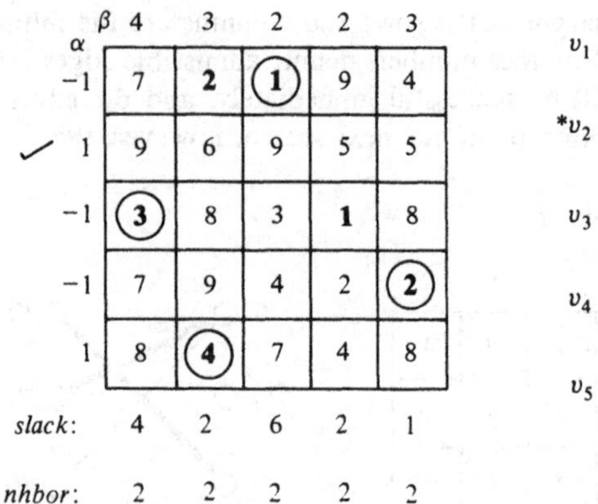

	α \ β	4	3	2	2	3
	−1	7	**2**	(1)	9	4
✓	1	9	6	9	5	5
	−1	(3)	8	3	**1**	8
	−1	7	9	4	2	(2)
	1	8	(4)	7	4	8
	slack:	4	2	6	2	1
	nhbor:	2	2	2	2	2

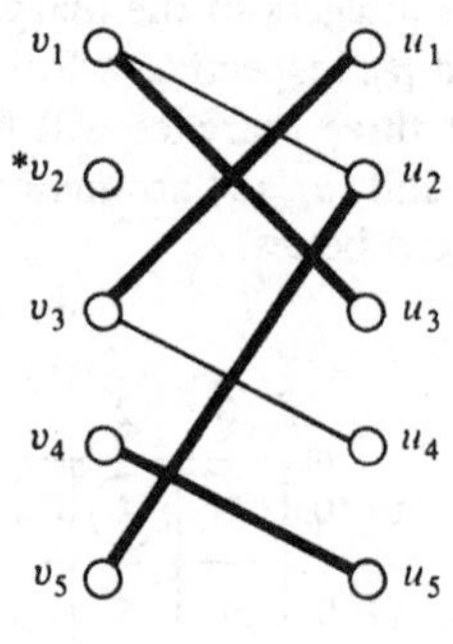

Now $\theta_1 = \frac{1}{2}$, and the modification results in the following table and admissible graph.

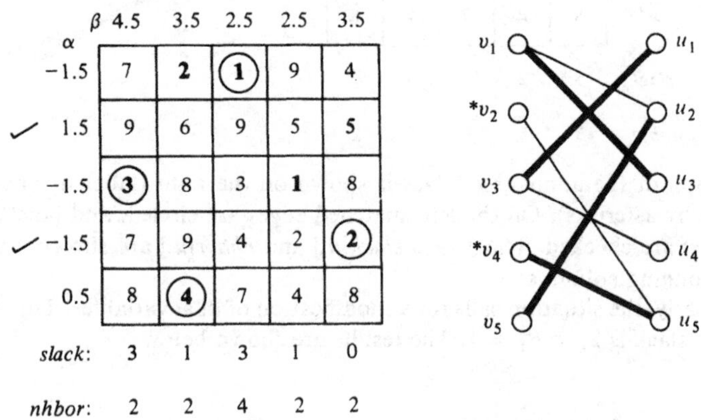

	α \ β	4.5	3.5	2.5	2.5	3.5
	−1.5	7	**2**	(1)	9	4
✓	1.5	9	6	9	5	**5**
	−1.5	(3)	8	3	**1**	8
✓	−1.5	7	9	4	2	(2)
	0.5	8	(4)	7	4	8
	slack:	3	1	3	1	0
	nhbor:	2	2	4	2	2

Notice that v_4 was labeled during the call of the procedure modify. Again, $\theta_1 = \frac{1}{2}$, and we get

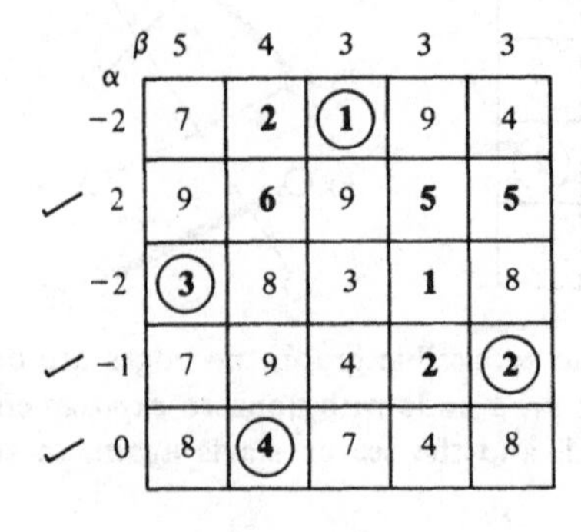

	α \ β	5	4	3	3	3
	−2	7	**2**	(1)	9	4
✓	2	9	**6**	9	**5**	**5**
	−2	(3)	8	3	**1**	8
✓	−1	7	9	4	**2**	(2)
✓	0	8	(4)	7	4	8

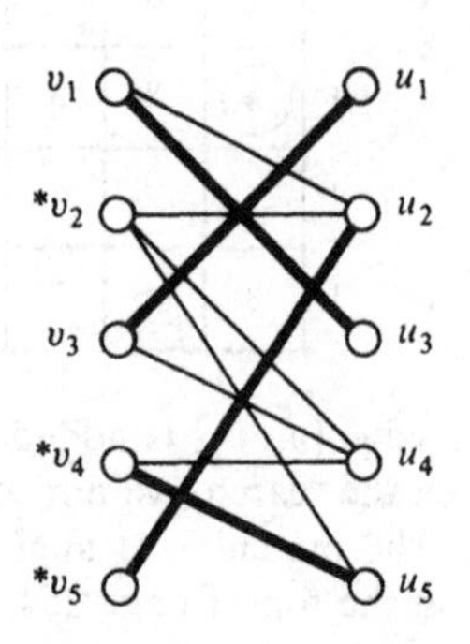

There is an augmenting path—namely, $[v_2, u_5, v_4, u_4]$. The optimal matching is shown below.

7	2	(1)	9	4
9	6	9	5	(5)
(3)	8	3	1	8
7	9	4	(2)	2
8	(4)	7	4	8

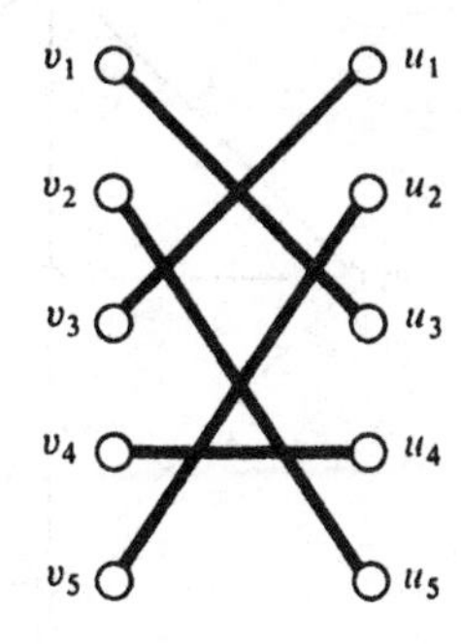

Its cost, 15, checks with

$$\sum_{j=1}^{5} \alpha_j + \sum_{j=1}^{5} \beta_j = 15 \quad \square$$

11.3 The Nonbipartite Weighted Matching Problem

We can formulate the general matching problem as follows:

$$\min \sum_{i,j} c_{ij} x_{ij}$$

subject to

$$\sum_{j=1}^{n} x_{ij} = 1 \qquad i = 1, \ldots, n \tag{11.1†}$$

$$x_{ij} \geq 0 \qquad 1 \leq i \leq j \leq n \tag{11.2}$$

n is assumed even. This linear program may have fractional optimal solutions that are meaningless as matchings, as was the case with the assignment problem. However, things are now much worse. These fractional solutions may now be *basic feasible solutions* of the LP above. This means that every technique similar to simplex would be doomed, for some set of costs, to be trapped into such a solution. An example is shown in Fig. 11–3. The *uniquely* optimal solution corresponding to these costs (the missing edges have a cost of, say, 100) is $x_{12} = x_{28} = x_{83} = x_{37} = x_{71} = x_{45} = x_{56} = x_{46} = \frac{1}{2}$. The optimum is therefore obtained by setting the variables of all edges in odd circuits equal to $\frac{1}{2}$. On the other hand, we know from the previous section that such problems never arise in the absence of odd circuits. We therefore justly suspect that it is the odd circuits that make the LP above inadequate as a description of the matching problem. The following theorem states that this is exactly the case, and in a very strong way indeed:

†By convention, $x_{ii} = 0$ and $x_{ij} = x_{ji}$.

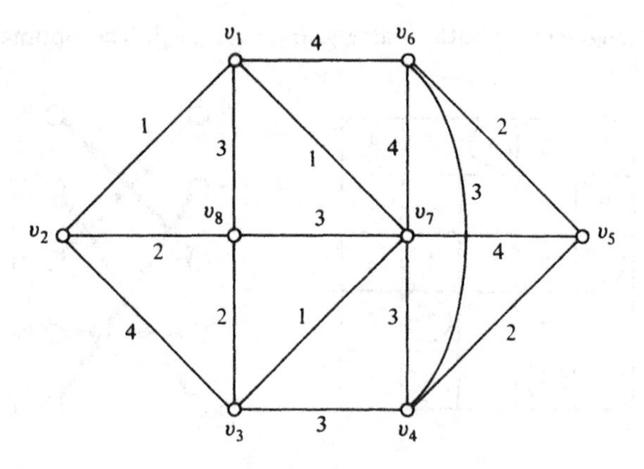

Figure 11–3

In order to describe the general matching problem, we have only to add certain constraints that prevent pathologies like the one in Fig. 11–3 from being feasible.

Recall that we are assuming that n is even. Consider all subsets of $\{1, 2, \ldots, n\}$ which have cardinality that is odd and greater than one. It is easy to check that there are $N = 2^{n-1} - n$ such subsets. Let $S_1, S_2, \ldots, S_N$ be an enumeration of these subsets; the cardinality of S_j will be denoted by $|S_j| = 2s_j + 1$.

Theorem 11.2 (Edmonds) *The general matching problem for a set of costs $\{c_{ij}: 1 \leq i < j \leq n\}$ is equivalent to the LP*

$$\text{minimize} \sum_{i,j} c_{ij}x_{ij}$$

subject to (11.1) *and* (11.2), *and also to a new class of constraints, one for each subset S_k,*

$$\sum_{i,j \in S_k} x_{ij} + y_k = s_k, \qquad y_k \geq 0 \quad \text{for} \quad k = 1, 2, \ldots, N \qquad (11.3) \quad \square$$

At first, it is somewhat disappointing that the LP formulation we are proposing has an exponentially large number of rows. However, we shall see that this is not important. Now (11.3) simply says that no odd set of nodes can be overstocked with more matched edges than appropriate—y_k is a slack variable. Clearly, all feasible matchings satisfy this new class of constraints. Notice that these constraints would disallow the pathological optimum of the example in Fig. 11–3. Indeed, the optimum of the LP in (11.1), (11.2), and (11.3) for the costs in Fig. 11–3 is $x_{12} = x_{38} = x_{45} = x_{67} = 1$, a legitimate matching. In order to prove the theorem, therefore, it suffices to prove that the LP consisting of (11.1), (11.2), and (11.3) always has an integral optimal solution—one that corresponds to a matching. Our proof will be constructive. We shall simply show that the primal-dual method of Chapter 5 converges to an integral solution—and we know that it converges to the optimum. What is even more exciting, this application of primal-dual to the LP above will lead us to an efficient algorithm for the general matching problem.

The dual D of the LP in (11.1), (11.2), and (11.3) is

$$\max \sum_{i=1}^{n} \alpha_i + \sum_{k=1}^{N} s_k \gamma_k$$

subject to

$$\alpha_i + \alpha_j + \sum_{i,j \in S_k} \gamma_k \leq c_{ij} \qquad \text{for all } i, j \leq n \tag{11.4}$$

$$\gamma_k \leq 0 \qquad \text{for all } k \leq N \tag{11.5}$$

where the α's correspond to the constraints in (11.1) and the γ's to (11.3).

The primal-dual algorithm will solve the weighted matching problem as follows: We start with the feasible dual solution $\gamma_k = 0$ for all k and $\alpha_j = \frac{1}{2} \min_i \{c_{ij}\}$. We then perform several iterations of solving a *restricted primal* (RP) and changing the dual variables accordingly. The RP will depend, as usual, on the set of *admissible* variables in the dual problem (11.4) and (11.5). The set of admissible variables is a set $J = J_e \cup J_b$, where J_e is a set of *admissible edges*—those for which (11.4) holds as an equality—and J_b is a set of *admissible odd sets*—those S_k for which $\gamma_k = 0$. We let $\bar{J}_b$ denote the set of all odd sets not in J_b. The RP is therefore

$$\min \xi = \sum_{j=1}^{n} x_j^a + 2 \sum_{k=1}^{N} x_{n+k}^a$$

subject to

$$\left.\begin{aligned} \sum_{j=1}^{n} x_{ij} + x_i^a &= 1 \qquad i = 1, \ldots, n \qquad (11.6) \\ \sum_{i,j \in S_k} x_{ij} + y_k + x_{n+k}^a &= s_k \qquad k = 1, \ldots, N \qquad (11.7) \end{aligned}\right\}$$

$$\left.\begin{aligned} x_{ij} \geq 0, \quad \text{and} \quad [v_i, v_j] \notin J_e &\Longrightarrow x_{ij} = 0 \qquad (11.8) \\ y_k \geq 0, \quad \text{and} \quad S_k \notin J_b &\Longrightarrow y_k = 0 \qquad (11.9) \end{aligned}\right\}$$

(The factor of 2 in the cost function allows some of the γ to be 2 in the dual of RP.) The dual of the RP (DRP) is, therefore,

$$\max \sum_{i=1}^{n} \alpha_i + \sum_{k=1}^{N} s_k \gamma_k$$

subject to

$$\alpha_i + \alpha_j + \sum_{i,j \in S_k} \gamma_k \leq 0 \qquad [v_i, v_j] \in J_e \tag{11.10}$$

$$\gamma_k \leq 0 \qquad S_k \in J_b \tag{11.11}$$

$$\alpha_i \leq 1 \qquad \text{for all } i$$

$$\gamma_k \leq 2 \qquad \text{for all } k$$

where the α's correspond to the constraints in (11.6) and the γ's to (11.7).

The rest of this section will be devoted to establishing that the dual variables α_j and γ_k maintained throughout the primal-dual algorithm are such that RP and DRP are "combinatorial" in nature and hence always yield integral solutions to RP. Because the final solution of RP is the solution of (11.1), (11.2), and (11.3), this will prove Theorem 11.2. Let us assume, for the moment, that the current α's and γ's of D and the optimal solution x_{ij} of the corresponding RP are such that the following holds.

Assumption

(a) The x_{ij}'s are 0-1, and they form a matching of the graph (V, J_e).

(b) If $S_k \in \bar{J}_b$ (that is, $\gamma_k < 0$) then the graph (V, J_e) restricted to S_k contains s_k matched edges. (Note that this implies that S_k is *full*; that is, $\sum_{i,j \in S_k} x_{ij} = s_k$.)

(c) If $S_l, S_k \in \bar{J}_b$ and $S_l \cap S_k \neq \varnothing$, then either $S_l \subseteq S_k$ or $S_k \subseteq S_l$.

With this assumption, we may forget about the constraints of (11.7) in the RP of (11.6), (11.7), (11.8), and (11.9) because if $\gamma_k = 0$, then y_k may be positive, and hence it can be used to fill the gap between $\sum_{i,j \in S_k} x_{ij}$ and s_k at no cost; on the other hand, if $\gamma_k < 0$—that is, $S_k \in \bar{J}_b$—there is no such gap to fill by (b) of our assumption. Thus we can take $x^a_{n+k} = 0$ for $k = 1, \ldots, N$, which we shall do in what follows.

Let us now define the *admissible graph* G_J corresponding to J_e and J_b. The graph G_J consists of the graph (V, J_e) *after shrinking all odd sets in* $\bar{J}_b$. For example, if $n = 12$ and J_e and J_b are as shown in Fig. 11–4(a), G_J is shown in Fig. 11–4(b). Notice that (c) of our assumption is needed in order for G_J to be well defined. Next, we call a matching of (V, J_e) *proper* if all odd sets in $\bar{J}_b$ are full. For example, in Fig. 11–4(a), we show the maximum proper matching of (V, J_e). Notice that it is not the maximum matching of (V, J_e).

Lemma 11.2 *There is a matching in G_J with d exposed nodes iff there is a proper matching in (V, J_e) with d exposed nodes.*

Proof The reader can follow the correspondence in Fig. 11–4. From a proper matching in (V, J_e) we can go to a matching in G_J by simply shrinking maximal odd sets in $\bar{J}_b$; this does not change the number of exposed nodes, since we started with a proper matching. For the other direction, a matching in G_J can be turned into a proper matching of (V, J_e) by expanding the odd sets and filling them with matched edges, as appropriate. An expanded odd set will contain an exposed node if and only if it was itself an exposed node in G_J; the lemma follows. □

One of the consequences of the lemma is that we can find the maximum cardinality proper matching in (V, J_e) by finding a maximum cardinality matching in G_J. This is quite important in view of our next claim.

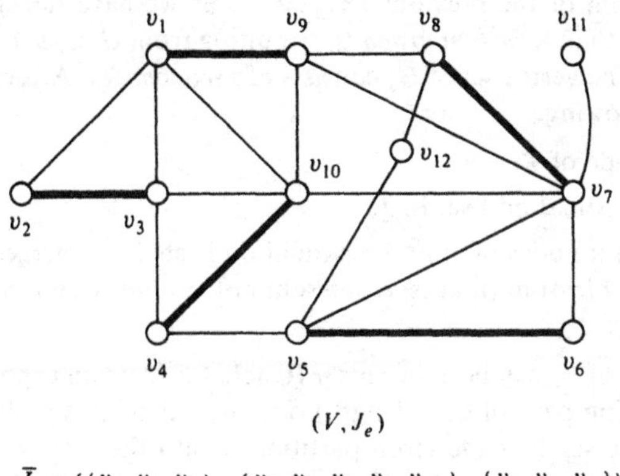

(V, J_e)

$\bar{J}_b = \{\{v_1, v_2, v_3\}, \{v_1, v_2, v_3, v_4, v_{10}\}, \{v_7, v_8, v_9\}\}$

(a)

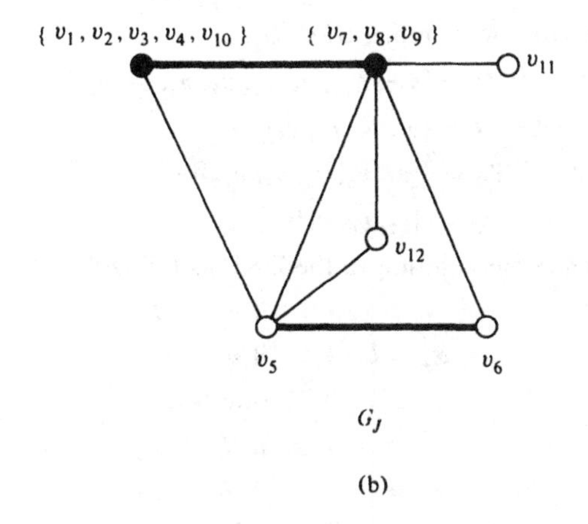

G_J

(b)

Figure 11–4

Claim The optimal solution $\{x_{ij}\}$ *of* RP *is the maximum proper matching of* (V, J_e).

The value of the claimed optimum is therefore, by (11.6),

$$\sum_{j=1}^{n} x_j^a = \sum_{j=1}^{n} 1 - \sum_{i,j} x_{ij}$$

which is the number d of exposed nodes in the maximum proper matching in (V, J_e). We shall prove our claim by exhibiting a solution of DRP that achieves the same cost d. Suppose that we have found the maximum matching of G_J

by the algorithm of the previous chapter. Thus we have failed to discover an augmenting path in a *current graph* G_c, resulting from G_J by shrinking a number of blossoms. The vertex set of G_c consists of *pseudonodes*. A pseudonode can be any of the following.

1. A node of V.
2. A maximal odd set in $\bar{J}_b$.
3. Several nodes of V and maximal odd sets in $\bar{J}_b$ merged into an outermost blossom (that is, a blossom not contained in any other blossom) of G_J.

A pseudonode of G_c can be either *outer* (reachable from an exposed pseudonode by an alternating path of even length) or *inner* (mate of an outer pseudonode) or neither. The set V is therefore partitioned into the sets O of *outer* vertices (those that are in outer pseudonodes), I (vertices in inner pseudonodes), and the remaining vertices. Similarly, the nodes of G_c that are not actual nodes of V are partitioned into Ψ_O (maximal odd sets corresponding to outer pseudonodes or to blossoms), Ψ_I (maximal odd sets corresponding to inner pseudonodes), and the rest.

In Fig. 11–4(b), for example, $G_c = G_J$,

$$O = \{v_{11}, v_{12}, v_6, v_1, v_2, v_3, v_4, v_{10}\}$$

$$I = \{v_5, v_7, v_8, v_9\}$$

$$\Psi_O = \{\{v_1, v_2, v_3, v_4, v_{10}\}\}$$

$$\Psi_I = \{\{v_7, v_8, v_9\}\}$$

We now define our solution to the DRP as follows.

$$\bar{\alpha}_j = \begin{cases} 1 & \text{if } v_j \in O \\ -1 & \text{if } v_j \in I \\ 0 & \text{otherwise} \end{cases}$$

$$\bar{\gamma}_k = \begin{cases} -2 & \text{if } S_k \in \Psi_O \\ 2 & \text{if } S_k \in \Psi_I \\ 0 & \text{otherwise} \end{cases}$$

It is easy to see that these values satisfy (11.11), because $\Psi_I \subseteq \bar{J}_b$. Inequality 11.10 is also satisfied, because if $[v_i, v_j] \in J_e$, then (11.10) can be violated only if $v_i, v_j \in O$. But this means that v_i and v_j belong to the same outer pseudonode (they cannot belong to different outer pseudonodes, because then the two pseudonodes would constitute a blossom), and hence $\bar{\gamma}_k = -2$ is added to (11.10) to salvage its validity. The cost of this solution is, therefore,

$$|O| - |I| - 2 \sum_{S_k \in \Psi_O} s_k + 2 \sum_{S_k \in \Psi_I} s_k \tag{11.12}$$

We can now see that (11.12) is the number of outer pseudonodes in G_c minus that of inner pseudonodes; this, however, is the number of exposed vertices in G_c, which is—by Lemma 11.2—equal to the number of exposed vertices in the maximum proper matching of (V, J_e). This is exactly the claimed optimal cost of RP. Optimality has been established and the Claim proved. Consequently, all solutions of RP will be maximum proper matchings. This includes the last iteration, at which the optimal solution of RP is that of the LP of (11.1), (11.2), and (11.3).

It remains to prove our Assumption, which we have been accepting for the longest moment now. This is very easy to do by induction on the number of iterations. It certainly holds at the first iteration, because initially all γ's and x_{ij}'s are zero. And the discussion above proves that if the assumption holds at the beginning of an iteration, it also holds at the beginning of the next iteration. This is because only the blossoms of the admissible graph may be added to $\bar{J}_b$, and blossoms are certainly full. Also, the odd sets that were already in $\bar{J}_b$ remain full, because the maximum matching of RP is always a proper matching. Finally, (c) of the assumption is also maintained, since only blossoms of G_J are added to $\bar{J}_b$, and these certainly satisfy (c). The theorem is proved.

In order to give a detailed description of the algorithm, we have to show how the dual variables are to be modified. Recall that the parameter θ_1 is calculated (Eq. 5.16) as the minimum of the quantities

$$\frac{c_{ij} - \alpha_i - \alpha_j}{\bar{\alpha}_i + \bar{\alpha}_j + \sum_{i, j \in S_k} \bar{\gamma}_k} \quad \text{and} \quad \frac{-\gamma_k}{\bar{\gamma}_k},$$

where the minimum is taken over all such quantities for which the denominator is positive.

The first denominator can be positive in two cases:

1. $v_i, v_j \in O$, but not in the same pseudonode of G_c, in which case it is 2;
2. $v_i \in O, v_j \in V - I - O$. Then $\bar{\alpha}_i + \bar{\alpha}_j + \sum_{i, j \in S_k} \bar{\gamma}_k = 1$.

Finally, $\bar{\gamma}_k > 0$ whenever $S_k \in \Psi_I$. Thus we take $\theta_1 = \min(\delta_1, \delta_2, \delta_3)$, where

$$\begin{aligned} \delta_1 &= \min\left\{\frac{c_{ij} - \alpha_i - \alpha_j}{2} : v_i, v_j \in O, \text{ not in the same pseudonode}\right\} \\ \delta_2 &= \min\{c_{ij} - \alpha_i - \alpha_j : v_i \in O, v_j \in V - I - O\} \\ \delta_3 &= \min\{-\gamma_k/2 : S_k \in \Psi_I\} \end{aligned} \tag{11.13}$$

The algorithm is shown in Fig. 11–5.

Theorem 11.3 *The algorithm of Fig. 11–5 correctly solves the weighted matching problem in $O(n^4)$ time.*

Proof By our Claim, the algorithm is a version of the primal-dual method applied to the LP of (11.1), (11.2), and (11.3), and is therefore correct by Theorem 11.2. For the time bound, the algorithm consists of a number of searches for

WEIGHTED MATCHING ALGORITHM

Input: A $n \times n$ symmetric matrix $[c_{ij}]$ of nonnegative integers; n is even.
Output: The complete matching M which has the smallest total cost under c_{ij}.

```
begin
for all vi∈V do αi:=½min{cij};
                         j
for all k do γk:=0;
M:=∅; (comment: initialize)
Jb:=∅; (comment: J̄b contains all odd sets Sk with γk<0)
while |M|<n/2 do
  begin
  construct the admissible graph GJ by including all edges [vi,vj] with αi+αj+   Σ   γk=cij,
                                                                               i, j∈Sk
  and shrinking all sets Sk in J̄b;
  find the maximum matching in GJ starting from the current matching M;
  let Gc be the current graph at the conclusion of the (unweighted) maximum matching
  algorithm for GJ;
  let O be the set of outer vertices in Gc, I the set of inner ones, ΨO the set of outer pseudonodes,
  and ΨI the set of inner pseudonodes;
  calculate θ1=min(δ1,δ2,δ3) (comment: Equations 11.13)
  for all vj∈O do αj:=αj+θ1;
  for all vj∈I do αj:=αj−θ1;
  for all Sk∈ΨO do γk:=γk−2θ1;
  for all Sk∈ΨI do γk:=γk+2θ1;
  recover the maximum proper matching M of (V,Je) from the maximum matching of GJ;
  let J̄b:={Sk∈J̄b∪ΨO : γk<0}
  end
end
```

Figure 11–5 The weighted matching algorithm.

augmenting paths interleaved with either augmentations or dual variable modifications. Call each search a *step* and the sequence of steps between two successive augmentations a *stage*. Obviously, there can be no more than $n/2$ stages; we shall next bound the number of steps in each stage.

A step can be of one of three types, depending on whether, at its conclusion, $\theta_1 = \delta_1, \theta_1 = \delta_2$, or $\theta_1 = \delta_3$. In the first case, two outer nodes of G_J are joined by a new admissible edge. This means that in the next step we shall discover either an augmenting path or a blossom, depending on whether the exposed nodes from which these outer vertices are reachable by an augmenting path of even length are different or identical, respectively. Since each blossom decreases the size of the current graph by at least 2, we cannot have more than $n/2 + 1$ steps of Type 1 in one stage.

If $\theta_1 = \delta_2$, a new outer vertex is discovered in the next step. This means that the number of steps of this type is also bounded by $n/2$. Finally, if $\theta_1 = \delta_3$, we remove from $\bar{J}_b$ an odd set S_k that corresponds to an inner pseudovertex of G_J. However, when S_k was first added to G_J it was an outer pseudovertex. Consequently, an augmentation must have happened in between. The number of such steps in a stage is therefore bounded by the number of odd sets in $\bar{J}_b$ at the conclusion of the previous stage. However, this number is bounded by $n/2$ by (c) of our assumption.

It follows that we have a total of $O(n^2)$ steps. Each search can be carried out in $O(n^2)$ time by the cardinality matching procedure. (A step involving augmentation, which might take n^3 time, can occur only once per stage.) Furthermore, the determination of O, I, Ψ_O, and Ψ_I can be carried out in $O(n^2)$ time while finding the optimal matching of G_J. Finally, the construction of G_J, the calculation of θ_1 and the variable modification can also be done in $O(n^2)$ time. This proves the theorem. □

Example 11.2

Let us apply our weighted matching algorithm to the example shown below.

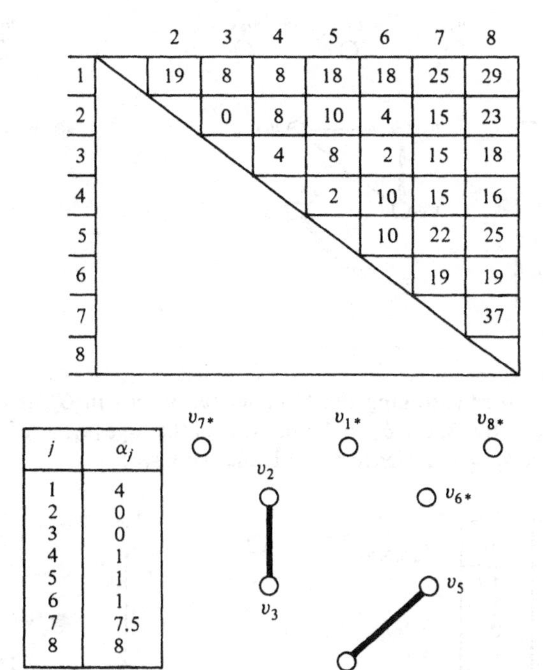

	2	3	4	5	6	7	8
1	19	8	8	18	18	25	29
2		0	8	10	4	15	23
3			4	8	2	15	18
4				2	10	15	16
5					10	22	25
6						19	19
7							37
8							

j	α_j
1	4
2	0
3	0
4	1
5	1
6	1
7	7.5
8	8

The initial values of the α_j's are also shown. We also depict the admissible graph G_J and its maximum matching. Outer vertices are denoted by an asterisk, and there are no inner nodes or blossoms. To calculate θ_1, we see that $\delta_1 = 5$ because of the edge $[v_8, v_6]$. $\delta_2 = 1$ because of $[v_3, v_6]$, and $\delta_3 = \infty$ because $\Psi_I = \varnothing$. We therefore have $\theta_1 = 1$. The resulting changes are shown on the top of page 264.

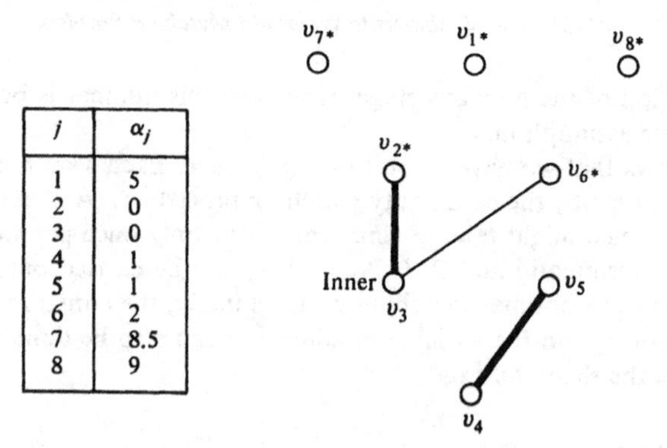

j	α_j
1	5
2	0
3	0
4	1
5	1
6	2
7	8.5
8	9

The maximum matching of G_J is shown, $\delta_1 = 1$ because of $[v_2, v_6]$, $\delta_2 = 2$, and $\delta_3 = \infty$. So $\theta_1 = 1$, and we have the following.

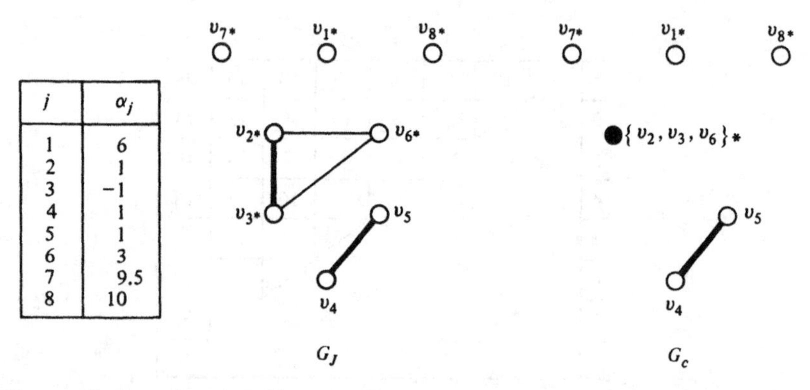

j	α_j
1	6
2	1
3	−1
4	1
5	1
6	3
7	9.5
8	10

The graph G_c, after shrinking the blossom $\{v_2, v_3, v_6\}$ in G_J, is shown above. Here $\bar{J}_b = \Psi_O = \{\{2, 3, 6\}\}$, $\delta_1 = 1.5$ because of the edge $[v_1, v_3]$, $\delta_2 = 1$ because of $[v_1, v_4]$, and $\delta_3 = \infty$. Hence $\theta_1 = 1$ and we have:

j	α_j
1	7
2	2
3	0
4	1
5	1
6	4
7	10.5
8	11

$\gamma_{\{2,3,6\}} = -2$

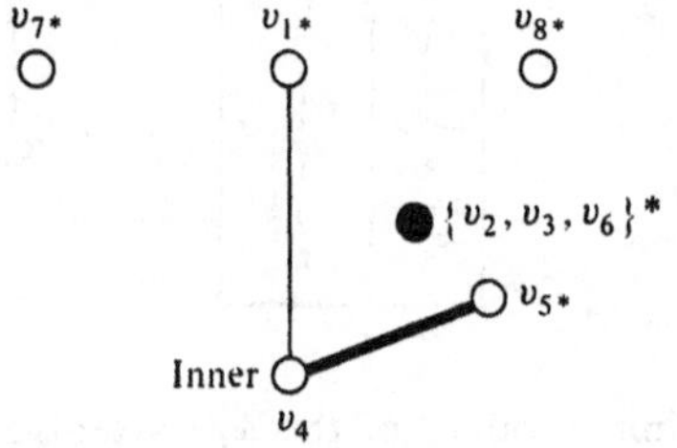

Notice that we modified the variable $\gamma_{\{2,3,6\}}$ corresponding to the set in Ψ_O. The optimal matching of G_J is shown, $\bar{J}_b = \Psi_O = \{\{2, 3, 6\}\}$, $O = \{v_1, v_2, v_3, v_5, v_6, v_7, v_8\}$, and $I = \{v_4\}$. Again, no augmentation is possible. We have $\delta_1 = 0.5$, $\delta_2 = \delta_3 = \infty$, and hence:

j	α_j
1	7.5
2	2.5
3	0.5
4	0.5
5	1.5
6	4.5
7	11.0
8	11.5

$\gamma_{\{2,3,6\}} = -3$

Thus v_1 is matched with $\{v_2, v_3, v_6\}$ in G_J. Now $\delta_1 = 7.25$ because of the edge $[v_8, v_7]$, $\delta_2 = 1.5$ because of the edge $[v_2, v_7]$, and $\delta_3 = \infty$ because $\Psi_I = \varnothing$. Hence we have $\theta_1 = 1.5$, and:

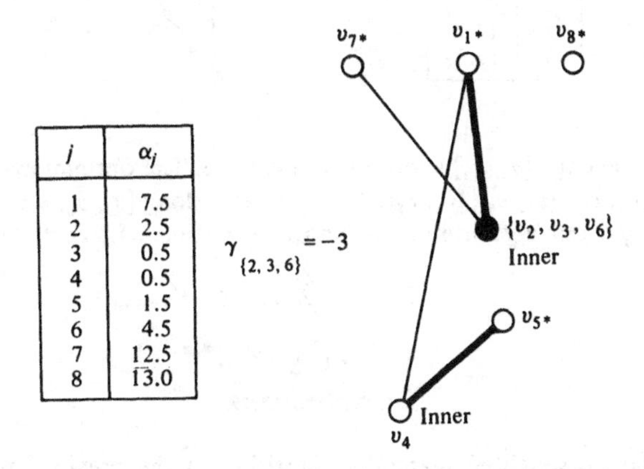

j	α_j
1	7.5
2	2.5
3	0.5
4	0.5
5	1.5
6	4.5
7	12.5
8	13.0

We show the optimal matching in G_J; $\bar{J}_b = \Psi_I = \{\{2, 3, 6\}\}$, $O = \{v_1, v_7, v_8, v_5\}$, and $I = \{v_2, v_3, v_4, v_6\}$. Because of $[v_1, v_7]$, $\delta_1 = 2.5$, and $\delta_2 = \infty$. Finally, $\theta_1 = \delta_3 = 1.5$ because $\{2, 3, 6\} \in \Psi_I$. Our modification gives

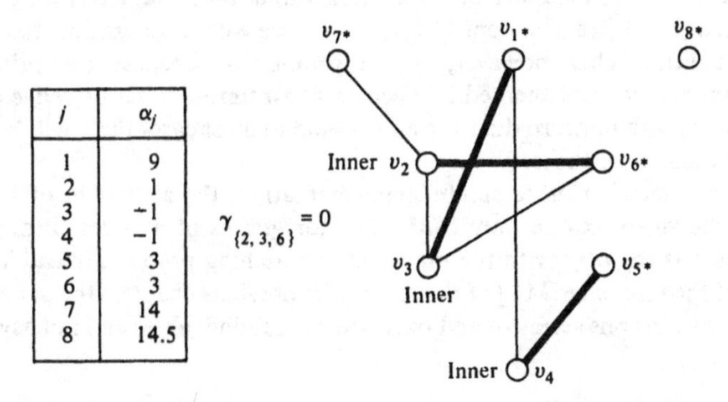

j	α_j
1	9
2	1
3	−1
4	−1
5	3
6	3
7	14
8	14.5

Because $\gamma_{\{2,3,6\}} = 0$, and hence $\{2, 3, 6\}$ is removed from $\bar{J}_b$, we do not shrink $\{v_2, v_3, v_6\}$ in G_J any more. At the conclusion of the search for an augmenting path in G_J the situation is as shown above. We find that $\theta_1 = \delta_1 = 0.75$ (because of the edge $[v_6, v_8]$), and the dual variables are modified thus:

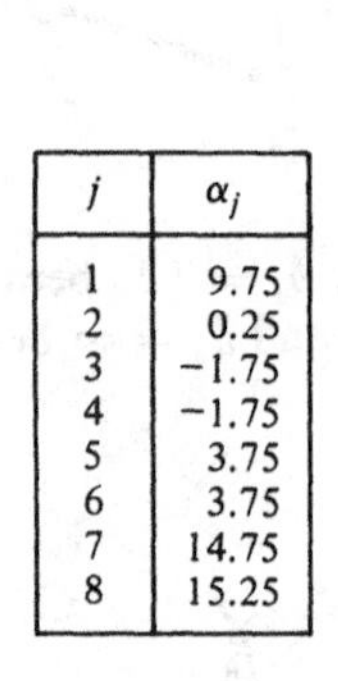

j	α_j
1	9.75
2	0.25
3	−1.75
4	−1.75
5	3.75
6	3.75
7	14.75
8	15.25

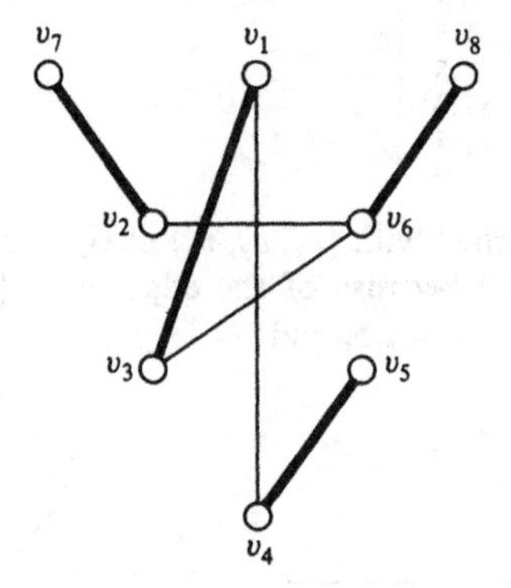

Notice that the edge $[v_2, v_3]$ is not in G_J anymore. The optimal matching in G_J is $\{[v_1, v_3], [v_2, v_7], [v_4, v_5], [v_6, v_8]\}$ (we augment along $[v_8, v_6, v_2, v_7]$). Because $|M| = 4 = n/2$, our algorithm stops and M is optimal. Its cost checks with

$$\sum_{j=1}^{n} \alpha_j + \sum_{k=1}^{N} \gamma_k s_k = 44 \quad \square$$

11.4 Conclusions

The nonbipartite weighted matching algorithm of the previous section is an excellent example of an application of the primal-dual algorithm. It reveals the nature of the primal-dual algorithm as a *general method* for *reducing* weighted problems to their unweighted counterparts. To carry out this reduction, we must first guess the complete set of linear constraints that characterize exactly the problem at hand (see Theorem 11.2). Often there will be an exponential number of inequalities. This, however, is not prohibitive, because the primal-dual algorithm can be implemented by keeping at all times the set of *active* inequalities—those with nonzero dual variables—and in most cases there will be a manageable number of those.

By a somewhat more careful implementation, the algorithm of Fig. 11–5 can be shown to require only $O(n^3)$ time for graphs of n nodes (Problem 14). Thus, as was the case with the unweighted matching problem (recall Theorem 10.3 and the reference [MV] at the end of the previous chapter) the generality of nonbipartite graphs seems to add only conceptual difficulties and not asymptotic inefficiency.

PROBLEMS

1. Solve the assignment problem shown below.

1	3	2	4	5
4	1	3	2	5
1	3	2	5	4
5	2	1	4	3
4	5	1	3	2

2. Show that if the costs c_{ij} are integers, then the values of the dual variables during the Hungarian method are always multiples of $\frac{1}{2}$.

3. Show that if the costs c_{ij} are integers, then the values of the dual variables during the nonbipartite weighted matching algorithm are always multiples of $\frac{1}{4}$.

4. Find the minimum cost matching for the graph shown below (all missing edges have cost ∞).

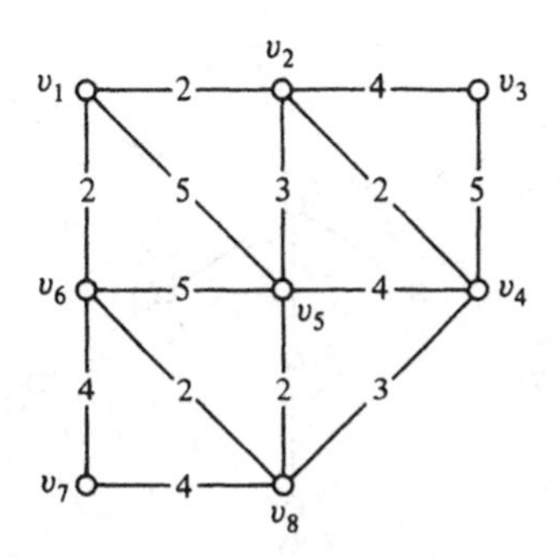

5. (a) Show by an example that the *maximum weight* matching of a (not necessarily complete) graph may not be of maximum cardinality, even for positive weights.
   (b) Give an algorithm for finding the maximum weight *complete* matching of a given graph.
   *(c) Give an algorithm for finding the maximum weight matching *with k edges*, for fixed k. (*Hint:* add the equation $\sum x_{ij} = k$ to (11.1), (11.2) and (11.3). How does this affect the primal-dual algorithm?)

*6. Show that the Hungarian method for weighted (not necessarily complete) bipartite graphs can be implemented in $O(|V||E|\log|V|)$ time.

7. Consider a set of costs $c_{ij} \geq 0$ on the edges of the complete graph K_n with n nodes, and let $S \subseteq \{v_1, \ldots, v_n\}$. We wish to find a subgraph of K_n that has an odd degree at nodes in S, an even (possibly zero) degree at all other nodes, and as little total cost as possible. Show that this is a matching problem.

***8.** (*Chinese Postman Problem*) We are given a connected graph $G = (V, E)$ and a cost function $c\colon E \longrightarrow Z^+$. We wish to find a *walk*, traversing each edge of E at least once, such that the total cost of the walk (with multiple traversals of an edge charged multiply) is as small as possible. Give a polynomial-time algorithm for this problem by using the result of Problem 7.

***9.** Repeat Problem 8 when G is *directed.*

10. Give a polynomial-time algorithm for the *minimum-cost edge cover* problem for weighted graphs (recall Problem 3 of Chapter 10) by reducing it to weighted matching.

11. Given a set of costs c_{ij} on the edges of the complete graph K_{2n+1} of $2n + 1$ nodes, we wish to find a subgraph of K_{2n+1} that is isomorphic to the "double star" shown below and which has the smallest possible total cost. Show that this problem is equivalent to weighted matching.

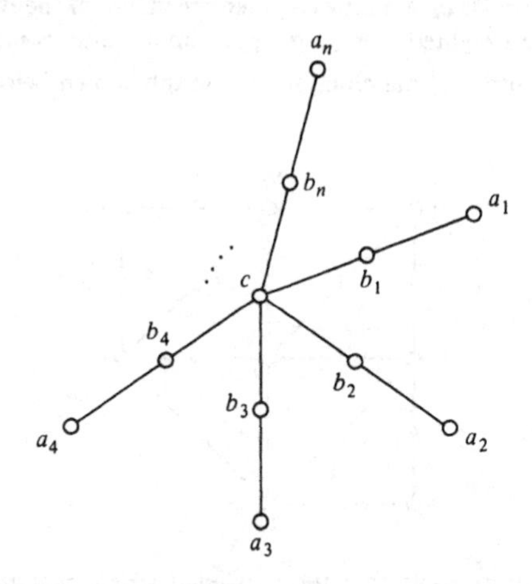

***12.** Given a set of costs c_{ij} on the edges of the complete graph K_n of n nodes and n integers $b(v_i)$, $i = 1, \ldots, n$, find a subgraph G of K_n that has degree $b(v_i)$ at node v_i, $i = 1, \ldots, n$, such that G has the smallest total cost possible. (*Hint:* Recall the construction of Problem 13 in Chapter 10.)

***13.** (a) Consider an instance of the assignment problem with $B = (V, U, V \times U)$, and $c_{ij}\colon V \times U \longrightarrow Z^+$ such that $|V| = |U| = n$, and the c_{ij}'s have the following special structure: There are real numbers $\alpha_1 \geq \alpha_2 \geq \cdots \geq \alpha_n$ and $\beta_1 \geq \beta_2 \geq \cdots \geq \beta_n$ such that, for all $1 \leq i, j \leq n$, we have

$$c_{ij} = \max(0, \alpha_i - \beta_j).$$

Show that the optimal matching is always

$$M = \{[v_i, u_i]\colon i = 1, \ldots, n\}.$$

(b) Generalize Part (a) for the case in which

$$c_{ij} = \begin{cases} \int_{\alpha_i}^{\beta_j} f(t)\,dt & \text{if } \alpha_i \le \beta_j \\ \int_{\beta_j}^{\alpha_i} g(t)\,dt & \text{otherwise} \end{cases}$$

for some nonnegative functions f and g.

14. In this problem we show that the nonbipartite weighted matching algorithm of Fig. 11–5 can be implemented in $O(n^3)$ time. The algorithm proceeds in $n/2$ *stages*, that is, portions between two successive augmentations. Each stage may have $O(n)$ *steps*, that is, portions between two successive modifications of dual variables. One aspect of the wastefulness of the algorithm of Fig. 11–5 is that G_J is constructed anew at each new step, and G_c is discarded.

(a) Describe an implementation of the algorithm in which we update the same graph during a stage. Do not forget the case of the step before the last of Example 11.2.

A step is said to be of Type 1, 2 or 3 depending on whether, at its conclusion, $\theta_1 = \delta_1, \delta_2$, or δ_3.

(b) Show that there are $O(n)$ steps of each type during a stage.

(c) Show how to implement in $O(n)$ time each step of any type, excluding the calculation of θ_1.

As for the calculation of θ_1, δ_3 can be computed in $O(n)$ time, since $|\Psi_J| = O(n)$.

(d) Show how δ_2 can be computed $O(n)$ time in a manner similar to that of the Hungarian method (that is, using arrays *slack* and *nhbor*).

Thus it remains to compute δ_1. We can do this by maintaining a list L of $O(n)$ sets of nodes, namely, (1) all singletons $\{v_i\}$, (2) all sets in $\bar{J}_b$, (3) all blossoms of G_c, possibly not yet in $\bar{J}_b$. For each outer vertex v_i and each $S \in L$, we have a node *closest* $[v_i, S] \in S$ such that

(i) *closest* $[v_i, S]$ was not outer when v_i became outer;

(ii) *closest* $[v_i, S]$ has the smallest $c_{ij} - \alpha_i - \alpha_j$ among all $v_j \in S$ satisfying (i).

Let *dist*$[v_i, S]$ be the value of $c_{ij} - \alpha_i - \alpha_j$ at the time that v_i became outer or S was inserted in L, whichever came later.

Also, for each $S \in L$, we maintain two quantities: $\delta[S] = \min\{c_{ij} - \alpha_i - \alpha_j$: $v_i \in S, v_j \notin S$, outer$\}$, and $\Delta[S]$, which is the total increment in the α's of vertices in S since S was added to L.

*(e) Show that this structure can be maintained at a cost of $O(n^2)$ time per stage.

(f) Conclude that the nonbipartite weighted matching algorithm of Fig. 11–5 can be implemented in $O(|V|^3)$ time.

NOTES AND REFERENCES

The Hungarian method is due to Kuhn.

[Ku] KUHN, H. W., "The Hungarian Method for the Assignment Problem," *Naval Research Logistics Quarterly*, 2 (1955), 83–97.

Theorem 11.2 and the primal-dual algorithm for weighted nonbipartite matching are due to Jack Edmonds.

[Ed] EDMONDS, J., "Matching and a Polyhedron with 0-1 Vertices," *J. Res. NBS*, 69B (1965), 125–30.

The $O(n^3)$ matching algorithm (Problem 14) is from

[La] LAWLER, E. L., *Combinational Optimization: Networks and Matroids*. New York: Holt, Rinehart & Winston, 1976.

Problem 6 is due to Mihalis Yannakakis. Problems 8 and 9 are from

[Kw] KWAN, MEI-KO, "Graphic Programming Using Odd and Even Points," *Chinese Math.*, 1 (1962), 273–77.

[EJ] EDMONDS, J., and E. L. JOHNSON, "Matching, Euler Tours and the Chinese Postman," *Math. Prog.*, 5 (1973), 88–124.

It turns out that the *mixed* version of the Chinese postman problem (on weighted graphs with *both* directed and undirected edges) is much harder. See Chapter 15 and

[Pa] PAPADIMITRIOU, C. H., "The Complexity of Edge Traversing," *J. ACM*, 23 (1976), 544–54.

Problem 11 is from

[PY] PAPADIMITRIOU, C. H., and M. YANNAKAKIS, "The Complexity of Restricted Spanning Tree Problems," pp. 460–70 in *Automata, Languages and Programming*, ed. H. A. Maurer. Berlin: Springer-Verlag, 1979.

It turns out that, for the cost matrices of the special type discussed in Problem 13, not only is the assignment problem easier, but so is the TSP. See

[GG] GILMORE, P. C., and R. E. GOMORY, "Sequencing a One-State Variable Machine: A Solvable Case of the Traveling Salesman Problem," *OR*, 12 (1964), 655–79.

12

Spanning Trees and Matroids

In this chapter we solve the minimum spanning tree problem. In doing so, we observe that this problem has a fundamental structural property that makes it susceptible to fast algorithmic solution. This property is shared by a class of optimization problems known as *matroidal problems*, because the underlying structure is a generalization of graphs called *matroids*.

We also show that a class of more complicated problems can be solved efficiently. These problems call for, roughly, finding the best solution that is feasible in *two different* matroidal problems. These problems, together with the matching problem of the previous chapter, are in some sense the hardest combinatorial problems that are known to be solvable by polynomial-time algorithms. They are to be contrasted to a plethora of much harder problems, such as the TSP and integer linear programming, which we are going to discuss in subsequent chapters.

12.1 The Minimum Spanning Tree Problem

We have referred to the MST as a classical, well-solved combinatorial optimization problem several times in earlier chapters. In this problem we are given a symmetric $|V| \times |V|$ matrix of positive numbers $[d_{ij}]$, the distance matrix, and our task is to find the shortest spanning tree of $K_{|V|}$, the complete graph of $|V|$ nodes, under this metric.

Sometimes the MST problem is posed in terms of a connected graph $G = (V, E)$—not necessarily complete—and a positive distance d_{ij}† for each edge $[v_i, v_j] \in E$. Again, we are looking for the shortest spanning tree of G; because G is connected, we know that it has at least one. To convert an instance of this version to an instance of the original problem, we simply set $d_{ij} = \infty$ for all $[v_i, v_j] \notin E$.

There are several efficient algorithms that solve the MST problem. They all exploit, in one way or another, the following fact.

Theorem 12.1 *Let* $\{(U_1, T_1), (U_2, T_2), \ldots, (U_k, T_k)\}$ *be a forest spanning* V, *and let* $[v, u]$ *be the shortest of all edges with only one endpoint in* U_1. *Then among all spanning trees containing all edges in* $T = \bigcup_{j=1}^{k} T_j$, *there is an optimal one containing* $[v, u]$.

Proof Suppose that there is a spanning tree (V, F) with $F \supseteq T$ and $[v, u] \notin F$, which is shorter than all spanning trees containing all of T and $[v, u]$. Let us add the edge $[v, u]$ to F; by Proposition 1.2 a unique cycle results. This cycle does not consist solely of nodes in U_1 because $v \notin U_1$ (see Fig. 12–1). Consequently, there is an edge $[v', u']$ on this cycle, different from $[v, u]$, with $u' \in U_1$ and $v' \in V - U_1$. By hypothesis, this edge is no shorter than $[v, u]$ and does not belong to T. Consequently, if we remove $[v', u']$, we obtain a new spanning tree (V, F'), $F' = F \cup \{[v, u]\} - \{[v', u']\}$, containing T and $[v, u]$ and with cost not larger than that of (V, F). However, this contradicts our assumption that (V, F) is shorter than any spanning tree that contains $[v, u]$ and T, and the theorem is proved. □

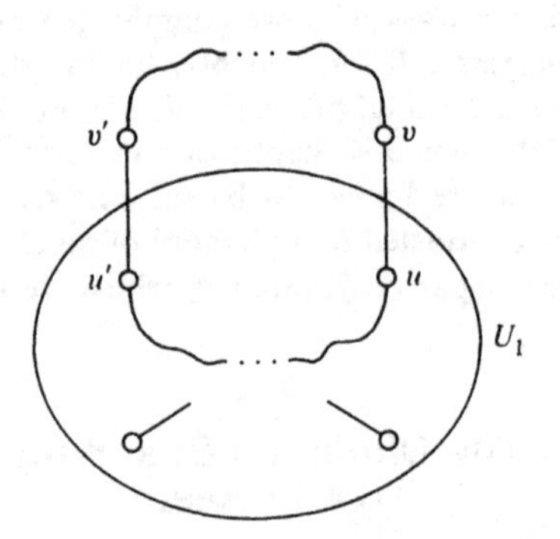

Figure 12–1

†We shall use both d_{ij} and $d(v_i, v_j)$ to denote the distance between v_i and v_j.

Let us see how this result leads us almost directly to an efficient algorithm for the MST problem. Because the theorem holds for all spanning forests T, it would certainly hold when $U_j = \{v_j\}, j = 1, \ldots, n$, and $T = \varnothing$. Applied to this case, the theorem states that there exists a minimum spanning tree containing the shortest edge leaving v_1. Hence we can decide at this point to include this edge in the tree without giving up optimality. Suppose that this edge is $[v_1, v_2]$. We next apply the theorem to the sets $U_1 = \{v_1, v_2\}, U_3, \ldots, U_n$ as before, and $T = \{[v_1, v_2]\}$. Now the theorem states that we may add the shortest edge leaving v_1 or v_2—not $[v_1, v_2]$—to the tree, and we shall be able to complete the tree formed into an optimal MST. An algorithm is now transparent: We start from a set $U = \{v_1\}$, for instance, and we recursively add to T the shortest edge leaving U, until all vertices have been added to U and a tree has been formed.

This algorithm for the MST problem is shown in Fig. 12–2. In order to

```
MINIMUM SPANNING TREE ALGORITHM
Input:  A set V of vertices, and the distances d_ij between any
        two vertices v_i, v_j of V.
Output: The shortest spanning tree (V, T).
begin
U:={v1}, T:=∅;
for all v∈V−{v1} do closest[v]:=v1; (comment: initialize)
while U≠V do
  begin
  min:=∞;
  for all v∈V−U do
    if d(v,closest[v])<min then min:=d(v,closest[v]), next:=v;
    (comment: find the node in V−U that is closest to U)
  U:=U ∪ {next}, T:=T ∪ {[next, closest[next]]};
  for all v∈V−U do
    if d(v, closest[v])>d(v,next) then closest[v]:=next;
  end
end
```

Figure 12–2 The minimum spanning tree algorithm.

facilitate the main computational task, that is, finding the shortest edge leaving U, we maintain an array *closest*[v]. For each vertex $v \in V - U$, *closest*[v] equals the vertex of U that is closest to v. So, in order to find the shortest edge leaving U, we need only find the shortest edge among those of the form [v, *closest*[v]], $v \notin U$. Naturally, the array *closest* must be updated each time a new vertex is added to U. The rest of the algorithm is quite straightforward.

Theorem 12.2 *The algorithm of Fig. 11–2 correctly solves the MST problem in* $O(|V|^2)$ *time.*

Proof Because $|U| = |V|$ at termination and because we always add to T edges leaving U, the resulting graph (V, T) is a spanning tree of G. It remains to show that it is an optimal one. We shall show by induction on the cardinality of the set U that there is always an optimal spanning tree of G containing the corresponding set T. This is certainly true when $U = \{v_1\}$ and $T = \varnothing$. Suppose it is true for some value j of $|U|$, $1 \leq j < |V|$. We can view the partial spanning tree (U, T) as part of a forest $\{(U_1, T_1), \ldots, (U_k, T_k)\}$ with $U_1 = U$, $k = |V| - |U| + 1$, and $T_2 = \cdots = T_k = \varnothing$. Hence Theorem 12.1 applies and among all trees containing T, there is a shortest one containing both T and the shortest edge leaving U. However, by the induction hypothesis, there is a globally optimal tree containing T. Hence there is also an optimal spanning tree containing T and the shortest edge leaving U, which is exactly T at the next stage when $|U| = j + 1$. The induction step is completed.

For the time bound, we note that the algorithm runs in $|V| - 1$ stages; one edge is added in each stage. The initialization takes $O(|V|)$ time, and finding *next* also requires $O(|V|)$ time. Finally, we can also update the array *closest* in $O(|V|)$ time. The $O(|V|^2)$ bound for the overall algorithm follows. □

Example 12.1

Let us find the minimum spanning tree of the graph $G = (V, E)$ shown in Fig. 12–3 by the algorithm of Fig. 12–2. (We consider that whenever $[v_i, v_j] \notin E$, $d_{ij} = \infty$.) The nine stages starting from the vertex v_1 are shown in Fig. 12–4. Beside each node v not yet in the set U, the value of *closest*[v] is shown. The minimum spanning tree is shown in Fig. 12–4(i). □

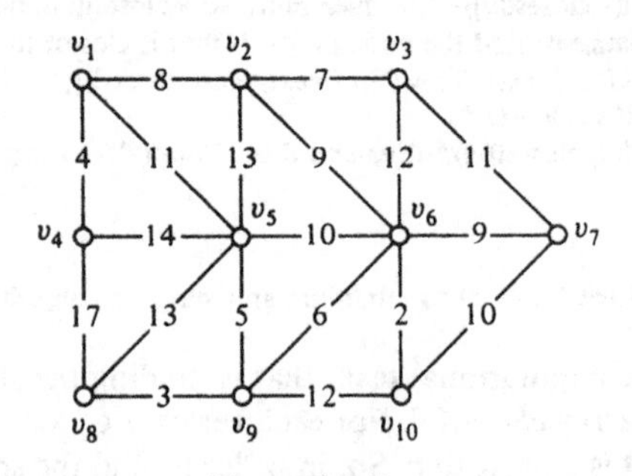

Figure 12–3

12.2 An O(|E| log |V|) Algorithm for the Minimum Spanning Tree Problem

The $O(|V|^2)$ algorithm of the previous section for finding the MST cannot be easily improved upon if our data is a $|V| \times |V|$ distance matrix. The reason is very simple: Any algorithm that is supposed to construct the MST must examine each entry of the distance matrix at least once. Otherwise, if an entry d_{ij} is never

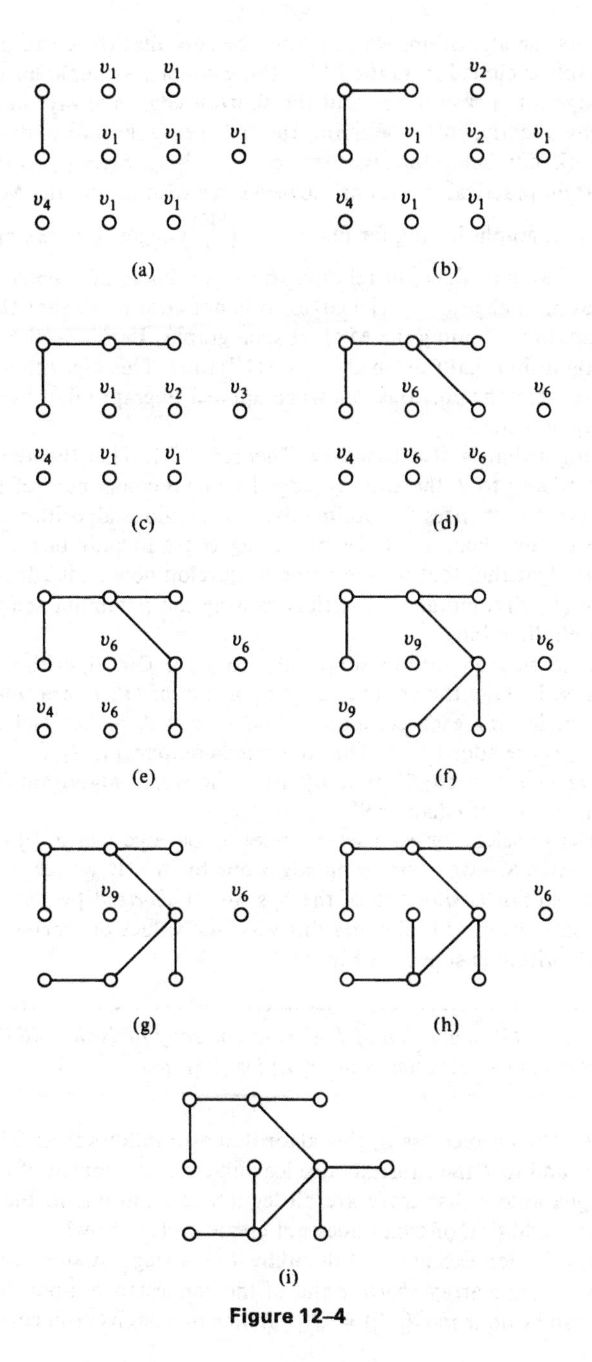
v1 v1
v1 v1 v1
v4 v1 v1
(a)
v2
v1 v2 v1
v4 v1 v1
(b)
v1 v2 v3
v4 v1 v1
(c)
v6 v6
v4 v6 v6
(d)
v6 v6
v4 v6
(e)
v9 v6
v9
(f)
v9 v6
(g)
v6
(h)
(i)

Figure 12–4

looked at by the algorithm, we can never be sure that the corresponding edge was justifiably excluded from the MST. For example, we might have skipped the shortest edge—it is easy to see that the shortest edge is always included in the MST. Consequently, just examining the indispensable information is already $\Theta(|V|^2)$ work, and hence the algorithm of Fig. 12–2 is *asymptotically optimal.*

In certain practical situations, however, we wish to find the MST of *sparse* graphs, that is, graphs having far fewer than $\binom{|V|}{2}$ edges. For example, in Problems 4 and 5 we introduce, in relation to certain kinds of *geometric* problems, graphs that have at most $3 \cdot |V|$ edges. It is not clear at all that $O(|V|^2)$ is the best we can do for finding the MST of such graphs. In this section we develop an MST algorithm that runs in $O(|E| \log |V|)$ time. This algorithm is asymptotically better than the previous one when applied to graphs that have fewer than $\Theta(|V|^2/\log |V|)$ edges.

This algorithm is also based on Theorem 12.1. That theorem says essentially that adding to T the shortest edge leaving a component of (V, T) is not going to ruin our chances for optimality. Our previous algorithm takes limited advantage of this theorem: It keeps adding edges to only one component of (V, T). The algorithm that we are going to develop now adds edges to all components of (V, T) simultaneously, thus causing the minimum spanning tree to "grow" in all directions.

Let us see how to put this simple idea to work. Our algorithm will work in stages. At each stage the connected components of (V, T) are found, and the shortest edge leaving each component is determined. At the end of the stage, all these edges are added to T. The connected components $S_1, \ldots, S_k$ of (V, T) are then calculated in $O(|E|)$ time by using the *search* algorithm (see Example 9.1 and Problem 1 of Chapter 9).

In order to select for each set of nodes S_i the edge *shortest*[i]—the shortest edge that leaves S_i—we examine all edges one by one. If we see an edge that is the shortest so far leaving one of the S_i's we set *shortest*[i] equal to this edge. When we have examined all edges this way, the values of *shortest* are final.

The algorithm is shown in Fig. 12–5.

Theorem 12.3 *The algorithm of Fig. 12–5 correctly finds the MST for a graph (V, E) and a distance function d in $O(|E| \log |V|)$ time.*

Proof The correctness of this algorithm also follows from Theorem 12.1. Because we add to T the shortest edge leaving each component of (V, T), Theorem 12.1 guarantees that these are all legitimate additions to the MST. (The simultaneous addition of edges does not create cycles. Why?)

Let us call each execution of the **while** loop a *stage.* A stage consists of the computation of the array *shortest* and of the connected components of (V, T). The latter can be done in $O(|T|) = O(|V|)$ time by search (Problem 1 of Chapter

```
SECOND MST ALGORITHM
Input: A connected graph G=(V,E) and, for each edge [v, u]∈E,
     a distance d(u, v).
Output: The shortest spanning tree (V,T) of G.
begin
T:=∅, C:={{v1}, . . . , {vn}};
(comment: initialize; C is the collection of connected
          components of (V,T))
while |C|≠1 do
  begin
  for all Si∈C do min[i]:=∞;
  for all [u, v]∈E do
    begin
    let Si, Sj be the sets containing v and u, respectively;
    if i≠j then
      begin
      if d(u, v)<min[i] then min[i]:=d(u, v), shortest[i]:=[u, v];
      if d(u, v)<min[j] then min[j]:=d(v, u), shortest[j]:=[u, v];
      end
    end;
    (comment: shortest[j] is the shortest edge leaving Sj)
  for all Sj∈C do T:=T∪{shortest[j]};
  find the set C of connected components of (V,T)
  end
end
```

Figure 12–5 The second MST algorithm.

9); in fact, we may produce in $O(|V|)$ time an array *component* containing the name of the connected component to which each node belongs (at the first stage, *component*$[v_j] = S_j$ for all j). With the help of the array *component* we can compute the array *shortest* in $O(|E|)$ time by examining the edges of G one by one, as indicated in Fig. 12–5.

How many stages are there? We claim that k, the number of connected components of (V, T), is divided by at least two at each stage. This is true because each connected component of the present (V, T) contains at least two connected components of the graph (V, T) at the previous stage, since each component is connected to another via its *shortest* edge. At the first stage $k = |V|$, and k is divided by two or more at each stage. So, after at most $\log |V|$ stages, k is reduced to one and the algorithm terminates with the MST. Hence, there may be at most $\log |V|$ stages in the algorithm, and the theorem is proved. □

Example 12.1 (Continued)

In Fig. 12–6, we solve the MST problem for the graph of Fig. 12–3 using the algorithm developed in this section. The graph (V, T) at the end of each of the three stages is shown in (a), (b), and (c) of Fig. 12–6. The vertex sets that constitute the connected components are labeled with the corresponding variable

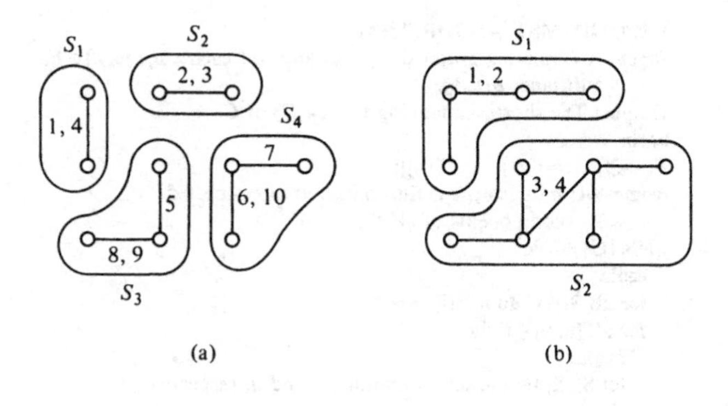

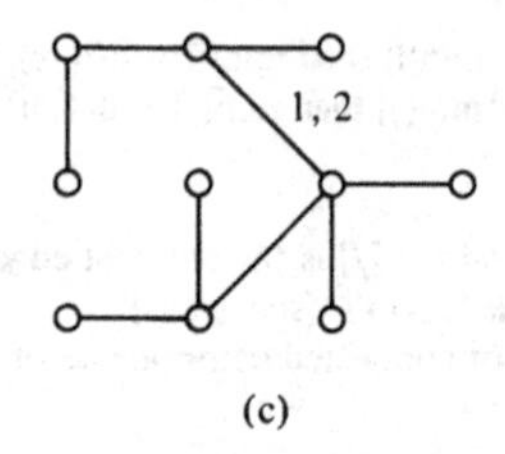

Figure 12–6

names—for example, S_1, S_2, and so on—and each newly added edge e is labeled with the integers i such that *shortest*$[i] = e$. □

12.3 The Greedy Algorithm

Let us look at a slightly different problem. As in the matching problem, we are given a graph $G = (V, E)$ and a *weight* $w_{ij} \geq 0$ for each edge $[v_i, v_j] \in E$. We wish to find a *forest*—an acyclic subgraph of G—that has the maximum total weight. It is not hard to see that this *maximum weight forest* (MWF) problem is very closely related to the MST. First, suppose that G is connected. Because we have assumed that $w_{ij} \geq 0$, any optimal forest can be made maximal, and hence a spanning tree of G. Furthermore, all spanning trees of G have $|V| - 1$ edges (see Proposition 1.2). So we may construct a distance function for E by letting $d_{ij} = W - w_{ij}$ for all $[v_i, v_j] \in E$, where $W = \max_{i,j} \{w_{ij}\}$. The sum of distances $d(T)$ of any spanning tree T will be related to the total weight $w(T)$ of T by $w(T) = (|V| - 1) \cdot W - d(T)$. It is immediately obvious that the MST of G under d is the same as the maximum weight forest of G under w.

Now, if G is disconnected, the situation is not much more complicated. We leave it as an exercise for the reader to verify that the following proposition governs this case:

Proposition 12.1 The maximum weight forest of $G = (V, E)$ under w is the union of the MST's of the connected components of G under d, where d is defined as above.

We are thus justified to think of the MST and the MWF problems as essentially the same problem. The counterpart of Theorem 12.1 for the MWF problem reads:

Lemma 12.1 *Let $\{(U_1, T_1), (U_2, T_2), \ldots, (U_k, T_k)\}$ be a forest spanning V, and let $[v, u]$ be the edge leaving U_1 that has the largest weight. Then among all forests containing $T = \bigcup_{i=1}^{k} T_k$, there is an optimal one also containing $[v, u]$.*

We could solve the MWF problem by using either of the two algorithms for the MST discussed in the previous sections. Nevertheless, we shall develop another very natural algorithm—natural to the extent that greed is.

```
THE GREEDY ALGORITHM
Input: A graph G=(V,E) with weights w_ij on the edges.
Output: The maximum-weight forest F of G.
begin
F:=∅;
while E≠∅ do
  begin
  let [u,v] be the edge in E that has maximum weight;
  remove [u,v] from E;
  if u and v are not in the same component of (V,F)
     then F:=F∪{[u,v]}
  end
end
```

Figure 12-7 The greedy algorithm for MWF.

The greedy algorithm clearly deserves its name. It constantly tries to include the heaviest possible edge in F. The only case in which it excludes a heavy edge is when this edge makes F infeasible. It is quite surprising—and indicative of the simplicity of the structure of the MWF problem—that such a naïve approach, with no elements of look-ahead or backtracking, works. The next theorem states that it does.

Theorem 12.4 *The greedy algorithm correctly solves the MWF problem.*

Proof The correctness of the algorithm follows directly from Lemma 12.1. □

Example 12.1 (Continued)

We now solve the MWF problem for the graph and *weights* shown in Fig. 12–3 by using the greedy algorithm. The various stages are shown in Fig. 12–8. □

12.4 Matroids

We may view the MWF problem as one in a wide class of combinatorial optimization problems.

Definition 12.1

A *subset system* $S = (E, \mathscr{I})$ is a finite set E together with a collection $\mathscr{I}$ of subsets of E closed under inclusion (that is, if $A \in \mathscr{I}$ and $A' \subseteq A$, then $A' \in \mathscr{I}$). The elements of $\mathscr{I}$ are called *independent*. The *combinatorial optimization problem associated with* a subset system $(E, \mathscr{I})$ is the following: Given a *weight* $w(e) \geq 0$ for each $e \in E$, find an independent subset that has the largest possible total weight. □

In this section, as well as in the next, we are going to find algorithms to solve the combinatorial optimization problems associated with certain subset systems. Toward this goal, two comments are in order. First, we need to describe a way by which a subset system $(E, \mathscr{I})$ is to be represented as the input to a computer algorithm. One could propose a listing of all independent subsets of E as an appropriate representation. However, such a way of representing $\mathscr{I}$ could be very inefficient, because $\mathscr{I}$ may contain up to $2^{|E|}$ subsets. For all subset systems that we are going to consider, our representation will be in terms of an *algorithm* $\mathscr{A}_{\mathscr{I}}$ which, given a subset I of E, decides whether $I \in \mathscr{I}$. For example, in the MWF problem for a graph (V, E), it would be very inefficient to list all forests of G. Even if we list only maximal forests, taking advantage of the fact that $\mathscr{I}$ is closed under inclusion, we may end up with as many as $|V|^{|V|-2}$ subsets. What we can do instead, is supply an algorithm $\mathscr{A}_{\mathscr{I}}$ which, given any subset F of E, tests whether F is acyclic; $\mathscr{A}_{\mathscr{I}}$ was the object of Problem 2 in Chapter 9.

This brings us to our second point. The MWF problem (that is, the set of *all* instances thereof) does not correspond to a single combinatorial optimization problem associated with a subset system, but to *many*, one for each graph, the MWF of which we may wish to find. In general, the combinatorial optimization problems corresponding to several subset systems $(E, \mathscr{I})$ with a *common* algorithm $\mathscr{A}_{\mathscr{I}}$ will be thought of as subcases of the same computational problem. Thus, although each combinatorial optimization problem associated with a subset system has an infinity of instances—one for each set of weights—and hence is an optimization problem in its own right, we shall usually refer to a *class* of problems with the same $\mathscr{A}_{\mathscr{I}}$ as a "problem."

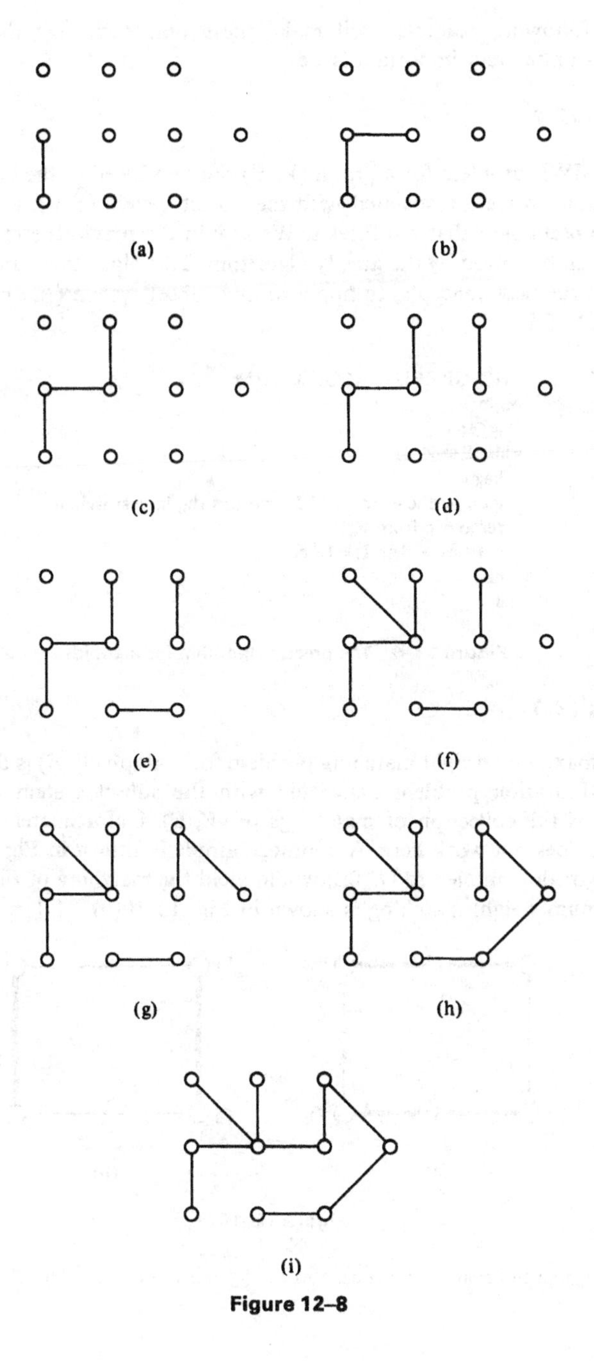

Figure 12–8

The following examples will make these points clearer; they will also manifest another very important issue.

Example 12.2

The MWF problem for a graph (V, E) can be viewed as the combinatorial optimization problem associated with the subset system $(E, \mathcal{F})$, where $\mathcal{F}$ is the class of subsets of E that are forests. We saw in the previous section that this problem can be solved by the greedy algorithm. This algorithm can be restated, in view of the new concepts, to apply to any subset system $(E, \mathcal{I})$ as shown in Fig. 12–9.† □

```
THE GREEDY ALGORITHM
begin
I:=∅;
while E≠∅ do
   begin
   let e be the element of E that has the largest weight;
   remove e from E;
   if I+e∈𝓘 then I:=I+e
   end
end
```

Figure 12–9 The greedy algorithm for matroids.

Example 12.3

The maximum weight matching problem for a graph (V, E) is the combinatorial optimization problem associated with the subset system $S = (E, \mathcal{M})$, where $\mathcal{M}$ is the collection of matchings of (V, E). Unfortunately, the greedy algorithm does not work here. A counterexample is shown in Fig. 12–10. The greedy algorithm applied to $(E, \mathcal{M})$ would yield the matching of Fig. 12–10(a); the maximum weight matching is shown in Fig. 12–10(b). □

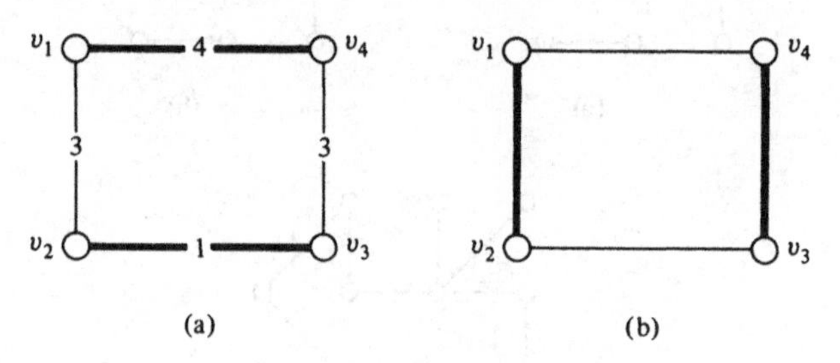

Figure 12–10

†Throughout this chapter, $I + e$ denotes $I \cup \{e\}$ and $I - e = I - \{e\}$.

Example 12.4

Given a digraph $D = (V, A)$ and a weight $w(a) \geq 0$ for each $a \in A$, we may wish to find a subset B of A with the largest possible weight, such that no two arcs of B have the same head. This is the combinatorial optimization problem associated with the subset system $(A, \mathcal{B})$, where a subset B of A is in $\mathcal{B}$ if and only if no two arcs of B have the same head. Despite the superficial similarity to the matching problem for undirected graphs, this problem is a much easier one. Referring to the digraph of Fig. 12–11, we may observe that in choosing

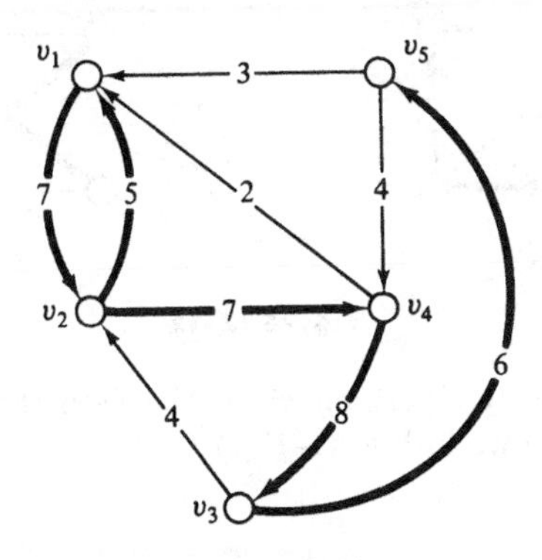

Figure 12–11

the arc to enter v_1, one has no reason to pick any but the heaviest one. This is because this choice is a local one, in that it does not affect in any way the feasibility of other choices at different nodes. The same reasoning holds for all other nodes. Thus, the greedy algorithm will yield an optimal solution in this problem. □

Example 12.5

Consider in a graph $G = (V, E)$ those sets of edges that are unions of node-disjoint paths; examples are shown in Figs. 12–12(a) and 12–12(b). Call the collection of these edge sets $\mathcal{I}$. It is not hard to see that the combinatorial optimization problem associated with the system $(E, \mathcal{I})$ is the TSP, slightly disguised. The difference is that now we are interested in *paths* instead of tours—it turns out that the two versions are equivalent (see Problem 8). Also, we have a maximization instead of a minimization problem. This, as in the case of the MWF versus MST, has no real significance (see Problem 9).

Once we know that this problem is essentially the same as the notorious TSP, we may suspect that the greedy algorithm fails to solve it—as do far more serious efforts—and this is precisely so. In the counterexample of Figs. 12–12(c)

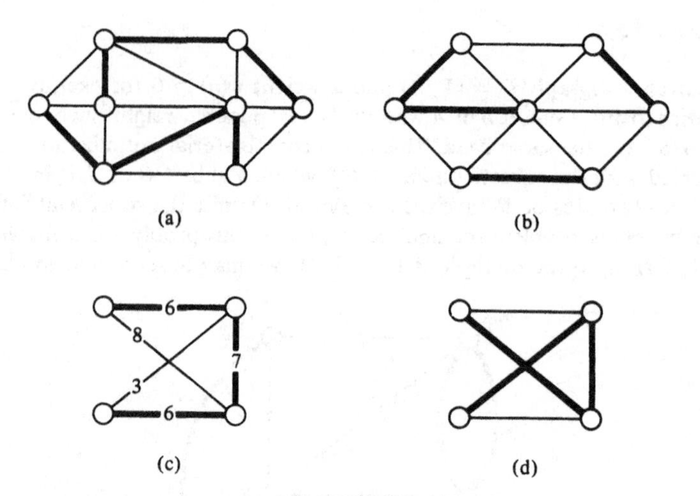

Figure 12–12

and 12–12(d) the greedy algorithm yields the path shown in Fig. 12–12(d); the optimal path is shown in Fig. 12–12(c). □

Example 12.6

Our next example of a system $(E, \mathcal{I})$ is of a somewhat different flavor. Here E is the set of columns of an $n \times |E|$ matrix A, and $\mathcal{I}$ is the collection of *linearly independent* sets of columns of A. In our example of Fig. 12–13, we have $\{e_1, e_2, e_3, e_4\} \in \mathcal{I}$, whereas $\{e_1, e_4, e_5, e_6\} \notin \mathcal{I}$ and $\{e_4, e_6, e_7\} \notin \mathcal{I}$.

3	1	3	0	2	1	1	1
0	2	1	1	0	0	1	2
1	1	2	0	1	0	0	0
2	0	0	−1	0	2	1	0
e_1	e_2	e_3	e_4	e_5	e_6	e_7	e_8

Figure 12–13

Would the greedy algorithm solve the combinatorial optimization problem associated with this curious subset system? Surprisingly enough, it would. This fact is the object of Example 12.9 and will follow easily from subsequent developments in this section and an elementary fact from linear algebra. □

From the previous examples an interesting issue emerges. Some of the subset systems discussed (for example, those in Examples 12.2, 12.4, and 12.6), although greatly different in nature and origin, have an important property in common:

The greedy algorithm solves the corresponding combinatorial optimization problems. In contrast, other subset systems are not equally gifted. We have shown by counterexample that the combinatorial optimization problems associated with the subset systems of Examples 12.3 and 12.5 cannot be solved by the greedy algorithm. This important distinction can be formalized by the following definition.

Definition 12.2

Let $M = (E, \mathcal{I})$ be a subset system. We say that M is a *matroid* if the greedy algorithm correctly solves any instance of the combinatorial optimization problem associated with M. $\square$

The subset system $(E, \mathcal{F})$ of Example 12.2 is thus a matroid; it is called a *graphic* matroid because of its interpretation as the system of forests of a graph. The system $(A, \mathcal{B})$ of Example 12.4 is also a matroid; we shall later classify it as a *partition* matroid (see Example 12.8). Finally, matroids like those in Example 12.6 are known as *matric* matroids. Historically, the notion of a matroid was first introduced as a generalization of this subclass, using certain simple axioms. The following result demonstrates the equivalence of the two approaches.

Theorem 12.5 *Let $M = (E, \mathcal{I})$ be a subset system. Then the following statements are equivalent.*

1. *M is a matroid.*
2. *If $I_p, I_{p+1} \in \mathcal{I}$ with $|I_p| = p$ and $|I_{p+1}| = p + 1$, then there is an element $e \in I_{p+1} - I_p$ such that $I_p + e \in \mathcal{I}$.*
3. *If A is a subset of E and I and I' are* maximal *independent subsets of A, then $|I'| = |I|$.*

Proof (1) $\Rightarrow$ (2) Suppose that M is a matroid, but (2) is false. That is, there are two independent subsets I_p, I_{p+1} with $|I_p| = p$ and $|I_{p+1}| = p + 1$ such that for no $e \in I_{p+1} - I_p$ is $I_p + e$ independent. Let us consider the following weights w of E:

$$w(e) = \begin{cases} p + 2 & e \in I_p \\ p + 1 & e \in I_{p+1} - I_p \\ 0 & e \notin I_p \cup I_{p+1} \end{cases}$$

Let us first observe that I_p is suboptimal, because $w(I_{p+1}) \geq (p + 1)^2 > p(p + 2) = w(I_p)$. The greedy algorithm, if applied to this instance, will start by picking all elements of I_p, since they are the ones with maximum weight. After this, the greedy algorithm will not improve the total weight, because for all other elements e, either $I_p + e \notin \mathcal{I}$ (if $e \in I_{p+1}$) or, otherwise, $w(e) = 0$.

Hence, the greedy algorithm yields the suboptimal solution I_p, and consequently M is not a matroid, as assumed. Hence the implication.

(2) $\Rightarrow$ (3). Suppose that (2) holds, and let I and I' be two maximal independent subsets of $A \subseteq E$. Suppose that $|I| < |I'|$. We can find a set $I'' \subseteq I'$ such that $|I''| = |I| + 1$ by discarding $|I'| - |I| - 1$ elements of I' (recall that $\mathcal{I}$ is closed under inclusion). By Property 2, we can find an element $e \in I'' - I$ such that $I + e \in \mathcal{I}$. Hence I is not a maximal independent subset of A, a contradiction that proves the desired implication.

(3) $\Rightarrow$ (1). We assume that Property 3 is true of M, and we shall show that the greedy algorithm solves M. Suppose it does not; that is, suppose that for some set of weights $w(e)$, $e \in E$, the greedy algorithm yields the independent set $I = \{e_1, e_2, \ldots, e_i\}$ whereas there is a set $J = \{e'_1, e'_2, \ldots, e'_j\} \in \mathcal{I}$ such that $w(J) > w(I)$. The elements of I and J are assumed to be ordered in such a way that $w(e_1) \geq w(e_2) \geq \cdots \geq w(e_i)$, and $w(e'_1) \geq w(e'_2) \geq \cdots \geq w(e'_j)$. We may assume that J is a maximal independent subset of E. By its construction I is maximal, and therefore, from Property 3 (by taking $E = A$), it follows that $i = j$. We shall show that, for $m = 1, 2, \ldots, i$, $w(e_m) \geq w(e'_m)$, which will contradict our assumption that $w(J) > w(I)$. The proof is by induction on m; for $m = 1$, the result holds. Suppose that for some $m > 1$ we have $w(e_m) < w(e'_m)$ and $w(e_s) \geq w(e'_s)$ for $s = 1, \ldots, m - 1$. Consider the set $A = \{e \in E: w(e) \geq w(e'_m)\}$. Now $\{e_1, \ldots, e_{m-1}\}$ is a maximal independent subset of A because, if $\{e_1, e_2, \ldots, e_{m-1}, e\} \in \mathcal{I}$ and $w(e) \geq w(e'_m) > w(e_m)$, the greedy algorithm would have chosen e instead of e_m as the next element of I. This contradicts Property 3, because $\{e'_1, \ldots, e'_m\}$ is another independent subset of A of larger cardinality. This completes the induction and the proof. □

Definition 12.3

Let $M = (E, \mathcal{I})$ be a matroid and $A \subseteq E$. The *rank* of A in M, $r(A)$, is the cardinality of the maximal independent subsets of A. Notice that, by Property 3 of Theorem 12.5, all such subsets of A have the same cardinality, and hence *rank* is always well defined. Maximal independent subsets of E are called *bases*. Notice that since $\mathcal{I}$ is closed under inclusion, M is completely specified by the collection $\mathcal{B}$ of its bases. Given $\mathcal{B}$, $\mathcal{I}$ can be recovered as $\mathcal{I} = \{I: I \subseteq B$ for some $B \in \mathcal{B}\}$. □

Definition 12.4

A subset D of E not in $\mathcal{I}$ is called *dependent*. A minimal dependent subset C of E is called a *circuit*. A *span* of A is a maximal superset S of A satisfying $r(S) = r(A)$. □

Two interesting properties of matroids are shown below. The first is a generalization of Proposition 1.2.

Theorem 12.6 *Let $I \in \mathcal{I}$ and $e \in E$. Then either $I + e \in \mathcal{I}$ or $I + e$ contains a unique circuit.*

Proof Suppose that $I + e \notin \mathcal{I}$. Let $C = \{c: I + e - c \in \mathcal{I}\}$. We claim that C is a circuit. First, it is dependent. This is because otherwise we could augment it into a basis of $I + e$ (notice that $C \subseteq I + e$), which would have to be of cardinality $|I|$ and thus of the form $I + e - d$; this is absurd, because then $d \in C$. Secondly, C is minimal, because removing any of its elements, say c, makes $C - c$ a subset of $I + e - c$, which is independent. To show uniqueness, suppose that D is another circuit in $I + e$ and there is a c in $C - D$. Then D is a subset of $I + e - c$, and hence independent, which is a contradiction. □

Theorem 12.7 *Any $A \subseteq E$ has a unique span, defined as follows:*

$$sp(A) = \{e: r(A + e) = r(A)\}.$$

Proof If S is a span of A and $e \in S$, then $r(A + e) = r(A)$. This is because otherwise, if $r(A + e) > r(A)$, we would have $r(S) \geq r(A + e) > r(A)$, a contradiction. Therefore $S \subseteq sp(A)$. It remains to show that $r(sp(A)) = r(A)$. It is not hard to see that a common basis of two sets is a basis of their union. A basis of A is therefore a basis of $sp(A)$, since it is a basis of $A + e$ for every $e \in sp(A)$. □

Corollary 1 *$sp(A)$ is the union of A and all circuits that have all but one of their elements in A.*

Corollary 2 *If $I \in \mathcal{I}$, $I + e \notin \mathcal{I}$, and c belongs to the circuit in $I + e$, then $sp(I) = sp(I + e - c)$.*

Example 12.7

If $\mathfrak{F}$ is the set of forests of the graph $G = (V, E)$, $M_G = (E, \mathfrak{F})$ is a *graphic* matroid.

Circuits of M_G are the (graph-theoretic) circuits of G.

The *rank* of a subset E' of E is easily seen to be $r(E') = |V| - c(E')$ where $c(E')$ is the number of connected components of $G' = (V, E')$.

The *span* of a subset E' of E, sp(E'), is the largest superset of E' with the same rank as E'. Hence

$$\text{sp}(E') = \{[v, u] \in E: v \text{ and } u \text{ are in the same component of } G' = (V, E')\}$$

□

Example 12.8

Let E be a finite set and Π a *partition* of E, that is, a collection of disjoint subsets of E covering E; $\Pi = \{E_1, E_2, \ldots, E_p\}$. Let us call a subset I of E independent ($I \in \mathcal{I}$) if and only if no two elements of I are in the same set of Π; that is, $|I \cap E_j| \leq 1, j = 1, 2, \ldots, p$. Then the system $M_\Pi = (E, \mathcal{I})$ is a matroid, called a *partition matroid.*

To show that M_Π is a matroid, it suffices to show that it has a well-defined *rank* function. Let $J(A) = \{j \leq p: E_j \cap A \neq \varnothing\}$. The reader may easily verify that $r(A) = |J(A)|$ is the desired rank function; this is because given A, we can construct a maximal independent subset of A by selecting one element of A from all E_j's that A hits. For example, if $E = \{e_1, \ldots, e_8\}$ and $\Pi = \{\{e_1\}, \{e_2, e_3\}, \{e_4, e_5\}, \{e_6, e_7, e_8\}\}$, then the rank of $A = \{e_1, e_2, e_3, e_6, e_7\}$ in M_Π is 3; a maximal independent subset of A is $I = \{e_1, e_2, e_7\}$.

The *span* of A is $\text{sp}(A) = \bigcup_{j \in J(A)} E_j$.

A *circuit* of M_Π is any set of two elements from the same E_j; thus $\{e_2, e_3\}$ and $\{e_6, e_8\}$ are circuits.

For another example of a partition matroid, let us look back to the subset system discussed in Example 12.4. We called a set B of arcs of the digraph D of Fig. 12–11 independent if no two arcs of B had the same head. This is equivalent to partitioning the arcs of D according to which node is the head of each. For our example, the partition is $\Pi = \{\{(v_2, v_1), (v_4, v_1), (v_5, v_1)\}, \{(v_1, v_2), (v_3, v_2)\}, \{(v_4, v_3)\}, \{(v_5, v_4)\}, \{(v_3, v_5)\}\}$. Hence this system is a partition matroid M_Π—called the *head-partition matroid* of the digraph D†—and the fact that the greedy algorithm solved that problem was only natural. □

Example 12.9

In a *matric matroid* $M_A = \{E, \mathcal{I}\}$, the set E is the set of columns of an $n \times |E|$ matrix A, and $\mathcal{I}$ is the set of *linearly independent* subsets of E. Here A may have entries from any field K. The subset system M_A is a matroid, because Property 3 of Theorem 12.5 holds. This follows from a well-known fact from linear algebra:

> All maximal linearly independent subsets of a set of vectors E' have the same cardinality.

In linear algebra this cardinality is, in fact, called the *rank* of the submatrix A' defined by E'. □

Matric matroids are a more general class than graphic and partition matroids. It can be shown that any graphic or partition matroid can be formulated as a matric matroid by a suitable choice of a matrix A over a field K (see

†The *tail-partition matroid* of D is defined similarly.

Problem 12). Nevertheless, this is not true of all matroids (see Problem 13); consequently, matroid theory is a *proper* generalization of linear algebra.

12.5 The Intersection of Two Matroids

We start by revisiting the bipartite matching problem.

Example 12.10

Consider the bipartite graph $B = (V, U, E)$ of Fig. 12–14, and the set $\mathfrak{M}$ of matchings of B. Is $(E, \mathfrak{M})$ a matroid?

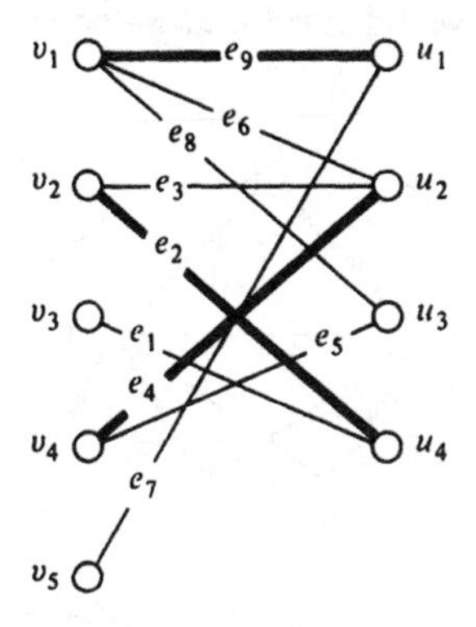

Figure 12–14

If $(E, \mathfrak{M})$ were a matroid, we would have a much simpler algorithm for bipartite matching than the one we developed in the previous chapters. The very essence of the matroidal structure is that any *maximal* independent set is also *maximum*. Finding maximal independent subsets is a trivial task: We keep adding elements until no further additions are possible. Unfortunately, $(E, \mathfrak{M})$ is not a matroid. This is simply demonstrated by the fact that the matching shown in Fig. 12–14 is *maximal* but not *maximum*.

Nevertheless, $\mathfrak{M}$ does have quite a nice structure: It is the *intersection* of two matroids. In other words, we can find two matroids $M = (E, \mathcal{I})$ and $N = (E, \mathcal{K})$ such that $\mathfrak{M} = \mathcal{I} \cap \mathcal{K}$. Both M and N are partition matroids. The partition of M is Π_V, the subdivision of E into subsets according to which vertex of V each is incident upon. In Fig. 12–14, for example, $\Pi_V = \{\{e_9, e_6, e_8\}, \{e_3, e_2\}, \{e_1\}, \{e_4, e_5\}, \{e_7\}\}$. Similarly, N is the partition matroid corresponding to the partition $\Pi_U = \{\{e_9, e_7\}, \{e_6, e_3, e_4\}, \{e_8, e_5\}, \{e_1, e_2\}\}$. It is not hard to see that a subset I of E is a matching if and only if it is independent in M (that is, no two edges in I share a node of V) and in N (that is, no two edges in I

share a node of U). Consequently, $\mathcal{M} = \mathcal{I} \cap \mathcal{K}$. Thus the bipartite matching problem can be viewed as the problem of finding a maximum subset of E that is independent in two matroids. □

Example 12.11

A digraph $D = (V, A)$ may be used to represent the flow of control in a group of people. We have a node for each person, and an arc (v, u) means that v "influences" u. Usually, such a situation will be chaotic (see Fig. 12–15). One person may receive more than one influence and may possibly receive contradictory ones (like the node v_2, controlled by all of v_1, v_3, v_6). Furthermore, we may have loops in our structure (the directed circuit $(v_1, v_3, v_7, v_9, v_1)$ for example) or short-circuits (like the arc (v_1, v_4), which is a shortcut of the path (v_1, v_2, v_4)). There may be no well-defined *leader*—node with no incoming arcs.

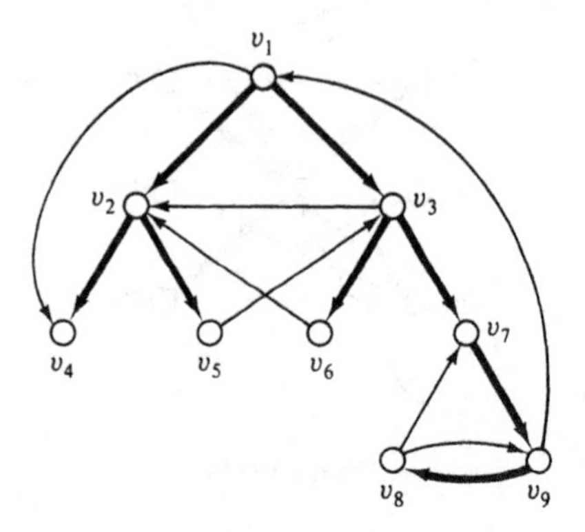

Figure 12–15

It is sometimes desirable to organize such a group of people into a *branching*. A branching is a digraph (V, B), $B \subseteq A$, with no loops or shortcuts: The graph $G = (V, E_B)$, which results from disregarding the directions of the arcs in B, is a tree. Furthermore, each vertex—except for the leader—has exactly one incoming arc. For example, the heavy lines in Fig. 12–15 constitute a branching. The problem is: Given a digraph $D = (V, A)$ is there a subset B of A that is a branching?

This problem may also be formulated as a problem of finding the maximum set that is simultaneously independent in two matroids. One of these matroids is the graphic matroid of the digraph D when we disregard direction. The second is the head-partition matroid of the digraph D. It should be clear that any forest of branchings will be in the intersection of these matroids. Thus A contains a branching B if and only if the maximum set that is independent in these two matroids has cardinality $|V| - 1$. □

Example 12.12

We are given two graphs G and G' such that their edges have names taken from the same set E. Such a situation is shown in Fig. 12–16. We are looking for the largest subset of E that is acyclic in both G and G'. Obviously, this is again a problem of finding the set of largest cardinality that is independent in two matroids, namely, the graphic matroids M_G and $M_{G'}$. □

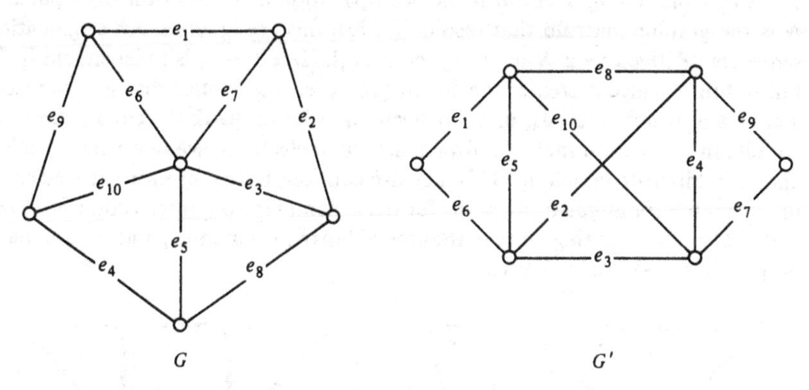

Figure 12–16

In this section we shall develop an algorithm for finding, given any two matroids $M = (E, \mathcal{I})$ and $N = (E, \mathcal{K})$, the largest set in $\mathcal{I} \cap \mathcal{K}$. Since the bipartite matching problem is a special case of the *matroid intersection* problem, it is well to first recall the methodology that we developed in Sec. 10.2 in order to solve that problem.

The bipartite matching problem is solved by repeated augmentations, using augmenting paths to find larger and larger matchings—in our new terminology, sets independent in both the $M = (E, \mathcal{I})$ and $N = (E, \mathcal{K})$ matroids of Example 12.10. For example, $S = [e_1, e_2, e_3, e_4, e_5]$ is an augmenting path relative to the matching I of Fig. 12–14. In light of our matroid formulation of the problem, we can obtain a new understanding of how S acts on I, finally augmenting it to $I \oplus S$. First I becomes $I + e_1$, then $I + e_1 - e_2$, $I + e_1 - e_2 + e_3$, $I + e_1 - e_2 + e_3 - e_4$; finally, $I \oplus S = I + e_1 - e_2 + e_3 - e_4 + e_5$. The first element e_1 of S starts from an exposed vertex of V; hence $I + e_1$ is independent in M. However, $I + e_1$ is not independent in N—otherwise, our augmentation would already be over. The circuit of N, $\{e_1, e_2\}$, is formed. Now, S "breaks" this circuit by removing e_2 to obtain $I + e_1 - e_2$. The new set is independent in both M and N, but it has the same cardinality as the set I with which we started. Thus we try again to find an edge, e_3, that creates no dependence in M but possibly creates a circuit in N, and then break this circuit (by deleting e_4). The hope is that eventually we shall be able to come up with an edge that creates no dependence in either M or N, thereby augmenting I. This happens in the next step, and thus e_5 is the last edge of S.

Augmenting sequences can be defined as above when M and N are general matroids, as shown in the following example.

Example 12.13

Suppose that we wish to augment the branching forest I of Fig. 12–17(a) to a single branching. Here M is the head-partition matroid of the digraph, and N is the graphic matroid that results if directions are ignored. An augmenting sequence relative to I is $S = [e_1, e_2, e_3, e_4, e_5]$. Here $I + e_1$ is independent in M but not in N, since it creates the circuit $\{e_1, e_2, e_4, e_9\}$. Notice that now we have a choice of which of e_2, e_4, or e_9 to delete in order to break the circuit. Because circuits in a partition matroid always have two elements, we never have such a choice in bipartite matching. This fact also cautions us that even-numbered steps in our search for augmenting paths for the general matroid intersection problem will not be trivial, as they were in the case of bipartite matching, and our auxiliary graph will not skip even levels.

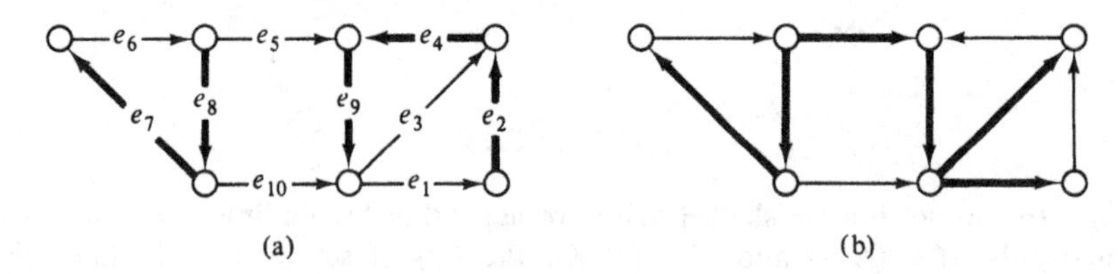

Figure 12–17

We choose e_2 as the next element of S. If e_3 is added to $I + e_1 - e_2$, it creates a circuit only in N, namely, $\{e_3, e_4, e_9\}$. We break this one by deleting e_4. Now e_5 can be added to $I + e_1 - e_2 + e_3 - e_4$ without creating any circuits in either M or N, and so S is an augmenting sequence. The resulting branching is shown in Fig. 12–17(b). □

Let $S = [e_1, e_2, \ldots, e_m]$ be a sequence of elements of E. We denote by $S_{i,j}$ $(i \leq j)$ the sequence $[e_i, e_{i+1}, \ldots, e_j]$. The span of a set $A \subseteq E$ with respect to the matroid M (or N) will be denoted by $\text{sp}_M(A)$ (or $\text{sp}_N(A)$). Similarly, the rank of a set A will be $r_M(A)$ or $r_N(A)$, depending on the underlying matroid.

Definition 12.5

A sequence $S = [e_1, e_2, \ldots, e_m]$ is called *alternating* with respect to a set $I \in \mathcal{I} \cap \mathcal{K}$ if:

Alt. 1. $I + e_1 \in \mathcal{I}$, and $e_1 \notin I$.

Alt. 2. For all even i, $2 \leq i \leq m$, $e_i \in I$ and $\text{sp}_N(I \oplus S_{1i}) = \text{sp}_N(I)$.

Alt. 3. For all odd i, $3 \leq i \leq m$, $e_i \notin I$ and $\text{sp}_M(I \oplus S_{1i}) = \text{sp}_M(I + e_1)$.
Furthermore, if m is odd and $I \oplus S \in \mathcal{K}$, S is called *augmenting*. $\square$

Lemma 12.2 *Let S be an alternating sequence. Then*

1. *$I \oplus S_{1i} \in \mathcal{K}$ for i even.*
2. *$I \oplus S_{1i} \in \mathcal{I}$ for i odd.*

Proof 1. Since $e_i \notin I$ for odd i, it follows that $I \oplus S_{1i} \neq I$ for i even. Since $\text{sp}_N(I \oplus S_{1i}) = \text{sp}_N(I)$, I and $I \oplus S_{1i}$ have the same rank in N. Since I is independent, that is, has full rank in N, it follows that $I \oplus S_{1i}$ is also independent.

2. The case for odd i is very similar. $\square$

Thus the alternating sequences with respect to an independent set I are sequences of elements in E whose odd-numbered elements create circuits in N; each circuit is then broken by the next element. Our goal is to search for alternating sequences using our familiar auxiliary digraph technique, for some appropriately defined auxiliary digraph with E as its node set. Unfortunately, this is not possible if we rely on the concept of the alternating sequence alone. The reason is that the auxiliary digraph immediately suggested by the definition of alternating sequences is *dynamic*, that is, it cannot be fixed beforehand. The choices available at each step for the continuation of the construction of an alternating sequence may depend heavily on previous choices. This is illustrated in Figure 12–18 in terms of the intersection of two graphic matroids for the graphs G and H. Here M is the matroid of G and N that of H. The current independent set is $\{e_2, e_2', e_4, e_5\} = I$. The alternating sequences $[e_1, e_2, e_3]$ and $[e_1, e_2', e_3]$ have the same endpoint, e_3, but totally different continuations. The first can be continued by the edges e_2' and e_4, whereas the second can be con-

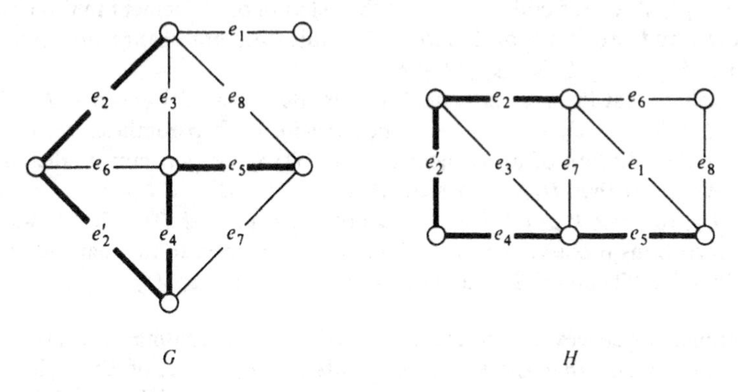

Figure 12–18

tinued by e_2 and e_5. No fixed auxiliary graph can capture the choices available to both.

This dynamic nature of path problems is sometimes a symptom of intractability. (See the next section and Chapter 15 for a discussion of the *Hamilton path* problem.) Fortunately, in the case of the matroid intersection problem, it suffices to restrict somewhat the definition of an alternating sequence in order to obtain an appropriate *static* auxiliary digraph. For a given $I \in \mathcal{I} \cap \mathcal{K}$, suppose that $I + e_i \notin \mathcal{I}$. We shall denote by C_i the unique M-circuit in $I + e_i$. If $I + e_i \notin \mathcal{K}$, let D_i denote the corresponding N-circuit.

Definition 12.6

A sequence $S = [e_1, e_2, \ldots, e_m]$ is *proper* with respect to $I \in \mathcal{I} \cap \mathcal{K}$ if:

Pr. 1 $I + e_1 \in \mathcal{I}$ and $e_1 \notin I$.

Pr. 2 For all even i, $2 \leq i \leq m$, $e_i \in I$ and $e_i \in D_{i-1}$. Furthermore, $e_i \notin D_{k-1}$ for any even $k < i$.

Pr. 3 For all odd i, $3 \leq i \leq m$, $e_i \notin I$ and $e_{i-1} \in C_i$. Furthermore, $e_{k-1} \notin C_i$ for any odd $k < i$.

Furthermore, if m is odd and $I \oplus S \in \mathcal{K}$, S is called a *proper augmenting sequence.* □

Lemma 12.3 *Proper sequences are alternating.*

Proof Pr. 1 clearly implies Alt. 1 (see Definition 12.5). We shall show that Pr. 3 implies Alt. 3.

We write $I \oplus S_{1i}$ for odd i as $I + e_1 + (e_3 - e_2) + \cdots + (e_i - e_{i-1})$. Each parenthesized addition and deletion of an element $(e_j - e_{j-1})$ is an addition of an element to $I \oplus S_{1,j-2}$ that creates an M-circuit—namely, C_j, because $e_{k-1} \notin C_j$ for odd $k < j$—and the deletion of an element in this circuit. However, by Corollary 2 of Theorem 12.7, this does not change the span of a set. Hence $\mathrm{sp}_M(I \oplus S_{1i}) = \mathrm{sp}_M(I + e_1)$.

To show that Pr. 2 implies Alt. 2, we rewrite $I \oplus S_{1i}$ for i even as $I + (e_{i-1} - e_i) + (e_{i-3} - e_{i-2}) + \cdots + (e_1 - e_2)$. Again, each parenthesis $(e_{j-1} - e_j)$ contains the addition of an element e_{j-1} to $I \oplus S_{j+1,i}$ that creates the circuit D_{j-1}. We know that D_{j-1} is formed when e_{j-1} is added to $I \oplus S_{j+1,i}$ because $e_k \notin D_{j-1}$ for even $k > j$. We also delete the element e_j of D_{j-1}. It follows that these operations preserve the span of the set with respect to the matroid N (see Corollary 2 of Theorem 12.7), and hence $\mathrm{sp}_N(I \oplus S_{1i}) = \mathrm{sp}_N(I)$. □

Proper sequences are indeed a restriction of alternating sequences. The reader may verify that the sequence $S = [e_1, e_2, e_3, e_4, e_5]$ of Example 12.13 is alternating but not proper. This is because $e_4 \in D_1$, in violation of Pr. 2.

There are two reasons for defining this restricted class of alternating sequences:

1. They can be discovered by searching a "static" auxiliary digraph.
2. They still guarantee that optimality will be reached.

For example, there *is* a proper augmenting sequence with respect to the branching forest of Fig. 12–17(a). It is $S' = [e_1, e_4, e_5]$.

Our auxiliary digraph (E, A) will have E as its set of nodes. Each element $e_i \in E - I$ will be a candidate for an odd numbered element in a proper augmenting sequence S. If $I + e_i \notin \mathcal{I}$, we find C_i and we add to A the arc (e_j, e_i) for each $e_j \in C_i - e_i$, according to Pr. 3. If $I + e_i \in \mathcal{I}$, then e_i is a candidate for the first element in S. An element of $E - I$ such that $I + e_i \in \mathcal{K}$ may be the last element of S; those elements are our *targets*, and reaching one of them will signal the discovery of a proper augmenting path. However, if $I + e_i \notin \mathcal{K}$, we add, according to Pr. 2, the arc (e_i, e_j) to A for each e_j in $D_i - e_i$.

Example 12.14

Let us construct the auxiliary graph (E, A) for the branching forest I shown in Fig. 12–17(a). We do this by examining one by one the elements of $E - I$. We start with e_1. $I + e_1 \in \mathcal{I}$, where $\mathcal{I}$ is the head partition matroid; hence e_1 is a possible starting element of S. Also, $I + e_1 \notin \mathcal{K}$, where $\mathcal{K}$ is the graphic matroid. Because $D_1 = \{e_1, e_2, e_4, e_9\}$, we add the arcs (e_1, e_2), (e_1, e_4), and (e_1, e_9) to A. In the case of e_3, $I + e_3 \notin \mathcal{I}$. Because $C_3 = \{e_2, e_3\}$, we add the arc (e_2, e_3) to A; $I + e_3 \notin \mathcal{K}$ and $D_3 = \{e_3, e_4, e_9\}$, so the arcs (e_3, e_4) and (e_3, e_9) are added to A. For e_6, we note that $I + e_6 \in \mathcal{I}$, so e_6 is a possible starting element of S. Because $I + e_6 \notin \mathcal{K}$ and $D_6 = \{e_6, e_7, e_8\}$, the arcs (e_6, e_7) and (e_6, e_8) are added to A. Also, $I + e_5 \notin \mathcal{I}$ and $C_5 = \{e_5, e_4\}$; and $I + e_{10} \notin \mathcal{I}$ and $C_{10} = \{e_9, e_{10}\}$; thus (e_4, e_5) and (e_9, e_{10}) are added to A. Finally, $I + e_5, I + e_{10} \in \mathcal{K}$, so both e_5 and e_{10} are target elements. The auxiliary digraph is shown in Fig. 12–19. Notice that (E, A) is bipartite, since all arcs are between I and $E - I$, in either direction. □

It is not sufficient to search for any path from S to T in the auxiliary digraph. Certain such paths may not correspond to augmenting sequences. This is illustrated in Fig. 12–20, where another branching problem is considered. The set I to be augmented is shown in heavy lines in Figure 12–20(a). The corresponding auxiliary digraph is shown in Fig. 12–20(b). We observe, however, that the path $S = [e_1, e_2, e_3, e_4, e_5]$ in the auxiliary digraph is not an augmenting path. Also, $I \oplus S = \{e_1, e_3, e_5, e_6\}$ is not a branching—it is not independent in the graphic matroid. The reason for this is that there is a *shortcut* of this path, namely, $S' = [e_1, e_4, e_5]$. This means that S violates Pr. 2—because $e_4 \in D_1$—and thus it is not proper. We therefore must use breadth-first search in order to

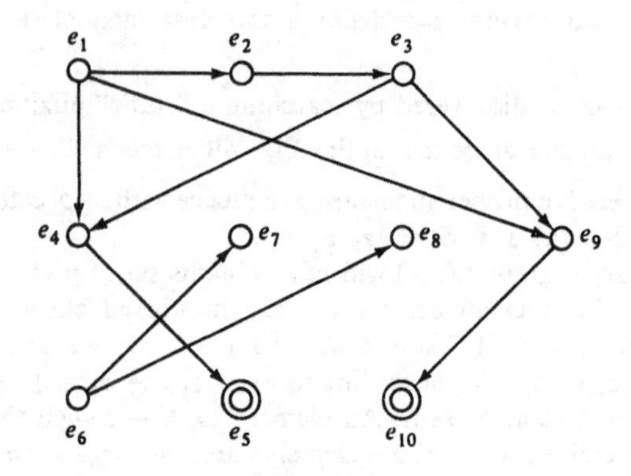

Figure 12–19

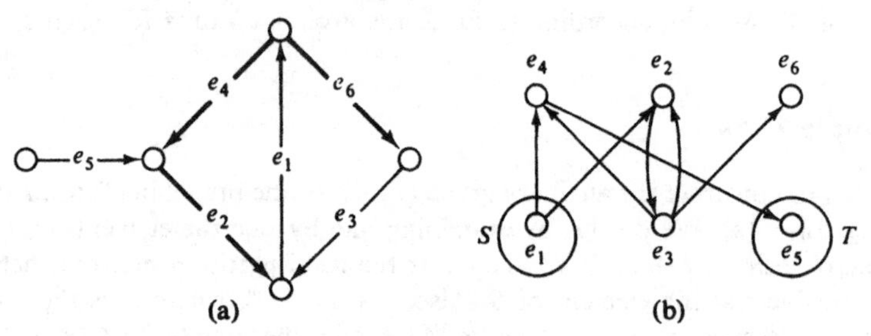

Figure 12–20

search the auxiliary digraph for *shortest* paths from S to T. It turns out that such paths will always be proper and therefore augmenting.

We show the complete algorithm in Fig. 12–21.

To show the correctness of our algorithm, we shall first need some preliminary work. Let E_1 and E_2 be such that $E_1 \cup E_2 = E$, and let us call the *rank* of this pair of sets (with the matroids M and N fixed) the number $r(E_1, E_2) = r_M(E_1) + r_N(E_2)$. Let J be any set in $\mathscr{I} \cap \mathscr{K}$.

Lemma 12.4 $r(E_1, E_2) \geq |J|$.

Proof Let $J_1 = J \cap E_1$ and $J_2 = J \cap E_2$. Since J_1 and J_2 are independent subsets of E_1 and E_2, respectively, we have that

$$r(E_1, E_2) = r_M(E_1) + r_N(E_2) \geq |J_1| + |J_2| \geq |J| \qquad \square$$

The two matroids M and N are given in terms of the algorithms $\mathfrak{a}_{\mathscr{I}}$ and $\mathfrak{a}_{\mathscr{K}}$. Let $C(|E|)$ be an upper bound on the complexity of these algorithms when operating on problems of size $|E|$.

```
MATROID INTERSECTION ALGORITHM
Input: Two matroids M=(E,𝓘) and N=(E,𝒦), given in terms of
       the algorithms 𝒜_𝓘 and 𝒜_𝒦.
Output: A set I ∈ 𝓘∩𝒦 of maximum cardinality.
begin
   I:=∅; (comment: initialize)
stage: A:=∅, Q:=∅, T:=∅;
   (comment: initialize for digraph search)
   for all eᵢ ∈ E−I do
   begin (comment: construction of the auxiliary digraph)
      if I+eᵢ ∈ 𝓘,𝒦 then I:=I+eᵢ, go to stage;
      if I+eᵢ ∈ 𝓘 then Q:=Q∪{eᵢ}, label[eᵢ]:=0,
         else for all eⱼ ∈ Cᵢ−eᵢ do A:=A∪{(eⱼ,eᵢ)};
      if I+eᵢ ∈ 𝒦 then T:=T∪{eᵢ}
         else for all eⱼ ∈ Dᵢ−eᵢ do A:=A∪{(eᵢ,eⱼ)}
      (comment: Cᵢ and Dᵢ are circuits that are defined in terms of
                eᵢ and I as in the text preceding Definition 12.6)
   end;
   while Q≠∅ do
   begin
      let e be an element of Q;
      remove e from Q; (comment: Q must be a queue)
      for all unlabeled e′ ∈ E such that (e,e′) ∈ A do
      begin
         label[e′]:=e, Q:=Q∪{e′};
         if e′ ∈ T then I:=I⊕ path(e′), go to stage
      end
   end
end

procedure path(e)
(comment: it returns a proper augmenting sequence from some
          element of S to e; path is recursive)
if label[e]=0 then return [e]
   else return path(label[e]) || [e]
```

Figure 12–21 The matroid intersection algorithm.

Theorem 12.8 *The algorithm of Fig. 12–21 correctly solves the matroid intersection problem in $O(|E|^3 C(|E|))$ time.*

Proof First, we claim that a sequence $S = [e_1, e_2, \ldots, e_m]$ constructed by the algorithm is proper and hence alternating. Pr. 1 of Definition 12.6 is obvious. To show Pr. 2, we note that elements of even rank in S are in I because of the bipartite property of the auxiliary graph. Also, the construction of the auxiliary digraph guarantees that $e_i \in D_{i-1}$ for even i. Now, assume that $e_i \in D_{k-1}$ for some $k < i$. Then the arc (e_k, e_i) is in the auxiliary graph. However this contradicts the fact that Q is a queue (and therefore our algorithm finds shortest paths). Pr. 3 is shown similarly. Finally, since $e_m \in T$, it follows that m is odd and $I + e_m \in \mathcal{K}$; thus $I \oplus S \in \mathcal{K}$, and S is an augmenting sequence.

Hence our algorithm augments along augmenting sequences, and consequently all sets I produced are in $\mathcal{I} \cap \mathcal{K}$, as required. We shall now show that the algorithm does not terminate unless I is as large as possible. Let L and U be the labeled and unlabeled elements of I at termination, respectively, and take $E_1 = \text{sp}_M(U)$ and $E_2 = \text{sp}_N(L)$. Each element of I belongs to one of E_1, E_2. Take any element e_i of $E - I$; we shall show that it also belongs to one of E_1, E_2. If e_i is labeled, then in our search we label all elements in the corresponding D_i. So if e_i is added to L, it creates an N-circuit, namely, D_i. Thus $e_i \in \text{sp}_N(L)$. Now, suppose that e_i is unlabeled. It must be the case that $I + e_i \notin \mathcal{I}$, because otherwise e_i would have a label—namely, zero. So $I + e_i$ contains an M-circuit C_i. No element e_j of C_i could be labeled, however, because it if were, e_i would have been labeled in the next step through the arc (e_j, e_i). So if we add e_i to U, an M-circuit results. Hence $e_i \in \text{sp}_M(U)$.

So, $E_1 \cup E_2 = E$, and hence $r(E_1, E_2) \geq |J|$ for all $J \in \mathcal{I} \cap \mathcal{K}$ (Lemma 12.4). However, it is easy to see that $r(E_1, E_2) = |L| + |U| = |I|$, and hence $|I| \geq |J|$ for all $J \in \mathcal{I} \cap \mathcal{K}$. Consequently, I is the desired maximum intersection.

For the time bound, we can have at most $|E|$ augmentations. For each augmentation, the time requirements are dominated by the construction of the auxiliary digraph. For each $e_i \in E - I$, we find the circuit D_i or C_i or both. This can be done by applying the tests $\mathcal{A}_{\mathcal{I}}$ and $\mathcal{A}_{\mathcal{K}}$ to $I + e_i - e_j$ for each $e_j \in E$. The construction of the auxiliary digraph, Q, and T can also be carried out in $O(|E|^2 \cdot C(|E|))$ time. Finally, searching the digraph takes $O(|E|^2)$ time. The time bound follows. □

12.6 On Certain Extensions of the Matroid Intersection Problem

12.6.1 Weighted Matroid Intersection

Suppose that we are given two matroids $M = (E, \mathcal{I})$ and $N = (E, \mathcal{K})$ and also a *weight* $w(e)$ for each $e \in E$. We wish to find a subset $I \in \mathcal{K} \cap \mathcal{I}$ such that $\sum_{e \in I} w(e)$ is as large as possible. There is a polynomial-time algorithm for this problem, which is yet another application of the primal-dual method. Once again, the weighted problem is *reduced* to the unweighted one with the help of a linear programming formulation of the former. This formulation—the counterpart of Theorem 11.2—is as follows.

Theorem 12.9 (Edmonds) *The weighted matroid intersection problem is equivalent to the LP*

$$\max \sum_{e \in E} w(e) \cdot x(e)$$

subject to

$$\left.\begin{array}{l} \sum_{e \in A} x(e) \le r_M(A) \\ \sum_{e \in A} x(e) \le r_N(A) \end{array}\right\} \quad \text{for all } A \subseteq E$$

It is obvious that if $x(e)$, $e \in E$, represents a set in $\mathcal{I} \cap \mathcal{K}$, then the constraints of Theorem 12.9 are satisfied. It remains to show that, for all weights, the optimal solution of this LP is always 0-1, and hence indeed corresponds to the heaviest set in $\mathcal{I} \cap \mathcal{K}$. A constructive proof of this can be obtained by applying the primal-dual method to this LP and observing that the restricted primal is at all times an *unweighted* matroid intersection problem for a modified pair of matroids, M' and N'. The details are quite involved and are not presented here (see [La1]).

12.6.2 Matroid Parity

We are given a matroid $M = (E, \mathcal{I})$, and a partition Π of E into disjoint pairs, $\Pi = \{[e_1, e_2], [e_3, e_4], \ldots, [e_{n-1}, e_n]\}$. We are asked to find a subset P of Π such that $\bigcup_{p \in P} p \in \mathcal{I}$ and P is as large as possible. This is the *matroid parity problem.*

For example, if M is a partition matroid, then the matroid parity problem can easily be seen to degenerate to the matching problem (Problem 19). If M is a graphic matroid, then the matroid parity problem is restated as follows: Given a graph G and a partition Π of its edges in pairs (see Fig. 12–22), find a spanning

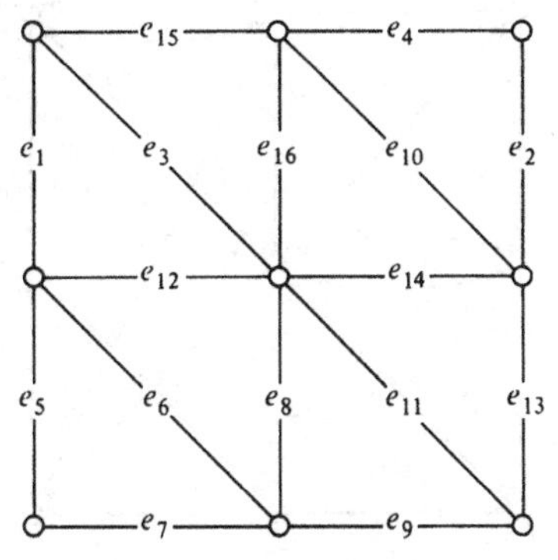

$\Pi = \{[e_1, e_2], [e_3, e_4], [e_5, e_6], [e_7, e_8], [e_9, e_{10}], [e_{11}, e_{12}], [e_{13}, e_{14}], [e_{15}, e_{16}]\}$

Figure 12–22

tree T such that if an edge e is in T, then so is its mate in Π. This problem is a common generalization of the two-graphic-matroid intersection problem (Example 12.12) and matching (see Problem 20).

It was recently shown by Lovász [Lo] that the parity problem *cannot* be solved for general matroids. Lovász gave an algorithm for the case of graphic matroids—in fact, for the more general *matric* matroids.

12.6.3 The Intersection of Three Matroids

Another direction in which we may wish to generalize the results that we obtained for the intersection of two matroids would be to develop an algorithm for finding the largest set that is independent in *three* matroids. Unfortunately, there is no known polynomial algorithm that performs this task. Furthermore, as we shall see in Chapter 15, there is evidence that this problem will remain forever beyond polynomial-time solution, no matter how advanced our understanding of its structure may be in the future. For the time being, we shall show that if we had a way to solve this problem in polynomial time—polynomial in $|E|$ and the complexities C_1, C_2, and C_3 of the algorithms that describe the three matroids in question—we would also be able to solve a well-studied combinatorial optimization problem related to the TSP, namely the DIRECTED HAMILTON PATH problem:

> Given a digraph $D = (V, A)$, does D have a *Hamilton path*, that is, a directed path h traversing each node in V exactly once?

For example, D of Fig. 12–23 does have a Hamilton path, namely $(v_1, v_2, v_3, v_9, v_7, v_6, v_4, v_5, v_8)$.

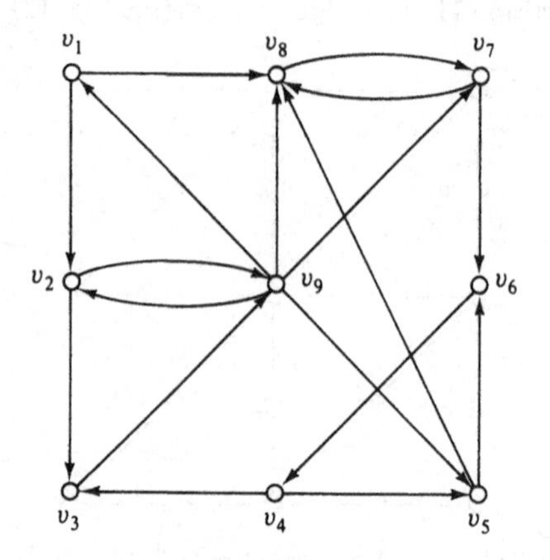

Figure 12–23

Theorem 12.10 *There are three matroids $M_1 = (A, \mathcal{I}_1)$, $M_2 = (A, \mathcal{I}_2)$, and $M_3 = (A, \mathcal{I}_3)$ with polynomially bounded algorithms $\mathcal{A}_{\mathcal{I}_1}$, $\mathcal{A}_{\mathcal{I}_2}$, and $\mathcal{A}_{\mathcal{I}_3}$ such that a digraph $D = (V, A)$ has a Hamilton path if and only if there exists a set $I \in \mathcal{I}_1 \cap \mathcal{I}_2 \cap \mathcal{I}_3$ with $|I| = |V| - 1$.*

Proof It suffices to observe that a Hamilton path is just a branching, such that no two arcs have the same tail. Thus any Hamilton path is by definition a subset of A that is simultaneously independent in the graphic matroid corresponding to the graph (V, E_A), the head-partition matroid of D, and the tail-partition matroid of D, and that also has cardinality $|V| - 1$. The theorem follows. □

PROBLEMS

1. Find the MST of the graph below by using both algorithms of Figure 12–2 and 12–5. Find the MWF of the same graph.

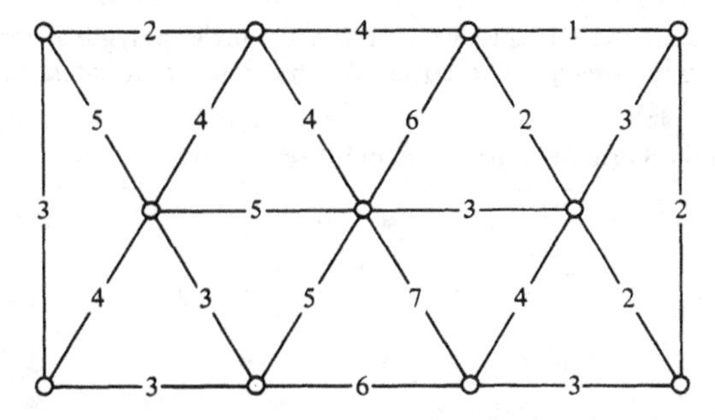

***2.** The MST algorithm of Figure 12–5 can be implemented in $O(|E| \log\log |V|)$ time by the following procedure: First, we do not compute the connected components of (V, T) from scratch at each stage. Instead, we *merge* the appropriate components.

(a) Show how to maintain a partition $C = \{S_1, \ldots, S_k\}$ of the vertices in terms of four arrays ***first***, ***next***, ***size***, and ***set*** so that we can identify in one step which set contains a given node, and we can merge two sets S_i and S_j in $O(\min(|S_i|, |S_j|))$ steps.

The ***median*** of a set of ***n*** numbers can be computed in $O(n)$ time by the clever algorithm in [BFPRT].

(b) Using this information, show that a set S of n numbers can be partitioned in $O(n \log p)$ time into its ***p-quantiles***, that is, p sets $S^1, S^2, \ldots, S^p$, each of cardinality either $\lfloor n/p \rfloor$ or $\lceil n/p \rceil$, such that $a \in S^i, b \in S^j, i < j$ imply $a \leq b$.

Assume now that the adjacency list of each vertex v has been subdivided into its p-quantiles, for some p to be determined. We compute *shortest* (see Figure 12–5) by examining the adjacency list of each node v in *batches* of whole quantiles. If such a search of a quantile finds at least one edge $[v, u]$ with u in a different set of C, then the search is a *success*; otherwise it is a *failure*.

(c) Show how *shortest* can be computed in such a way that (i) the total time due to successes is $O(|E|/p)$ *per stage*, and (ii) the total time due to failures is $O(|E|)$ *for all stages together*.

(d) Show that the algorithm described above operates in time $O(\log|V| \cdot (|E|/p + |V|) + |E| \log p)$. By choosing p accordingly, conclude that the MST in a graph with $|E| \geq |V| \log |V|$ can be found in $O(|E| \log\log |V|)$ time.

(e) Generalize the result in (d) for graphs with $|E| < |V| \log |V|$. (*Hint:* First apply the algorithm of Figure 12–5 for $\log\log |V|$ stages.)

3. Show that the greedy algorithm for MST can be implemented in $O(|E| \log |V|)$ time.

4. Let $P = \{p_1, p_2, \ldots, p_n\} \subseteq R^2$ be a finite set of points on the plane. Let $d_{ij} = \text{dist}(p_i, p_j)$ be the Euclidean distance between p_i and p_j. The *Dirichlet cell* of $p_i \in P$ is the set of all $q \in R^2$ such that $\text{dist}(p_i, q) \leq \text{dist}(p_j, q)$ for all $j \neq i$.

(a) Show that the Dirichlet cells of P are convex polygonal regions—some of them unbounded—that cover all the plane and intersect only at their boundaries.

(b) Find the Dirichlet cells of the point set P shown below.

(c) Show that the Dirichlet cell of p_j is bounded iff p_j is the convex combination of three points in P that are not colinear.

5. Two points $p_i, p_j \in P$ are *Dirichlet neighbors* if the intersection of their Dirichlet cells is neither empty nor a single point. The *Dirichlet graph* of P is the graph $G_D = (P, E_D)$, where $E_D = \{[p_i, p_j]: p_i \text{ and } p_j \text{ are Dirichlet neighbors}\}$.

(a) A graph is *planar*, informally, if it can be drawn on the plane so that no two edges intersect. Show that the Dirichlet graph of any point set P is planar.

If a graph (V, E) is connected and planar, then it can be shown that $|E| \leq 3|V| - 6$ for $|V| \geq 3$.

*(b) Prove that the Dirichlet graph of a point set of n points can be constructed in $O(n \log n)$ time.

(c) Show that the MST of a point set P under d_{ij} is a subgraph of the Dirichlet graph of P. Conclude that the MST of P can be computed in $O(n \log n)$ time.

(d) Show by an example of six points that the shortest *matching* of a point set is not a subgraph of the Dirichlet graph.

*(e) Is the shortest TSP tour of a set of points a subgraph of the Dirichlet graph?

6. Let P be a set of points on the plane. Show that no two edges of the MST of P (with respect to d_{ij}) cross, when drawn as straight lines on the plane.

7. Let P be a point set. Show that there is an MST of P with no degree greater than 5.

8. The *wandering salesman problem* (WSP) is the TSP, except that the salesman can start wherever he wishes and does not have to return to the starting city after visiting all cities.

(a) Show how to transform in polynomial time any instance of the WSP to an equivalent instance of the TSP.

(b) Show how to transform in polynomial time any instance of the TSP to an equivalent instance of the WSP. (By *equivalent*, we mean that the optimal tour of one can be easily derived from the optimal tour of the other).

9. Which of the following problems remain essentially unchanged when we turn them from minimization to maximization problems? Why?

(a) TSP.

(b) Shortest path from s to t.

(c) Minimum weight complete matching.

(d) MST.

10. (Matching Problem with Node Weights) Given a graph $G = (V, E)$ and weights $w: V \longrightarrow Z^+$ on the nodes of G, find a matching M of G that maximizes $\sum_M w(v)$, where the sum is taken over all nodes v incident upon some edge of M. Show that the greedy algorithm solves this problem:

(a) By a direct argument.

(b) By showing that a certain independence system $(V, \mathfrak{M})$ satisfies Property 2 of Theorem 12.5.

(c) By showing that $(V, \mathfrak{M})$ satisfies Property 3 of Theorem 12.5.

(d) What are the circuits, spans, and rank function of $(V, \mathfrak{M})$? (*Note:* Matroids such as $(V, \mathfrak{M})$ are called *matching* matroids.)

11. Let E be a finite set, $C = \{S_1, \ldots, S_m\}$ a collection of subsets of E, and let $T = \{e_1, \ldots, e_t\} \subseteq E$. We say T is a *transversal* of C if there exist *distinct* integers $j(1), \ldots, j(t)$ such that $e_i \in S_{j(i)}$, $i = 1, \ldots, t$. Let $\mathfrak{I}$ be the set of all transversals of E. Show that $M_C = (E, \mathfrak{I})$ is a matroid. What are the circuits, spans, and rank function of M_C?

12. For each of the following matroids M, choose a field K, a dimension d, and a vector $L(e) \in K^d$ for each element e of E, to show that M is matric.
    (a) A graphic matroid M_G.
    (b) A partition matroid M_Π.
    (c) A transversal matroid M_C.

13. Consider the system of seven points and seven lines (six line segments and a circle) shown below.

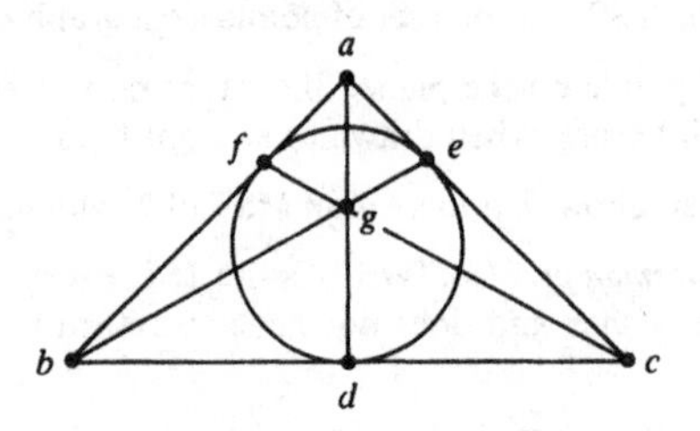

Define the independence system $M = (E, \mathcal{I})$ where $E = \{a, b, c, d, e, f, g\}$ and a subset S of E is in $\mathcal{I}$ iff one of the following is true.
    (i) $|S| < 3$.
    (ii) $|S| = 3$ and none of the seven lines passes through all three points in S.

    (a) Show that M is a matroid. What are the circuits, spans, and rank function of M?
    *(b) Show that M is *not* a matric matroid. That is, there is no system of seven vectors over any field such that the independent sets have exactly the same structure as this $\mathcal{I}$.

14. Let $M = (E, \mathcal{I})$ be a matroid, and let $\mathfrak{C}$ be its set of circuits.
    (a) Show each of the following.
        (i) If $C_1, C_2 \in \mathfrak{C}$ and $C_1 \subseteq C_2$, this implies $C_1 = C_2$.
        (ii) If $C_1, C_2 \in \mathfrak{C}$, $e \in C_1 \cap C_2$, and $e' \in C_1 - C_2$, then there is a $C_3 \in \mathfrak{C}$ such that $C_3 \subseteq (C_1 \cup C_2) - e$ and $e' \in C_3$.
    (b) What is the interpretation of (ii) for graphic matroids? Matching matroids?
    *(c) Conversely, show that if a system $(E, \mathfrak{C})$ satisfies (i) and (ii), then $(E, \mathcal{I})$ is a matroid, where $\mathcal{I} = \{I \subseteq E: C \nsubseteq I \text{ for all } C \in \mathfrak{C}\}$.

15. We are given a finite set E and a class of subsets of E, $C = \{S_1, S_2, \ldots, S_n\}$. We ask whether there is a set $H \subseteq E$ with $|H| = n$ and $|H \cap S_i| = 1$ for $i = 1, \ldots, n$. Formulate this as the intersection problem of a transversal and a partition matroid. Then solve it in a much simpler way.

16. We are given a graph $G = (V, E)$ and a set $L \subseteq V$. We wish to determine whether there is a spanning tree T of G such that the nodes in L are all leaves of T. Formulate this as the intersection problem of a graphic and a partition matroid. Then solve it in a much simpler way.

17. Find the maximum intersection of the two graphic matroids of Figure 12–16.

*18. (Matroid Partition Problem) Given a matroid $M = (E, \mathcal{I})$ and $S \subseteq E$, can S be partitioned into two sets $I_1, I_2 \in \mathcal{I}$? Show that this problem is equivalent to matroid intersection.

19. Show that the matroid parity problem (Subsec. 12.6.2) for partition matroids is a matching problem.

20. Show that the matroid parity problem for graphic matroids is a common generalization of the two-graphic-matroid intersection problem and of matching. (*Hint:* You may have to use *multigraphs*, that is, graphs with repetitions of edges allowed.)

21. Let $M = (E, \mathcal{I})$ be a matroid, and let $\mathcal{B}$ be the set of bases of M. Let $\bar{\mathcal{I}} = \{I \subseteq E: E - I \supseteq B$ for some $B \in \mathcal{B}\}$. Show that $\bar{M} = (E, \bar{\mathcal{I}})$ is a matroid. (*Note:* $\bar{M}$ is called the *dual matroid* of M.)

22. Given a graph $G = (V, E)$, a vertex $v \in V$, and an integer k, we wish to find whether there is a spanning tree of G in which v has degree k or less. Show that this is a matroid intersection problem.

NOTES AND REFERENCES

It seems that there is no earliest reference for MST. One of the earlier ones is a Czechoslovokian paper:

[Bo] BORUVKA, O., "On a Minimal Problem," *Prace Morawske Predovedecke Spolecrosti*, 3 (1926).

The algorithm of Figure 12–2 is usually attributed to

[Pr] PRIM, R. C., "Shortest Connection Networks and Some Generalizations," *BSTJ*, 36 (1957), 1389–1401.

[Di] DIJKSTRA, E. W., "A Note on Two Problems in Connexion with Graphs," *Numerische Mathematik*, 1 (1959), 269–71.

The algorithm of Figure 12–5 is described in

[BG] BERGE, C., and A. GHOUILLA-HOURI, *Programming, Games, and Transportation Networks*. New York: John Wiley & Sons, Inc., 1965.

The $O(|E| \log\log |V|)$ implementation of Problem 2 is from

[Ya] YAO, A.C-C., "An $O(|E| \log\log |V|)$ Algorithm for Finding Minimum Spanning Trees," *Inf. Proc. Letters*, 4 (1975), 21–25.

Many related algorithms, including an $O(|E|)$ algorithm for dense graphs, are described in

[CT] CHERITON, D., and R. E. TARJAN, "Finding Minimum Spanning Trees," *J. SIAM Comp.*, 5 (1976), 724–42.

The greedy algorithm for MST is presented in

[Kr] KRUSKAL, J. B., "On the Shortest Spanning Subtree of a Graph and the Traveling Salesman Problem," *Proc. Amer. Math. Soc.*, 7 (1956), 48–50.

Matroids (also known as pregeometries for reasons related to Problem 13) were first studied in

[Wh] WHITNEY, H., "On the Abstract Properties of Linear Dependence," *Amer. J. Math.*, 57 (1935), 509–33.

For in-depth reading on the subject, see also

[Tu] TUTTE, W. T., "Lectures on Matroids," *J. Res. NBS*, 69B (1965), 1–48.

[CR] CRAPO, H. H., and G-C. ROTA, *On the Foundations of Combinatorial Theory: Combinatorial Geometries*. Cambridge, Mass.: M.I.T. Press, 1970.

The relation of matroid theory to combinatorial optimization and the algorithm of Kruskal was first shown in

[Ed1] EDMONDS, J., "Matroids and the Greedy Algorithm," *Math. Prog.*, 1 (1971), 127–36.

The polynomial-time solution of the two-matroid intersection problem was also announced in that paper. The algorithm of Figure 12–21 was first published in

[La1] LAWLER, E. L., "Matroid Intersection Algorithms," *Math. Prog.*, 9 (1975), 31–56.

The equivalent matroid *partition* problem (Problem 18) was solved in

[Ed2] EDMONDS, J., "Minimum Partition of a Matroid into Independent Subsets," *J. Res. NBS*, 69B (1965), 67–77.

Problems 4 and 5 are from

[Sh] SHAMOS, M. I., *Computational Geometry*, (Unpublished Ph.D. thesis, Yale University, 1978).

Problems 10 and 11 are from

[EF] EDMONDS, J., and D. R. FULKERSON, "Transversals and Matroid Partition," *J. Res. NBS*, 69B (1965), 147–53.

and Problems 13–14 from Whitney's original paper [Wh]. For efficient algorithms for finding the *shortest branching* in a weighted digraph, see

[Ed3] EDMONDS, J., "Optimum Branchings," *J. Res. NBS*, 71B (1967), 233–40.

[Ka] KARP, R. M., "A Simple Derivation of Edmonds' Algorithm for Optimum Branching," *Networks*, 1 (1971), 265–72.

[Ta] TARJAN, R. E., "Finding Optimum Branchings," *Inf. Proc. Letters*, 3 (1974), 25–35.

The matroid parity problem (Subsec. 12.6.2) for matric matroids is solved in

[Lo] LOVÁSZ, L., "The Matroid Matching Problem," *Proc. Conf. on Algebraic Graph Theory*, Szeged, Hungary, 1978.

The fact used in Problem 2 is from

[BFPRT] BLUM, M., R. W. FLOYD, V. R. PRATT, R. L. RIVEST, and R. E. TARJAN, "Time Bounds for Selection," *JCSS*, 7 (1973), 448–61.

13

Integer Linear Programming

13.1 Introduction

Integer linear programming (ILP) is the following optimization problem.

$$\begin{aligned} &\min c'x \\ &Ax = b \\ &x \geq 0 \\ &x \text{ integer} \end{aligned}$$

Also, ILP in canonical form or in general form can be defined similarly, and the observation (Sec. 2.1) that they are all equivalent is still valid. The entries of A, b, and c are integer.

The necessity of formulating and solving ILP's came initially from the fact that in some applications of LP, fractional solutions were undesirable—imagine an application in which x_j is, for example, the number of aircraft assigned to route j. The usual first reaction to the ILP problem is, "Why not solve the corresponding LP and round the solutions to the closest integer?" This is certainly a plausible strategy in many cases, especially when the solution x is expected to contain large integers and therefore to be insensitive to rounding. There are problems here too: rounding to a feasible integer solution may not always be straightforward (see Figs. 13–1 and 14–1). There are, however, very serious limitations to this approach. Most applications of ILP are not just LP's with solution variables that happen to come in quanta of one unit. Most frequently, an ILP is the result of a *conscious use* of the integer constraint to *model combinatorial constraints* or *nonlinearities* of different sorts. These ILP's are, by their very nature, not susceptible to the rounding approach, essentially because rounding defeats the purpose of the ILP formulation, and it is as hard to perform as solving the original combinatorial problem from scratch. We examine several such examples.

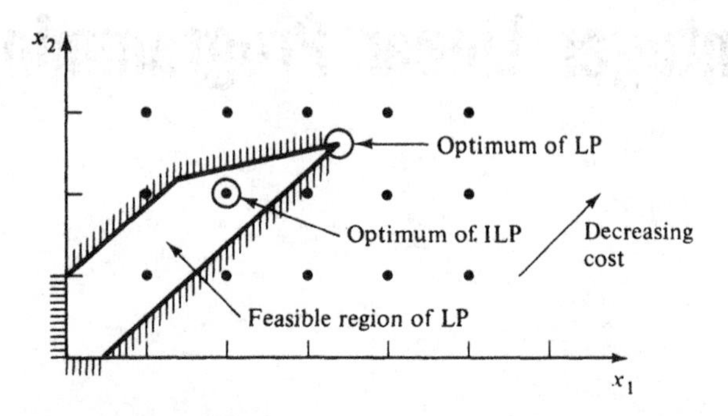

Figure 13–1 The four integer points closest to the LP optimum are infeasible.

Example 13.1 (Formulation of the Traveling Salesman Problem as an Integer Linear Programming Problem)

Consider the traveling salesman problem with $n + 1$ cities, nodes 0, 1, . . . , n, and intercity distances $[c_{ij}]$ (not necessarily a symmetric matrix). If we associate the variable x_{ij} with the arc (i, j) and let $x_{ij} = 1$ if arc (i, j) is in a tour and zero if not, we can begin to formulate the TSP by writing

$$\min z = \sum_{\substack{i,j=0 \\ i \neq j}}^{n} c_{ij} x_{ij}$$

$$0 \leq x_{ij} \leq 1 \qquad \text{all } i, j$$

$$x_{ij} \text{ integer} \qquad \text{all } i, j \tag{13.1}$$

$$\text{(a)} \quad \sum_{i=0}^{n} x_{ij} = 1 \qquad j = 0, \ldots, n$$

$$\text{(b)} \quad \sum_{j=0}^{n} x_{ij} = 1 \qquad i = 1, \ldots, n$$

The inequalities in (a) express the fact that exactly one arc enters each node and (b) that exactly one arc leaves each of nodes 1, . . . , n (From (a) and (b) together, it follows that one arc leaves node 0 as well; see Problem 7.) All this does not capture all the constraints of TSP, however, since any set of disjoint cycles will satisfy (13.1), as illustrated in Fig. 13–2(a). In fact, (13.1) describes the *assignment problem* (Sec. 11.1), which is certainly easier than the TSP. What we need are constraints that rule out disjoint cycles, or *subtours*, as they are called in this context. One way to do this is as follows [DFJ]: Let $(S, \bar{S})$ be a nontrivial partition of $\{0, \ldots, n\}$, and for each such partition demand that

$$\sum_{\substack{i \in S \\ j \in \bar{S}}} x_{ij} \geq 1 \tag{13.2}$$

Figure 13–2(b) illustrates the idea—all such constraints taken together ensure that the graph is a single tour. This is hardly a practical formulation for even moderately large problems, however, since there are $2^{n+1} - 2$ nontrivial partitions, each contributing a constraint.

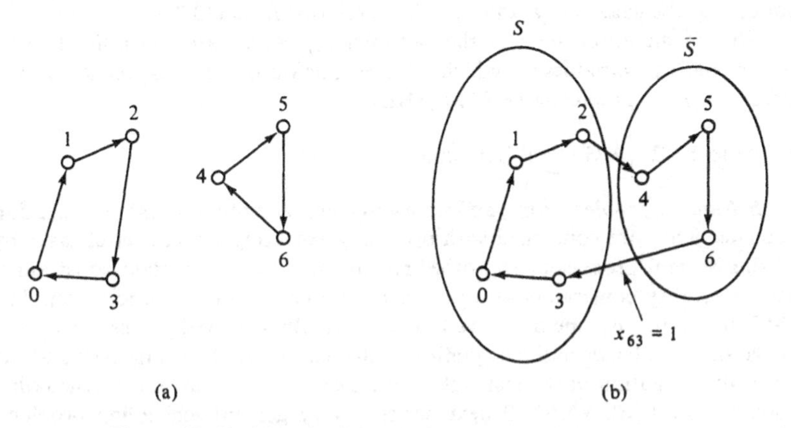

Figure 13–2 (a) A feasible solution to Eq. 13.1 that is not a tour. (b) A subtour elimination constraint.

A more concise formulation, due to A.W. Tucker [MTZ], replaces the subtour elimination constraints in (13.2) by

$$u_i - u_j + nx_{ij} \leq n - 1 \qquad 1 \leq i \neq j \leq n \tag{13.3}$$

where the u_i, $i = 1, \ldots, n$, are unrestricted real variables. We next show that these inequalities not only restrict solutions to be tours, but do not exclude any tours.

Proposition The constraints in (13.1) and (13.3) define the TSP.

Proof We first show that every feasible solution is a tour. To see this, we show that every circuit passes through city 0. Suppose not: Suppose that the sequence of cities $i_1, \ldots, i_k$ is a circuit that excludes city 0. We can write the inequalities in (13.3) along the tour as

$$u_{i_j} - u_{i_{j+1}} + n \leq n - 1 \qquad j = 1, \ldots, k-1$$
$$u_{i_k} - u_{i_1} + n \leq n - 1$$

Adding these, we get a contradiction.

To finish the proof, we show that for every legitimate tour, there are values of u_i that satisfy (13.3). In fact, let $u_0 = 0$ and $u_i = t$ if city i is the tth city visited on the tour, $t = 1, \ldots, n$. If $x_{ij} = 0$, we have

$$u_i - u_j \leq n - 1 \qquad 1 \leq i \neq j \leq n$$

which always holds since the case $u_i = n$ and $u_j = 0$ is ruled out because the arc $(i, 0)$ is in the tour and hence $x_{i0} = 1$. If $x_{ij} = 1$, we have

$$u_i - u_j + n \leq n - 1$$

which holds because $u_i - u_j = -1$ if i and j are successive cities on the tour (notice that the case $u_i = n$ and $u_j = 0$ is excluded from (13.3)).

This formulation involves the variables x_{ij}, which are constrained to be integer, and the variables u_i, which are not. Such a problem is called a *mixed integer linear programming problem* (MILP). □

Example 13.2 (A General Scheduling Problem)

Scheduling problems form an important class of combinatorial optimization problems. They are concerned with optimally executing a given set of tasks by utilizing several processors and other resources, subject to certain constraints, such as priority constraints of a task over another, deadlines, and so on. The goal is to minimize some objective function, usually the total processing time—that is, the time between the beginning of the execution of the first task and the end of the execution of the last task. For an extensive discussion of scheduling problems, see [Co]. We shall next define a very general scheduling problem, which contains most scheduling problems appearing in the literature as special cases.

We are given a *task system* $\mathcal{J} = \{J_1, \ldots, J_n\}$. For each task J_i in $\mathcal{J}$, we have its integer *processing time* τ_i, and the amount R_{ji} of the *jth resource* it needs, $j = 1, \ldots, r$. There are B_j units of the *jth* resource available at all times, $j = 1, \ldots, r$. These tasks are to be executed on *m processors*; a bit a_{ij}, $i = 1, \ldots, n$, $j = 1, \ldots, m$, indicates whether or not J_i can be executed on the *jth* processor. There may be a *deadline* d_i associated with J_i; J_i must have completed execution d_i time units after the beginning of execution of the first task. Finally, we have a *precedence relation* $(\mathcal{J}, A)$ (that is, a digraph with no cycles) such that $(J_i, J_j) \in A$ means that J_i must complete its execution before J_j starts its own.

We are going to show how any instance of the scheduling problem defined above can be formulated as an ILP. The main integer variables will be s_i and p_{ij}, $i = 1, \ldots, n, j = 1, \ldots, m$; s_i is the starting time of J_i, (we can assume that the s_i's are integer, and that $\min \{s_i\} = 0$), whereas $p_{ij} = 1$ if i is executed on machine j and zero otherwise. Naturally, we require that

$$\sum_{j=1}^{m} p_{ij} = 1 \qquad i = 1, \ldots, n$$

Therefore p_{ij} and s_i completely specify a solution of the scheduling problem. Let us now see how to describe the different scheduling constraints in terms of linear equations and inequalities.

1. We can have p_{ij} equal to 1 only in cases in which the task J_i can be executed on processor j; thus $p_{ij} \leq a_{ij}, j = 1, \ldots, m, i = 1, \ldots, n$.
2. No two jobs can be executed simultaneously on the same processor. We want to express the fact that whenever $p_{ij} = p_{kj} = 1$ and job J_i executes before J_k, we have $s_i + \tau_i \leq s_k$. To do this, let $T = \sum_{j=1}^{n} \tau_j$, and let δ_{ik} be a nonnegative integer variable which is 1 if $s_i \leq s_k$ and zero otherwise. We can guarantee that δ_{ik} has its prespecified meaning by writing
$$s_k - s_i \leq \delta_{ik} \cdot T \qquad i \neq k, \quad i, k = 1, \ldots, n$$
$$\delta_{ik} + \delta_{ki} = 1 \qquad i < k, \quad i, k = 1, \ldots, n$$
Since $s_k - s_i$ is never equal to or greater than T in an optimal solution, the inequality above is restrictive only if $\delta_{ik} = 0$. So, if $\delta_{ik} = 0$, s_k starts no later than s_i, and vice versa. We can now express the constraint that two jobs do not execute simultaneously on the same processor by insisting that
$$s_i + \tau_i - s_k \leq T \cdot (1 - \delta_{ik} + 2 - p_{ij} - p_{kj}) \qquad j = 1, \ldots, m$$
$$i \neq k, \quad i, k = 1, \ldots, n$$
The above inequality is restrictive only if $\delta_{ik} = 1$ and $p_{ij} = p_{kj} = 1$; that is, J_i executes before J_k on the jth processor. In this case it guarantees that $s_i + \tau_i \leq s_k$, as it should.
3. If $(J_i, J_k) \in A$, then J_k starts executing after the end of J_i, independent of processors. This can be written as
$$\delta_{ik} = 1 \qquad (J_i, J_k) \in A$$
$$s_i + \tau_i - s_k \leq 0 \qquad (J_i, J_k) \in A$$
4. We can restate the resource constraints by saying that whenever a job starts, the sum of the resource requirements of the active jobs (all the jobs which started at the same time or before and have not yet completed) does not exceed the bounds. To express this by linear constraints, we first define a new set of nonnegative integer variables $\epsilon_{ik}, i, k = 1, \ldots, n$. If J_i is active when J_k starts, $\epsilon_{ik} = 1$; otherwise it is zero.
$$\epsilon_{ik} \leq \delta_{ik}$$
$$1 + (\epsilon_{ik} - 1)T \leq s_i + \tau_i - s_k \leq (\epsilon_{ik} + \delta_{kj})T$$
$$i \neq k, \quad i, k = 1, \ldots, n$$

The resource constraints are therefore

$$R_{jk} + \sum_{i \neq k} (\epsilon_{ik} + \epsilon_{ki}) R_{ji} \leq B_j \qquad j = 1, \ldots, r, \quad k = 1, \ldots, n$$

5. Deadline constraints can be expressed as

$$s_i + \tau_i \leq d_i \qquad i = 1, \ldots, n$$

6. Finally, let f be the finishing time of the schedule; naturally

$$f \geq s_i + \tau_i \qquad i = 1, \ldots, n$$

Our goal is to minimize f subject to all these constraints. Consequently any instance of our general scheduling problem with n jobs, m machines, r resources, and $|A|$ precedence constraints can be expressed as an ILP with $n(2n + m - 1) + 1$ variables and $n(n-1)(m + \frac{9}{2}) + n(r + m + 3) + 2|A|$ equations and inequalities. □

Example 13.3 (Nonlinearities)

The constraint x *integer* is said to be a *nonlinear* constraint because it cannot be replaced by linear constraints (however, see the next section). It appears, furthermore, that it is a fundamental nonlinear constraint, because many different nonlinear constraints can be expressed in terms of it. We examine here a few specific examples.

A. *Nonlinear Costs* The linear cost $c'x$ is not always an acceptable first approximation of the costs occurring in real situations. An important type of a nonlinear cost is the one with a *set-up cost* component defined by

$$c(x) = \begin{cases} ax + b & \text{if } x > 0 \\ 0 & \text{if } x = 0 \end{cases}$$

and depicted in Fig. 13–3(a). We can construct an MILP model of this by bringing in an integer variable δ, $1 \geq \delta \geq 0$. Here δ is 1 iff $x > 0$; we can ensure this by adding the constraints

$$\delta \leq x \leq U\delta,$$

where U is an upper bound on x which we assume to be available. The cost becomes the linear functional

$$c(x, \delta) = ax + b\delta$$

Similar methods can be used in order to find ILP models for more complicated nonlinearities in the cost function, such as the one shown in Fig. 13–3(b)—which in most practical situations will be, no doubt, a piecewise linear approximation of an even more complex curved cost function. (See Problem 6.)

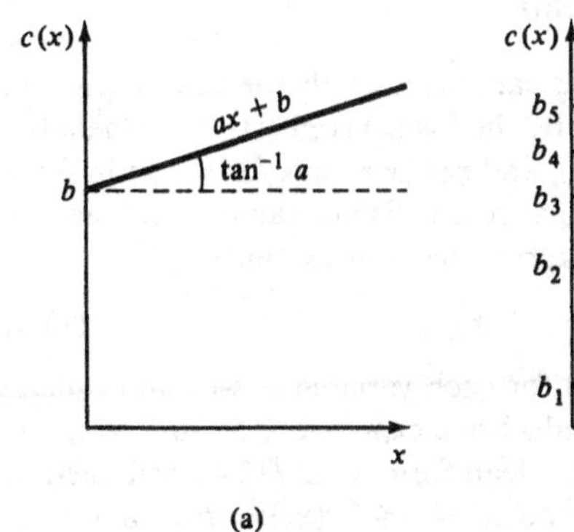

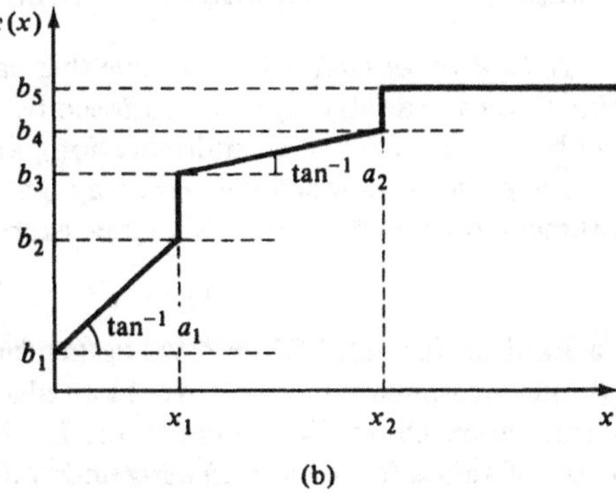

Figure 13–3

B. *Dichotomies* Suppose that in an application the feasible solutions must obey the disjunction of two constraints:

$$x \geq a \quad or \quad y \geq b$$

We can write a portion of MILP which is equivalent to this by introducing an integer variable δ, $1 \geq \delta \geq 0$. We now rewrite this disjunctive constraint as

$$x \geq \delta a$$
$$y \geq (1 - \delta)b$$

Generalizations are possible to disjunctions of *conjunctions* of constraints and disjunctions of more than two constraints. This trick is especially useful when one is faced with conditional constraints of the form

$$\textit{if } x > a \textit{ then } y \geq b$$

This is, however, equivalent to

$$x \leq a \quad or \quad y \geq b$$

and hence the previous method applies.

C. *Discrete Variables* Suppose that we have a constraint of the form $x \in \{s_1, \ldots, s_m\}$; in other words x is restricted to assume only a finite set of values, possibly otherwise unrelated. We can write this constraint as

$$x = s_1\delta_1 + s_2\delta_2 + \cdots + s_m\delta_m$$

where the variables δ_j obey

$$\delta_1 + \delta_2 + \cdots + \delta_m = 1$$
$$\delta_j \geq 0, \qquad \text{integer}, \ j = 1, \ldots, m. \ \square$$

Example 13.4 (The Satisfiability Problem)

A *Boolean variable* x is a variable that can assume only the values *true* and *false*. Boolean variables can be combined by the logical connectives *or* (denoted here by $+$), *and* (shown as multiplication), and *not* ($\bar{x}$ stands for *not* x) to form *Boolean formulas* in much the same way that real variables can be combined by arithmetic operations to form algebraic expressions. For example,

$$\bar{x}_3 \cdot (x_1 + \bar{x}_2 + x_3) \tag{13.4}$$

is a Boolean formula. Given a value $t(x)$ for each variable x, we can evaluate a Boolean formula, just as we would an algebraic expression, by following the interpretation above. For example, the Boolean formula in (13.4) evaluated at the set of values (called a *truth assignment*) $t(x_1) =$ *true*, $t(x_2) =$ *true*, and $t(x_3) =$ *false* gives the value *true*.

The formula above can be made *true* by some truth assignment: Such Boolean formulas are called *satisfiable*. Not all formulas are satisfiable; there are some that cannot be made *true* by any truth assignment, essentially because they are encodings of "contradiction." For example, consider

$$(x_1 + x_2 + x_3) \cdot (x_1 + \bar{x}_2) \cdot (x_2 + \bar{x}_3) \cdot (x_3 + \bar{x}_1) \cdot (\bar{x}_1 + \bar{x}_2 + \bar{x}_3) \tag{13.5}$$

For (13.5) to be *true*, all subformulas within parentheses (called *clauses*) that contain *literals* (that is, variables or negations) must be *true*. The first clause says that at least one of the variables must be *true*. The next three clauses force all variables to be the same. To see this, suppose that x_1 is *false*. Because the second clause must be made *true*, x_2 must be *false*; finally, from the third clause, x_3 must be *false* too. If a variable assumes any value, the second, third, and fourth clauses will force the other variables to take the same value. So all variables must have the same value, and because we know that at least one must be *true*, they all must be *true*. However, the last clause requires that not all of them be *true*, and hence Formula 13.5 contains a contradiction. It is unsatisfiable. The SATISFIABILITY problem is as follows:

> Given m clauses $C_1, \ldots, C_m$ involving the variables $x_1, \ldots, x_n$, is the formula $C_1 \cdot C_2 \ldots C_m$ satisfiable?

SATISFIABILITY is a central problem in mathematical logic, and there has been great interest in devising efficient algorithms for its solution. Of course, one solution is to try all possible truth assignments to see if one satisfies the formula. This is not an efficient algorithm, however, since there are 2^n truth assignments to try—we have a binary choice for each variable. To date there is no known efficient algorithm that solves SATISFIABILITY. It is interesting, therefore, that SATISFIABILITY can be formulated as an ILP. The formulation is immediate if one identifies *true* with 1 and *false* with zero. Thus *or* becomes ordinary addition, $\bar{x}$ is expressed as $1 - x$, and we require that for each clause

C we have at least one *true* literal. In other words,

$$\sum_{x \in C} x + \sum_{\bar{x} \in C} (1 - x) \geq 1$$

For example, the full ILP for the formula in (13.5) would be

$$\begin{aligned} x_1 + x_2 + x_3 &\geq 1 \\ x_1 + (1 - x_2) &\geq 1 \\ x_2 + (1 - x_3) &\geq 1 \\ x_3 + (1 - x_1) &\geq 1 \\ (1 - x_1) + (1 - x_2) + (1 - x_3) &\geq 1 \\ x_1, x_2, x_3 &\leq 1 \\ x_1, x_2, x_3 &\geq 0, \qquad \text{integer} \end{aligned} \tag{13.6}$$

It is easy to see how the satisfiability of (13.5) is captured by the constraints in (13.6). In general, a formula consisting of many clauses connected by *and*'s—a formula in *conjunctive normal form*, as we say—is satisfiable iff the corresponding ILP has a feasible point. Thus the formulation in (13.6) is not an ILP in that we do not seek to minimize a linear functional. However, it can be easily transformed to an equivalent ordinary ILP. For example, in (13.6) we could replace the first inequality by $x_1 + x_2 + x_3 \geq y$ and maximize y. The formula is satisfiable iff the optimal value $\hat{y}$ exists and satisfies $\hat{y} \geq 1$. The ILP of (13.6) belongs to an important class of ILP's: Those that admit solutions which are zero or 1 only. Such ILP's are called *binary linear programs* or *zero-one linear programs* (ZOLP). The last two rows of the ILP of (13.6) are usually written $x_j \in \{0, 1\}, j = 1, \ldots, n$. The equality $x_j^2 = x_j$ is a fancier way of saying the same thing.

The ILP formulation (13.6) of satisfiability illustrates a previous point: We cannot in general solve ILP's in an acceptable manner just by rounding the solution of the corresponding LP. The constraints corresponding to the satisfiability problem for any formula in conjunctive normal form with at least two literals in each clause—a condition easy to guarantee (see Problem 5)—is satisfied by the fractional values $x_j = \frac{1}{2}$ for all variables x_j. Hence a feasible solution to the LP is always trivially available; rounding this solution in an acceptable manner (in fact just deciding whether it *can* be rounded) is as hard as the satisfiability problem itself, which is very hard indeed (see Chapter 15). □

We see, therefore, that ILP is a very general framework, within which many diverse problems can be formulated. Alas, it is this generality that brings about its main weakness. After more than two decades of intensive effort there is no known practical algorithm for solving large ILP's. Although we describe two plausible methods for attacking ILP's in this book, we also point out that excessive time requirements make them practically infeasible for instances of dissappointingly small size. In Chapter 15, in fact, we shall introduce the notion of *NP-completeness* in order to characterize those problems which, like ILP, appear to be intrinsically difficult exactly because of their generality.

13.2 Total Unimodularity

Recall that in the max-flow and weighted bipartite matching problems, for example, solutions to the linear program without special integer constraints were, nevertheless, always integer. It is natural to ask what is at the root of our good fortune in such cases, so that we can take full advantage of such a mechanism. To answer this question, we first need the following definition of a central concept.

Definition 13.1

A square, integer matrix B is called *unimodular* (UM) if its determinant $\det(B) = \pm 1$. An integer matrix A is called *totally unimodular* (TUM) if every square, nonsingular submatrix of A is UM. □

If B is formed from a subset of m linearly independent columns of A, it determines the basic solution

$$x = B^{-1}b = \frac{B^{adj}b}{\det(B)}$$

where B^{adj} is the adjoint of B, and so if B is UM and b is integer (which we always assume), x is integer. If we define the polytope

$$R_1(A) = \{x: Ax = b, \quad x \geq 0\}$$

to be the usual feasible set for the standard form LP, we have proved the following theorem.

Theorem 13.1 *If A is TUM, then all the vertices of $R_1(A)$ are integer for any integer vector b.*

Thus a standard form LP with TUM matrix will always lead to an integer optimum when solved by the simplex algorithm.

When an LP is formulated with inequality constraints, the same result holds. Let the corresponding polytope be

$$R_2(A) = \{x: Ax \leq b, \quad x \geq 0\}$$

Then we have the next theorem.

Theorem 13.2 *If A is TUM, then all the vertices of $R_2(A)$ are integer for any integer vector b.*

Proof This amounts to showing that if A is TUM, so is $(A \mid I)$, for then we can add slack variables and apply Theorem 13.1. Let C be a square, nonsingular submatrix of $(A \mid I)$. The rows of C can be permuted so that it can be written

$$C = \left(\begin{array}{c|c} B & O \\ \hline D & I_k \end{array}\right)$$

where I_k is an identity matrix of size k and B is a square submatrix of A, possibly with its rows permuted. Therefore

$$\det(C) = \det(B) = \pm 1$$

because A is TUM and C is nonsingular. □

We shall now show that the cases we have observed in previous chapters where integer solutions were automatic were in fact cases where the constraint matrix was TUM. The convenient sufficient (but not necessary) condition is given by Theorem 13.3.

Theorem 13.3 *An integer matrix A with $a_{ij} = 0, \pm 1$ is TUM if no more than two nonzero entries appear in any column, and if the rows of A can be partitioned into two sets I_1 and I_2 such that:*

1. *If a column has two entries of the same sign, their rows are in different sets;*
2. *If a column has two entries of different signs, their rows are in the same set.*

Proof The proof is by induction on the size of submatrices. For the basis, we need only observe that any submatrix of one element is TUM. Let C be any submatrix of size k. If C has a column of all zeros, it is singular. If C has a column with one nonzero entry, we can expand its determinant along that column, and the result follows from the induction hypothesis.

The last case occurs when C has two nonzero entries in every column. Then Conditions 1 and 2 of the theorem imply that

$$\sum_{i \in I_1} a_{ij} = \sum_{i \in I_2} a_{ij} \qquad \text{for every } j$$

That is, a linear combination of rows is zero, and hence $\det(C) = 0$. □

We then have the desired result.

Corollary *Any LP in standard or canonical form whose constraint matrix A is either*

1. *The node-arc incidence matrix of a directed graph, or*
2. *The node-edge incidence matrix of an undirected bipartite graph,*

has only integer optimal vertices. This includes the LP formulations of shortest path, max-flow, the Hitchcock problem, and weighted bipartite matching.

Proof The matrices in Case 1 satisfy the condition of Theorem 13.3 with $I_2 = \varnothing$; those in Case 2 with $I_1 = U$ and $I_2 = V$, where the bipartite graph is $B = (V, U, E)$. □

It is also true that the converse of Theorem 13.2 (but not 13.1) holds: If all the vertices of $R_2(A)$ are integer for any integer b, then A must be TUM (see Problems 1 and 2). In some sense, therefore, we have discovered why certain LP's lead to optimal basic solutions which are automatically integer while others do not. In the latter cases we are forced to look beyond the simplex algorithm for solutions to the corresponding ILP's.

13.3 Upper Bounds for Solutions of ILP's

We showed in Lemma 2.1 that any bfs of an LP is bounded in absolute value by an upper bound depending on the dimensions of the LP and the size of the integers appearing in it. There is sound geometric intuition behind this result. What it really says is that the faces of the polytope cannot be extended for "too long" to create "big" bfs's, unless huge integers are involved to create very small angles (see Fig. 13–4(a)). A similar argument seems to be valid for ILP. It is very difficult for the feasible region to avoid all integer points except for

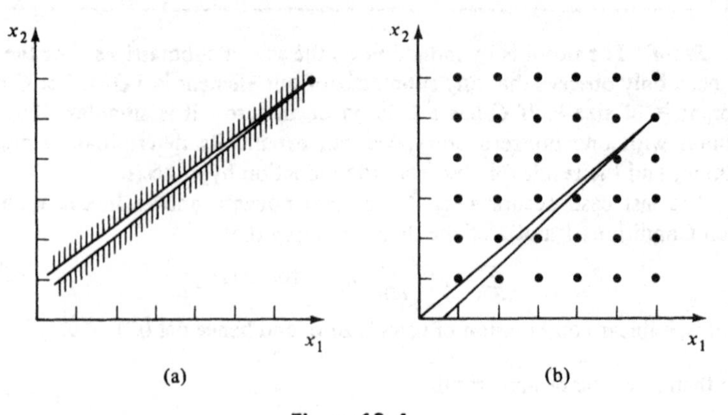

Figure 13–4

a very large one, unless again extremely small angles and huge coefficients are employed (Fig. 13–4(b)). In this section we give a formal argument to this effect.

We consider the ILP with constraints

$$Ax = b$$
$$x \geq 0, \quad \text{integer} \tag{13.7}$$

where A is an $m \times n$ integer matrix and b is an integer m-vector. Let $a_1 = \max_{i,j} \{|a_{ij}|\}$, $a_2 = \max_i \{|b_i|\}$.

The following is a very intuitive geometric fact. Consider three directions on the plane (Fig. 13–5). Then *exactly* one of the following is true.

1. The directions belong to the *same half-plane* (Fig. 13–5(a)).
2. The directions can be the directions of three *balanced forces* (that is, there are three nonnegative multipliers, not all zero, that annul them).

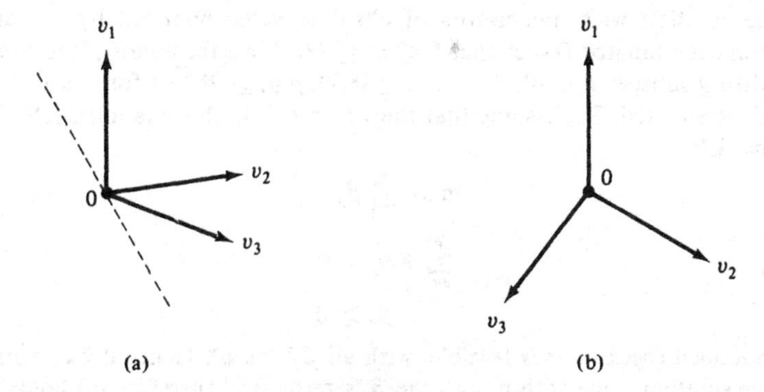

Figure 13–5

We next prove a multi-dimensional, finite-precision generalization of this fact, which can be considered as a finite-precision version of Farkas' lemma (Sec. 3.3).

Lemma 13.1 *Let $v_1, v_2, \ldots, v_k$ be $k > 0$ vectors in $\{0, \pm 1, \pm 2, \ldots, \pm a_1\}^m$, and let $M_1 = (ma_1)^{m+1}$. Then the following statements are equivalent.*

(a) *There exist k reals $\beta_1, \ldots, \beta_k \geq 0$, not all zero, such that $\sum_{j=1}^{k} \beta_j v_j = 0$.*

(b) *There exist k* integers *$0 \leq \beta_1, \ldots, \beta_k \leq M_1$, not all zero, such that $\sum_{j=1}^{k} \beta_j v_j = 0$.*

(c) *There is no vector $h \in R^m$ such that $p_j = h'v_j > 0$ for $j = 1, \ldots, k$.*

(d) *There is no vector $h \in \{0, \pm 1, \pm 2, \ldots, \pm M_1\}^m$ such that $h'v_j \geq 1$ for $j = 1, \ldots, k$.*

Proof (a) $\Rightarrow$ (b) This follows directly from Lemma 2.1 and Theorem 2.1.

(b) $\Rightarrow$ (c) Suppose that (b) holds, and yet a vector h satisfying the conditions of (c) still exists. Then we have

$$0 = h' \sum_{j=1}^{k} \beta_j v_j$$
$$= \sum_{j=1}^{k} \beta_j p_j > 0$$

which is absurd.

(c) $\Rightarrow$ (d) This is trivial.

(d) $\Rightarrow$ (a) Suppose that (d) holds, and consider the linear program

$$\min h'0 \tag{13.8}$$
$$h'v_j \geq 1, \quad j = 1, \ldots, k$$

If this LP has a feasible solution, then, by Cramer's rule, it has a rational feasible solution with numerators of absolute value bounded by M_1 and a common denominator D such that $1 \leq D \leq M_1$. If g is the vector of the numerators, then g satisfies $g \in \{0, \pm 1, \ldots, \pm M_1\}^m$, $g'v_j \geq D \geq 1$ for $j = 1, \ldots, k$, and (d) is violated. So, assume that the LP in (13.8) above is infeasible. Then the dual LP

$$\max \sum_{j=1}^{k} \beta_j$$
$$\sum_{j=1}^{k} \beta_j v_j = 0$$
$$\beta_j \geq 0$$

is unbounded (because it *is* feasible, with all β_j's zero). Hence it has a strictly positive solution—one with not all the β_j's zero—and therefore (a) holds. □

We can now show the main result of this section.

Theorem 13.4 *If the ILP of (13.7) has a feasible solution, then it has a feasible solution $x \in \{0, 1, \ldots, M_2\}^n$, where $M_2 = n(ma_1)^{2m+3}(1 + a_2)$.*

Proof Consider the *smallest* (in terms of sum of components) feasible solution x of (13.7). If all components of x are smaller than or equal to $M_1 = (ma_1)^{m+1}$, then we are done, since $M_1 < M_2$. Otherwise, let us assume, without loss of generality, that the components of x that are larger than M_1 are the first k ones. Let $v_1, v_2, \ldots, v_k$ be the corresponding columns of A. We distinguish between two cases:

Case 1 There exist integers $\beta_1, \ldots, \beta_k$ between zero and M_1, not all zero, such that $\sum_{j=1}^{k} \beta_j v_j = 0$. In this case we immediately have that the vector

$x' = (x_1 - \beta_1, x_2 - \beta_2, \ldots, x_k - \beta_k, x_{k+1}, \ldots, x_n)$ is also a solution of (13.7) and, in fact, one with sum of components smaller than that of x, a contradiction.

Case 2 There are no such integers $\beta_1, \ldots, \beta_k$. By Lemma 13.1, there is a vector $h \in \{0, \pm 1, \ldots, \pm M_1\}^m$ such that $h'v_j \geq 1$ for $j = 1, \ldots, k$. Let us premultiply the equation $Ax = b$ by h'. We obtain

$$\sum_{j=1}^{k} h'v_j x_j = h'b - \sum_{j=k+1}^{n} h'v_j x_j$$

Thus

$$\sum_{j=1}^{k} x_j \leq \sum_{j=1}^{k} h'v_j x_j = h'b - \sum_{j=k+1}^{n} h'v_j x_j$$

The right-hand side is bounded from above by $ma_2M_1 + (n - k)ma_1M_1^2 \leq M_2$, and hence no component of x can be larger than M_2. ☐

Corollary *If the ILP*

$$Ax \leq b$$
$$x \geq 0, \qquad \textit{integer}$$

has a feasible solution, then it has one with components bounded by $(n + m)(ma_1)^{2m+3}(1 + a_2)$.

Proof Reduce this ILP to the form of (13.7) by using slack variables, and then apply the theorem. ☐

So far we have been ignoring the cost component of the ILP's. Let us therefore consider the ILP

$$\begin{aligned} &\min c'x \\ &Ax = b \\ &x \geq 0, \qquad \text{integer} \end{aligned} \tag{13.9}$$

and its linear programming "relaxation"

$$\begin{aligned} &\min c'x \\ &Ax = b \\ &x \geq 0 \end{aligned} \tag{13.10}$$

Lemma 13.2 *If (13.10) is unbounded and (13.9) has a feasible solution, then (13.9) is also unbounded.*

Proof Let x be a feasible solution of (13.9). Obviously, x is a feasible solution of (13.10). Since (13.10) is unbounded, we know (see Problem 17 in Chapter 2) that there is a rational vector α such that

(a) $c'\alpha < 0$;

(b) $x + k\alpha$ is feasible in (13.10) for all $k > 0$.

Let P be the product of the denominators of the components of α. Then the points $\{x + jP\alpha : j = 1, 2, \ldots\}$ are integer, feasible, and have cost that is unbounded from below. □

Theorem 13.5 *Suppose that the ILP in (13.9) has a finite optimal feasible solution* $\hat{x}$. *Then* $|c'\hat{x}| \leq M_2 \sum_{j=1}^{n} |c_j|$.

Proof The value $c'\hat{x}$ is bounded from above by the value of the objective function at the feasible solution $\bar{x}$ of bounded size guaranteed by Theorem 13.4. Hence

$$c'\hat{x} \leq c'\bar{x}$$

and consequently

$$c'\hat{x} \leq M_2 \sum_{j=1}^{n} |c_j|$$

We now need to bound $c'\hat{x}$ from below. Such a lower bound is the optimum of the LP in (13.10). And we know that (13.10) has a bounded minimum, because otherwise the ILP in (13.9) would be unbounded by Lemma 13.2. Suppose that the minimum of (13.10) is some bfs $\tilde{x}$. The components of $\tilde{x}$ are bounded in absolute value by $m!a_1^m$ (Lemma 2.1); hence

$$c'\hat{x} \geq c'\tilde{x} \geq -m!a_1^m \sum_{j=1}^{n} |c_j| \geq -M_2 \sum_{j=1}^{n} |c_j|$$

Because $-M_2 \sum_{j=1}^{n} |c_j| \leq c'\hat{x} \leq M_2 \sum_{j=1}^{n} |c_j|$, we conclude that

$$|c'\hat{x}| \leq M_2 \sum_{j=1}^{n} |c_j|.$$

□

Let $a_3 = \max(\{a_1, a_2\} \cup \{|c_j| : j = 1, \ldots, n\})$.

Corollary *If the ILP in (13.9) has a finite optimum, then it has an optimal solution* x *such that* $|x_j| \leq n^3[(m+2)a_3]^{4m+12} = M_3, j = 1, \ldots, n$.

Proof Let d be the value of the optimum of (13.9). Then any optimal feasible solution x satisfies

$$\begin{aligned} c'x &\geq d \\ -c'x &\geq -d \\ Ax &\geq b \\ x &\geq 0, \quad \textit{integer}. \end{aligned}$$

Hence the corollary to Theorem 13.4 applies, with m replaced by $m + 2$, a_1 by a_3, and a_2 by $\max(a_2, d)$. It follows that, for all j,

$$|x_j| \leq M_3.$$ □

The important feature of the bounds we proved above is that their *logarithms* are polynomial in the size of the input. Recall that the size of an LP—or of an ILP for that matter—can be taken to be $L = mn + \log|P|$, where P is the product of the nonzero entries of A, b, and c. Hence $a_3 \leq |P|$, and therefore $\log a_3, mn \leq L$. Consequently,

$$\log M_3 = 3\log n + (4m + 12)[\log(m + 2) + \log a_3] = O(L^2)$$

We shall use this fact in order to prove that ILP and ZOLP are *polynomially equivalent* problems; that is, there is a polynomial-time algorithm for one iff there is one for the other.

Theorem 13.6 *There is a polynomial-time algorithm for ILP iff there is a polynomial-time algorithm for ZOLP.*

Proof If we have a polynomial-time algorithm for ILP, then we can certainly solve the special case of ZOLP in polynomial time.

For the other direction, suppose that we have a polynomial-time algorithm for ZOLP. We shall show how to rewrite every ILP as an equivalent ZOLP. Let $l = \lceil \log M_3 \rceil$, as defined in Theorem 13.5. Because we know that we are interested in integers between 0 and M_3, we can replace each variable x_j by $\sum_{i=0}^{l} x_{ji} 2^i$, where x_{ji} are now 0-1 variables. Any such solution of the ILP has a unique representation as a binary integer with up to l binary digits, and hence it uniquely corresponds to a set of x_{ji}'s. Thus the original ILP is equivalent to the resulting ZOLP. Furthermore, if L is the size of the original ILP, it is easy to see that the size of the ZOLP is $O(l^2 L) = O(L^5)$. The theorem follows. □

We shall soon see, however, that neither problem has good chances of being solvable by a polynomial algorithm.

PROBLEMS

*1. Show by counterexample that the converse of Theorem 13.1 is false.

*2. [VD] Prove the converse of Theorem 13.2.

3. [VD] Show that if any one of the matrices A, A^T, $-A$, $(A \mid A)$, or $(A \mid I)$ is TUM, then so are all the others.

*4. Show that if there is a polynomial-time algorithm for ILP, then there is a polynomial-time algorithm for *mixed*-ILP. (*Hint:* See Subsec. 8.7.1.)

5. Show that any instance of SATISFIABILITY can be transformed in polynomial time into an equivalent instance in which all clauses have more than one literal.

6. Generalize A of Example 13.3 by showing how to use integer variables to formulate the objective function of Fig. 13–3(b).

7. Show that we are justified in leaving out the case $i = 0$ in Eq. 13.1(b).

8. Show that the constraint matrix of the min-cost flow problem is TUM.

9. Consider the integer linear program

$$\max c'x$$
$$Ax \leq b$$
$$x \geq 0, \qquad \text{integer}$$

where A, b, and c are all composed of *positive* integers. Let LP be the linear program obtained by relaxing the constraint that x be integer. Call the solution to ILP x_0 and the solution to LP x_1. Show that in ILP $\lfloor x_1 \rfloor$ is feasible, and that its cost can be no farther from optimal than

$$\sum_{i=1}^{n} c_i$$

NOTES AND REFERENCES

[Co] COFFMAN, E. G., JR., ed., *Computer and Job-Shop Scheduling Theory*. New York: John Wiley & Sons, 1976.

The formulations of the TSP are from

[DFJ] DANTZIG, G. B., D. R. FULKERSON, and S. M. JOHNSON, "Solution of a Large Scale Traveling Salesman Problem," *OR*, 2 (1954), 393–410.

[MTZ] MILLER, C. E., A. W. TUCKER, and R. A. ZEMLIN, "Integer Programming Formulation and Traveling Salesman Problems," *J. ACM*, 7 (1960), 326–29.

Theorems 13.1 and 13.2 are due to

[HK] HOFFMAN, A. J., and J. B. KRUSKAL, "Integral Boundary Points of Convex Polyhedra," pp. 223–46 in *Linear Inequalities and Related Systems*, ed. H. W. Kuhn and A. W. Tucker. Princeton, N.J.: Princeton University Press, 1956.

See also

[VD] VEINOTT, A. F., JR., and G. B. DANTZIG, "Integral Extreme Points," *SIAM Rev.*, 10, no. 3 (July 1968), 371–72.

Theorem 13.3 is from

[HT] HELLER, I., and C. B. TOMPKINS, "An Extension of a Theorem of Dantzig's," pp. 247–52 in *Linear Inequalities and Related Systems*, ed. H. W. Kuhn and A. W. Tucker. Princeton, N.J.: Princeton University Press, 1956. (See also the appendix to this paper, by A. J. Hoffman and D. Gale.)

Other characterizations of total unimodularity are given in

[Ca] CAMION, P., "Characterization of Totally Unimodular Matrices," *Proc. Amer. Math. Soc.*, 16 (1965), 1068–73.

[Pad] PADBERG, M. W., "A Note on the Total Unimodularity of Matrices," *Discrete Math.*, 14, no. 3 (March 1976), 273–78.

Example 13.3 is from

[Da] DANTZIG, G. B., "On the Significance of Solving Linear Programming Problems with Some Integer Variables," *Econometrica* (1960), 30–44.

Theorems 13.4 and 13.5 are from

[Pa] PAPADIMITRIOU, C. H., "On the Complexity of Integer Programming," *JACM* (1981), in press.

Many other proofs of similar results exist in the literature; see, for example,

[BT] BOROSH, I., and L. B. TREYBIG, "Bounds on Positive Integral Solutions to Linear Diophantine Equations," *Proc. Amer. Math. Soc.*, 55 (1976), 299–304.

14

A Cutting-Plane Algorithm for Integer Linear Programs

14.1 Gomory Cuts

We have seen that certain ILP's, those corresponding to totally unimodular constraint matrices, are really no more difficult to solve than the corresponding LP's. In such problems an optimal basic feasible solution to the corresponding LP is guaranteed to be integer. What do we do, however, when no such convenient property holds? There is, unfortunately, no simple answer to this question. The general ILP, as we shall see later on, seems to be inherently difficult, and a wide variety of algorithms have been developed for it. Some such algorithms are especially effective on certain classes of integer programs, but it is safe to say that no general algorithm is known that is practical for large problems (say a few dozen constraints and variables), whereas the solution of LP's of this size by the simplex algorithm is routine.

General algorithms for ILP fall into two categories (which overlap to some extent): the *cutting-plane algorithms*, which are derived from the simplex algorithm, and *enumerative algorithms*, which are based on intelligent enumeration of all possible solutions. This chapter is devoted to a description of a simple cutting-plane algorithm, which belongs to a class of methods that are useful on problems of modest size and are interesting from a theoretical point of view as well.

To fix our notation, we shall consider the ILP in standard form

$$\begin{aligned} &\min c'x \\ &Ax = b \\ &x \geq 0, \quad \text{integer} \end{aligned} \tag{14.1}$$

where A, b, and c are integer. We shall call the LP without integer constraints

$$\begin{aligned} &\min c'x \\ &Ax = b \\ &x \geq 0 \end{aligned} \tag{14.2}$$

the *relaxation* of the ILP in Eq. 14.1.

Suppose we solve the relaxation of an ILP, by a simplex algorithm, for example, to obtain a basic feasible solution x^*. In general, of course, x^* is not integer. But it is natural to try to obtain a solution to the original ILP by rounding off the coordinates of x^* to the nearest integers. Figure 14–1 shows why this does not work: There may, in fact, be no feasible point near x^* at all. The reader should have no trouble constructing such two-dimensional examples in which the continuous and discrete optima are arbitrarily far apart in planar distance and in cost.

We note here two simple but useful facts: If the continuous optimum x^*, the solution to LP, is integer, then it solves the corresponding ILP. Second, the

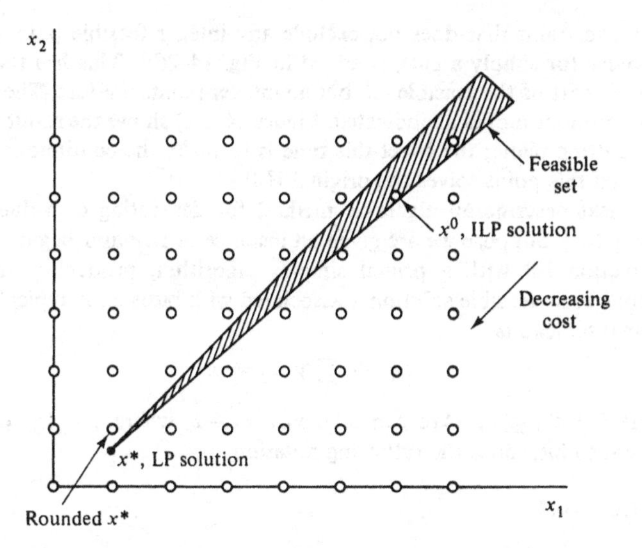

Figure 14–1 A hypothetical ILP with optimum x^0 and its relaxation with optimum x^*.

cost of the continuous optimum, $c(x^*)$, is a lower bound on the cost of the discrete optimum, $c(x^0)$.

We now come to the main idea of cutting-plane algorithms. If we add a constraint to an ILP *that does not exclude integer feasible points*, then the solution is unchanged. Our strategy will be to add such linear constraints to an ILP, one at a time, until the solution to the LP relaxation is integer. Because we have excluded no integer feasible points, this final solution to the relaxed ILP with added constraints will solve the original ILP. This process is illustrated in Fig. 14–2. Figure 14–2(a) shows the original ILP and the continuous optimum x^*.

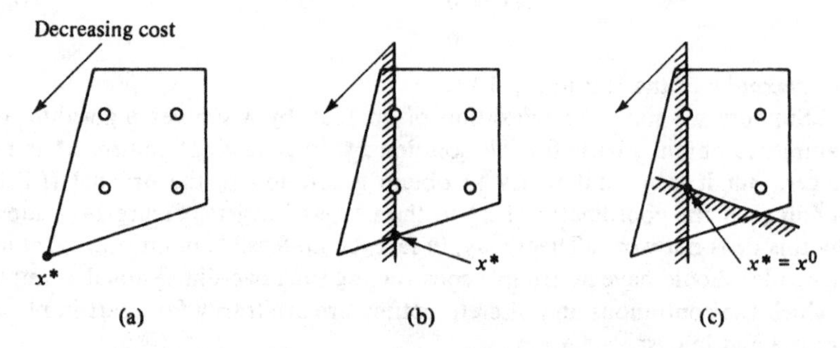

Figure 14–2 Illustration of a cutting-plane algorithm. (a) The continuous optimum x^*. (b) The new x^* after one cut. (c) The solution of the original ILP after two cuts.

A linear constraint that does not exclude any integer feasible points, called a *cutting-plane* (or simply a *cut*), is added in Fig. 14–2(b). This has the effect of lopping off part of the feasible set, but no integer points are lost. The new continuous optimum moves as indicated. Figure 14–2(c) shows the result of adding another cutting plane; the effect this time is to make the continuous optimum integer, and this point solves the original ILP.

We next describe an algebraic method for generating cuts due to R. E. Gomory [Go1]. Suppose we are given an instance of ILP and begin by solving the relaxation LP with a primal simplex algorithm, producing an optimal continuous basic feasible solution x associated with basis $\mathcal{B}$. A typical equation in the final tableau is

$$x_{B(i)} + \sum_{j \notin B} y_{ij} x_j = y_{i0} \tag{14.3}$$

for some i, $0 \leq i \leq m$. (We can take $x_{B(0)} = -z$, where z is the cost.) It is convenient to introduce the following notation.

Definition 14.1

Given a real number y, $\lfloor y \rfloor$ (called the *integer part* of y) is defined to be the largest integer q such that $q \leq y$. □

Example 14.1

$$\lfloor 2.7 \rfloor = 2$$
$$\lfloor -8.1 \rfloor = -9$$
$$\lfloor 0 \rfloor = 0 \quad \square$$

The variable x in Eq. 14.3 is constrained to be nonnegative, so

$$\sum_{j \notin B} \lfloor y_{ij} \rfloor x_j \leq \sum_{j \notin B} y_{ij} x_j \tag{14.4}$$

Therefore Eq. 14.3 becomes

$$x_{B(i)} + \sum_{j \notin B} \lfloor y_{ij} \rfloor x_j \leq y_{i0} \tag{14.5}$$

In ILP, the problem we are trying to solve, x is constrained to be integer, and so the left-hand side of Eq. 14.5 is integer. The right-hand side can therefore be replaced by its integer part without disturbing the relation, yielding

$$x_{B(i)} + \sum_{j \notin B} \lfloor y_{ij} \rfloor x_j \leq \lfloor y_{i0} \rfloor \tag{14.6}$$

Subtracting Eq. 14.6 from Eq. 14.3 gives

$$\sum_{j \notin B} (y_{ij} - \lfloor y_{ij} \rfloor) x_j \geq y_{i0} - \lfloor y_{i0} \rfloor \tag{14.7}$$

Let

$$f_{ij} = y_{ij} - \lfloor y_{ij} \rfloor \qquad i = 0, \ldots, m \tag{14.8}$$

The number f_{ij} is called the *fractional part* of y_{ij} and satisfies

$$0 \leq f_{ij} < 1 \tag{14.9}$$

Finally we get the constraint

$$\sum_{j \notin B} f_{ij} x_j \geq f_{i0} \tag{14.10}$$

called the *Gomory cut* corresponding to the *source* or *generating row i.*

Our plan is to add the Gomory cut in (14.10) to our tableau. To keep a basic solution, we multiply Eq. 14.10 by -1 and add the slack variable s, yielding

$$-\sum_{j \notin B} f_{ij} x_j + s = -f_{i0} \tag{14.11}$$

The following summarizes the effect of adding the cut in (14.11) to an optimal tableau.

Lemma 14.1 *If the cut in (14.11) is added to an optimal tableau of an LP, no integer feasible points are excluded, and the new tableau is basic, primal infeasible if y_{i0} is not integer, and dual feasible.*

Proof That Eq. 14.11 excludes no integer feasible points follows from the fact that the cut is implied by the integer constraints of the original ILP. The new slack variable s is a new basic variable and forms a new basis when added to the optimal basis $\mathcal{B}$. When y_{i0} is not integer, $f_{i0} > 0$ and therefore the basic solution includes

$$s = -f_{i0}$$

which shows it to be primal infeasible. It remains dual feasible because the zeroth row is not changed. □

It should now be clear how to proceed. Given a tableau with a basic solution which is primal infeasible and dual feasible, it is most natural to apply the dual simplex algorithm described in Sec. 3.6. One or more pivots will either bring us to a new continuous optimum or tell us that the primal is infeasible. This last possibility must mean that the original ILP had no integer feasible points. Figure 14.3 shows a sketch of the overall algorithm, which is usually called a

```
procedure fractional dual
begin
solve the relaxation of ILP, obtaining optimal solution x*;
feasible:="yes";
while x* is not integer and feasible="yes" do
  begin
    choose a source row i;
    add the generated Gomory cut and a
    corresponding basic variable s;
    apply the dual simplex algorithm;
    if dual is unbounded then feasible := "no";
    let x* be the new optimum
  end
end
```

Figure 14–3 A fractional dual algorithm for ILP.

fractional dual algorithm, because the tableaux involved have fractional entries and dual feasibility is maintained. We can now work through a complete example.

Example 14.2

Consider the ILP

$$\max x_2 \tag{14.12}$$

$$3x_1 + 2x_2 \leq 6 \tag{14.13}$$

$$-3x_1 + 2x_2 \leq 0 \tag{14.14}$$

$$x \geq 0, \quad \text{integer} \tag{14.15}$$

Figure 14–4 shows the constraints in the x_1–x_2 plane. The problem has been chosen to be very simple, with an integer optimum at $x = (1, 1)$. Adding slack variables x_3 and x_4, we get the standard form tableau for the relaxation of the problem:

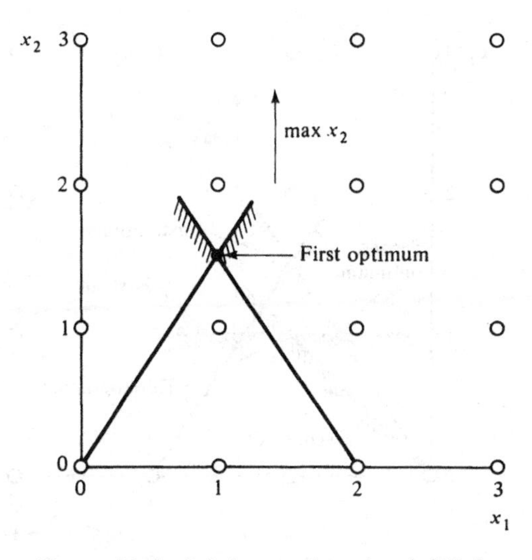

Figure 14–4 Solution to the relaxed ILP in Example 14.2.

		x_1	x_2	x_3	x_4
$-z =$	0	0	-1	0	0
$x_3 =$	6	3	2	1	0
$x_4 =$	0	-3	2	0	1

(14.16)

The optimal tableau is obtained after two pivots:

$-z =$	$\frac{3}{2}$	0	0	$\frac{1}{4}$	$\frac{1}{4}$
$x_1 =$	1	1	0	$\frac{1}{6}$	$-\frac{1}{6}$
$x_2 =$	$\frac{3}{2}$	0	1	$\frac{1}{4}$	$\frac{1}{4}$

(14.17)

corresponding to $x = (1, \frac{3}{2})$ and a cost $z = -x_2 = -\frac{3}{2}$, as shown in Fig. 14–4. Because the solution to the relaxed problem is not integer, we generate cuts. The zeroth equation is

$$-z + \tfrac{1}{4}x_3 + \tfrac{1}{4}x_4 = \tfrac{3}{2} \tag{14.18}$$

which yields the cut

$$\tfrac{1}{4}x_3 + \tfrac{1}{4}x_4 \geq \tfrac{1}{2} \tag{14.19}$$

Substituting for x_3 and x_4 from the constraints in Tableau 14.16, this is seen to be equivalent to

$$x_2 \leq 1 \tag{14.20}$$

which is shown in Fig. 14–5 as the first cut.

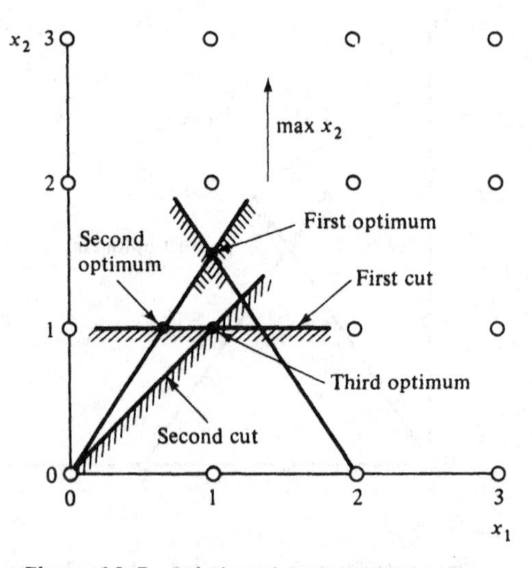

Figure 14–5 Solution of the ILP by two cuts.

Had we chosen the first or second row as source row, we would have obtained for our cut

$$\tfrac{1}{6}x_3 + \tfrac{5}{6}x_4 \geq 0 \tag{14.21}$$

or

$$\tfrac{1}{4}x_3 + \tfrac{1}{4}x_4 \geq \tfrac{1}{2} \tag{14.22}$$

respectively. The first of these is equivalent to

$$-2x_1 + 2x_2 \leq 1 \tag{14.23}$$

and the second is the same as the one obtained above from the zeroth row.

Choosing Row 0 as the source row, we add the equation

$$-\tfrac{1}{4}x_3 - \tfrac{1}{4}x_4 + s_1 = -\tfrac{1}{2} \tag{14.24}$$

to Tableau 14.17, yielding

(14.25)

		x_1	x_2	x_3	x_4	s_1
$-z =$	$\frac{3}{2}$	0	0	$\frac{1}{4}$	$\frac{1}{4}$	0
$x_1 =$	1	1	0	$\frac{1}{6}$	$-\frac{1}{6}$	0
$x_2 =$	$\frac{3}{2}$	0	1	$\frac{1}{4}$	$\frac{1}{4}$	0
$s_1 =$	$-\frac{1}{2}$	0	0	$\boxed{-\frac{1}{4}}$	$-\frac{1}{4}$	1

Executing the indicated dual pivot, we obtain the second optimal tableau:

$-z =$	1	0	0	0	0	1
$x_1 =$	$\frac{2}{3}$	1	0	0	$-\frac{1}{3}$	$\frac{2}{3}$
$x_2 =$	1	0	1	0	0	1
$x_3 =$	2	0	0	1	1	-4

(14.26)

with the corresponding second optimum $x = (\frac{2}{3}, 1)$, still not integer.

The zeroth row is integer and so generates a trivial cut. The first row generates the cut

$$\tfrac{2}{3}x_4 + \tfrac{2}{3}s_1 \geq \tfrac{2}{3} \tag{14.27}$$

which is equivalent to

$$x_1 \geq x_2 \tag{14.28}$$

Adding the corresponding row

$$-\tfrac{2}{3}x_4 - \tfrac{2}{3}s_1 + s_2 = -\tfrac{2}{3} \tag{14.29}$$

to Tableau 14.26 and reoptimizing with the dual simplex algorithm, we arrive at the final optimal tableau

		x_1	x_2	x_3	x_4	s_1	s_2
$-z =$	1	0	0	0	0	1	0
$x_1 =$	1	1	0	0	0	1	$-\frac{1}{2}$
$x_2 =$	1	0	1	0	0	1	0
$x_3 =$	1	0	0	1	0	-5	$\frac{3}{2}$
$x_4 =$	1	0	0	0	1	1	$-\frac{3}{2}$

(14.30)

corresponding to the solution $x = (1, 1)$, as shown in Fig. 14–5. □

We now need to consider the finiteness of this procedure: How do we know that this procedure terminates with an integer solution after a finite number of steps? First, however, we shall study a method for dealing with degeneracy that is very important in the analysis of many simplex-based ILP algorithms.

14.2 Lexicography

Bland's method for ensuring finiteness of the simplex algorithm (see Sec. 2.7) is a relatively late development in the history of linear programming. The classical method is based on lexicographic ordering of vectors in the tableau, for which we now need a definition.

Definition 14.2

A nonzero vector $v \in R^n$ is said to be *lexicographically positive* (*lex-positive*, for short) if its first nonzero component is greater than zero. If $v = 0$, we say it is *lex-zero,* and v is said to be *lex-negative* if $-v$ is lex-positive. We write these conditions as

$$v \overset{L}{>} 0$$

$$v \overset{L}{=} 0$$

$$v \overset{L}{<} 0$$

respectively.

A vector $v \in R^n$ is said to be *lex-greater-than* $w \in R^n$ (written $v \overset{L}{>} w$) if and only if $v - w$ is lex-positive; similar definitions can be made for *lex-less-than* and *lex-equal.* The terms *lex-min* and *lex-max* also are defined in the obvious ways. □

Example 14.3

$$\text{col}\,(0 \quad 0 \quad 1 \quad 0) \overset{L}{>} \text{col}\,(0 \quad 0 \quad 0 \quad 2)$$

$$\text{col}\,(0 \quad 3 \quad 1 \quad 2) \overset{L}{<} \text{col}\,(1 \quad 2 \quad 4 \quad 8) \quad \square$$

Example 14.4

Assign the numbers 1, 2, 3, . . . to the letters $a, b, c, \ldots,$ and assign 0 to an absent letter. Code a word as a vector in Z^n, where n is some number larger than the length of any word. Then lexicographic ordering of the code vectors for words corresponds exactly to the ordering of words used in a dictionary. For example,

$$\text{bad} \longrightarrow \text{col}\,(2 \quad 1 \quad 4 \quad 0 \quad 0 \quad 0 \quad \ldots)$$

and

$$\text{good} \longrightarrow \text{col}\,(7 \quad 15 \quad 15 \quad 4 \quad 0 \quad 0 \quad 0 \quad \ldots)$$

so that in our code

$$\text{bad} \overset{L}{<} \text{good}$$

Notice also that

$$\text{bad} \overset{L}{<} \text{bade} \quad \square$$

The idea is to break ties in the simplex algorithm by using the lexicographic criterion. Thus, in the primal simplex algorithm, when we come to choose a row by

$$\min_{\substack{i \text{ such that} \\ a_{is}>0}} \left[\frac{a_{i0}}{a_{is}}\right] \tag{14.31}$$

we resolve ties by using the criterion

$$\underset{\substack{i \text{ such that} \\ a_{is}>0}}{\text{lex-min}} \left[\frac{a_i}{a_{is}}\right] \tag{14.32}$$

where as usual a_i is the vector of numbers in the ith row of the tableau. This is equivalent to Eq. 14.31 when no ties occur, since in that case the first component of a_i, a_{i0}, will determine the choice of pivot row. When there is a tie, however, Eq. 14.32 will base the decision on the first component for which ties do not occur.

Example 14.5

In the section of tableau shown below,

	$s \downarrow$			
1	4	1	3	...
2	1	10	1	...
3	12	1	2	...

we have

$$\frac{a_1}{a_{1s}} = \text{col}\left(\tfrac{1}{4},\ 1,\ \tfrac{1}{4},\ \tfrac{3}{4},\ \ldots\right)$$

$$\frac{a_2}{a_{2s}} = \text{col}\left(2,\ 1,\ 10,\ 1,\ \ldots\right)$$

$$\frac{a_3}{a_{3s}} = \text{col}\left(\tfrac{1}{4},\ 1,\ \tfrac{1}{12},\ \tfrac{1}{6},\ \ldots\right)$$

so the lex-min operation in Eq. 14.32 resolves the tie between Row 1 and Row 3 by choosing Row 3. □

Notice that in an LP, the lex-min choice will be unique, because no two rows will be proportional. (Why not?)

The main application to LP can be stated as the following theorem, which provides an alternative to Bland's anticycling rules.

Theorem 14.1 (Lexicographic Anticycling Rules for Primal Simplex) *Let us begin the primal simplex algorithm with all the rows after the cost row lex-positive:*

$$a_i \overset{L}{>} 0 \qquad i = 1, \ldots, m$$

Then the following pivoting rules imply that these rows stay lex-positive, that Row zero strictly lex-increases (written $a_0 \uparrow L$), and that the primal simplex algorithm terminates after a finite number of pivots:

(*a*) *Choose any column s such that $a_{0s} < 0$.*

(*b*) *Choose row $i = r$ by*

$$\underset{\substack{i \text{ such that} \\ a_{is}>0}}{\text{lex-min}} \left[\frac{a_i}{a_{is}}\right] \tag{14.33}$$

Proof We first check that the a_i, $i = 1, \ldots, m$, remain lex-positive after pivoting. The rth row becomes

$$\hat{a}_r = \frac{a_r}{a_{rs}}$$

with $a_{rs} > 0$, so $\hat{a}_r \overset{L}{>} 0$ if $a_r \overset{L}{>} 0$. When $i \neq r$ and $a_{is} > 0$, we have

$$\hat{a}_i = a_i - \frac{a_{is}a_r}{a_{rs}}$$

$$= a_{is}\left[\frac{a_i}{a_{is}} - \frac{a_r}{a_{rs}}\right] \overset{L}{>} 0$$

by the lexicographic choice. Finally, when $i \neq r$ and $a_{is} \leq 0$, we have

$$\hat{a}_i = a_i - \frac{a_{is}a_r}{a_{rs}}$$

$$= a_i + \frac{|a_{is}|a_r}{a_{rs}}$$

$$\overset{L}{\geq} a_i \overset{L}{>} 0$$

so the a_i in fact stay lex-positive after each pivot.

The effect on a_0 of a pivot is

$$\hat{a}_0 = a_0 - \frac{a_{0s}a_r}{a_{rs}}$$

$$= a_0 + \frac{|a_{0s}|a_r}{a_{rs}}$$

$$\overset{L}{>} a_0$$

since $a_{0s} < 0$ and $a_r \overset{L}{>} 0$ by assumption. Therefore the zeroth row *strictly increases* lexicographically and hence can never return to a previous value during pivoting. The zeroth row is determined completely by the basis, however, so the fact that

$$a_0 \uparrow L$$

implies that no basis can ever be repeated. This in turn implies that the simplex algorithm terminates after a finite number of pivots. □

Of course, we can expect an analogous result for the dual simplex algorithm, because it is really a primal simplex in disguise (see Section 3.7).

Theorem 14.2 (Lexicographic Anticycling Rules for Dual Simplex) *Let us begin the dual simplex algorithm with all the columns after the zeroth column lex-positive:*

$$A_j \overset{L}{>} 0 \qquad j = 1, \ldots, n$$

Then the following pivoting rules imply that these columns stay lex-positive, that Column zero strictly lex-decreases

$$A_0 \downarrow L,$$

and that the dual simplex algorithm terminates after a finite number of pivots:

(*a*) *Choose any row r such that* $a_{r0} < 0$.

(*b*) *Choose column* $j = s$ *by*

$$\underset{\substack{j \text{ such that} \\ a_{rj}<0}}{\text{lex-max}} \left[\frac{A_j}{a_{rj}}\right] \tag{14.34}$$

Proof The proof of this is left as an exercise; it uses the same techniques as the proof of Theorem 14.1 (see Problem 8). □

We also ask the reader to explain how to begin the primal algorithm with lex-positive rows, to explain how to begin the dual algorithm with lex-positive columns, and to show that the rule in (14.34) yields a unique result (see Problems 9 and 10).

14.3 Finiteness of the Fractional Dual Algorithm [Go1, Go2]

The lexicographic method guarantees the finiteness of the fractional dual algorithm, as well as the ordinary simplex algorithm. We state this in the form of a theorem.

Theorem 14.3 *Make the following choices in the fractional dual algorithm for ILP (see Fig.* 14–3)*:*

(*a*) *Choose the source row to be the first row with a noninteger* a_{i0}.

(*b*) *Use the lexicographic version of the dual simplex algorithm.*

Then, assuming that the original problem has an upper bound on the cost z of a feasible solution, the algorithm terminates with an integer solution in a finite number of steps or finds that there is no feasible integer solution to the original problem.

Proof [Go1, Go2] Call the main loop in Fig. 14–3 a *re-optimization* and call the tableau after the lth re-optimization A^l, with corresponding columns A^l_j. The algorithm then produces a sequence of zeroth columns that are monotonically lex-decreasing, because rows are always added to the bottom of the tableau:

$$A_0^1 \overset{L}{>} A_0^2 \overset{L}{>} A_0^3 \overset{L}{>} \cdots \tag{14.35}$$

We have assumed that the first component $a_{00} = -z$ is bounded from below, so the sequence a_{00} converges to some number, say w_{00}, which can be written

$$w_{00} = \lfloor w_{00} \rfloor + f_{00} \tag{14.36}$$

After some finite number of re-optimizations, a_{00}^l falls below $\lfloor w_{00} \rfloor + 1$, and for some $l = k$, we can write

$$a_{00}^k = \lfloor w_{00} \rfloor + f_{00}^k \tag{14.37}$$

By Rule (a), the next source row is Row 0, and the cut

$$-f_{00}^k = -\sum_{j \notin \mathcal{B}} f_{0j}^k x_j + s \tag{14.38}$$

is added to the tableau. We then pivot using the dual simplex algorithm, choosing column p, say, to enter the basis. After this pivot we obtain

$$a_{00}^{k+1} = a_{00}^k - \frac{a_{0p}^k}{f_{0p}^k} f_{00}^k \tag{14.39}$$

Now at an optimal tableau of the dual simplex algorithm,

$$a_{0p}^k \geq 0 \tag{14.40}$$

and therefore it is larger than its fractional part

$$a_{0p}^k \geq f_{0p}^k \tag{14.41}$$

Equation 14.39 then tells us that

$$a_{00}^{k+1} \leq a_{00}^k - f_{00}^k = \lfloor a_{00}^k \rfloor = \lfloor w_{00} \rfloor \tag{14.42}$$

Because the sequence a_{00}^l converges to w_{00}, this shows that from this point on $a_{00}^l = \lfloor w_{00} \rfloor$, an integer.

The vectors A_0^l are lex-decreasing, and we have shown that after some point the first component becomes fixed (at an integer), so it must be that the second component is monotonically nonincreasing. It is bounded from below by zero, for otherwise the dual simplex algorithm would have failed. The argument above can then be repeated for a_{10}^l. We need to show however that the element

$$a_{1p}^k \geq 0 \tag{14.43}$$

so that the steps following Eq. 14.40 still go through. This follows because a_{00}^k must remain fixed, which implies that $a_{0p}^k = 0$, which in turn implies Eq. 14.43 because $A_p^k \overset{L}{>} 0$. Hence a_{10}^l becomes integer after a finite number of steps.

We can continue in this way down Column 0, showing that all components eventually reach integer values, at which point the algorithm terminates. The only other possible termination occurs when the dual simplex algorithm finds that the dual is unbounded, and hence that the original ILP is infeasible. □

It is usual to avoid accumulating an indefinite number of rows and columns in the fractional dual method by dropping the slack variable s_t associated with a Gomory cut if such a variable should enter the basis at any point in the algorithm. This means that the number of rows never exceeds n, the number of original variables, so that the number of cuts in use at any one time is no more than $n - m$. This refinement does not affect the proof of finiteness (see Problem 4). In fact, with this modification it is not hard to establish an exponential bound on the number of iterations taken by the algorithm (see Problem 11).

14.4 Other Cutting-Plane Algorithms

The fractional dual algorithm has two main problems associated with it: one because it is fractional, another because it is dual. Dealing with these problems has led to other classes of algorithms, which we now discuss briefly. (See Notes and References.)

The fact that the fractional dual algorithm necessitates dividing one integer by another means that an implementation in a digital computer will in general store tableau entries to only a finite precision (recall the discussion accompanying the ellipsoid algorithm, Subsec 8.7.4). If 1 is divided by 3, the result is stored as a binary number to some finite number of bits; if that is then multiplied by 3, we do not return to the value 1. (The reader can try this on a pocket calculator; we get 0.9999999.) This effect accumulates from stage to stage, with the result that it may become difficult to decide whether a given entry is or is not an integer, a distinction which is necessary for the generation of cuts. Hence all-integer algorithms have been developed that ensure that pivot elements always have the value one.

The fractional dual algorithm is after all a dual algorithm and does not produce a primal feasible solution until it reaches optimality. This would ordinarily not be a problem, but it is an unfortunate fact that algorithms for ILP are frequently terminated before optimality has been reached, because it is difficult to predict their running times and because their running times may vary tremendously with the problem data. When this happens in a dual algorithm, we are in the embarrassing position of having invested a certain amount of computer time with no useful result; we have neither an integer nor a feasible solution to our original problem. A cutting-plane algorithm that maintains suboptimal solutions that are both integer and feasible in the original constraints would, on the other hand, enable the user to use the result of a premature termination. Such algorithms have in fact been developed and are generally called *primal integer* cutting-plane algorithms.

PROBLEMS

1. Show that the cutting-plane solution described for Example 14.2 violates the rules for the lexicographic dual algorithm given later in Theorem 14.3. Rework the problem, adhering to the lexicographic rules. Express new cuts in terms of the original variables x_1 and x_2.

2. Suppose an integer linear program is formulated by adding slack variables to

$$\min c'x$$
$$Ax \leq b$$
$$x \geq 0, \qquad \text{integer}$$

where A, b, and c are integer. Call x the *original* variables. Let s be the slack variable corresponding to a Gomory cut at any point in the cutting-plane algorithm. Prove that s can be expressed as a linear combination with integer coefficients of the original variables, plus an integer constant.

3. Show that any element in the lth optimal noninteger tableau A^l of the cutting-plane algorithm can be written as L/D, where L is an integer and D is the determinant of the corresponding optimal basis.

*4. Extend the finiteness proof of the fractional dual method to the case where the row and column corresponding to slack variable s are dropped whenever s enters the basis.

5. Describe how the fractional dual cutting-plane algorithm can detect and distinguish between the cases when the relaxation is unbounded, and the integer program is either unbounded or infeasible. (When the relaxation is unbounded, the integer problem cannot have a finite optimum, as shown in Lemma 13.2.)

6. Show that the assumption in Theorem 14.3, that the cost z of a feasible solution has an upper bound, is not restrictive. (*Hint:* Recall Theorem 2.2.)

7. Prove the following properties of the lexicographic ordering relation.

   (a) $x \overset{L}{<} y$ and $y \overset{L}{<} z \Rightarrow x \overset{L}{<} z$

   (b) $x \overset{L}{<} y \Rightarrow \neg (y \overset{L}{<} x)$

   (c) $\neg (x \overset{L}{<} x)$

   (d) $x \overset{L}{<} y$ and $z \overset{L}{<} w \Rightarrow x + z \overset{L}{<} y + w$

8. Prove Theorem 14.2: the lexicographic dual simplex algorithm is finite.

9. Show how to begin the lexicographic algorithms with lex-positive rows or columns.

10. Explain how the lex-max choices in the operation of the dual lexicographic simplex algorithm can be made unique.

11. Show that the fractional dual cutting-plane algorithm, dropping slacks corresponding to cuts should they become basic, takes no more than an exponential number of iterations. (By *exponential* here we mean $2^{p(L)}$, where $p(L)$ is a polynomial in the input length.)

NOTES AND REFERENCES

The fractional dual algorithm is due to Gomory.

[Go1] GOMORY, R. E., "Outline of an Algorithm for Integer Solution to Linear Programs," *Bulletin Amer. Math. Soc.*, 64, no. 5 (1958).

[Go2] ———, "An Algorithm for Integer Solutions to Linear Programs," pp. 269–302 in *Recent Advances in Mathematical Programming*, ed. R. L. Graves and P. Wolfe. New York: McGraw-Hill Book Company, 1963.

The lexicographic anticycling method is from

[DOW] DANTZIG, G. B., A. ORDEN, and P. WOLFE, "The Generalized Simplex Method for Minimizing a Linear Form under Linear Inequality Constraints," *Pacific J. Math.*, 5, no. 2 (1955), 183–95.

Several textbooks are available that describe varieties of cutting-plane algorithms in detail; among them are

[GN] GARFINKEL, R. S., and G. L. NEMHAUSER, *Integer Programming*. New York: John Wiley & Sons, Inc., 1972.

[Hu] HU, T. C., *Integer Programming and Network Flows*. Reading, Mass.: Addison-Wesley Publishing Co., Inc., 1969.

[Sa] SALKIN, H. M., *Integer Programming*. Reading, Mass.: Addison-Wesley Publishing Co., Inc., 1975.

H. W. Lenstra described a remarkable result in a lecture on Dec. 17, 1980 in Amsterdam (recorded by A. J. J. Talman and transmitted by H. E. Scarf and P. van Emde Boas). Consider the ILP feasibility constraints $Ax \leq b, x \in Z^n$. Then for any *fixed n*, Lenstra described an algorithm that is polynomial in the input data and determines if these constraints are feasible, and if so, provides a feasible point x. Of course n appears in the exponent of the polynomial time bound. See

[Le] LENSTRA, H. W., JR., "Integer Programming with a Fixed Number of Variables," Report 81-03, University of Amsterdam, April 1981.

15

NP-Complete Problems

15.1 Introduction

Our main objective in the previous chapters has been the development of *efficient* algorithms for the solution of combinatorial optimization problems of various sorts. In Chapter 8, in fact, we accepted the thesis that it is reasonable by "efficient algorithm" to understand "an algorithm requiring a number of steps that grows as a *polynomial* in the size of the input." So far, we have witnessed the main successes of research in the area of combinatorial optimization algorithms: We have developed efficient (in our technical sense) algorithms that solve such involved problems as weighted matching, matroid intersection, and LP. The time has now come to meet the most prominent *failures* of this approach; problems, that is, for which no efficient algorithm is known. In doing so, we shall develop a beautiful theory that unifies these failures into a deep mathematical conjecture.

The principal result of this investigation is the notion of an *NP-complete* problem. Intuitively, an *NP*-complete problem is a computational problem that is *as hard as any reasonable problem,* all these concepts subject to a precise formulation. The traveling salesman problem (TSP), which has haunted two generations of mathematicians with its stubborn resistance to all kinds of attacks, will be shown to be an *NP*-complete problem. So will such hard nuts as integer linear programming (ILP), the satisfiability problem, three-matroid intersection, and a host of other well-known, difficult computational problems, many of them of optimizational flavor. This class of *NP*-complete problems has the following very interesting properties.

1. No *NP*-complete problem can be solved by any known polynomial algorithm (and this is despite persistent efforts by many brilliant researchers for many decades).
2. If there is a polynomial algorithm for *any NP*-complete problem, then there are polynomial algorithms for *all NP*-complete problems.

Based on these two facts, many people have conjectured that there can be no polynomial algorithm for any *NP*-complete problem; however, nobody has been able to prove this. In fact, it is now believed that the proof of this conjecture will not come about without the development of entirely new mathematical techniques.

The practical significance of the notion of an *NP*-complete problem lies exactly in the widespread belief that such problems are *inherently intractable* from the computational point of view; that they are not susceptible to efficient algorithmic solution; and that any algorithm that correctly solves an *NP*-complete problem will require in worst case an *exponential* amount of time, and hence will be impractical for all but very small instances. So, a researcher who faces a new combinatorial optimization problem and cannot find an efficient algorithm for it now has the option of trying to prove that the problem at hand is *NP*-complete. Of course, in most cases this will not be nearly as exciting as finding an efficient algorithm. Nevertheless, this approach has certain definite merits. First, it will save the researcher in question—and most probably some of his or her colleagues—from further futile efforts to solve this problem algorithmically. Moreover, once a problem is known to be *NP*-complete, one is generally willing to settle for goals that are less ambitious than developing an algorithm that *always* finds an *exact* solution and has time requirements that *never* exceed a given polynomial growth. One may wish to take one of several possible alternative approaches. It is these alternative approaches that motivate the remaining chapters of this book.

15.2 An Optimization Problem is Three Problems

In Chapter 1, an optimization problem was defined as a set $\mathscr{I}$ of instances. Each instance is a pair (F, c), where F is the set of *feasible solutions* and c is a *cost function*: $c: F \longrightarrow R$. Since in this chapter we shall study optimization problems from the viewpoint of the theory of computation, it is well first to fix ideas on how an instance of a combinatorial optimization problem will be represented as input to a computer. Naturally, we can list all feasible solutions and the value of c for each. However, most interesting problems will have instances with a disproportionately large number of feasible solutions, and hence this method of representation is especially undesirable. The reader may recall the TSP and MST as examples of such situations.

In what follows, we shall assume that F and c are given implicitly in terms of two algorithms $\mathcal{A}_F$ and $\mathcal{A}_c$. The algorithm $\mathcal{A}_F$, given a combinatorial object f and a set S of parameters, will decide whether f is an element of F, the set of feasible solutions specified by the given parameters. On the other hand $\mathcal{A}_c$, given a feasible solution f and another set of parameters Q, returns the value of $c(f)$. An *instance* of the combinatorial problem can now be defined as the representation of the parameters in S and Q—using a fixed finite alphabet and some standard, reasonable encoding as discussed in Chapter 8 and later in this chapter.

Example 15.1

An instance of the TSP with n cities has as parameter S the integer n; its parameters Q are the entries of the $n \times n$ symmetric distance matrix $[d_{ij}]$. The algorithm $\mathcal{A}_F$, given an object f and the parameter n, will examine whether f is a tour of n cities—a cyclic permutation, that is, of the set $\{1, 2, \ldots, n\}$. The algorithm $\mathcal{A}_c$, given a tour f and $[d_{ij}]$, calculates the cost $c(f)$ by summing all entries of $[d_{ij}]$ that correspond to distances traversed by the tour. □

Example 15.2

Consider the following combinatorial optimization problem, called the *maximum clique problem*:

Given a graph $G = (V, E)$ find the largest subset $C \subseteq V$ such that for all distinct $u, v \in C$, $[v, u] \in E$.

The parameter S is in this case a graph G. Given a graph $G = (V, E)$, $\mathcal{A}_F$ determines whether f constitutes a *clique* of G, that is, a fully connected subset of V. The cost evaluator $\mathcal{A}_c$ now simply calculates the cardinality of f; Q is in this case empty. □

Example 15.3

Let us formulate integer linear programming (ILP) in a similar manner. We shall be interested in the form

$$\begin{aligned} &\min c'x \\ &Ax = b \\ &x \geq 0, \quad \text{integer} \end{aligned}$$

The algorithm $\mathcal{A}_F$ has as parameters S, the matrix A, and the vector b. Given any integer vector x, $\mathcal{A}_F$ tests whether $Ax = b$ and $x \geq 0$ are satisfied. The algorithm $\mathcal{A}_c$ takes the vector c as the parameter set Q and evaluates $c'x$ for each given feasible solution x. □

We notice that in all three examples above, $\mathcal{A}_F$ and $\mathcal{A}_c$ are polynomial-time algorithms. This will be a frequent and pragmatic assumption. For an interesting exception, in which c is hard to compute (more precisely, $c(f)$ is hard to compare with an integer) see Part (a) of Problem 17.

A *combinatorial optimization problem* is thus the following form of computational problem.

> Given representations of the parameters S and Q for the algorithms $\mathcal{A}_F$ and $\mathcal{A}_c$, find the optimal feasible solution.

We call this the *optimization version* of the problem. However, a combinatorial optimization problem can also be posed in the following, more relaxed, form.

> Given S and Q, find the cost of the optimal solution.

This will be referred to as the *evaluation version* of a combinatorial optimization problem. It is immediate that, under the assumption that $\mathcal{A}_c$ is a polynomial-time algorithm—in other words the cost c is not too hard to compute—the evaluation version of a combinatorial optimization problem cannot be much harder than the optimization version.

A third version of a combinatorial optimization problem is particularly important in studying the complexity of the problem, because it is closest to the prototype of computational problems traditionally studied by the theory of computation. This version—called the *recognition version*—is of the following form.

> Given an instance—a representation, that is, of S and Q—and an integer L, is there a feasible solution $f \in F$ such that $c(f) \leq L$?†

Unlike the two previously introduced versions, the recognition version is in fact a *question*, which can be answered by *yes* or *no*. Obviously, answering this question is not much harder than solving the evaluation problem above, because, once we have done this, we need only compare the optimal cost $c(\hat{f})$ to L, and report *yes* iff $c(\hat{f}) \leq L$. We have thus established that each of the optimization, evaluation, and recognition versions, in this order, is not harder than the previous ones, using only the assumption that c is easy to compute. One natural question now arises: Is it the case that all these versions are roughly of the same complexity? In other words, can we solve the evaluation version by making efficient use of a hypothetical algorithm that solves the recognition version, and can we do the same with the optimization and evaluation versions, respectively?

†If the original problem is in fact a maximization problem, such as the maximum clique problem, the inequality becomes $c(f) \geq L$.

Under very general and realistic assumptions—namely, that the cost of the optimal solution is an integer with logarithm bounded by a polynomial in the size of the input—we can show that the evaluation version can be solved efficiently whenever the recognition version can. To show this, we must somehow evaluate the optimal cost $c(f)$ by asking the question "Is $c(f) \leq L$?" for many values of L. This, however, can be accomplished by the *binary search* technique of Lemma 8.4. By our assumption that $\log c(f)$ is bounded by a polynomial in the size of the input, it follows that we can make efficient use of any algorithm that solves the recognition problem in order to solve the evaluation problem. Notice that our assumption on the logarithm of the cost holds for all optimization problems that we have seen or shall introduce subsequently in this book, and of course is implied by our assumption that $\mathcal{A}_c$ is polynomial.

There is no known general method for solving the optimization version of a problem by making use of an algorithm for the evaluation version. However, certain ad hoc variations of a "dynamic programming" technique seem to be applicable to some problems.

Example 15.4

Let us consider the maximum clique problem introduced in Example 15.2, and assume that we have a procedure *cliquesize* which, given any graph G, will evaluate the size of the maximum clique of G. In other words cliquesize solves the evaluation version of the maximum clique problem. We can then make efficient use of this routine in order to solve the optimization version by the procedure *maxclique* of Fig. 15–1. In this recursive procedure we first find a node

```
procedure maxclique(G)
(comment: it returns the largest clique of G; it is recursive,
          and it uses the assumed procedure cliquesize).
if G has no nodes then return ∅
else
   begin
   let v be a node such that cliquesize(G(v)) = cliquesize(G),
   where G(v) is the subgraph of G consisting of v and all
   of its adjacent nodes;
   return {v} ∪ maxclique(G(v) − v);
   end
```

Figure 15–1

v of G that certainly participates in a maximum clique. We do this by checking that the cliquesize of G is not reduced if we omit all vertices nonadjacent to v; this means that v is a vertex of some maximum clique. We then recursively find the maxclique of the subgraph of G consisting of all vertices adjacent to v and the pertinent edges, and add v to this clique. The result is guaranteed to be a maximum clique of G. If cliquesize has a time bound $C(n)$ when invoked on graphs with n nodes, the time bound $T(n)$ of maxclique obeys

$$T(0) = O(1)$$
$$T(n) \leq (n + 1)C(n) + T(n - 1) + O(n)$$

and therefore

$$T(n) = O(n^2 \cdot C(n))$$

Thus, if cliquesize operates within a polynomial bound, maxclique will also. Unfortunately, we shall demonstrate shortly that it is highly unlikely that any of the three versions of the maximum clique problem can be solved efficiently. □

We can also show that similar techniques apply to the TSP (Problem 2) as well as many other combinatorial optimization problems. Thus, the three versions of such combinatorial optimization problems are all equivalent, at least as far as the existence of efficient algorithms is concerned.

15.3 The Classes *P* and *NP*

We saw in the previous section how, given an optimization problem, we can define a closely related *recognition problem*, that is, a question that can be answered by *yes* or *no*. Several well-known computational problems, however, are recognition problems to begin with. Such are the problems traditionally studied by the theory of computation. In Sec. 8.1, for example, we mentioned the HALTING PROBLEM:

Given an algorithm and its input, will it ever halt?

We also introduced the SATISFIABILITY problem in Chapter 13:

Given a Boolean formula, is it satisfiable?

In Section 12.6, we introduced a version of the HAMILTON CIRCUIT problem:

Given a graph G, is there a circuit in G visiting all nodes exactly once?

These are all recognition problems. Our definition of *recognition* versions of *optimization* problems allows us now to study both kinds of problems in a uniform setting. Furthermore, since we have pointed out that a recognition version is no harder than the original optimization problem, any *negative* results proved about the complexity of the recognition version will apply to the optimization version as well.

We are interested in classifying recognition problems according to their complexity. We shall denote by P the class of recognition problems that can be solved by a polynomial-time algorithm. The class P can be defined very precisely in terms of any mathematical formalism for algorithms, such as the *Turing machine* [Tu]. It turns out, however, that all such reasonable models of computation have a remarkable property: if a problem can be solved in polynomial time by one of them, it can be solved in polynomial time by all the rest. This class P, therefore, is *extremely stable* under variations in the details of our assumptions. Thus we are satisfied to define P informally, as the class of recognition problems that have polynomial-time algorithms; in other words, P is the class of relatively easy recognition problems, those for which efficient algorithms exist. We have already seen many representatives of P. We list some below.

GRAPH CONNECTEDNESS

Given a graph G, is G connected? (Sec. 9.1)

PATH IN A DIGRAPH

Given a digraph $D = (V, A)$, and two subsets $S, T \subseteq V$, is there a path from a vertex of S to a vertex of T in D? (Sec. 9.1)

MAXIMUM MATCHING

Given a graph G and an integer k, is there a matching in G with k or more edges? (Chapter 10)

MINIMUM SPANNING TREE

Given a connected graph $G = (V, E)$, a cost function d defined on E, and an integer L, is there a spanning tree of G with cost L or less? (Chapter 12)

All these problems belong to the class P. Given an instance of each, we have an efficient way for telling whether the answer is *yes* or *no*.

We shall now introduce NP, a seemingly richer class of recognition problems. For a problem to be in NP, we do not require that every instance can be answered in polynomial time by some algorithm. We simply require that, if x is a *yes* instance of the problem, then there exists a *concise* (that is, of length bounded by a polynomial in the size of x) *certificate* for x, which can be checked in polynomial time for validity.

Example 15.5

Consider the recognition version of the maximum clique problem:

CLIQUE

Given a graph $G = (V, E)$ and an integer k, is there a clique (that is, a completely connected subset of V) of size k?

It is not clear that CLIQUE is in P. The obvious way to solve this problem would be to subject all $\binom{|V|}{k}$ subsets of V with cardinality k to the test of whether they fulfill the requirement of the problem. The catch is, of course, that there is an exponentially large number of such sets. In fact, no polynomial algorithm is known for this problem.

Nevertheless, CLIQUE is in *NP*. To see this, suppose that we are given a *yes* instance of CLIQUE; in other words, a graph $G = (V, E)$ and an integer k such that G does have a clique C of cardinality k. Then this instance has a quite concise certificate, namely a list of the nodes in the clique C. This certificate can be checked efficiently for validity, because one has only to verify that C indeed consists of k nodes, and that all these nodes are in fact connected by edges in E. □

We can formalize these ideas as follows: Let Σ be a fixed finite alphabet and \$ be a *distinguished symbol* in Σ. (The symbol \$ marks the end of the input and the beginning of the certificate.) If x is a string of symbols from Σ, then its length—the number of symbols in it—is denoted by $|x|$.

Definition 15.1

We say that a recognition problem A is in the class *NP* if there exists a polynomial $p(n)$ and an algorithm $\mathcal{A}$ (the *certificate-checking algorithm*) such that the following is true:

> The string x is a *yes* instance of A if and only if there exists a string of symbols in Σ, $c(x)$, (the *certificate*), $|c(x)| \leq p(|x|)$, with the property that $\mathcal{A}$, if supplied with the input $x\$c(x)$, reaches the answer *yes* after at most $p(|x|)$ steps. □

Example 15.5 (Continued)

In our CLIQUE example, if x is the encoding of a *yes* instance of the clique problem (G, k), then the certificate of x, $c(x)$, would be an encoding of the list of the nodes in an appropriate clique C. The certificate-checking algorithm checks whether x is indeed the encoding of a graph G and an integer k, whether $c(x)$ is a set of vertices C, whether $|C| = k$, and whether there is an edge in G for each pair u, v of vertices of C. The polynomial $p(n)$ may be taken to be $2n^2$. □

Example 15.6

Consider our well-known

TRAVELING SALESMAN PROBLEM (TSP)

Given an integer n, an $n \times n$ symmetric matrix of nonnegative integers $[d_{ij}]$, and an integer L, find whether there exists a cyclic permutation (tour) τ such that $\sum_{j=1}^{n} d_{i\tau(j)} \leq L$.

This problem is in *NP* because, given an instance $(n, [d_{ij}], L)$ of the TSP with a *yes* answer, we can demonstrate this fact quite handily by exhibiting an appropriate tour. The certificate $c(x)$ of such an instance would then be an encoding of a tour τ satisfying $\sum_{i=1}^{n} d_{i,\tau(i)} \leq L$. The algorithm $\mathcal{A}$ would check whether n, L, and $[d_{ij}]$ are appropriate, whether τ is indeed a tour, and whether its total length is L or less. □

Example 15.7

The SATISFIABILITY problem examined in Chapter 13 is in *NP*. Given a set of clauses $C_1, \ldots, C_m$ involving the Boolean variables $x_1, \ldots, x_n$ which is indeed satisfiable, an appropriate certificate would be a truth assignment, represented as vector in $\{0, 1\}^n$. The certificate-checking algorithm would simply make sure the C_i's are legitimate clauses involving n variables and that all clauses come out true under the truth assignment of the certificate. □

Example 15.8

Consider the following recognition version of the integer linear programming (ILP) problem:

ILP
Given an $m \times n$ integer matrix A and an integer m-vector b, is there an integer n-vector x such that $Ax = b$, $x \geq 0$?

Notice that, in transforming integer linear programming to its recognition version above, we did not need to include explicitly the integer L and the inequality $c'x \leq L$, as we usually do in such a transformation. This is because this inequality can be transformed into an equality by adding a slack variable and thus incorporated in the system $Ax = b$.

The recognition problem ILP is in *NP*. To see this, recall Theorem 13.4, which states that if the integer program has a feasible solution, then it has one with all components bounded by $M_2 = n \cdot (ma_1)^{2m+3}(1 + a_2)$. Such a concise solution could, therefore, serve as a succinct certificate for a *yes* instance of ILP, because the length of its representation in binary is polynomial in the size of the input, as required. □

It is important to note that, in order to establish that a problem is in *NP*, one does not have to explain how the certificate $c(x)$ can be computed efficiently starting from an input x. One simply has to show the *existence* of *at least one* such string for each x.

Next notice that *P* is a subset of *NP*. In other words, every efficiently solvable problem is also succinctly certifiable. To see this, suppose that there is a polynomial-time algorithm $\mathcal{A}_A$ for Problem A. Given any *yes* instance x of A, $\mathcal{A}_A$ will operate on x for a polynomial number of steps and answer *yes*. Now, the *record* of this operation of $\mathcal{A}_A$ on x is a valid certificate $c(x)$. Indeed, one can check $c(x)$ easily by simply checking that it is a valid execution of $\mathcal{A}_A$; recall that $c(x)$

is, by its definition, polynomial in length. So, whenever A $\in$ *P*, it is also the case that A $\in$ *NP*.

But why is *NP* an interesting class of recognition problems? First, it contains *P*, the problems with which we are happy. Secondly, it contains many problems of interest to us, and whose membership in *P* is in doubt. Examples are the TSP, ILP, CLIQUE, and HAMILTON CIRCUIT. In fact we may say that the recognition versions of *all reasonable* combinatorial optimization problems are in *NP*. To argue for this, recall that combinatorial optimization problems aim at the *optimal design* of objects, such as tours, routes, sets of nodes, partitions, and lists of integers. Hence it is reasonable to expect that, once found, the optimal solution can be written down concisely and thus serve as a certificate for the recognition version. One cannot expect to design something optimally that cannot be designed within reasonable time bounds at all. So the class *NP* arises very naturally in the study of the complexity of combinatorial optimization problems.

But is *P* a proper subset of *NP*, or is it the case that *P* and *NP* are the same? Were the latter true, several problems now notorious for their difficulty—the TSP, CLIQUE, ILP, and SATISFIABILITY, to name a few—would be in *P*, and hence (presumably) easy. It is widely accepted today that *P is* a proper subset of *NP*. No formal proof of this is known yet, and the so-called $P = NP$ problem is now the most prominent theoretical question facing computer scientists. However, despite the absence of a proof either way, some light has been shed recently on this mystery. The principal mathematical tool of these investigations has been the notion of a *reduction*, which we examine next.

15.4 Polynomial-Time Reductions

It is often the case that solving a computational problem becomes easy once we assume that we have an efficient algorithm for solving a second problem. For example, the algorithm of Fig. 9–13 for finding the maximum flow in a network uses as a subroutine a clever algorithm for finding a *maximal* flow in a *layered* network. Also, in Sec. 12.6 we saw that the directed version of HAMILTON PATH can be solved in polynomial time, given an efficient algorithm for the three-matroid intersection problem. Finally, we noticed in Chapter 13 that if we had a polynomial-time algorithm for ILP, we would be able to solve SATISFIABILITY—and many other problems—efficiently. This is a pattern common enough to justify the following formalism.

Definition 15.2

Let A_1 and A_2 be recognition (that is, *yes-no*) problems. We say that A_1 *reduces in polynomial time* to A_2 if and only if there exists a polynomial-time algorithm $\mathcal{A}_1$ for A_1 that uses several times as a subroutine *at unit cost* a (hypothetical) algorithm $\mathcal{A}_2$ for A_2. We call $\mathcal{A}_1$ a *polynomial-time reduction* of A_1 to A_2. □

A point of central importance in the definition above is the *at unit cost* clause. By this we mean that algorithm $\mathcal{A}_2$ is considered as a single instruction, taking unit time to execute. Naturally, in almost all cases this will be a very unrealistic assumption. It is interesting, therefore, that the following proposition relates this purely hypothetical situation with the realities of computational complexity.

Proposition 15.1 If A_1 polynomially reduces to A_2 and there is a polynomial-time algorithm for A_2, then there is a polynomial algorithm for A_1.

Proof Suppose that $p_1(n)$ and $p_2(n)$ are the polynomials bounding the complexity of algorithm $\mathcal{A}_1$ (again, with the assumption of unit-cost invocation of $\mathcal{A}_2$) and $\mathcal{A}_2$. Then the *real* complexity of $\mathcal{A}_1$ on an input of size n, with each invocation of $\mathcal{A}_2$ costing as much time as it takes to run $\mathcal{A}_2$ with the current parameters, is bounded by

$$p(n) = p_1(n) \cdot p_2(p_1(n))$$

To see this, just notice that in the worst case $\mathcal{A}_1$ will consist of continuous calls of $\mathcal{A}_2$, each with the longest possible input. There can be at most $p_1(n)$ such invocations; but how long can their input be? Even if we assume that the algorithm spends all of its $p_1(n)$ steps just writing inputs for $\mathcal{A}_2$, these inputs cannot be longer than $p_1(n)$. (Here we have assumed that the algorithm can manipulate only one symbol at a time; the argument is identical if we allow simultaneous manipulation of up to $k > 1$ symbols, or even a polynomial—in the length of the input—number of symbols.) Consequently, there is a polynomial-time algorithm for A_1. □

We shall find a special kind of polynomial-time reduction to be of particular interest.

Definition 15.3

We say that a recognition problem A_1 *polynomially transforms* to another recognition problem A_2 if, given any string x, we can construct a string y within polynomial (in $|x|$) time such that x is a *yes* instance of A_1 if and only if y is a *yes* instance of A_2. □

Polynomial-time transformations can be thought of as polynomial-time reductions with just one call of the subroutine for A_2, exactly at the end of the algorithm for A_1. The rest of the algorithm simply constructs y, the input to the algorithm for A_2. Examples of polynomial-time transformation are the one from SATISFIABILITY to ILP (Example 13.4) and the transformation from the general minimum-cost flow problem to Hitchcock. (See Sec. 7.5; it is easy to verify that the transformation applies to the *recognition* versions of both problems as well.)

Definition 15.4

A recognition problem A $\in$ *NP* is said to be *NP-complete* if all other problems in *NP* polynomially transform to A. □

By Proposition 15.1, if a problem A is *NP*-complete, then it has a formidable property: If there is an efficient algorithm for A, then there is an efficient algorithm for *every* problem in *NP*. This includes such old and hard nuts as the TSP, ILP, SATISFIABILITY, and CLIQUE.

The notion of *NP*-complete problems and our suspicion that $P \neq NP$ suggests the topography of the class *NP* shown in Figure 15–2; complexity is assumed to increase as we go up. Of course, it is not at all obvious at this point that *NP*-complete problems exist. But they do, as we shall see in the next section.

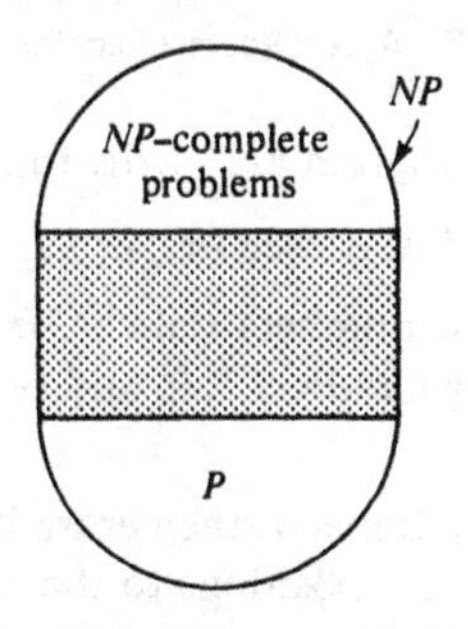

Figure 15–2 *P* and *NP*.

15.5 Cook's Theorem

In order to prove that a problem is *NP*-complete, we must show two things:

(a) That the problem is in *NP*.
(b) That *all* other problems in *NP* polynomially transform to our problem.

In practice, Part (b) is usually carried out by showing that a *known NP*-complete problem is polynomially transformable to the problem at hand. Since the property of polynomial transformability is *transitive*—that is, if A polynomially transforms to B and B polynomially transforms to C, then A polynomially transforms to C—this will be sufficient. However, our first *NP*-completeness proof must include an *explicit* proof of Part (b). In order to do this, we need to employ in our proof the common characteristics of all diverse problems in *NP*. These are, of course, the existence of at least one certificate $c(x)$ for each *yes* instance x, and of the certificate-checking algorithm $\mathcal{A}$.

Thus we must formalize our notion of a certificate-checking algorithm. This algorithm can be thought of as a device that reads and modifies the symbols of a string, one position at a time, via a read-write head; its operation, moves, and other decisions are governed by a program (see Fig. 15-3). Initially, the head is scanning the leftmost position of the string, and the program is about to execute its first instruction.

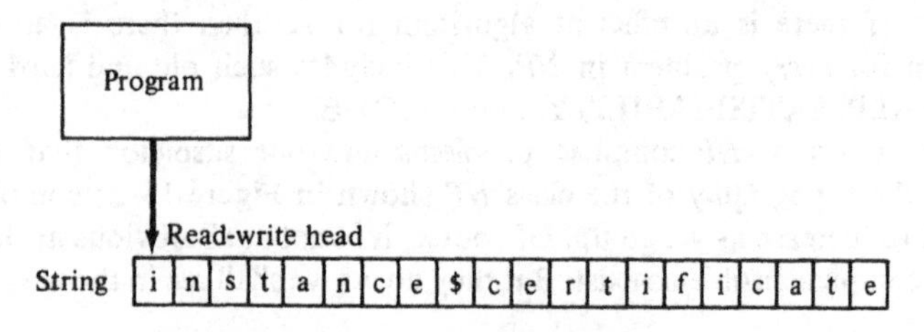

Figure 15-3 A certificate-checking algorithm.

The instructions of the program are of the form

$$l\text{: } \textbf{if } \sigma \textbf{ then } (\sigma'; o; l')$$

where l and l' are instruction numbers (labels), σ and σ' are symbols of the alphabet Σ, and o is one of the numbers 1, 0, and -1. The *meaning* of the above instruction is as follows.

> If the currently scanned symbol is σ, then erase it and write σ' in its place, move to the symbol which is o positions to the right, and execute next the instruction numbered l'; otherwise, continue to the next instruction.

Moving zero and -1 places to the right means staying at the same position and moving one to the left, respectively. The last statement of the program is

$$|\mathcal{A}|\text{: } \textbf{accept}$$

where $|\mathcal{A}|$ is the length of the program—the total number of its instructions.

We say that the string $x\$c(x)$ is *accepted* by $\mathcal{A}$ if the algorithm, properly started on the string, reaches the last instruction after at most $p(|x|)$ steps, where $p(n)$ is the polynomial bound of $\mathcal{A}$. If the algorithm executes $p(|x|)$ instructions without reaching the **accept** instruction or if the head falls off the string, we say that $x\$c(x)$ is *rejected*. Thus we can now define *NP* more formally to be the set of recognition problems for which there exists a certificate-checking algorithm $\mathcal{A}$ and a polynomial $p(n)$, such that x is a *yes* instance if and only if there exists a string $c(x)$, with $|c(x)| \leq p(|x|)$ such that $\mathcal{A}$ accepts $x\$c(x)$.

This model of a certificate-checking algorithm may seem, at first glance, somewhat primitive and restricted in its power. Let us examine several legitimate reservations.

1. The operation of the algorithm seems to be confined within the space provided by the string, and there is no extra "scratch paper." Nevertheless, this objection can be taken care of easily by noticing that the certificate may be defined so that it contains in itself enough extra space—for example, a long string of trailing "blanks"—for the algorithm to use. Because the algorithm is polynomial, it cannot need more than a polynomial amount of such space, and so the length of the certificate remains polynomial.

   With this idea in mind, we shall henceforth make the simplifying assumption that for all x's of the same size, $c(x)$ has length equal to $|c(x)| = p(|x|) - |x| - 1$. Consequently, the length of the whole input $x\$c(x)$ is exactly equal to $p(|x|)$. This is not a loss of generality, because if the certificate is shorter, it can always be "padded" by several blanks to reach this length. On the other hand, inputs longer than $p(|x|)$ are not meaningful for $\mathfrak{A}$, since $\mathfrak{A}$ cannot even inspect the whole length of such an input within $p(|x|)$ steps.
2. Another legitimate reservation as to the strength of our model of computation is that the repertoire of "instructions" available is quite limited. Still, it should be clear that our restricted model can be used to simulate more complicated instructions (for example: *Do the following until the scanned symbol is a* \$, *Move 7 positions to the right*, and *Delete the current symbol*) by introducing only a multiplicative constant factor in the total number of steps.
3. A more significant departure from this model would be to allow *addressing*; for example, instructions such as *Move to the fifteenth symbol of the string* (or even *Move as many positions to the right as indicated by the binary integer between the current symbol and the first* \$ *to its right*), in a fashion reminiscent of ordinary random access computers. However, we leave it to the reader to verify that this restriction can multiply the total number of steps only by approximately $p(|x|)$, since any position in the string can be accessed within this time bound.

We should like therefore to claim that our model for certificate-checking algorithms is sufficiently realistic and general, in that one can express in it efficient algorithms for any problem for which efficient algorithms are possible at all. Since this model is essentially a Turing machine, further confidence in our claim can be obtained by examining the elegant argument of A. M. Turing [Tu]. See Chapter 1 of [AHU] for a more technical discussion.

We shall use this formalism in order to prove the following important result due to S. A. Cook:

Theorem 15.1 *SATISFIABILITY is NP-complete.*

Proof We recall from Example 15.7 that the problem SATISFIABILITY is in *NP*. Thus, it remains to show that if A is a problem in *NP*, then A is polynomially transformable to SATISFIABILITY. In other words, given a string x we must construct formula $F(x)$—using only the fact that $A \in NP$—such that x is a *yes* instance of A if and only if $F(x)$ is satisfiable. To do this, we shall consider the certificate-checking algorithm $\mathcal{A}$ for A, which, by our hypothesis that $A \in NP$, operates within a polynomial bound $p(n)$.

The Boolean formula $F(x)$ will involve the following Boolean variables.

1. A Boolean variable $x_{ij\sigma}$ for all $0 \leq i, j \leq p(|x|)$ and $\sigma \in \Sigma$. The intended meaning for $x_{ij\sigma}$ is: *at time i, the jth position of the string contains the symbol* σ.
2. A Boolean variable y_{ijl} for all $0 \leq i \leq p(|x|), 0 \leq j \leq p(|x|) + 1$, and $1 \leq l \leq |\mathcal{A}|$—where $|\mathcal{A}|$ is the number of instructions in the algorithm $\mathcal{A}$. The meaning attached to this variable is: *at time i the jth position is scanned and the lth instruction is being executed.* If $j = 0$, or if $j = p(|x|) + 1$, this means that the head has fallen off the string, and hence the computation is unsuccessful.

We shall next combine these Boolean variables in a formula $F(x)$ such that $F(x)$ is satisfiable if and only if x is a *yes* instance of A. If the Boolean variables have the indicated meanings, $F(x)$ will state essentially that the algorithm $\mathcal{A}$, started with x in the leftmost part of the string, may eventually accept the string; this would mean, by definition, that there exists an appropriate certificate $c(x)$ and hence that x is a *yes* instance of A. The formula $F(x)$ is the conjunction of four parts: $F(x) = U(x) \cdot S(x) \cdot W(x) \cdot E(x)$.

1. The purpose of $U(x)$ is to make sure that at each time $i, 0 \leq i \leq p(|x|)$, each position of the string contains a *unique* symbol, the head scans a *unique* position within the bounds of the string, and the program executes a *single* statement

$$U(x) = \Bigg(\prod_{\substack{0 \leq i, j \leq p(|x|) \\ \sigma \neq \sigma'}} (\bar{x}_{ij\sigma} + \bar{x}_{ij\sigma'})\Bigg) \cdot \Bigg(\prod_{\substack{0 \leq i \leq p(|x|) \\ j \neq j' \text{ or } l \neq l'}} (\bar{y}_{ijl} + \bar{y}_{ij'l'})\Bigg)$$
$$\cdot \Bigg(\prod_{\substack{0 \leq i \leq p(|x|) \\ 1 \leq l \leq |\mathcal{A}|}} \bar{y}_{i0l} \cdot \bar{y}_{i,p(|x|)+1,l}\Bigg)$$
$$\cdot \Bigg(\prod_{0 \leq i \leq p(|x|)} \Bigg(\Bigg(\prod_{1 \leq j \leq p(|x|)} \sum_{\sigma \in \Sigma} x_{ij\sigma}\Bigg) \cdot \sum_{\substack{1 \leq j \leq p(|x|) \\ 1 \leq l \leq |\mathcal{A}|}} y_{ijl}\Bigg)\Bigg)$$

2. The formula $S(x)$ states that the operation of $\mathcal{A}$ *starts* correctly; in other words, at time zero the leftmost $|x| + 1$ symbols in the string

spell $x\$$, the read-write head scans the leftmost symbol of the string, and the first instruction of $\mathcal{A}$ is about to be executed.

$$S(x) = \left(\prod_{j=1}^{|x|} x_{0jx(j)}\right) \cdot x_{0,|x|+1,\$} \cdot y_{011}$$

(Here $x(j)$ stands for the jth symbol of the string x.)

3. The formula $W(x)$ states that $\mathcal{A}$ *works* right, according to the instructions of the program; $W(x)$ is the conjunction of the formulas $W_{ij\sigma l}$, one for each $0 \leq i < p(|x|)$, $1 \leq j \leq p(|x|)$, $\sigma \in \Sigma$, $1 \leq l < |\mathcal{A}|$, such that the lth instruction of $\mathcal{A}$ is

$$l\text{: } \mathbf{if}\ \sigma\ \mathbf{then}\ (\sigma'; o; l').$$

The formula $W_{ij\sigma l}$ is defined as follows:

$$W_{ij\sigma l} = (\bar{x}_{ij\sigma} + \bar{y}_{ijl} + x_{i+1,j,\sigma'}) \cdot (\bar{x}_{ij\sigma} + \bar{y}_{ijl} + y_{i+1,j+o,l'}) \cdot \prod_{\tau \neq \sigma} ((\bar{x}_{ij\tau} + \bar{y}_{ijl} + x_{i+1,j,\tau})(\bar{x}_{ij\tau} + \bar{y}_{ijl} + y_{i+1,j,l+1}))$$

This means that whenever $x_{ij\sigma}$ and y_{ijl} are both true, in the next time instant the x-and y-variables stating that $\mathcal{A}$ made the right move must also be true. For the last instruction of $\mathcal{A}$, we have for each i, j and σ

$$W_{ij\sigma|\mathcal{A}|} = (\bar{x}_{ij\sigma} + \bar{y}_{ij|\mathcal{A}|} + y_{i+1,j,|\mathcal{A}|})$$

stating that once the algorithm reaches the **accept** instruction, it stays there. Finally, $W(x)$ contains the clauses

$$\prod_{\substack{0 \leq i \leq p(|x|) \\ \sigma \in \Sigma \\ 1 \leq l \leq |\mathcal{A}| \\ j \neq j'}} (\bar{x}_{ij\sigma} + \bar{y}_{ij'l} + x_{i+1,j,\sigma})$$

meaning that whenever $\mathcal{A}$ is scanning a position different from the jth, the symbol of the jth position remains unchanged.

4. The last part of $F(x)$ states simply that the operation of $\mathcal{A}$ *ends* correctly, with the program executing the **accept** instruction. It consists of just one clause:

$$E(x) = \sum_{j=1}^{p(|x|)} y_{p(|x|),j,|\mathcal{A}|}$$

This completes the construction of $F(x)$. First, let us observe that this construction requires only a polynomial—in $|x|$—amount of time. This follows from the fact that the total length (number of occurrences of literals multiplied by the length of the subscripts of these literals of the formula F) is $O(p^3(|x|) \log p(|x|))$. Hence, to show that the construction is a polynomial transformation from A to SATISFIABILITY, it remains to prove the following claim.

Claim. $F(x)$ is satisfiable if and only if x is a *yes* instance of A.

To show the *only if* direction, suppose that $F(x)$ is satisfiable. Hence all of $U(x)$, $S(x)$, $W(x)$, and $E(x)$ are satisfied by the same truth assignment t. Because

t satisfies $U(x)$, it must be that for all i and j exactly one variable $x_{ij\sigma}$ is true; take this to mean that the jth cell contains σ at time i. Also, for all i, exactly one of the variables y_{ijl} is true; we assign to this the meaning that at time i, the jth position of the string is scanned and statement l is executed. Finally, no variable of the form y_{i0l} or $y_{i,p(|x|)+1,l}$ can be true, and this means that the head will never fall off the string. The truth assignment t, therefore, describes some sequence of strings, head positions, and instructions. We shall now show that this sequence is in fact a valid accepting computation of $\mathcal{A}$ on input $x\$c(x)$ for some certificate $c(x)$.

Since $S(x)$ must also be satisfied by t, this means that the sequence starts off correctly, with the first $|x| + 1$ places occupied by the correct string, $x\$$, and with the first symbol of x scanned while the first instruction is executed.

The fact that $W(x)$ is also satisfied by t means that the sequence changes according to the rules governing the execution of the algorithm $\mathcal{A}$. Finally, t satisfies $E(x)$ only if the algorithm ends up in its last accepting instruction. Consequently, if $F(x)$ is satisfiable, there exists a string $c(x)$ of appropriate length such that $\mathcal{A}$ accepts $x\$c(x)$; hence x is a *yes* instance of A.

For the *if* part, assume that x is a *yes* instance. Then there exists a string $c(x)$ of length $p(|x|) - |x| - 1$ such that $\mathcal{A}$ accepts $x\$c(x)$. This means that there exists a succession of $p(|x|)$ strings (with $x\$c(x)$ first), instruction numbers, and positions scanned, that are legal according to $\mathcal{A}$ and end up with the acceptance of $x\$c(x)$. This succession immediately defines a truth assignment t that necessarily satisfies $F(x)$. This completes the proof of the Claim.

The Claim says that $F(x)$ is a *yes* instance of SATISFIABILITY if and only if x was a *yes* instance of Problem A. Hence, what we have just described is a polynomial transformation from A to SATISFIABILITY. Because A was taken to be an arbitrary problem in *NP*, the theorem is now proved. □

Corollary *ILP is NP-complete.*

Proof In Example 15.8, we showed that ILP is in *NP*. To show that it is *NP*-complete, we simply recall the polynomial-time transformation from SATISFIABILITY to ILP described in Example 13.4. □

15.6 Some Other *NP*-Complete Problems: CLIQUE and the TSP

We shall next show that certain other problems in *NP* are *NP*-complete by showing that SATISFIABILITY—or some other problem previously shown to be *NP*-complete—is polynomial-time transformable to the new problem. The proof techniques employed are quite interesting in themselves, and we shall be pointing out certain basic elements of methodology.

If we restrict SATISFIABILITY to the special case in which each clause is restricted to consist of just *two* literals, the resulting problem can be solved in linear time (see Problem 6). However, we shall now show that if we allow the number of literals in each clause to be three, the resulting problem, called 3-SATISFIABILITY, remains as hard as SATISFIABILITY itself.

Theorem 15.2 *3-SATISFIABILITY is NP-complete.*

Proof Because 3-SATISFIABILITY is a special case of SATISFIABILITY, it is in *NP*. To show *NP*-completeness, we shall show that SATISFIABILITY polynomially transforms to 3-SATISFIABILITY. Consider any formula F consisting of clauses $C_1, \ldots, C_m$. We shall construct a new formula F' with three literals per clause, such that F' is satisfiable if and only if F is. We shall examine the clauses of F one by one and replace each C_i by an equivalent set of clauses, each with three literals. We distinguish among three cases.

1. If C_i has three literals, we do nothing.
2. If C_i has more than three literals, say $C_i = (\lambda_1 + \lambda_2 + \cdots + \lambda_k)$, $k > 3$, we replace C_i by the $k - 2$ clauses $(\lambda_1 + \lambda_2 + x_1)(\bar{x}_1 + \lambda_3 + x_2)(\bar{x}_2 + \lambda_4 + x_3) \cdots (\bar{x}_{k-3} + \lambda_{k-1} + \lambda_k)$, where $x_1, \ldots, x_{k-3}$ are new variables. It is not hard to see that these new clauses are satisfiable if and only if C_i is.
3. If $C_i = \lambda$, we replace C_i by $\lambda + y + z$, and if $C_i = \lambda + \lambda'$, we replace it by $\lambda + \lambda' + y$. We then add the clauses

$$(\bar{z} + \alpha + \beta)(\bar{z} + \bar{\alpha} + \beta)(\bar{z} + \alpha + \bar{\beta})(\bar{z} + \bar{\alpha} + \bar{\beta})(\bar{y} + \alpha + \beta)$$
$$(\bar{y} + \alpha + \bar{\beta})(\bar{y} + \bar{\alpha} + \beta)(\bar{y} + \bar{\alpha} + \bar{\beta})$$

to the formula, where y, z, α, and β are new variables. This addition forces the variables z and y to be *false* in any truth assignment satisfying F', so that the clauses λ and $\lambda + \lambda'$ are equivalent to their replacements.

We have thus eliminated all occurrences of clauses with other than three literals, and furthermore we have shown that the resulting formula F' is satisfiable if and only if F is. Finally, the construction of F' can obviously be carried out in polynomial time. Consequently, what we have just described is a polynomial-time transformation from SATISFIABILITY to 3-SATISFIABILITY. Hence 3-SATISFIABILITY is *NP*-complete. □

We shall use the *NP*-completeness of 3-SATISFIABILITY in order to show the following theorem.

Theorem 15.3 *CLIQUE is NP-complete.*

Proof We already know that CLIQUE $\in NP$. Hence it will suffice to show that 3-SATISFIABILITY polynomially transforms to CLIQUE. Given any Boolean formula F with three literals per clause, we shall construct a graph $G = (V, E)$ and an integer k such that G has a clique of size k if and only if F is satisfiable.

Let us first introduce some terminology and notation. A *partial truth assignment* t assigns the value *true* or *false* to certain variables only; the rest of the variables have truth value $t(x) = d$, a special value meaning *undefined.* Let us fix $x_1, \ldots, x_n$ as our set of variables. A partial truth assignment will be denoted as a sequence of zeros, 1's and d's. Thus $t = d01d0$ ($n = 5$) is the partial truth assignment in which $t(x_2) =$ *false*, $t(x_3) =$ *true*, $t(x_5) =$ *false*, and the rest are undefined. Two partial truth assignments t and t' are called *compatible* if, for all variables x for which $t(x) \neq d$ and $t'(x) \neq d$, we always have $t(x) = t'(x)$.

Let $C_1, \ldots, C_m$ be the clauses of F. The vertex set V contains seven nodes for each clause C_i, namely, the nodes $t_{ij}, j = 1, \ldots, 7$. We can think of those nodes as partial truth assignments with values defined only on the three variables involved in C_i. Of the eight such partial truth assignments, we omit only the one assignment that makes all three literals of C_i false. An illustration of this is given in Fig. 15–4. The edges of G are those joining all pairs of compatible truth assignments. We let k be equal to m.

We now claim that G, as constructed above, has a clique of size k if and only if F is satisfiable. For suppose that G has a clique of size k. Because $k = m$ and because there are no edges joining two nodes in the same column (two distinct partial truth assignments defined on the same variables must necessarily be incompatible), it follows that the clique consists of one node in each of the m columns. Since all these partial truth assignments are mutually compatible, it follows that they come from the same complete truth assignment t (in the example of Fig. 15–4, $t = 1011$). Now t must satisfy all clauses in F because, for each clause, the only partial truth assignment that falsifies it is omitted from the corresponding column. Hence t satisfies F.

For the *if* part, suppose that there is a truth assignment t satisfying F; then the restriction of this assignment to the variables appearing in each clause must be an existing node in the column corresponding to this clause. Since these partial assignments are restrictions of the same truth assignment, they are pairwise compatible, and consequently they constitute a clique of size k.

Thus we have transformed 3-SATISFIABILITY to CLIQUE and our construction can be carried out in polynomial time; hence CLIQUE is *NP*-complete. □

There are two graph-theoretic problems that are very closely related to CLIQUE;

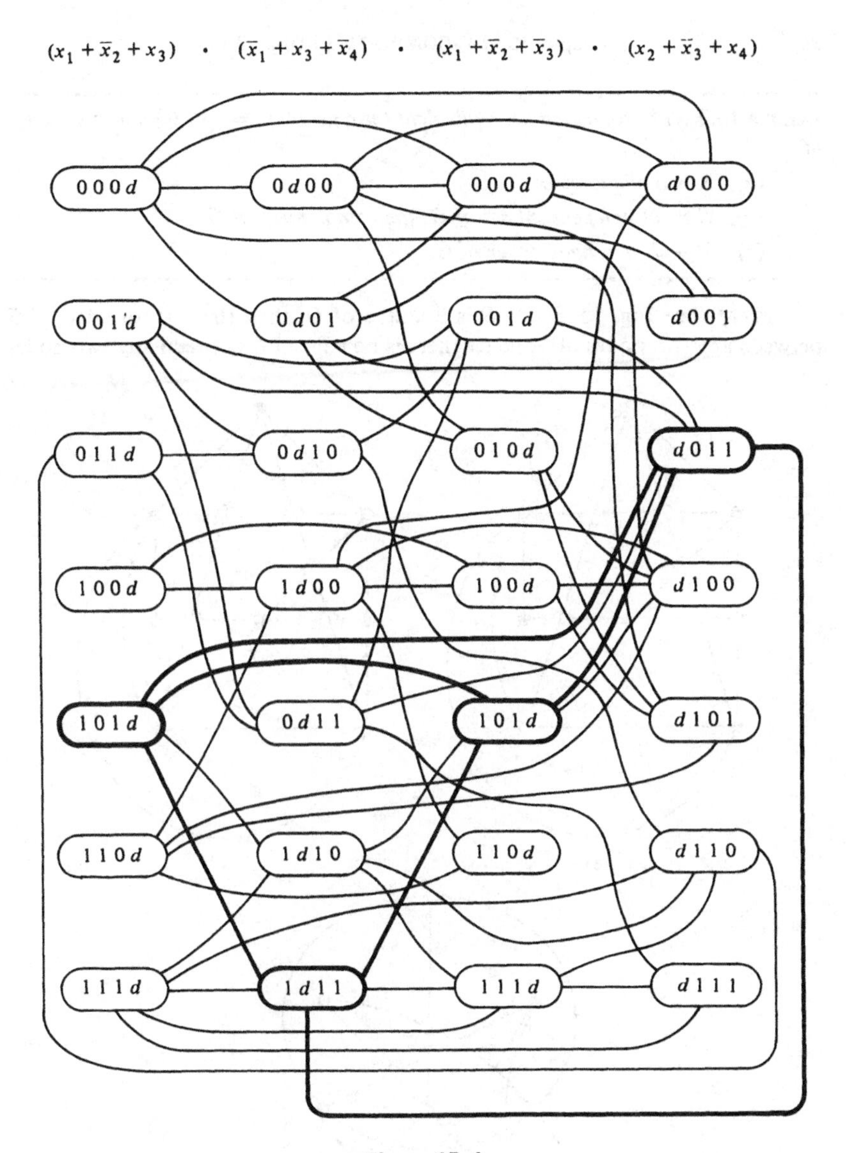

Figure 15–4

NODE COVER

Given a graph G and an integer k, is there a set C of k vertices such that all edges of G are adjacent to at least one node of C?

INDEPENDENT SET

Given a graph G and an integer k, is there a set I of k vertices such that no two nodes in I are connected by an edge?

Lemma 15.4 *The following are equivalent for a graph $G = (V, E)$ and a subset S of V:*

(*a*) *S is a clique of G.*
(*b*) *S is an independent set of $\bar{G}$, the complement of G.*
(*c*) *$V - S$ is a node-cover of $\bar{G}$.*

Proof (See Fig. 15–5) If S is a clique of G, then there is an edge of G between any two nodes of S; hence there is no edge of $\bar{G}$ between any two nodes

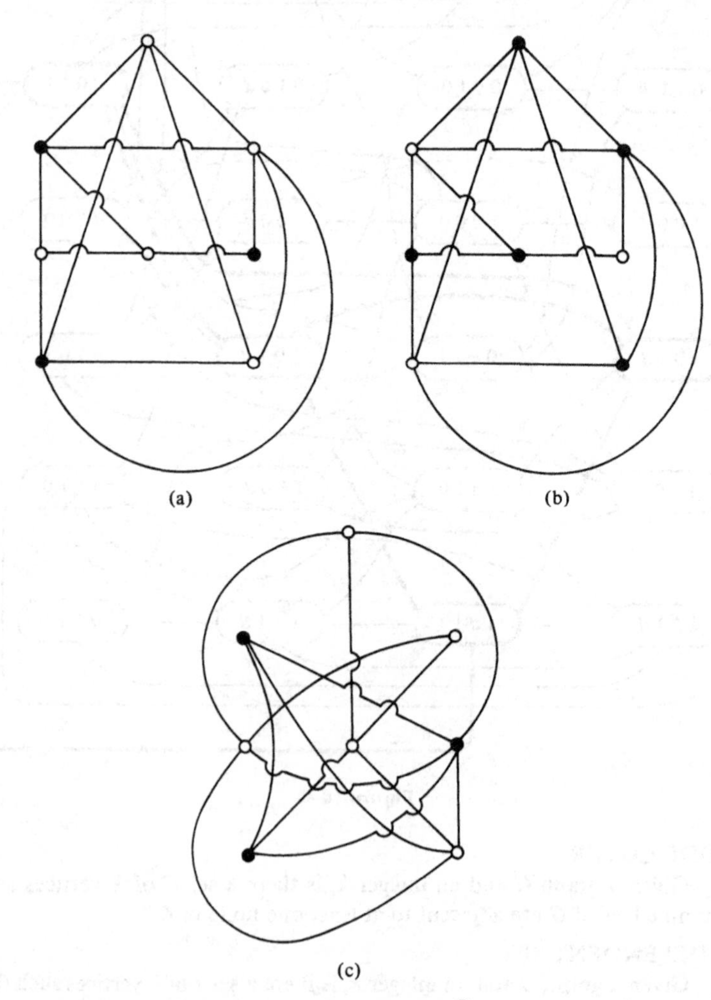

Figure 15–5 (a) An independent set in $\bar{G}$. (b) A node cover in $\bar{G}$. (c) A clique in G.

of S, and S is independent in $\bar{G}$. Now, if S is independent in $\bar{G}$, all edges in $\bar{G}$ must have an endpoint out of S; therefore, $V - S$ is a node cover of $\bar{G}$. Finally, if $V - S$ is a node cover in $\bar{G}$, this means that every edge missing from G does not have both endpoints in S; S is therefore a clique in G. □

Corollary *INDEPENDENT SET and NODE COVER are NP-complete.*

Proof Obviously G has a clique of size k if and only if $\bar{G}$ has an independent set of size k, if and only if $\bar{G}$ has a node cover of size $|V| - k$. Hence all three polynomially transform to one another. □

We shall now prove that an interesting scheduling problem (recall Example 13.2) is *NP*-complete. The problem involves scheduling jobs having equal—say, unit—execution times on a number of identical processors subject to some *precedence constraints*, stating that certain jobs can execute only if some other jobs have already been processed.

MULTIPROCESSOR SCHEDULING

Given a set of jobs $\mathcal{J} = \{J_1, \ldots, J_n\}$, a directed acyclic graph $P = (\mathcal{J}, A)$ (the *precedence* partial order), an integer m (the number of *machines*) and an integer T (the *deadline*), is there a function S (the *schedule*) mapping $\mathcal{J}$ to $\{1, 2, \ldots, T\}$ such that the following hold?

1. For all $j \leq T$, $|\{J_i: S(J_i) = j\}| \leq m$.
2. If $(J_i, J_j) \in A$, then $S(J_i) < S(J_j)$.

Example 15.9

Consider the instance of MULTIPROCESSOR SCHEDULING shown in Fig. 15–6(a). A schematic representation of a feasible schedule S is shown in Fig. 15–6(b). □

Theorem 15.5 *MULTIPROCESSOR SCHEDULING is NP-complete.*

Proof The problem is in *NP* because a feasible schedule, if it exists, can be exhibited and checked for feasibility quite easily. We shall now transform the CLIQUE problem to MULTIPROCESSOR SCHEDULING. Given any graph $G = (V, E)$ and integer k, we shall construct a set $\mathcal{J}$ of jobs, a partial order $P = (\mathcal{J}, A)$, and integers m and T such that there exists a feasible schedule for $\mathcal{J}$ if and only if G has a clique of size k. We shall assume here that G has no isolated vertices; this is no loss of generality, since isolated vertices can be removed from a graph without affecting the size of cliques. We define $\mathcal{J} = V \cup E \cup B \cup C \cup D$; B, C, and D are nonempty disjoint sets $B = \{b_1, \ldots, b_{|B|}\}$, $C = \{c_1, \ldots, c_{|C|}\}$, and $D = \{d_1, \ldots, d_{|D|}\}$, where $|B|$, $|C|$, and $|D|$ satisfy

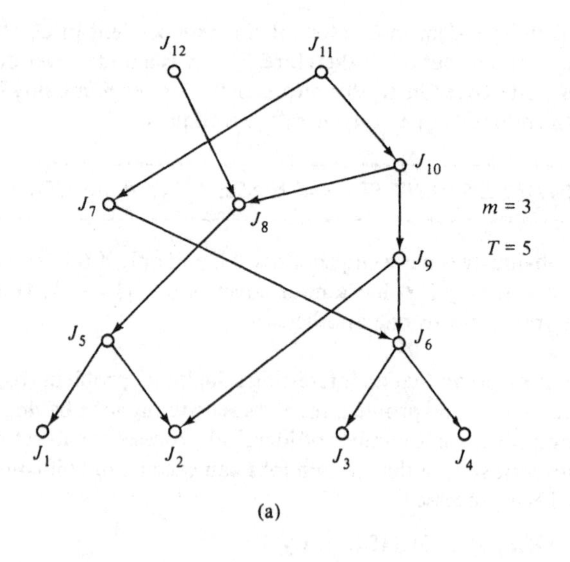

(a)

Time →

	1	2	3	4	5
Machine 1	J_{11}	J_7	J_9	J_6	J_3
Machine 2	J_{12}	J_8	J_5	J_1	J_4
Machine 3		J_{10}		J_2	

(b)

Figure 15–6

1. $m = k + |B| = \binom{k}{2} + |V| - k + |C| = |E| - \binom{k}{2} + |D|$, and
2. $\min(|B|, |C|, |D|) = 1$.

It is not hard to see that such cardinalities exist; m is defined by (1); T is taken to be 3.

For the construction of $P = (\mathcal{J}, A)$, we start by adding to A all possible arcs of the form (b_i, c_j) or (c_j, d_k). Finally, we add to A the arcs (v, e) for all $v \in V$, $e \in E$ such that e is incident upon v. This completes the construction of the instance of MULTIPROCESSOR SCHEDULING. An illustration is shown in Parts (a) and (b) of Fig. 15–7.

Let us now argue that there exists a feasible schedule for $\mathcal{J}$ under A, m, and T if and only if G has a clique of size k. Suppose that a feasible schedule S exists. Because $T = 3$ and all jobs in B must be executed before those in C, and those in C before the jobs in D, it necessarily follows that $S(b_i) = 1$, $S(c_i)$

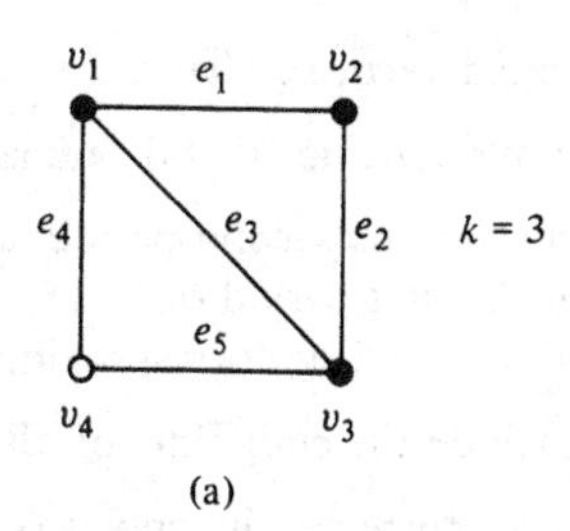

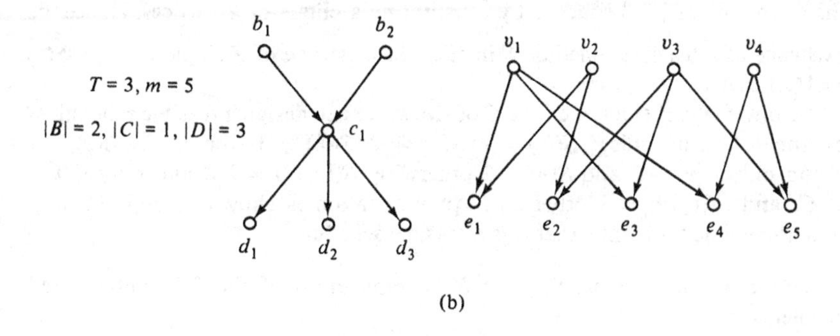

Time →

	1	2	3
Machine 1	b_1	c_1	d_1
Machine 2	b_2	e_1	d_2
Machine 3	v_1	e_2	d_3
Machine 4	v_2	e_3	e_4
Machine 5	v_3	v_4	e_5

(c)

Figure 15–7 (a) The graph G. (b) The precedence relation P. (c) The feasible schedule corresponding to the node cover $\{v_1, v_2, v_3\}$.

$= 2$, and $S(d_i) = 3$ for all i. Secondly, one notices that $|\mathcal{J}| = T \cdot m$; consequently all m machines must be executing some job during all three periods. Now, in the last period, besides the d-jobs, only $m - |D| = |E| - \binom{k}{2}$ jobs corresponding to edges can be executed. This is because if a job corresponding to a vertex is executed in the third period, the jobs corresponding to the edges incident upon it cannot be executed until after the third period; here we use the fact that G has no isolated vertices. Hence the schedule would not be feasible with $T = 3$, as assumed. We conclude that the remaining $\binom{k}{2}$ jobs corresponding to edges must be executed in the second period, and the jobs corresponding to vertices are executed in the first period ($m - |B| = k$ of them) and the second period. Furthermore, the k vertices corresponding to jobs that are executed in the first period must include the endpoints of all $\binom{k}{2}$ edges corresponding to jobs in the second period. However, the only way that k vertices can include all the endpoints of $\binom{k}{2}$ edges is by constituting a clique of k vertices. Hence the existence of a feasible schedule S implies the existence of a clique $C = \{v: S(v) = 1\}$ of size k.

Conversely, if G has a clique C of size k, we can design a feasible schedule S by the following rules: $S(b_i) = 1$; $S(c_i) = 2$; $S(d_i) = 3$, for all i; $S(v) = 1$ if and only if $v \in C$ and $S(v) = 2$ otherwise; $S([v, u]) = 2$ if and only if $v, u \in C$, and $S([v, u]) = 3$ otherwise. An illustration is shown in Fig. 15–7(c). It is immediately obvious that S is a feasible schedule. □

Next we turn to proving the *NP*-completeness of the TSP and related problems.

Theorem 15.6 *HAMILTON CIRCUIT is NP-complete.*

Proof We know that HAMILTON CIRCUIT is in *NP*; we shall now show that 3-SATISFIABILITY polynomially transforms to HAMILTON CIRCUIT. Given a Boolean formula F consisting of m clauses $C_1, \ldots, C_m$ and involving n variables $x_1, \ldots, x_n$, we shall construct a graph $G = (V, E)$ such that G has a Hamilton circuit if and only if F is satisfiable. Our construction involves the design of *special-purpose components*, a methodology that is commonplace in many interesting *NP*-completeness proofs.

Consider, for example, the graph A shown in Fig. 15–8(a). Suppose that A is a subgraph of some other graph G such that

1. No other edges (that is, edges not shown in Fig. 15–8(a)) are incident upon any node of A except for u, u', v, and v'.
2. G has a Hamilton circuit c.

Then the claim is that c traverses A in one of the ways shown in Fig. 15–8(b) and 15–8(c). To show this, first notice that the eight "vertical" edges of A must always be part of c, because only thus can we "pick up" the four nodes z_1, z_2, z_3, and z_4. Furthermore, any other combination of "horizontal" edges other than the ones shown in Fig. 15–8(b) and 15–8(c) cannot be a part of a Hamilton circuit, as can be checked easily. To summarize our observations, the graph A behaves *as if it were just a pair of edges* $[u, u']$ and $[v, v']$ of G with the additional

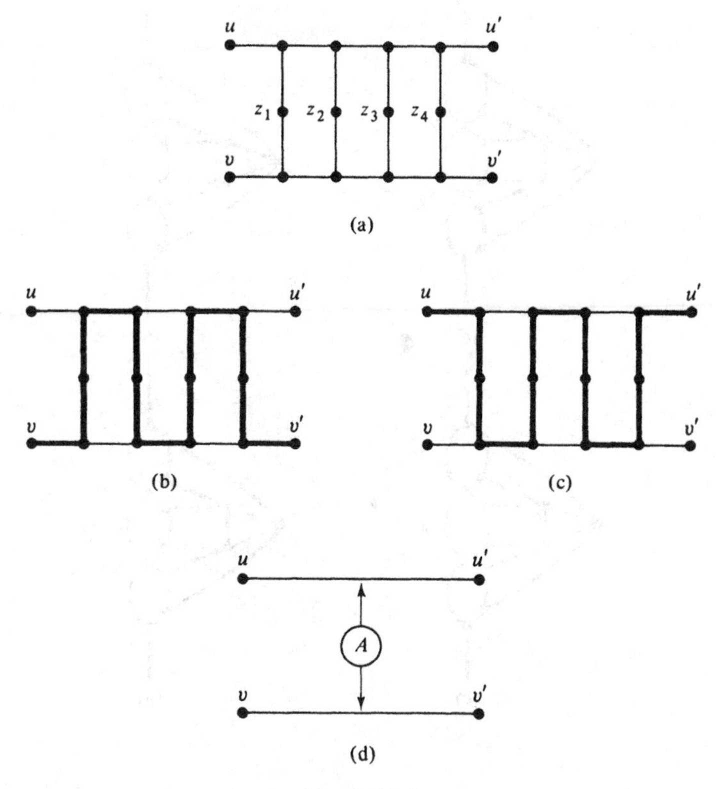

Figure 15–8

restriction that any Hamilton circuit of G must traverse *exactly* one of them. We shall represent this as shown in Fig. 15–8(d).

The graph B shown in Fig. 15–9(a) has a similar property. If B is a subgraph of G such that no other edges of G are incident upon any node of B other than u_1 and u_4 and G has a Hamilton circuit c, then c cannot traverse *all three* of the edges $[u_1, u_2]$, $[u_2, u_3]$, and $[u_3, u_4]$. Moreover, any proper subset of these edges *can* be a part of a Hamilton circuit of G, via the configurations shown in Figs. 15–9(b) through 15–9(d) (and more configurations not shown here). We shall represent this subgraph as in Fig. 15–9(e).

The graph $G = (V, E)$ will consist mainly of copies of the subgraphs A and B. For the m clauses $C_1, \ldots, C_m$, we have m copies of the subgraph B joined in series (see Fig 15–10 for an illustration of the construction of G). Also, for each variable x_i, we have two nodes v_i and w_i and *two copies* of the edge $[v_i, w_i]$, distinguished as the *right* and *left* copies of $[v_i, w_i]$, respectively. We also have the edges $[w_i, v_{i+1}]$ for $i = 1, \ldots, n - 1$ and $[u_{11}, v_1]$, $[u_{m4}, w_n]$, where u_{ij} denotes the ith copy of u_j.

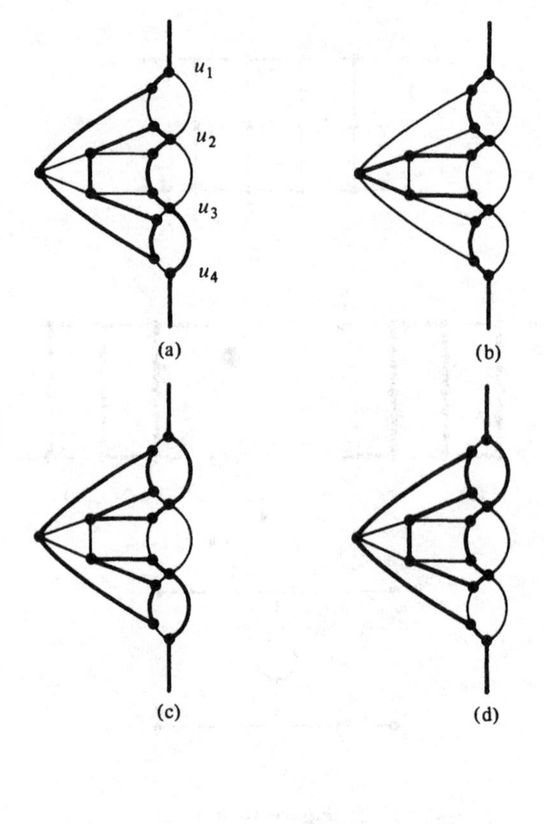

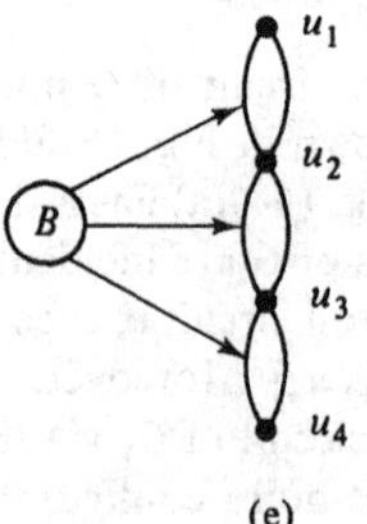

Figure 15–9

Notice that, so far, only the parameters m and n of our formula have entered into the construction of G. Since G is supposed to capture the intricacy of the satisfiability question for F, we must now take into account the exact nature of the clauses of F. Thus we connect (via the A-connector) the edge $[u_{ij}, u_{i,\,j+1}]$ with the left copy of $[v_k, w_k]$ in the case that the jth literal of C_i is x_k, and with the

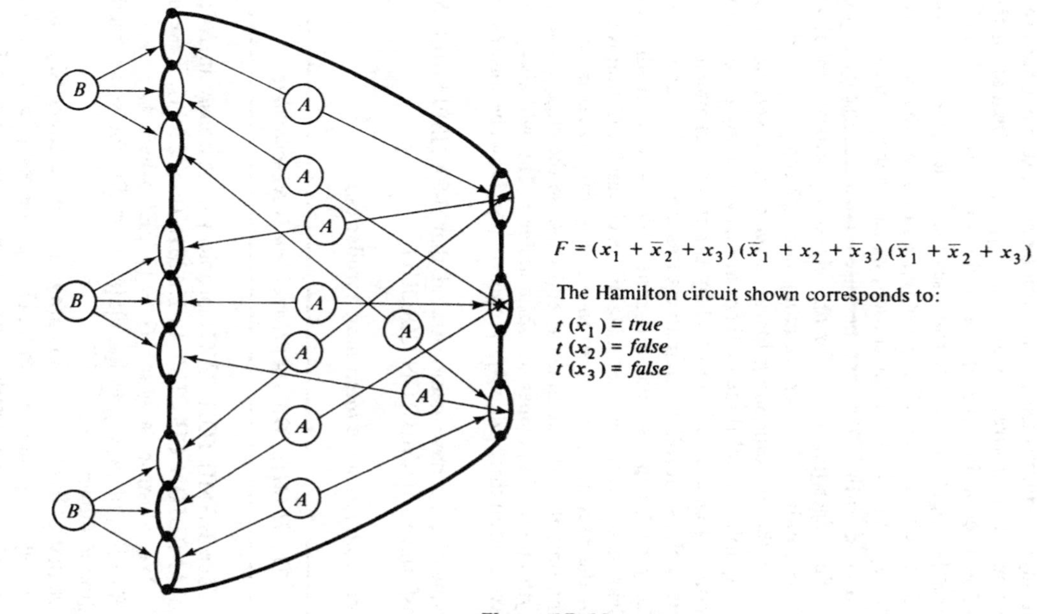
B
B
B
A
A
A
A
A
A
A
A
A
$F = (x_1 + \bar{x}_2 + x_3)(\bar{x}_1 + x_2 + \bar{x}_3)(\bar{x}_1 + \bar{x}_2 + x_3)$
The Hamilton circuit shown corresponds to:
$t(x_1) = \textit{true}$
$t(x_2) = \textit{false}$
$t(x_3) = \textit{false}$

Figure 15–10

right copy of $[v_k, w_k]$ if it is $\bar{x}_k$ (see Fig. 15–10). This completes the construction of G.

We shall now argue that the construction of G from F described above is such that G has a Hamilton circuit if and only if F is satisfiable. For the *only if* direction, suppose that G has a Hamilton circuit c. It is not hard to see that c must have a special structure: It traverses $[u_{11}, v_1]$ and then all of the v and w nodes top-down, choosing one of the copies of $[v_i, w_i]$ for all $i = 1, \ldots, n$, then traverses $[w_n, u_{m4}]$, and finally traverses the copies of B bottom-up. Think of the fact that c chooses the left copy of $[v_i, w_i]$ to mean that x_i takes the value *true*; otherwise, if the right copy is chosen, we say that x_i is *false*. The resulting function is a valid truth assignment, because c must traverse exactly one of the two copies. The edges $[u_{ij}, u_{i,j+1}]$ for any clause C_i behave similarly; they are traversed if and only if the corresponding copy of the $[v_k, w_k]$ edge is not; in other words, if the corresponding literal is *false*. However, because they are parts of a B graph, not all three of them are traversed by c—or, equivalently, the corresponding clause C_i is satisfied by the truth assignment. Because this holds for every clause, it follows that all clauses are satisfied, and F is satisfiable.

For the *if* part, suppose that F is satisfiable by some truth assignment t. It is then clear that we can construct a Hamilton circuit for G, simply by following the rules of the previous paragraph. In other words, traverse the left copy of $[w_i, v_i]$ if and only if x_i is *true* under t, and traverse the edge $[u_{ij}, u_{i,j+1}]$ if and only if the jth literal of C_i comes out *false* under t. This will always be possible without traversing all three $[u_{ij}, u_{i,j+1}]$ edges of any clause, because t was assumed to satisfy F.

Consequently, we have demonstrated that 3-SATISFIABILITY polynomially transforms to HAMILTON CIRCUIT. □

Theorem 15.6 has some interesting corollaries.

Corollary 1 *The HAMILTON PATH problem is NP-complete.*

Proof The HAMILTON PATH problem is obviously in NP; we shall transform HAMILTON CIRCUIT to HAMILTON PATH. Take any graph $G = (V, E)$. We construct a graph $G' = (V', E')$ where $V' = V \cup \{u, u', w\}$, and $E' = E \cup \{[u', u], [w, v_0]\} \cup \{[u, v]: [v_0, v] \in E\}$, for some fixed $v_0 \in V$. An illustration is shown in Fig. 15–11.

Suppose that G' has a Hamilton path p. The two extreme edges of p must be $[u', u]$ and $[v_0, w]$. Suppose now that $[u, v] \in p$; the rest of p is a path traversing each point in $V - \{v_0, v\}$ exactly once. Furthermore, because $[u, v] \in E'$, we also have $[v, v_0] \in E$. Therefore, this path—together with $[v_0, v]$—is a Hamilton circuit in G. Conversely, if G has a Hamilton circuit $c = [v_0, \ldots, v, v_0]$, we can find in G' the Hamilton path $p = [w, v_0, \ldots, v, u, u']$. Hence G has a Hamilton circuit if and only if G' has a Hamilton path; the proof is complete. □

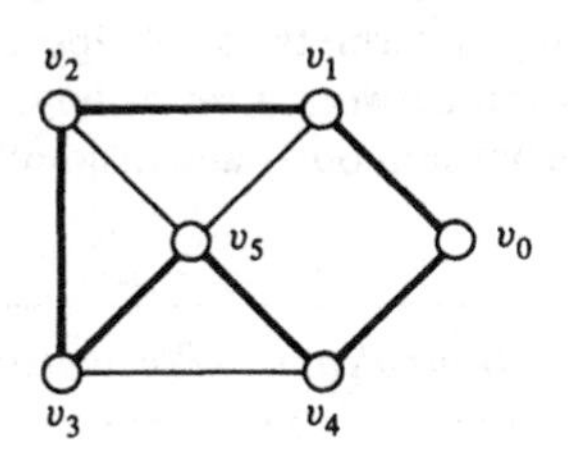

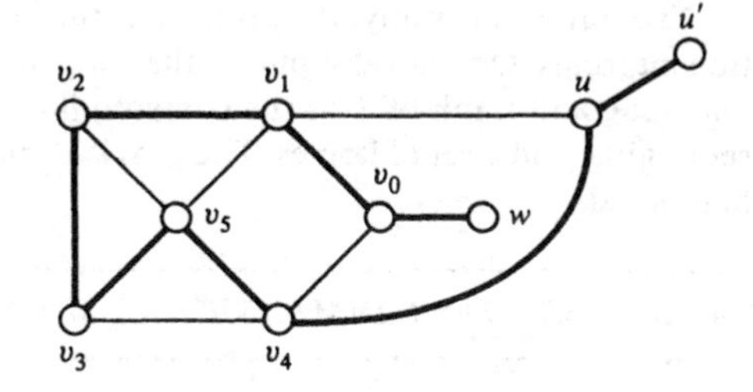

Figure 15–11

It is now easy to see that the *directed* versions of the problems in Theorem 15.6 and Corollary 1 above are also *NP*-complete. To show this, observe that, for the purposes of "path problems" such as the Hamilton circuit and Hamilton path problems, we can consider a graph as a special case of a digraph, namely, as one for which $(u, v) \in A$ if and only if $(v, u) \in A$. Thus the directed Hamilton problems are generalizations of the corresponding undirected problems, and hence they are all *NP*-complete. Also, according to the discussion of Chapter 12, we now observe that it is an *NP*-complete problem to tell, given three matroids and an integer k, whether there exists a set simultaneously independent in all three matroids with cardinality k. This is, again, because the DIRECTED HAMILTON PATH problem can be formulated as the intersection problem of a graphic matroid and two partition matroids. We shall see in the next section that the problem of intersecting three *partition* matroids is also *NP*-complete. Finally, we have the next corollary.

Corollary 2 *The TSP is NP-complete.*

Proof We shall show that, in effect, the HAMILTON CIRCUIT problem is a special case of the TSP. Towards this goal, given any graph $G = (V, E)$, we construct an instance of the $|V|$-city TSP by letting $d_{ij} = 1$ if $[v_i, v_j] \in E$, and 2 otherwise. We let the "budget" L be equal to $|V|$. It is immediate that there is a tour of length L or less if and only if there exists a Hamilton circuit in G. □

15.7 More *NP*-Complete Problems: Matching, Covering, and Partitioning

We now introduce a generalization of the bipartite matching problem discussed in Chapter 10.

3-DIMENSIONAL MATCHING

Given three sets U, V, and W of equal cardinality, and a subset T of $U \times V \times W$, is there a subset M of T with $|M| = |U|$ such that whenever (u, v, w) and (u', v', w') are distinct triples in M, $u \neq u'$, $v \neq v'$, and $w \neq w'$?

To put it differently, the problem is the intersection problem of three partition matroids. One can also pursue the "boys and girls" interpretation of bipartite matching and think of T as the compatibility relation between a set of boys, a set of girls, and a set of homes. The goal is to create harmonious—and disjoint—households.

Theorem 15.7 *The 3–DIMENSIONAL MATCHING problem is NP-complete.*

Proof The 3-DIMENSIONAL MATCHING problem is obviously in NP; furthermore, we can polynomially transform SATISFIABILITY to it, as follows. Let F be a Boolean formula involving the literals $x_1, \ldots, x_n$ and consisting of the clauses $C_1, \ldots, C_m$. We shall construct an instance (U, V, W, T) of 3-DIMENSIONAL MATCHING such that the required matching M exists if and only if F is satisfiable.

U consists of one copy of each literal for each clause:

$$U = \{x_i^j, \bar{x}_i^j : i = 1, \ldots, n; j = 1, \ldots, m\}$$

V contains three kinds of nodes:

$$V = \{a_i^j : i = 1, \ldots, n, \quad j = 1, \ldots, m\} \cup \{v_j : j = 1, \ldots, m\}$$
$$\cup \{c_i^j : j = 1, \ldots, m, \quad i = 1, \ldots, n-1\}$$

The structure of W is completely analogous to that of V:

$$W = \{b_i^j : i = 1, \ldots, n, \quad j = 1, \ldots, m\} \cup \{w_j : j = 1, \ldots, m\}$$
$$\cup \{d_i^j : j = 1, \ldots, m, i = 1, \ldots, n-1\}$$

T is the union of the following three kinds of triples (see Figure 15–12 for an illustration of the construction).

1. The triples (a_i^j, b_i^j, x_i^j), $i = 1, \ldots, n, j = 1, \ldots, m$, and $(a_i^{j+1}, b_i^j, \bar{x}_i^j)$, $i = 1, \ldots, m, j = 1, \ldots, m$ (here $a_i^{m+1} = a_i^1$). The a and b nodes do not participate in any other triples. Hence, for each i, these triples will force M to match either all of $\bar{x}_i^j$ (this means that x_i is *true*) or all of x_i^j (x_i is *false*) with a's and b's.
2. The second kind of triple of T is the set $\{(v_j, w_j, \lambda^j) : j = 1, \ldots, m$, λ a literal of $C_j\}$. These are the only triples involving the v and w nodes. Since at this point only *true*—according to our convention above—literals are available, the v and w nodes must be matched with true literals of their clauses; hence all clauses must be satisfied by this truth assignment.
3. The purpose of the c and d nodes, as well as of the last kind of triples (the "garbage collection" triples in the elegant terminology of [GJ]), is to "pick up" the remaining copies of different literals. These triples form the set $\{(c_i^j, d_i^j, \lambda^k) : i = 1, \ldots, n-1, j = 1, \ldots, m, k = 1, \ldots, m$, λ a literal$\}$.

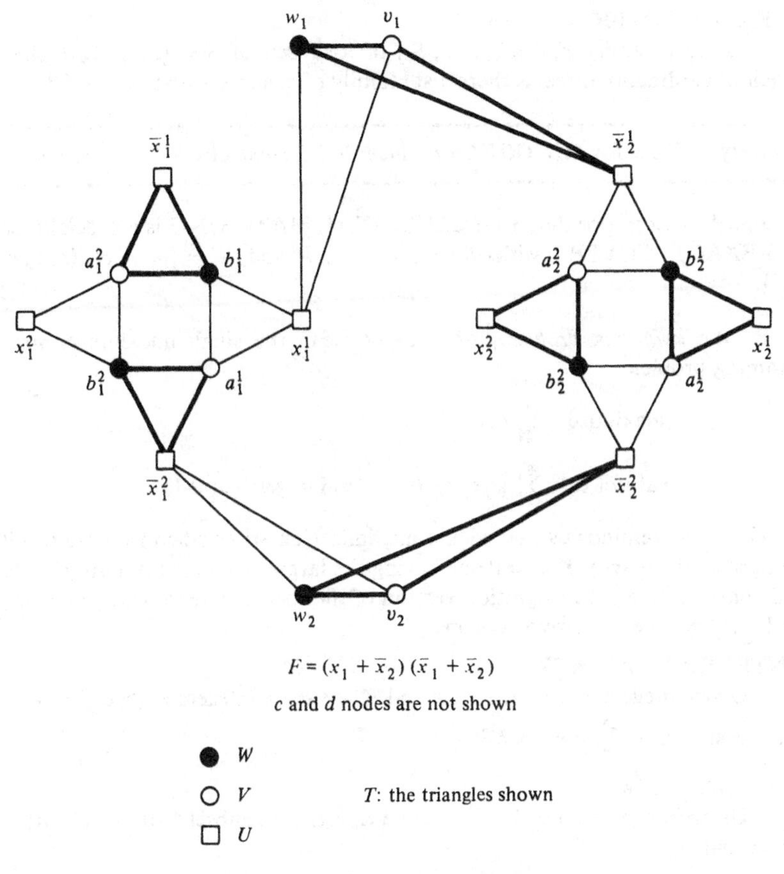

Figure 15–12

From the arguments above, it follows that whenever there exists a perfect matching M, F is satisfiable. Conversely, if F is satisfiable by a truth assignment t, we can match the a and b nodes with literals according to t; we can then match each of the v and w nodes with one true literal from the corresponding clause. Since t was assumed to satisfy F, such a literal will always exist. Finally, the rest of the literals are picked up by the c and d nodes. □

The following variation of 3-DIMENSIONAL MATCHING has been extremely useful as the point of departure for proving many *NP*-completeness results.

3-EXACT COVER

Given a family $F = \{S_1, \ldots, S_n\}$ of n subsets of $S = \{u_1, \ldots, u_{3m}\}$, each of cardinality three, is there a subfamily of m subsets that covers S?

Corollary *The 3-EXACT COVER problem is NP-complete.*

Proof Just notice that 3-DIMENSIONAL MATCHING is a special case of 3-EXACT COVER, with $S = U \cup V \cup W$ and $F = \{\{a, b, c\}: (a, b, c) \in T\}$. □

By the *knapsack problem*, we usually mean the single-line integer programming problem:

$$\text{maximize} \quad \sum_{j=1}^{n} c_j x_j$$

$$\text{subject to} \quad \sum_{j=1}^{n} w_j x_j \leq K, \qquad x_j \text{ integers (or 0-1).}$$

The name reminds us that such a maximization arises when we wish to fill a knapsack of capacity K with items having the largest possible total utility. We shall consider here the recognition version of the special case in which $c_j = w_j$, $j = 1, \ldots, n$. There are two versions.

INTEGER KNAPSACK

Given integers $c_j, j = 1, \ldots, n$ and K, are there integers $x_j \geq 0, j = 1, \ldots, n$ such that $\sum_{j=1}^{n} c_j x_j = K$?

0-1 KNAPSACK

Given integers $c_j, j = 1, \ldots, n$, and K, is there a subset S of $\{1, \ldots, n\}$ such that $\sum_{j \in S} c_j = K$?

Theorem 15.8 *The* 0-1 *KNAPSACK problem is NP-complete.*

Proof It is obviously in *NP*; we can show, furthermore, that 3-EXACT COVER polynomially transforms to 0-1 KNAPSACK. Starting with any family F of n sets of cardinality three, we shall construct integers $c_1, \ldots, c_n$ and K such that there is a subset of the c_i's summing to K if and only if there exists a subfamily of F covering exactly the universe $S = \{u_1, \ldots, u_{3m}\}$.

We can think of all sets in F as bit-vectors of length $3m$; for example, $\{u_1, u_5, u_6\}$ and $\{u_2, u_4, u_6\}$ become 100011 and 010101, respectively. Now interpret these bit vectors as *integers* written in the base-$(n + 1)$ system. In other words, the integer corresponding to a set S_j is

$$c_j = \sum_{u_i \in S_j} (n + 1)^{i-1}.$$

Also, let K be the integer corresponding to $\underbrace{11\ldots1}_{3m}$:

$$K = \sum_{j=0}^{3m-1} (n+1)^j$$

We now claim that there exists a subfamily of F covering $\{u_1, \ldots, u_{3m}\}$ exactly if and only if there is a subset of the c_j's adding up to K.

If Suppose that there exists a set $S \subseteq \{1, 2, \ldots, n\}$ such that $\sum_{j \in S} c_j = K$. In conducting this summation in base-$(n+1)$ arithmetic, we notice that only the digits 0 and 1 appear in the summands, and, moreover, that the number of summands is less than $n+1$, the base. Hence there is no "carry" in this addition. Consequently, if $\sum_{j \in S} c_j = K$, this means that there exists exactly one 1 in each of the $3m$ positions, or, equivalently, the subfamily $C = \{S_j : j \in S\}$ covers $\{u_1, u_2, \ldots, u_{3m}\}$ exactly.

Only if Starting from an exact cover C of $\{u_1, \ldots, u_{3m}\}$, we notice immediately that $\sum_{S_j \in C} c_j = K$.

This concludes the proof. □

We can also show that 0-1 KNAPSACK is *NP*-complete even if K is restricted to be $\frac{1}{2} \sum_{j=1}^{n} c_j$. This problem is better known as the PARTITION problem.

> Given integers $c_1, \ldots, c_n$, is there a subset $S \subseteq \{1, 2, \ldots, n\}$ such that $\sum_{j \in S} c_j = \sum_{j \notin S} c_j$?

Corollary 1 *PARTITION is NP-complete.*

Proof We shall polynomially transform 0-1 KNAPSACK to PARTITION. Given any instance $c_1, \ldots, c_n, K$ of 0-1 KNAPSACK, we construct the following instance of PARTITION: $c_1, \ldots, c_n, c_{n+1} = 2M, c_{n+2} = 3M - 2K$, where $M = \sum_{j=1}^{n} c_j > K$. We claim that there exists a subset S of $\{1, 2, \ldots, n\}$ with $\sum_{j \in S} c_j = K$ if and only if there exists a subset S' of $\{1, 2, \ldots, n+2\}$ such that

$$\sum_{j \in S'} c_j = \sum_{j \notin S'} c_j$$

If In any feasible partition S' of $\{c_1, \ldots, c_{n+2}\}$, c_{n+1} and c_{n+2} must be separated, because they add up to $5M - 2K > \sum_{j=1}^{n} c_j$. Hence we have, for $S = S' - \{n+1, n+2\}$,

$$\sum_{j \in S} c_j + c_{n+2} = \sum_{\substack{j \notin S \\ j \neq n+1, n+2}} c_j + c_{n+1}$$

It follows directly from the arithmetic that $\sum_{j \in S} c_j = K$.

Only if Suppose that $\sum_{j \in S} c_j = K$ for some $S \subseteq \{1, 2, \ldots, n\}$. Then immediately

$$\sum_{j \in S} c_j + c_{n+2} = \sum_{j \notin S} c_j + c_{n+1}$$ □

Corollary 2 *INTEGER KNAPSACK is NP-complete.*

Proof INTEGER KNAPSACK is in *NP*, since it is a special case of ILP. To show *NP*-completeness, we shall polynomially transform 0-1 KNAPSACK to INTEGER KNAPSACK. Suppose that we are given an instance $\{c_1, \ldots, c_n, K\}$ of 0-1 KNAPSACK. Naturally, we can assume that $1 \leq c_j \leq K$ for $j = 1, \ldots, n$ and $K > 0$. We shall construct an instance $\{d_1, \ldots, d_{2n}, L\}$ of INTEGER KNAPSACK such that there exist integers $y_1, \ldots, y_{2n} \geq 0$ with

$$\sum_{j=1}^{2n} d_j y_j = L$$

if and only if there exist $x_1, \ldots, x_n \in \{0, 1\}$ such that

$$\sum_{j=1}^{n} c_j x_j = K$$

Let $M = 2n(n + 1) \cdot K$. We define the d_j's as follows:

$$d_j = \begin{cases} M^{n+1} + M^j + c_j & \text{if } j \leq n \\ M^{n+1} + M^{j-n} & \text{otherwise} \end{cases}$$

Also, $L = n \cdot M^{n+1} + \sum_{j=1}^{n} M^j + K$. This completes the construction of the INTEGER KNAPSACK instance, starting from the instance of 0-1 KNAPSACK.

Suppose that the constructed instance of INTEGER KNAPSACK has a solution $\{y_1, \ldots, y_{2n}\}$; in other words,

$$\sum_{j=1}^{2n} d_i y_j = L$$

or

$$M^{n+1} \sum_{j=1}^{2n} y_j + \sum_{j=1}^{n} M^j (y_j + y_{j+n}) + \sum_{j=1}^{n} c_j y_j = n \cdot M^{n+1} + \sum_{j=1}^{n} M^j + K \quad (15.1)$$

Now, each y_j must be smaller than the quantity

$$\frac{L}{d_j} < n + 1 \leq \frac{M}{2nK}$$

Consequently

$$\sum_{j=1}^{2n} y_j, y_j + y_{j+n}, \sum_{j=1}^{n} c_j y_j \; < \; M$$

It follows that (15.1) is an equation involving positive linear combinations of the powers of a large integer M, with all coefficients less than M. It is immediate that the coefficients corresponding to like powers of M must be equal. Hence

$$\sum_{j=1}^{2n} y_j = n, \qquad y_j + y_{j+n} = 1 \qquad \text{for all } j$$

and

$$\sum_{j=1}^{n} c_j y_j = K$$

The first two equalities show that exactly one of y_j, y_{j+n} must be 1 for all j, and the other must be zero. Consequently, the third equality shows that $\{y_1, \ldots, y_n\}$ is a solution to the original 0-1 problem.

Conversely, starting with a solution $\{x_1, \ldots, x_n\}$ of the 0-1 KNAPSACK instance, we can construct a solution $\{y_1, \ldots, y_{2n}\}$ for INTEGER KNAPSACK by taking $y_j = x_j, j = 1, \ldots, n$, and $y_j = 1 - x_{j-n}, j = n + 1, \ldots, 2n$. Consequently, the constructed instance of INTEGER KNAPSACK has a solution iff the original instance of 0-1 KNAPSACK has one. Hence, INTEGER KNAPSACK is *NP*-complete. □

PROBLEMS

1. Formalize the maximum matching problem for weighted graphs as an optimization problem by describing appropriate algorithms $\mathcal{A}_F$ and $\mathcal{A}_c$.

2. Prove that if we had a polynomial-time algorithm for *computing the length* of the shortest TSP tour, then we would have a polynomial-time algorithm for *finding* the shortest TSP tour.

3. Consider the following problem.

   GRAPH COLORING
   Given a graph $G = (V, E)$ and an integer k, is there a mapping $\chi: V \Rightarrow \{1, 2, \ldots, k\}$ such that $[v, u] \in E$ implies $\chi(v) \neq \chi(u)$?

   Show that GRAPH COLORING is in *NP* by giving a detailed description of a certificate and a certificate-checking algorithm.

4. Give a *direct* proof that ILP is *NP*-complete by showing that any problem in *NP* polynomial-time transforms to ILP.

5. Show that SATISFIABILITY is *NP*-complete even if each variable is restricted to appear once negated and once or twice unnegated (a solution is given in the next chapter).

6. Let F be a formula consisting of clauses with two literals each. From F, let us construct a directed graph $D(F) = (X, A)$, as follows: X is the set of variables that appear in F and their negations. There is an arc $(\lambda_1, \lambda_2) \in A$ iff the clause $(\bar{\lambda}_1 + \lambda_2)$ is in F.

(a) Show that if, for some variable x, x and $\bar{x}$ are in the same *strongly connected component* (see Problems 4 and 5 of Chapter 9) of $D(F)$, then F is unsatisfiable.

*(b) Show the converse of (a).

(c) Give an $O(n)$ algorithm for solving the SATISFIABILITY problem restricted to formulas with 2-literal clauses.

7. Give an explicit construction of a certificate-checking algorithm (as defined in Sec. 15.5) that accepts strings of the form $\underbrace{00\ldots0}_{n}\ \$\underbrace{11\ldots1}_{2n}$ for $n \geq 1$.

8. Give a sequence of instructions in our model of certificate-checking algorithms that *simulates* each of the following more elaborate instructions.

(a) l: **if** σ **then** $(\sigma'; o'; l')$ **else** $(\sigma''; o''; l'')$

(b) l: **do** the following instruction **while** the scanned symbol is not a \$.

(c) l: **go to** l'.

9. Show how to avoid the possibility of the head of a certificate-checking algorithm falling out of bounds by making an appropriate first move, and also assuming that $c(x)$ has a special form.

***10.** Show that GRAPH COLORING (Problem 3) is *NP*-complete.

11. Show that the following six problems are *NP*-complete (compare (a) and (b) with Problem 12 of Chapter 12).

Given a graph $G = (V, E)$, a set $L \subseteq V$, and an integer k, is there a spanning tree T of G such that

(a) The set of leaves of T is L?

(b) There are no leaves of T outside L?

(c) T has k leaves?

(d) T has at most k leaves?

*(e) T has at least k leaves?

(f) The nodes of T have degree at most k?

(*Hint:* The HAMILTON PATH problem is useful for all except (e).)

***12.** Show that the following problems are *NP*-complete.

(a) FEEDBACK VERTEX SET
Given a digraph $D = (V, A)$ and an integer k, is there a subset F of V such that $|F| \leq k$ and the digraph resulting from D by omitting the vertices of F is acyclic?

(b) FEEDBACK ARC SET
This is the same as (a), except that F is now a set of arcs.

(*Hint:* Start from NODE COVER.)

***13.** Show that the following problem is *NP*-complete.

MAX-CUT

Given a graph $G = (V, E)$ and an integer k, is there a partition of V into V_1 and V_2 such that there are at least k edges in E between V_1 and V_2?

***14.** Show that the following problem is *NP*-complete.

MIXED CHINESE POSTMAN

Given a *mixed graph* $M = (V, E, A)$—where E is a set of edges and A a set of arcs on V—an integer L and a weight $w: E \cup A \longrightarrow Z^+$, is there a walk of M that includes every edge and arc and has total weight at most L?

Compare with Problems 8 and 9 in Chapter 11.

15. Show that the following problem is *NP*-complete.

MULTICOMMODITY FLOW

Given a digraph $D = (V, A)$ and $2k$ nodes $s_1, s_2, \ldots, s_k, t_1, t_2, \ldots, t_k$ in V, are there node-disjoint directed paths from s_1 to t_1, s_2 to $t_2, \ldots,$ and s_k to t_k?

***16.** (a) Show that the following simplified form of *quadratic programming* is at least as hard as SATISFIABILITY.

QUADRATIC PROGRAMMING

Given integer matrices A, Q, and integer vector b, find a real vector x such that

$$x'Qx = \min$$
$$Ax \leq b$$

(b) Give a polynomial-time algorithm for the problem in (a) in the special case that Q is positive-definite. (*Hint:* Use the ellipsoid algorithm of Sec. 8.7.)

17. An interesting special case of the TSP is the *geometric* (or *Euclidean*) case, in which we are given n points in the plane with integer coordinates, and we wish to find the shortest tour, when d_{ij} is the Euclidean distance

$$\sqrt{(x_i - x_j)^2 + (y_i - y_j)^2}$$

(a) Suppose that we formulate this case of the TSP as a recognition problem called ETSP. Explain why it is not easy to argue that ETSP $\in$ *NP*. (*Hint:* Calculate the length of the perimeter of the triangle below up to the fifth, and then the seventh, significant digit.)

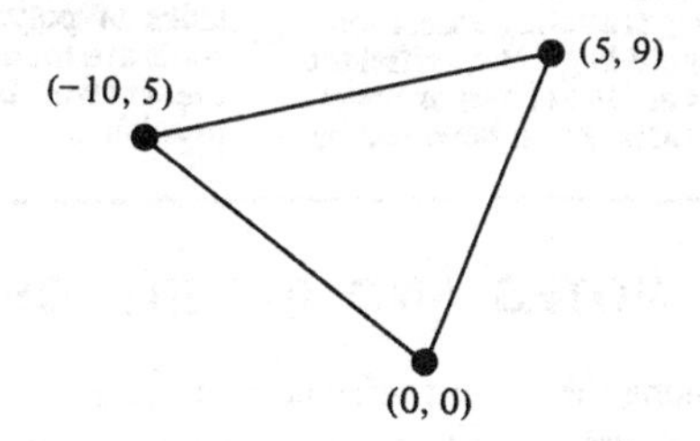

*(b) Define $d_{ij} = \lfloor \sqrt{(x_i - x_j)^2 + (y_i - y_j)^2} \rfloor$. Show that ETSP, under this definition, is *NP*-complete.

18. Consider the problem of finding the shortest path from s to t in a weighted digraph, when negative weights are allowed (recall Chapter 6).

(a) Show that there is a polynomial-time algorithm for this problem under the restriction that no cycles of negative total weight exist.

(b) Show that without this restriction, the problem is *NP*-complete. (*Hint:* Use the TSP.)

19. The following is from the *New York Times* of November 27, 1979.† Determine, when possible, whether each statement is (a) true, (b) false, (c) misleading, (d) equivalent to a well-known conjecture, the solution of which was probably not known to Mr. Browne.

An Approach to Difficult Problems

Mathematicians disagree as to the ultimate practical value of Leonid Khachiyan's new technique, but concur that in any case it is an important theoretical accomplishment.

Mr. Khachiyan's method is believed to offer an approach for the linear programming of computers to solve so-called "traveling salesman" problems. Such problems are among the most intractable in mathematics. They involve, for instance, finding the shortest route by which a salesman could visit a number of cities without his path touching the same city twice.

Each time a new city is added to the route, the problem becomes very much more complex. Very large numbers of variables must be calculated from large numbers of equations using a system of linear programming. At a certain point, the compexity becomes so great that a computer would require billions of years to find a solution.

In the past, "traveling salesmen" problems, including the efficient scheduling of airline crews or hospital nursing staffs, have been solved on computers using the "simplex method" invented by George B. Dantzig of Stanford University.

As a rule, the simplex method works well, but it offers no guarantee that after a certain number of computer steps it will always find an answer. Mr. Khachiyan's approach offers a way of telling right from the start whether or not a problem will be soluble in a given number of steps.

Two mathematicians conducting research at Stanford already have applied the Khachiyan method to develop a program for a pocket calculator, which has solved problems that would not have been possible with a pocket calculator using the simplex method.

Mathematically, the Khachiyan approach uses equations to create imaginary ellipsoids that encapsulate the answer, unlike the simplex method, in which the answer is represented by the intersections of the sides of polyhedrons. As the ellipsoids are made smaller and smaller, the answer is known with greater precision. MALCOLM W. BROWNE

NOTES AND REFERENCES

The following are among the earliest references to *P*, *NP*, and related concepts.

[Co] COBHAM, A., "The Intrinsic Computational Difficulty of Functions," pp. 24–30 in *Proc. 1964 Int. Congress for Logic Methodology and Phil. of Science*, ed. Y. Bar-Hillel. Amsterdam: North Holland, 1964.

[Ed1] EDMONDS, J., "Paths, Trees and Flowers," *Canad. J. Math.* 17 (1965), 449–67.

[Ed2] ———, "Minimum Partition of a Matroid in Independent Subsets," *J. Res. NBS*, 69B (1975), 67–72.

In the last two papers, Edmonds informally defines P and NP, he conjectures that $P \neq NP$, and in fact that TSP $\in NP - P$.

Reduction, without time bounds, has been a well-known technique in the theory of computation since the 1930s. Polynomial-time reductions have also been used by researchers in combinatorial optimization, albeit for the purpose of demonstrating that a problem is easy, not hard. For an early use of reduction in the other direction, see

[DBR] DANTZIG, G. B., W. O. BLATTNER, and M. R. RAO, "All Shortest Routes from a Fixed Origin in a Graph," pp. 85–90 in *Theory of Graphs: An International Symposium.* New York: Gordon & Breach, Inc., 1967.

The theory of *NP*-completeness started with Cook's paper

[Cook] COOK, S. A., "The Complexity of Theorem Proving Procedures," *Proc. 3rd ACM Symp. on the Theory of Computing*, ACM (1971), 151–158.

where Theorems 15.1 and 15.2 were proved. However, the wealth of the consequences of Cook's work and its close relationship to combinatorial optimization were made clear in the classical paper by Karp:

[Ka1] KARP, R. M., "Reducibility among Combinatorial Problems," pp. 85–103 in *Complexity of Computer Computations*, ed. R. E. Miller and J. W. Thatcher. New York: Plenum Press, 1972.

Among many other *NP*-completeness results, the original proofs to Theorems 15.3, 15.4, 15.6, 15.7, and 15.8 and the solutions of Problems 10, 12, and 13 can be found in this paper. An excellent tutorial on *NP*-completeness is the paper

[Ka2] KARP, R. M., "On the Complexity of Combinatorial Problems," *Networks*, 5 (1975), 45–68.

The encyclopedia of the subject is the delightful book by Garey and Johnson:

[GJ] GAREY, M. R., and D. S. JOHNSON, *Computers and Intractability: A Guide to the Theory of NP-completeness.* San Francisco: W. H. Freeman & Company, Publishers, 1979.

The reader is referred to this book for appreciating the richness and variety of the class of *NP*-complete problems. Theorem 15.5 was first proved in

[Ul] ULLMAN, J. D., "*NP*-complete Scheduling Problems," *JCSS*, 10 (1975), 384–93.

Our proof follows

[LR] LENSTRA, J. K., and A. H. G. RINOOY KAN, "Complexity of Scheduling under Precedence Constraints," *OR*, 26 (1978), 22–35.

Corollary 2 to Theorem 15.8 is from

[Lu] LUEKER, G. S., "Two *NP*-complete Problems in Nonnegative Integer Programming," TR 178, Princeton University, 1975.

Problem 11(e) is due to M. R. Garey and D. S. Johnson. Problem 14 is from

[Pa1] PAPADIMITRIOU, C. H., "On the Complexity of Edge Traversing," *J. ACM* 23 (1976), 544–54.

A solution to Problem 15 due to D. E. Knuth is reported in [Ka2]. Problem 17 is from

[Pa2] PAPADIMITRIOU, C. H., "The Euclidean TSP is *NP*-complete," *Theor. Comp. Sci.*, 4 (1977), 237–44.

Problem 18(b) is from [DBR]. The reader's confidence in the generality of our model of certificate-checking algorithms (Sec. 15.5) can be enhanced by examining Problems 7–9. Our model is a version of the Turing machine; see

[Tu] TURING, A. M., "On Computable Numbers, with an Application to the Entscheidungsproblem," *Proc. London Math. Soc. Ser. 2*, 47 (1936), 730–65.

The equivalence of this and other models is discussed in

[AHU] AHO, A. V., J. E. HOPCROFT, and J. D. ULLMAN, *The Design and Analysis of Computer Algorithms*, Chapter 1. Reading, Mass: Addison-Wesley Publishing Co., Inc., 1974.

16

More About NP-Completeness

In this chapter, we continue our discussion of *NP*-complete problems by addressing some important issues that surround the notion of *NP*-completeness.

16.1 The Class *co-NP*

Recall the following familiar *yes-no* problem:

HAMILTON CIRCUIT
Given a graph $G = (V, E)$, is G Hamiltonian? (That is, does G have a Hamilton circuit?)

It was routine to show that Hamilton circuit is in *NP*, since every *yes* instance has a concise certificate—namely, the circuit itself. Consider, however, the following *no-yes* version of the same problem:

HAMILTON CIRCUIT COMPLEMENT
Given a graph $G = (V, E)$, is G non-Hamiltonian?

It is not clear at all that this problem is in *NP*. In fact, we shall soon see that, most likely, it is not! The only general method known to date for demonstrating that a graph is non-Hamiltonian consists essentially of systematically listing all circuits of G and verifying that none contains all of the nodes. This list is certainly a certificate, but unfortunately one of exponential length.

In general, Problem $\bar{A}$ is the *complement* of Problem A if the set of strings with symbols in Σ that are encodings of *yes* instances of $\bar{A}$ are exactly those that are *not* encodings of *yes* instances of A. Thus, strictly speaking, HAMILTON CIRCUIT COMPLEMENT is *not* the complement of HAMILTON CIRCUIT, because there are strings that are neither encodings of *yes* instances of A nor of $\bar{A}$: These are the strings that fail to encode a graph at all—Hamiltonian or not. However, these other instances do not affect the complexity of the problem, since they can always be detected by a simple "syntactic check" and disposed of promptly. Consequently, it is an acceptable practice to disregard this point and consider the two problems above as complements of one another. For another example, the complement of TSP is the following problem:

TSP COMPLEMENT
Given n, $n \times n$ integer matrix $[d_{ij}]$, and integer L, is it true that for all cyclic permutations (tours) τ,

$$\sum_{j=1}^{n} d_{j\tau(j)} > L?$$

Again, it is not at all obvious how we can construct a certificate for a *yes* instance of TSP COMPLEMENT, short of exhibiting all tours, together with the corresponding costs, all larger than L. However, consider the following complement of a problem known to be in P:

CONNECTEDNESS COMPLEMENT
Given a graph G, is it disconnected?

The *search* algorithm that solves the connectedness problem in polynomial time obviously can be used to solve its complement. Thus, CONNECTEDNESS COMPLEMENT is in P as well. It is very easy to prove a general statement to this effect.

Theorem 16.1 *If* A *is a problem in P, then the complement* $\bar{A}$ *of* A *is also in P.*

Proof Since A is in P, then there is a polynomial algorithm that solves A. A polynomial algorithm for solving the complement of A is exactly the same algorithm, only with the substitution of *no* whenever *yes* was previously reported, and vice-versa. □

The same argument *cannot* be applied to show that the complement of any problem in NP is also in NP. This is because there is a certain *asymmetry* in the very definition of NP: If x is a *yes* instance of A $\in NP$, then it has a certificate so indicating; but if it is a *no* instance, it may not have a concise certificate to prove it. In fact, as we have pointed out in our examples, there are at least two problems in *NP* whose complements may very well *not* be in *NP*. This motivates the following definition.

Definition 16.1

The class *co-NP* is the class of all problems that are complements of problems in *NP*. □

Saying that the complements of all problems in *NP* are also in *NP* is thus equivalent to saying that $NP = co\text{-}NP$. However, there is reason to believe that $NP \neq co\text{-}NP$. The evidence is as circumstantial as that pointing to the conjecture $P \neq NP$: Many able researchers have tried for a long time, without success, to construct succinct proofs for the non-Hamiltonian property, as well as for many other complements of succinctly certifiable properties. Nevertheless, as in the case of the $P \neq NP$ conjecture, no proof is yet available. However, it can be shown (as with $P \neq NP$) that if this conjecture is true, then it is the *NP*-complete problems that will testify to its validity.

Theorem 16.2 *If the complement of an NP-complete problem is in NP, then* $NP = co\text{-}NP$.

Proof Suppose that the complement $\bar{C}$ of an *NP*-complete problem C is in *NP*; we shall show that then the complement $\bar{A}$ of *any* problem A in *NP* is also in *NP*.

Because C is *NP*-complete, we know that A is polynomially transformed to it; notice that this transformation is also a polynomial transformation from $\bar{A}$ to $\bar{C}$. We can thus exhibit a concise certificate for any *yes* instance of $\bar{A}$: It consists of (a) a record of the operation of this polynomial transformation, resulting in a *yes* instance of $\bar{C}$, and (b) a certificate for this instance of $\bar{C}$—such a certificate exists, because $\bar{C}$ is in *NP*. The whole certificate is polynomially concise, since the transformation is by hypothesis a polynomial-time one.

It follows that $\bar{A}$ is in *NP*. Since A was taken to be *any* problem in *NP*, this implies that $NP = co\text{-}NP$. □

Thus, of all problems in *NP*, the *NP*-complete problems are those with complements least likely to be in *NP*. Conversely, if the complement of a problem in *NP* is also in *NP*, this is evidence that the problem is not *NP*-complete. Recall, for example, the following recognition problem, shown in Sec. 8.7 to be equivalent to Linear Programming:

LINEAR INEQUALITIES (LI)
Given an $m \times n$ integer matrix A and an m-vector b with integer entries, is there a rational n-vector x such that $Ax \leq b$?

What is the complement of LI? Duality theory says that $Ax \leq b$ is infeasible if and only if the dual program

$$\begin{aligned} &\min y'b \\ &y'A = 0 \\ &y \geq 0 \end{aligned} \tag{16.1}$$

is unbounded (it *is* feasible, with $y = 0$). Here we applied the dual construction to the program

$$\begin{aligned} &\max 0 \cdot x \\ &Ax \leq b \\ &x \gtrless 0 \end{aligned}$$

However (16.1) is unbounded if and only if it has *any* solution with negative cost. Hence, the *no-yes* version of the instance $Ax \leq b$ of LI is *itself* another instance of LI, namely that with inequalities

$$\begin{aligned} y'A &\leq 0 \\ y'A &\geq 0 \\ y &\geq 0 \\ y'b &\leq -1 \end{aligned}$$

We conclude that LI *is its own complement*, and thus it is in both the classes *NP* and *co-NP*.

Naturally, this is not too impressive a result, because we now know that LI is in *P* (Chapter 8). However, in the years of uncertainty that preceded the ellipsoid algorithm, this fact—in view of Theorem 16.2—was the only available *theoretical* evidence that LP is not an intractable problem.

We can use our understanding of the issues and conjectures related to *co-NP* to make a more detailed chart of *NP* than the one of Fig. 15–2 (See Fig. 16–1.)

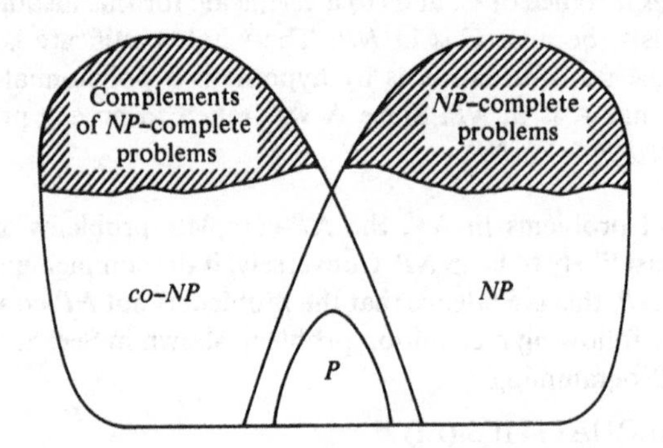

Figure 16–1 An updated conjectured topography of *NP*.

16.2 Pseudo-Polynomial Algorithms and "Strong" *NP*-Completeness

Recall the INTEGER KNAPSACK problem, which we proved in Chapter 15 to be *NP*-complete. Let $c_1, \ldots, c_n, K$ be an instance of this problem; the question is, of course, whether there exist integers $x_1, x_2, \ldots, x_n \geq 0$ such that

$$\sum_{j=1}^{n} c_j x_j = K$$

Given any such instance, we can construct a digraph $G(c_1, \ldots, c_n; K) = (V, A)$ as follows

$$V = \{0, 1, 2, \ldots, K\}$$
$$A = \{(m, k): 0 \leq m < k \leq K \quad \text{and} \quad k - m = c_j \text{ for some } j \leq n\}$$

Thus $G(c_1, \ldots, c_n; K)$ has $K + 1$ nodes and $O(nK)$ arcs. For example, the graph $G(3, 7; 13)$ is shown in Fig. 16–2.

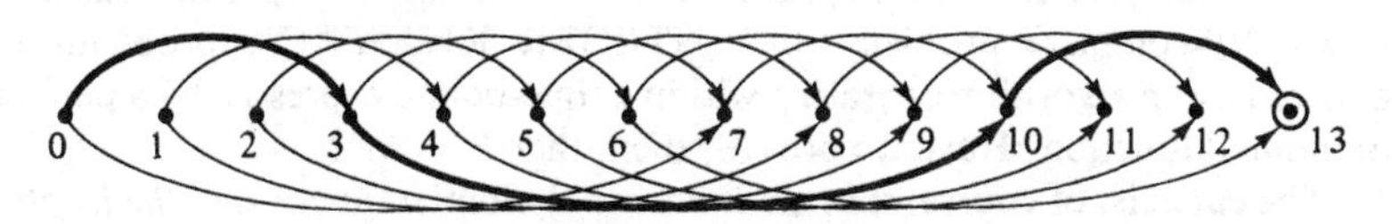

Figure 16–2

Lemma 16.1 *There is a path from* 0 *to* K *in* $G(c_1, \ldots, c_n; K)$ *if and only if the instance* $(c_1, \ldots, c_n, K)$ *of INTEGER KNAPSACK has a solution.*

Proof Suppose that $(0 \equiv i_0, i_1, \ldots, i_m \equiv K)$ is a path in G. Consider the sequence $(\delta_1, \ldots, \delta_m) = (i_1 - i_0, \ldots, i_m - i_{m-1})$. Now, $\delta_1, \ldots, \delta_m$ are all among $\{c_1, \ldots, c_n\}$, by the definition of G. Furthermore,

$$\sum_{i=1}^{m} \delta_i = K$$

It follows that

$$\sum_{j=1}^{n} c_j x_j = K$$

has a nonnegative integer solution, namely with x_j taken to be the number of times that c_j appears in $(\delta_1, \ldots, \delta_m)$. Conversely, if

$$\sum_{j=1}^{n} c_j x_j = K$$

for nonnegative integers $x_1, \ldots, x_n$, then we can recover a path from 0 to K in $G(c_1, \ldots, c_n; K)$ by taking

$$(\delta_1, \ldots, \delta_m) = (\underbrace{c_1, c_1, \ldots, c_1}_{x_1 \text{ times}}, \underbrace{c_2, \ldots, c_2}_{x_2 \text{ times}}, \ldots, \underbrace{c_n, \ldots, c_n}_{x_n \text{ times}})$$

The 0-K path is $(0 \equiv i_0, \ldots, i_m \equiv K)$, where

$$i_j = \sum_{l=1}^{j} \delta_l$$ □

This suggests the following result.

Theorem 16.3 *Any instance* $(c_1, \ldots, c_n, K)$ *of the INTEGER KNAPSACK problem can be solved in* $O(nK)$ *time.*

Proof Given $(c_1, \ldots, c_n, K)$, we construct in $O(nK)$ time the digraph $G(c_1, \ldots, c_n, K)$. We then determine in $O(nK)$ time whether there exists a path from 0 to K using our algorithm findpath of Chapter 9. By the above lemma, this solves the original instance of INTEGER KNAPSACK. □

Theorem 16.3 has the appearance of an extremely important result. A known *NP*-complete problem—the INTEGER KNAPSACK problem—can be solved by an algorithm operating within a time-bound expressed by a perfectly polynomial function! Have we proved, then, that $P = NP$?

The catch is, of course, that nK is *not* a polynomial function *of the length of the input*. It takes about $n \log K$ space to record the encoding of an instance of the INTEGER KNAPSACK problem, because integers can be written down in binary—or decimal, for that matter. Therefore, $O(nK)$ is *not* a polynomial bound. Nevertheless, Theorem 16.3 still stands as a very interesting curiosity. It makes our dichotomy of algorithms into practical and impractical (depending on whether they run in polynomial time in the length of the input) seem somewhat arbitrary. Under certain circumstances, the $O(nK)$ algorithm could be considered to be "practical," despite its exponential time performance. For example, suppose that in some application of INTEGER KNAPSACK, we seek to maximize the utility or money gained by an enterprise. What the $O(nK)$ time bound says is that in order to achieve this maximization, we have to spend a fixed—and, it is to be hoped, very small—percentage of the profits for purchasing computer time. And this may be considered a reasonable and practically feasible policy. To make another point, we shall see in the next chapter that "pseudo-polynomial" algorithms such as the $O(nK)$ algorithm above may be indicative of further positive algorithmic properties of the problem under examination, despite the fact that it is *NP*-complete. It appears, therefore, that certain exponential algorithms are not so catastrophic, after all.

What the above arguments suggest is that the length of the encoding may not be the *only* meaningful measure of the "size" of an instance. The example of INTEGER KNAPSACK leads to the following definition.

Definition 16.2

Let I be an instance of a computational problem—typically I will be a squence of combinatorial objects such as graphs, sets, or integers. Then number(I) is the *largest integer* appearing in I. □

For example, an instance $I = (c_1, \ldots, c_n, K)$ of INTEGER KNAPSACK has number(I) $= K$ (assuming, as is reasonable, that $c_1, \ldots, c_n \leq K$). If $I = (G, k)$ is an instance of the CLIQUE problem, then number(I) $= k$.† Let $I = (n, [d_{ij}], L)$ be an instance of the TSP. Then number(I) is the largest integer among $\{n, |d_{ij}|, i = 1, \ldots, n, j = 1, \ldots, n, L\}$. One very important issue manifests itself immediately after this definition: Certain problems, such as CLIQUE, have instances with naturally limited number; it does not make sense for an instance $(G = (V, E), k)$ of CLIQUE to have $k > |V|$. Some other hard problems, such as the TSP, have no natural limitations on the size of the integers appearing in their instances. However, these unlimited integers are not essential to the complexity of the problem: In order to show the *NP*-completeness of the TSP, the largest integer that we needed to construct was n (recall the proof of Corollary 2 to Theorem 15.6). On the other hand, we have problems such as INTEGER KNAPSACK, which owe their complexity exactly to the unlimited size of the integers appearing in their instances. In the example of INTEGER KNAPSACK, for instance, we needed to resort to the construction of huge integers in order to establish *NP*-completeness via Theorem 15.8 and its Corollaries. This seems to be necessary, especially in view of Theorem 16.3. These considerations lead to the following definition.

Definition 16.3

Let A be a computational problem and f a function mapping N to N. We use A_f to denote A restricted to instances I for which number(I) $\leq f(|I|)$, where $|I|$ is the length of the encoding of instance I. We say A is *strongly NP-complete* if A_p is *NP*-complete for some polynomial $p(n)$. □

Example 16.1

CLIQUE is strongly *NP*-complete, because CLIQUE_n is the same problem as CLIQUE—recall that $k =$ number(I) $\leq |V|$ for any meaningful instance $I = ((V, E), k)$ of CLIQUE. Similarly, TSP is strongly *NP*-complete, because, as the proof of Corollary 2 to Theorem 15.6 shows, it remains *NP*-complete even if we restrict the integers involved to be bounded by the number of cities. Also

†If $I = ((V, E), k)$ is an instance of CLIQUE, one may argue that number(I) $= |V|$, because one must use subscripts of size up to $|V|$ to represent the nodes of (V, E). This point, however, is of no consequence in this discussion.

strongly NP-complete are the problems HAMILTON CIRCUIT, 3-DIMENSIONAL MATCHING, MULTIPROCESSOR SCHEDULING, and 3-EXACT COVER, as well as all other NP-complete problems that can be shown to be NP-complete by transformations that do not use exponentially large integers. In contrast, 0-1 KNAPSACK, PARTITION, and INTEGER KNAPSACK do not fall in this category. □

Definition 16.4

An algorithm $\mathcal{A}$ for a problem A is *pseudo-polynomial* if it solves any instance I of A in time bounded by a polynomial in $|I|$ *and* number(I). □

Thus the $O(nK)$ algorithm that solves the INTEGER KNAPSACK problem is clearly a pseudo-polynomial algorithm. As we have already mentioned, the existence of a pseudo-polynomial algorithm for a problem is in many ways a positive fact. Our next theorem establishes that strong NP-completeness makes the existence of a pseudo-polynomial algorithm extremely unlikely, exactly as NP-completeness makes the existence of polynomial algorithms unlikely.

Theorem 16.4 *Unless $P = NP$, there can be no pseudo-polynomial algorithm for any strongly NP-complete problem.*

Proof Suppose that A is a strongly NP-complete problem; in other words, $A_{p(n)}$ is NP-complete for some polynomial $p(n)$. Furthermore, suppose that there exists a pseudo-polynomial algorithm $\mathcal{A}$ for A, which solves any instance I of A in time $q(|I|, \text{number}(I))$, for some bivariate polynomial q. It is then clear that $\mathcal{A}$ solves the NP-complete problem $A_{p(n)}$ in time $q(n, p(n))$, a polynomial. This is impossible unless $P = NP$. □

Thus, proving a problem to be *strongly* NP-complete rules out—modulo the $P \neq NP$ conjecture—not only the existence of polynomial algorithms, but that of pseudo-polynomial algorithms. In problems such as CLIQUE and HAMILTON CIRCUIT, in which numbers play a minimal role, this stronger variety of NP-completeness comes about naturally. Curiously, however, there are some strongly NP-complete problems in which integers seem to play a central part. A paradigm is the following problem.

THREE PARTITION
Given $3n$ integers $\{c_1, \ldots, c_{3n}\}$, is there a partition of these integers into n triples $T_1, \ldots, T_n$ such that

$$\sum_{c_j \in T_i} c_j = \sum_{c_j \in T_k} c_j$$

for all i, k?

This problem is strongly NP-complete (Problem 4).

We should conclude this section with an apology to the fundamentalists. Our "definition" of number(I) for an instance I of a problem was only an *indication* of the *intended meaning* of this quantity. Formally speaking, the input I will be a dry string of symbols, and in general it will not be easy to recover from it those parts that represent integers. This, however, does not diminish the importance of the concepts introduced and the results proved in this section. The critical observation is that our results hold if we take number to be *any fixed* (for a given problem) polynomially computable mapping from the set of instances to the positive integers satisfying number $(I) \leq 2^{|I|}$. The more "reasonable" our definition of the function number for a given problem, the more meaningful will be the implications of Theorem 16.4.

16.3 Special Cases and Generalizations of *NP*-Complete Problems

In this section we elaborate on the following obvious statement: *the more general a problem, the harder it is to solve.* We shall see that this statement can be very useful for proving *NP*-completeness results. We shall also illustrate it by pointing out that, on several occasions, restricted subproblems of *NP*-complete problems are in *P*. Finally, we shall give examples in which special cases of *NP*-complete problems remain hard, despite some very drastic restrictions.

16.3.1 *NP*-Completeness by Restriction

Consider the following problem.

MINIMUM COVER
Given a family $F = S_1, \ldots, S_n$ of subsets of a finite set U, and an integer $k \leq n$, is there a subfamily C of F containing k sets such that $\bigcup_{S_j \in C} S_j = U$?

What is the complexity of MINIMUM COVER? It is easy to see that this problem is a *generalization* of the 3-EXACT COVER problem shown to be *NP*-complete in the previous chapter. The 3-EXACT COVER problem is just the special case of MINIMUM COVER in which $|S_j| = 3$ for $j = 1, \ldots, n$, and $k = \frac{1}{3}|U|$. Consequently, any instance of 3-EXACT COVER can be trivially transformed into an instance of MINIMUM COVER. So, MINIMUM COVER is *NP*-complete, because it is a generalization of an *NP*-complete problem.

To take another example, we proved that the DIRECTED HAMILTON CIRCUIT problem is *NP*-complete by simply observing that it is a *generalization* of the undirected HAMILTON CIRCUIT problem. This is because, roughly speaking, as far as Hamilton circuits are concerned, undirected graphs are just special cases of digraphs, for which, whenever $(u, v) \in A$, then also $(v, u) \in A$.

Similarly, the ASYMMETRIC TSP problem is *NP*-complete, since it generalizes TSP. Also, consider the following problem.

SUBGRAPH ISOMORPHISM
Given two graphs G and G', is there a subgraph of G that is isomorphic (that is, identical up to renaming of the vertices) to G'?

This problem is *NP*-complete, because it contains the CLIQUE and the HAMILTON CIRCUIT problems as special cases. Indeed, when G' is a complete graph with k vertices, we have the CLIQUE problem. If G' is restricted to be a Hamilton circuit (that is, a connected graph with as many nodes as G and with degree of all nodes equal to 2) then we have the HAMILTON CIRCUIT problem.

Proving *NP*-completeness by restriction is not always as trivial as the examples above may have implied. Consider the following problem, for example.

SURVIVABLE NETWORK DESIGN
Given two $n \times n$ symmetric matrices, $[d_{ij}]$ (the distance matrix) and $[r_{ij}]$ (the redundancy matrix), and an integer L, is there a graph on n vertices having cost at most L such that between the ith and the jth vertex there exist at least r_{ij} node-disjoint paths?

We shall show that, in effect, this is a generalization of HAMILTON CIRCUIT. Consider the case in which d_{ij} is either 1 or 2, $L = n$, and $r_{ij} = 2$ for all $i \neq j$. Then it is not hard to see that the only n-edge graph that has two node-disjoint paths between any two nodes is the n-vertex circuit (and this requires some proof). So this special case of the SURVIVABLE NETWORK DESIGN problem has a solution if and only if the n-node graph in which an edge $[i,j]$ is present if and only if $d_{ij} = 1$ is Hamiltonian. Hence, SURVIVABLE NETWORK DESIGN *is* a generalization of the HAMILTON CIRCUIT problem, and consequently is *NP*-complete.

16.3.2 Easy Special Cases of *NP*-Complete Problems

The main point of didactic value in this subsection is that special cases of *NP*-complete problems *need not be hard.* This can be of considerable practical importance. Suppose that, in a practical situation, we are interested in obtaining exact optimal solutions to a given combinatorial optimization problem. Unfortunately, we soon realize that the problem is *NP*-complete, and hence there is no hope of solving the general problem in any efficient way. Should we give up?

Not immediately. It is possible (in fact, likely) that we have been the victims of *unnecessary generality*—like many researchers who have formulated every discrete optimization problem as an integer program and every sequencing problem as a TSP, only to give up once they realized that these general problems are too hard to solve exactly.

A better approach is to formulate problems in the least general terms possible and try to exploit any special features of the instances that interest us. For example, if we are looking at a routing problem that involves graphs, it may be that the graphs of interest have some nice properties, such as planarity, bounded degree of nodes, and so on. On the other hand, in order to prove the problem *NP*-complete, we may have used—as is usually the case—reductions that construct graphs that are highly nonplanar and feature very large degrees. So there is still hope that there is an efficient algorithm for solving the problem in the special case in which the graph is planar and the degrees low. Of course, this is not to say that the *NP*-completeness of the general problem is of no value in this case. Once such a result is proved, it is then up to the optimists to explain how they hope to solve the special case by taking advantage of its properties.

Example 16.2

Recall the CLIQUE problem, known to be *NP*-complete. Suppose that we consider PLANAR CLIQUE, its restriction to planar graphs. Now, Kuratowski's theorem [Ev] says that a planar graph can have no clique with five or more vertices. Consequently, the maximum clique of a planar graph $G = (V, E)$ can have up to four vertices, and hence it can be found by exhaustive search in $O(|V|^4)$ time. In fact an $O(|V|)$ algorithm is possible (see Problem 5). Hence PLANAR CLIQUE is indeed a polynomial special case of the CLIQUE problem. □

Example 16.3

Recall the MULTIPROCESSOR SCHEDULING problem that was shown to be *NP*-complete in Theorem 15.5. We are given a directed acyclic graph $(\mathcal{J}, A)$ describing the precedence requirements of the jobs, the number m of processors, and the length T of the schedule, and we are asked whether a feasible schedule exists. The following special cases, however, are polynomial.

1. $T = 2$. If we are to complete the schedule in 2 time units, the digraph $(\mathcal{J}, A)$ must have no path of length 2 or more; in other words, it must be a *bipartite* digraph with arcs going from a set S of *sources* to a set R of *sinks* and possibly a set I of isolated nodes. In this case, it is easy to see that a feasible schedule exists if and only if (a) $|R|, |S| \leq m$ and (b) $|R| + |S| + |I| \leq 2m$.
2. $m = 2$. It was shown in Problem 5 of Chapter 10 that the problem of scheduling two machines can be solved in $O(n^3)$ time.
3. $(\mathcal{J}, A)$ is a *branching*. In other words, indegree $(J) \leq 1$ for all $J \in \mathcal{J}$. For this case, too, we have a polynomial algorithm (see Problem 6 and [Hu]).

Thus several interesting special cases of the MULTIPROCESSOR SCHEDULING problem can be solved efficiently, despite the fact that the general problem is *NP*-complete. □

For another example, when SATISFIABILITY is restricted to clauses with 2 literals, the corresponding problem (called 2-SATISFIABILITY) can be solved in linear time (Problem 6 of Chapter 15). More examples of such situations can be found in the problems.

16.3.3 Hard Special Cases of *NP*-Complete Problems

The approach discussed in the previous subsection does not always work. Certain *NP*-complete problems remain *NP*-complete even if their instances are restricted very substantially. Very often the most interesting questions concerning *NP*-completeness involve understanding exactly which special cases of an *NP*-complete problem capture the complexity of the problem. Showing that a special case of an *NP*-complete problem is itself *NP*-complete usually involves a special kind of transformation. The purpose of this transformation is to modify any given instance of the general problem to eliminate the features that are disallowed in the special case under consideration without changing the *yes* or *no* nature of the instance. We have already seen such a demonstration: the proof that 3-SATISFIABILITY is *NP*-complete (Theorem 15.2), in which our goal was to replace clauses with a number of literals other than three with equivalent sets of 3-literal clauses. We next give another example of such a proof.

Theorem 16.5 *SATISFIABILITY remains NP-complete, even for formulas in which each variable is restricted to appear once or twice unnegated and once negated.*

Proof Consider any formula F, and any particular variable x appearing in F. Suppose that x occurs $k > 3$ times in total in F. We may substitute the first occurrence of x by a new variable x_1, the second with x_2, and so on, up to the kth. Now, we have to make sure that all k variables $x_1, \ldots, x_k$ agree in their truth value in any truth assignment satisfying the formula. This can be done by appending the clauses $(x_1 + \bar{x}_2)(x_2 + \bar{x}_3)\cdots(x_k + \bar{x}_1)$ to the formula; the reader can easily verify that the only way to satisfy these new clauses is by letting $x_1, \ldots, x_k$ all have the same truth value.

Consequently, if we repeat this for all variables appearing more than three times, we create a new formula in which all variables appear at most three times, and this formula is satisfiable if and only if F is satisfiable. Now we identify all variables that appear in the formula either always unnegated or always negated. Clearly, in any attempt to satisfy F', we can satisfy all clauses in which these variables appear by letting them be *true* or *false*, respectively, and no conflict arises from this. Hence we can delete all these clauses from F', and the resulting formula again is satisfiable if and only if F is satisfiable. If any variable x in the new formula appears twice negated and once unnegated—this is the only way left to violate the restriction of the theorem—we substitute $\bar{y}$ for x in the formula, where y is a new variable. It is easy to check that in the resulting formula each variable appears once or twice unnegated and once negated. □

Sometimes, in order to prove that a special case of an *NP*-complete problem is *NP*-complete, we can simply observe that our construction for proving the general problem *NP*-complete creates only instances of the problem that fall within the special case of interest.

Theorem 16.6 *HAMILTON CIRCUIT is NP-complete, even if the underlying graph is restricted to have nodes of degree four or less.*

Proof We have only to observe that in the proof of Theorem 15.6 we construct a graph with nodes of degree not exceeding four. □

Let us end by showing the following stronger result.

Theorem 16.7 *HAMILTON CIRCUIT for graphs with all nodes of degree three is NP-complete.*

Proof Consider a graph $G = (V, E)$ with degrees four or less. We shall show how to dispense with the nodes of degrees four and two. We start with the nodes of degree four. We replace every node v of degree four in G by the subgraph shown in Fig. 16–3(b), the nodes of which have degree three or two. Now, if this subgraph is a part of any graph, it can be traversed by a Hamilton circuit only as shown in Fig. 16–3(c) through 16–3(e) (or three more, completely symmetric, configurations). Hence each of these transformations preserves the existence of Hamilton circuits. To complete the proof, we notice that we can eliminate nodes of degree two by replacing them as shown in Fig. 16–4. □

16.4 A Glossary of Related Concepts

In this section we explore several issues pertaining to the theory of *NP*-completeness, and explain the relationship of these issues to the concepts introduced and discussed in previous sections.

16.4.1 Polynomial-Time Reductions

In Sec. 15.4 we defined the notion of a polynomial-time reduction from Problem A to Problem B to be a polynomial-time algorithm solving A by making several "calls" of a subroutine that solves B, at unit cost. However, we have focused our attention on the special case of a polynomial-time *transformation.* This "neater" variant was all we needed in order to develop the theory of

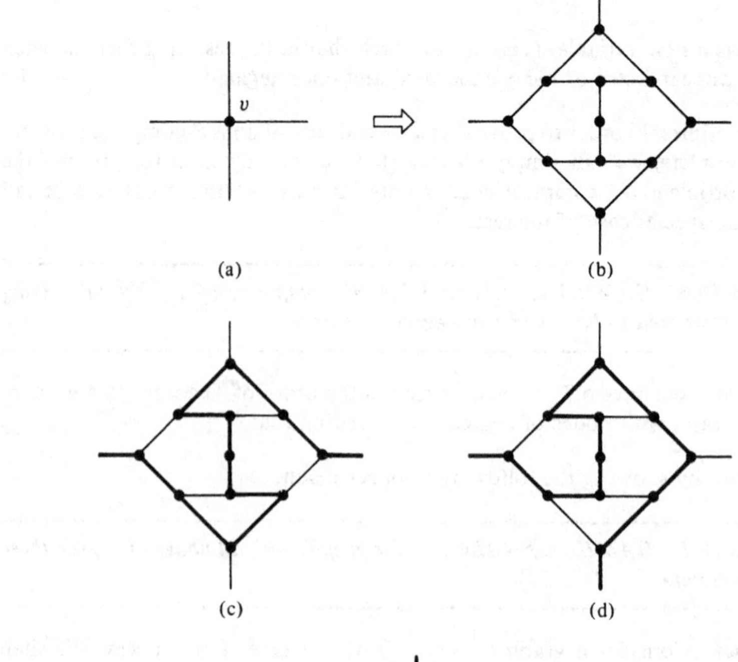

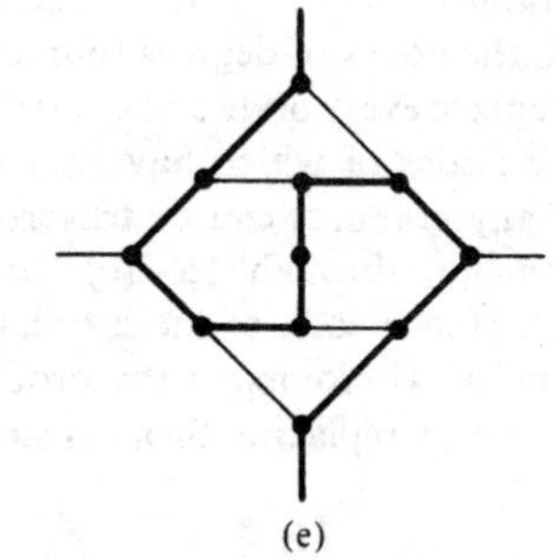

Figure 16–3

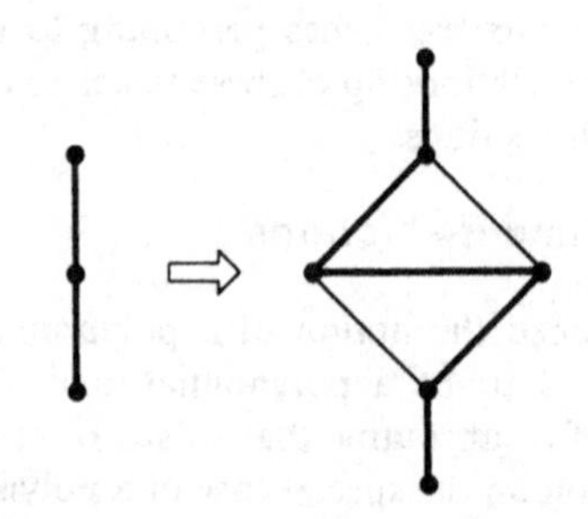

Figure 16–4

NP-completeness. Have we lost something by restricting our attention to polynomial-time transformations? It is not known whether relaxing our definition of *NP*-completeness to allow for general reductions would actually enlarge the class of *NP*-complete problems—even in the case that $P \neq NP$.

16.4.2 *NP*-Hard Problems

Sometimes we may be able to show that all problems in *NP* polynomially reduce to some problem A, but we are unable to argue that A $\in$ *NP*. So A does not qualify to be called *NP*-complete. Yet, undoubtedly A is as hard as any problem in *NP*, and hence most probably intractable. It is for these problems that we have reserved the term *NP-hard.* Here is an example.

> Kth HEAVIEST SUBSET
> Given integers $c_1, \ldots, c_n$, K, and L, are there K *distinct* subsets $S_1, \ldots, S_K \subseteq \{1, \ldots, n\}$ such that
> $$\sum_{j \in S_i} c_j \geq L \qquad \text{for } i = 1, \ldots, K?$$
> (In other words, is the Kth heaviest subset of $\{1, \ldots, n\}$ at least as heavy as L?)

Is Kth HEAVIEST SUBSET in *NP*? This is not at all clear. What shorter certificate is there for a *yes* instance $(c_1, \ldots, c_n, K, L)$ other than a listing of K subsets that are heavier than L? Naturally, this is not a succinct certificate, because K can be as large as 2^{n-1}, for example.

Nevertheless, all problems in *NP* are polynomially *reducible* (not transformable) to the Kth HEAVIEST SUBSET problem.

Theorem 16.8 *PARTITION is polynomially reducible to Kth HEAVIEST SUBSET.*

Proof Suppose that we have a subroutine $\mathfrak{A}$ that solves the Kth HEAVIEST SUBSET problem. We can use this to solve any instance $c_1, \ldots, c_n$ of the PARTITION problem as follows.

1. First, we determine the rank R within the subsets of $\{1, \ldots, n\}$ (ordered by weight) of the lightest subset with weight greater than or equal to
$$\tfrac{1}{2} \sum_{j=1}^{n} c_j$$
Naturally, if
$$\sum_{j=1}^{n} c_j$$
is odd, we immediately answer *no*. We do this by determining R bit by bit, with n calls of $\mathfrak{A}$.

2. If fewer than $2^n - R$ subsets have weight greater than or equal to

$$\frac{1}{2}\sum_{j=1}^{n} c_j + 1$$

this means that at least one—in fact, at least two—subsets have weight

$$\frac{1}{2}\sum_{j=1}^{n} c_j$$

and hence we answer *yes*. Otherwise, we answer *no*. □

Incidentally, this is one of the rare instances in which polynomial reducibility is not known to be replaceable by the stricter notion of polynomial transformability (recall the previous subsection).

Besides its use to describe recognition problems not known to be in *NP*, the term *NP-hard* is sometimes used in the literature to describe *optimization* problems (which, not being recognition problems, are certainly not in *NP*), the recognition versions of which are *NP*-complete. For example, we may say that the TSP (the optimization problem, that is) is *NP*-hard.

16.4.3 Nondeterministic Turing Machines

The initials in *NP* stand for *nondeterministic polynomial.* The reason is that, historically, *NP* was first introduced in terms of certain computing devices called *nondeterministic Turing machines.*

A *Turing machine* is a device operating in a manner similar to the certificate-checking algorithm of the previous chapter—only a Turing machine operates on the input alone, not a certificate, and has unbounded tape to work on. It is controlled by a program with statements such as

$$l\text{: } \mathbf{if}\ \sigma\ \mathbf{then}\ (\sigma'; o; l')$$

A Turing machine accepts its input if it reaches a designated statement l: **accept**. Thus P can be defined as the set of *yes-no* problems recognizable by Turing machines with the property that the computation always halts after a number of steps that is bounded by a fixed polynomial in the size of the input. It should be clear that this definition coincides with our informal notion of polynomial-time algorithm.

A *nondeterministic Turing machine M* is again such a device, only now the program has statements of a more general form

$$l\text{: } \mathbf{if}\ \sigma\ \mathbf{then\ one\ of}\ \{(\sigma_1'; o_1; l_1'), (\sigma_2'; o_2; l_2'), \ldots, (\sigma_k'; o_k, l_k')\}.$$

In executing this statement, M has the *choice* of taking *any one* of the indicated $k \geq 1$ actions; hence the term *nondeterministic*. Now *NP* can be defined to be the class of all *yes-no* problems A with the property that there exists a nondeterministic Turing machine M such that for each *yes* instance x of A there exists a sequence of legal "choices" of moves by M, of total length bounded by a fixed polynomial in $|x|$, eventually leading to the **accept** statement; furthermore, no such choice of moves exists for *no* instances.

We shall now argue that this definition of *NP* is equivalent to our definition, which made use of the notion of a certificate. Suppose that a problem has the "concise certificate" property. Then it can be recognized by a nondeterministic Turing machine that operates by first (nondeterministically) printing next to the input some alleged certificate of appropriate length, and then simulating the certificate-checking algorithm on the string that it has generated. Because our problem has the concise certificate property, for each *yes* instance (and for no *no* instance) there exists a "correct" certificate; that is, there exists a sequence of moves of the nondeterministic machine leading to an **accept** statement, and this sequence has a length bounded by the polynomial corresponding to the problem.

Conversely, suppose that a *yes-no* problem A can be recognized by a nondeterministic Turing machine with a polynomial bound $p(n)$. Then each state of the operation of this machine on an input of length n can be represented as a word of $p(n)$ symbols—possibly many of them blank—and certain extra information denoting the label of the statement about to be executed and the position of the read-write head on the tape. Hence, the whole operation of this machine on a *yes* instance x of A can be represented by juxtaposing $p(|x|)$ such words, with a total length of $O(p^2(|x|))$. Now this record of the operation can be checked for validity by making sure that each word in it comes from the previous one by an option that was legitimate at that time—and this check requires only a polynomial amount of time. Hence, all *yes* instances of A—and only these—have a succinct and efficiently checkable certificate, namely, a legal record of the operation of the nondeterministic Turing machine that terminates at the **accept** instruction. We conclude that the definition of *NP* in terms of nondeterministic Turing machines is equivalent to ours.

16.4.4 Polynomial-Space Complete Problems

We have the class *P* of problems solvable in polynomial time, and the class *NP* of problems that have polynomially certifiable *yes* instances. If we are even more generous and require only that problems have algorithms whose operation can be confined within an amount of *space* that is bounded by a polynomial in the size of the input, we arrive at a class of problems known as *PSPACE*. If an algorithm requires only a polynomial amount of time, it certainly cannot consume more than a polynomial amount of space, so *P* is definitely a subset of *PSPACE*. Also, *NP* is a subset of *PSPACE*. To see this, imagine a machine that, on some input, systematically generates *all* possible concise certificates one after the other, erasing the previous one each time, and simulates the certificate-checking algorithm on each. This machine takes an exponential amount of time, since there is a huge number of possible certificates of the appropriate length, but it uses only a polynomial amount of space. The same argument also proves that $co\text{-}NP \subseteq PSPACE$.

Still, *PSPACE* may very well be even richer. For example, the *K*th HEAVIEST SUBSET problem discussed earlier (and whose membership in *NP* is doubtful) is in *PSPACE*. This is because we can examine exhaustively all subsets of $\{1, \ldots, n\}$, each time erasing the previous one, while keeping on the side a count (in binary) of those subsets found to be heavier than L. All this can be done in polynomial space.

A problem is said to be *PSPACE-complete* if it is in *PSPACE* and all other problems in *PSPACE* are polynomially reducible to it. The paradigm of *PSPACE*-complete problems is a quantified version of SATISFIABILITY, which we do not define here.

PSPACE-complete problems are even less likely to be in *P* than *NP*-complete problems, since *PSPACE* contains *NP* (and hence $P = PSPACE$ would imply $P = NP$). In the light of these new concepts, we show below our final conjectured view of *NP* and its vicinity. It is indicative of our poor understanding of the area that, for all we know now, all these regions can conceivably be collapsed to one: *P*!

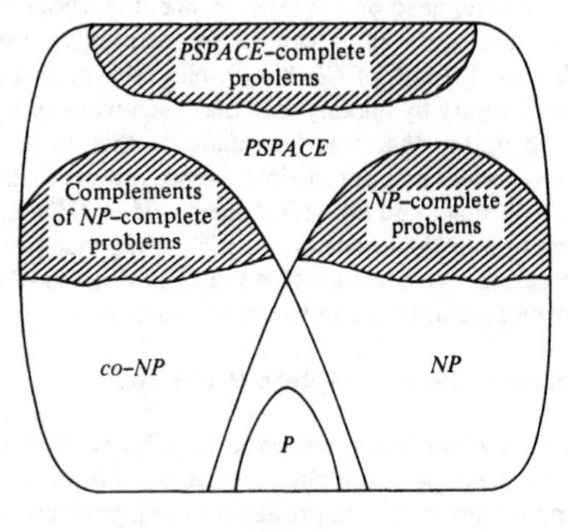

16.5 Epilogue

The final word on the intractability of *NP*-complete problems cannot be written until the $P \neq NP$ conjecture is resolved. Despite the great theoretical importance of this problem and the declared interest of many computer scientists, no proof of this conjecture appears to be in sight. In fact, it now seems very likely that the answer to this question will not come about without the development of entirely new mathematical methodology.

In compiling the small sample of problems that we proved *NP*-complete in the previous chapter, we selected only those that pertain to combinatorial optimization, or are fundamental enough to be useful as starting points for proving other problems *NP*-complete. There are many more *NP*-complete problems, however, arising in disciplines as diverse as graph theory, optimization, mathematical programming, logic, number theory, and the theory of computation. Some can be found in the problems following the previous chapter. The book by Garey and Johnson [GJ2]—the most thorough census attempted to date—contains many hundreds of *NP*-complete problems. Among these, there are several optimization problems whose solution is of great practical significance. The fact that these problems are *NP*-complete—together with the widespread and well-founded confidence that this implies intractability—has led many researchers to reassess their strategy for attacking these problems. There are several alternatives that are less bleak than attempting to solve *NP*-complete optimization problems exactly and efficiently. We list the principal ones below.

1. *Approximation algorithms* These are algorithms that produce not optimal solutions, but solutions that are guaranteed to be a fixed percentage away from the actual optimum. We study this approach and its limitations in the next chapter.
2. *Probabilistic algorithms* Sometimes it is possible to design algorithms that do not behave badly—in terms of the quality of the produced solutions or the time spent—too often, assuming some probabilistic distribution of the instances of the problem.
3. *Special cases* We saw in Sec. 16.3 that interesting special cases of an *NP*-complete problem may be easy. If we are interested only in these special instances, the fact that the general problem is *NP*-complete is more or less irrelevant.
4. *Exponential algorithms* Such algorithms may not be too bad, after all. We have already argued that pseudo-polynomial algorithms (Sec. 16.2) may sometimes be practical, despite the fact that they are, strictly speaking, exponential. Furthermore, some search techniques of exponential worst-case complexity, such as *branch-and-bound* of Chapter 18, can be applied successfully to instances of reasonable size.
5. *Local search* One of the most successful methods of attacking hard combinatorial optimization problems is the discrete analog of "hill climbing," known as *local* (or *neighborhood*) search. This is the subject of Chapter 19.
6. *Heuristics* Any of the approaches above without a formal guarantee of performance can be considered a "heuristic." However unsatisfying mathematically, such approaches are certainly valid in practical situations.

PROBLEMS

1. The following is a classical recognition problem.

PRIMES

Given the decimal representation of an integer, is it a prime?

(a) Give an algorithm for recognizing PRIMES. What is the complexity of your algorithm?

(b) Show that PRIMES $\in$ *co-NP*.

*(c) Show that an integer n is prime iff there exists an integer a such that

(i) $a^{n-1} \equiv 1 \pmod{n}$ and

(ii) $a^{(n-1)/p} \not\equiv 1 \pmod{n}$ for all prime divisors p of $n - 1$.

(d) Based on (c), show that PRIMES $\in$ *NP*.

2. A polynomial-time transformation T from Problem A to Problem B is called *parsimonious* if for any instance x of A, $T(x)$ has the same number of *solutions* (that is, different certificates $c(T(x))$ as x. Which of the transformations used in the proofs in the previous chapter are (or can easily be made) parsimonious?

***3.** Let m be any fixed integer and m-ILP the problem of ILP restricted to instances with m equations. Show that there is a pseudopolynomial algorithm for m-ILP. (*Hint:* Recall the bounds of Sec. 13.3.)

4. *(a) Give a polynomial-time transformation from 3-DIMENSIONAL MATCHING to the following problem.

FOUR PARTITION

Given $4n$ integers $\{c_1, \ldots, c_{4n}\}$, is there a partition of these integers into n quadruples $Q_1, \ldots, Q_n$ such that

$$\sum_{c_j \in Q_i} c_j = \sum_{c_j \in Q_k} c_j \qquad \text{for all } i, k?$$

The numbers c_j constructed in the transformation should obey $c_j \leqq p(n)$ for some polynomial p.

(b) Show that THREE PARTITION is strongly *NP*-complete.

***5.** Show that whether a planar graph (V, E) has a clique of size 4 can be determined in $O(|V|)$ time.

***6.** Consider the following algorithm for solving the MULTIPROCESSOR SCHEDULING problem when the precedence relation is an *antibranching* (that is, all outdegrees are 1, except for the root).

1. Assign to each task a *priority*, which equals its distance from the root.
2. Assign tasks to time slots and processors by assigning first the tasks that have no unexecuted predecessor and the highest priority.

(a) Show that this algorithm correctly solves this special case in polynomial time.

(b) Give polynomial-time algorithms for the cases in which the precedence relation is a *forest* of antibranchings, a branching, and a forest of branchings.

7. A graph is *chordal* if all cycles $[v_1, v_2, \ldots, v_k]$ of length four or more have a *chord*, that is, an edge $[v_i, v_j]$ with $j \not\equiv i \pm 1 \pmod{k}$.

 *(a) Show that the following is an equivalent (recursive) definition of chordal graphs:

 A graph $G = (V, E)$ is chordal iff either $G = (\varnothing, \varnothing)$ (the empty graph) or there is a $v \in V$ such that (i) the neighborhood of v (that is, v and its adjacent nodes) forms a clique, and (ii) recursively, $G - v$ is chordal.

 (b) Show the GRAPH COLORING, INDEPENDENT SET, and CLIQUE are polynomial-time problems for chordal graphs.

8. Let $G = (V, E)$ be a planar graph, and let F be the set of regions (*faces*) into which a planar "drawing" of G separates the plane. Two faces of G are *adjacent* if their intersection is a line (not a point). The *dual* graph of G is $G^D = (F, H)$, where $[f_1, f_2] \in H$ iff f_1 and f_2 are adjacent (G^D may have repeated edges).

 (a) Formulate MAX CUT (Problem 13 in Chapter 15) for a planar graph G as a matching-like problem in G^D, and thus solve it in polynomial time.

 (b) Repeat (a) for the *weighted* version of MAX CUT.

*9. Show that the (asymmetric) TSP with distances as those described in Part (a) (and (b)) of Problem 13 in Chapter 11 can be solved in polynomial time.

10. Show that the (asymmetric) TSP with nonnegative distances d_{ij} satisfying the condition $d_{ij} = 0$ if $i \geq j$ can be solved in polynomial time.

11. Show that the Hamilton circuit problem for planar graphs is *NP*-complete. (*Hint:* Recall Fig. 15–9 in the proof of Theorem 15.6. All that is required is to show how the A lines can be allowed to cross. Consider now the following idea.)

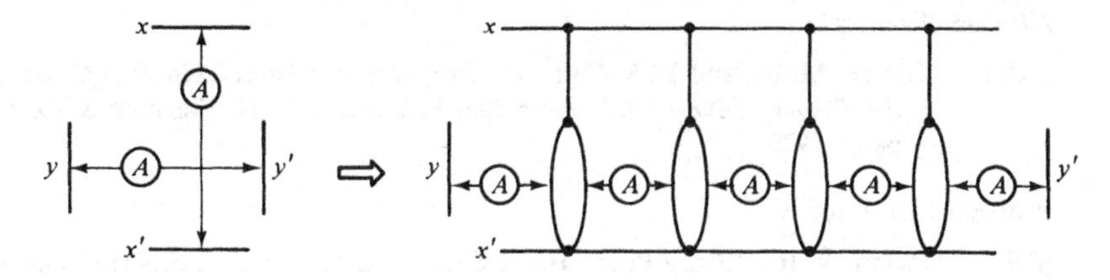

12. Prove by *restriction* (Subsec. 16.3.1) that the following problems are *NP*-complete.

 (a) SUBGRAPH HOMEOMORPHISM
 Given two graphs $G = (V, E)$ and $H = (U, F)$, is there a mapping h from U to V such that (i) $h(u) = h(v)$ implies $u = v$; (ii) for each $[u, v] \in F$, there is a path from $h(u)$ to $h(v)$ in G; and (iii) all paths in (ii) are node-disjoint.

 (*Hint:* HAMILTON CIRCUIT.)

 Given a digraph $D = (V, A)$ the *transitive closure* of D is the digraph $D^* = (V, A^*)$ such that $(u, v) \in A^*$ iff there is a path from u to v in A (or $u = v$).

(b) TRANSITIVE REDUCTION BY ELIMINATION
Given a digraph $D = (V, A)$ and an integer k, is there a subdigraph $F = (V, B)$ such that $D^* = F^*$ and $|B| \leq k$?

(*Hint:* DIRECTED HAMILTON CIRCUIT.)

(c) HITTING SET
Given a family $C = \{S_1, \ldots, S_n\}$ of finite sets and an integer k, is there a set H, with $|H| \leq k$, such that $H \cap S_j \neq \varnothing$ for $j = 1, \ldots, n$?

(*Hint:* NODE COVER.)

(d) HAMILTON COMPLETION
Given a graph $G = (V, E)$ and an integer k, is there a set B of edges such that $|B| \leq k$ and $(V, E \cup B)$ has a Hamilton circuit?

NOTES AND REFERENCES

The ideas discussed in Sec. 16.2 are from

[GJ1] GAREY, M. R., and D. S. JOHNSON, "Strong *NP*-completeness Results: Motivation, Examples, and Implications," *J. ACM*, 25 (1978), 499–508.

Theorem 16.8 is due to

[JK] JOHNSON, D. B., and S. D. KASHDAN, "Lower Bounds for Selection in $X + Y$ and other Multisets," *J. ACM*, 25 (1978), 556–70.

For further discussion of the issues mentioned in Sec. 16.4, see the chapter "Beyond *NP*-Completeness" in

[GJ2] GAREY, M. R., and D. S. JOHNSON, *Computers and Intractability: A Guide to the Theory of NP-completeness.* San Francisco: W. H. Freeman & Company, 1979.

Problem 1(d) is due to

[Pr] PRATT, V. R., "Every Prime Has a Succinct Certificate," *J. SIAM Comp.*, 4 (1975), 214–70.

Parsimonious reductions were introduced in

[Si] SIMON, J., *On Some Central Problems in Computational Complexity*, (Unpublished Ph.D. Thesis, Cornell University, 1977).

and in

[Va1] VALIANT, L. G., "A Polynomial Reduction of Satisfiability to Hamiltonian Circuits that Preserves the Number of Solutions," (unpublished manuscript).

Using such reductions, Valiant has developed a complexity theory for enumeration problems (that is, problems related to *counting* the solutions) which parallels that of recognition and optimization problems. See

[Va2] VALIANT, L. G., "The Complexity of Enumeration and Reliability Problems," Report CSR-15-77, Univ. of Edinburgh, 1977.

Problem 3 is from

[Pa] PAPADIMITRIOU, C. H., "On the Complexity of Integer Programming," M.I.T. Lab. for Computer Science, TM-152, 1980. Also, *J. ACM*, 28 (1981, in press).

Problem 4 is from [GJ2]. Problem 6 is from

[Hu] HU, T. C., "Parallel Sequencing and Assembly Line Problems," *OR*, 9 (1961), 841–48.

Problem 7 is from

[Ga] GAVRIL, F., "Algorithms for Minimum Coloring, Maximum Clique, Minimum Covering by Cliques, and Maximum Independent Set of a Chordal Graph," *J. SIAM Comp.*, 1 (1972), 180–87.

Problem 8 is from

[OD] ORLOVA, G. I., and Y. G. DORFMAN, "Finding the Maximum Cut in a Graph," *Engnrg. Cybernetics*, 10 (1972), 502–06.

Problem 5 is from

[PY] PAPADIMITRIOU, C. H., and M. YANNAKAKIS, "The Clique Problem for Planar Graphs," *Inf. Proc. Letters* (1981, in press).

Problems 9 and 10 are from

[GG] GILMORE, P. C., and R. E. GOMORY, "Sequencing a One State-Variable Machine: A Solvable Case of the Traveling Salesman Problem," *OR*, 12 (1964), 655–79.

and

[La] LAWLER, E. L., "A Solvable Case of the Traveling Salesman Problem," *Math. Prog.*, 1 (1971), 267–69.

For Kuratowski's theorem see, for example,

[Ev] EVEN, S., *Graph Algorithms*. Potomac, Maryland: Computer Science Press, 1979.

17

Approximation Algorithms

17.1 Heuristics for Node Cover: An Example

Consider the NODE COVER problem.

> Given a graph $G = (V, E)$, find the smallest possible set C of nodes such that $[u, v] \in E \Rightarrow v \in C$ or $u \in C$.

This is a very practical problem, arising, for example, whenever one wants to monitor the operation of a large network by monitoring as few nodes as possible. Certainly the fact that this problem is *NP*-complete (Corollary to Lemma 15.4) is quite disappointing in this regard. Nevertheless, there are certain plausible techniques for obtaining "good" (but perhaps not optimal) solutions to this problem. For example, the "greedy" heuristic in Fig. 17–1 appears quite promising.

Because our goal in this problem is to cover all edges of G with as few nodes as possible, selecting each time the single node that by itself covers as many of the remaining edges as possible is an attractive strategy. In view of the fact that NODE COVER is *NP*-complete, of course, we do not expect this efficient

```
Input: A graph G=(V,E)
Output: A node cover C of G, presumably not much larger than
        the optimal one.
begin
  C:=∅;
  while E≠∅ do
    choose the node in V that has the largest degree,
    (comment: break ties arbitrarily)
    remove it from G and add it to C
end
```

Figure 17–1 Algorithm 1

algorithm to yield the smallest node cover all the time. But how close does it come?

Let us consider this scheme applied to the graph of Fig. 17–2. We first choose one of the degree-5 nodes a_1, a_2, or a_3—say, a_1—then a_2, then a_3, and finally c_1, c_2, c_3, c_4 and c_5. The resulting node cover consists of 8 nodes. The optimal node cover, however, has just 5: $\{b_1, b_2, b_3, b_4, b_5\}$. Furthermore, if we generalize this graph to one with n a-nodes, $n + 2$ b-nodes, and $n + 2$ c-nodes (together with the $[c_j, b_j]$ and $[a_i, b_j]$ edges for all i, j), we find that Algorithm 1 finds a cover of $2n + 2$ nodes, whereas the optimal cover has only $n + 2$. Since n can be arbitarily large, we see that the error can become arbitrarily close to 100 percent.

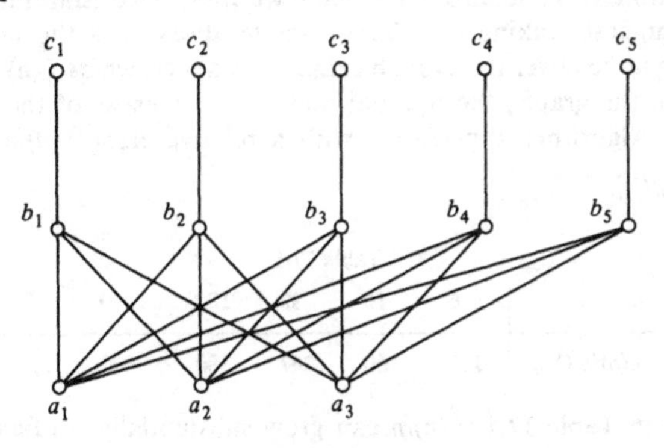

Figure 17–2

Is this the worst possible performance for this heuristic, or can the error get even higher? The example of Fig. 17–3 shows that it can. In the beginning, node a_7 has the maximum degree, namely 5. After the removal of a_7, a_6 has the largest degree, then a_5, and so on. At each stage, an a-node is among those of highest degree, and so it is removed. Finally, the node cover is completed with, say, all the c-nodes. This solution has a total of 13 nodes, whereas the optimal one, $\{b_1, b_2, b_3, b_4, b_5, b_6\}$, has only 6. The deviation from the optimum exceeds 100 percent.

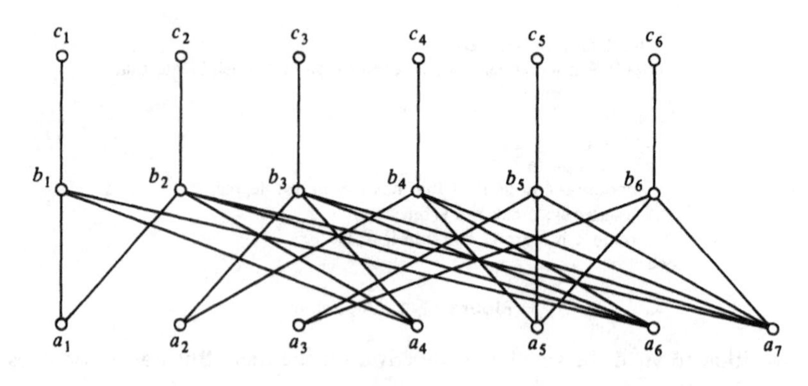

Figure 17–3

In order to generalize the counterexample of Figure 17–3, we have first to understand its structure. We can think of it as 6 edges $[c_i, b_i]$, $i = 1, \ldots, 6$, to which the a-nodes are appended in the following fashion: First we partition the 6 b-nodes into 3 pairs and join the nodes in each pair with an a-node. Then we partition the b-nodes into 2 triples, and again join all nodes in a triple with a new a-node. We repeat the same with quadruples and quintuples, and so on, possibly leaving out some b-nodes and always adding a new a-node for each set in each partition. It is not hard to see that if we apply Algorithm 1 to the resulting graph, the highest-ranking remaining a-node always has the highest degree. The resulting node-cover therefore has $L(n) + n$ nodes, where $L(n)$ is the number of a-nodes in the graph; the optimal node cover consists of the n b-nodes. It follows that Algorithm 1 performs with a relative error $L(n)/n$. Notice that $L(n) = \sum_{j=2}^{n-1} \lfloor n/j \rfloor$.

Table 17.1

n	6	10	30	100	1000	...
$L(n)/n$ (%)	117	160	267	380	600	...

As indicated in Table 17.1, $L(n)/n$ can grow substantially. In fact, it grows as fast as $\ln n$. Consequently, the seemingly plausible Algorithm 1 for node cover has no fixed bound on the relative (that is, percentage) error that it introduces. We can construct instances in which it behaves as badly as desired!

Can we design a heuristic for NODE COVER with bounded relative error? Consider the algorithm in Fig. 17–4.

The resulting set C of nodes is certainly a node cover. Now, any node cover must by definition cover all the edges chosen by the algorithm. However, these edges have no node in common, and so each must be covered by a different node of the node cover. Consequently, any node cover must contain one node

```
Input and output: As in Algorithm 1.
begin
  C:=∅;
  while E≠∅ do
    Choose any edge [v,u] of E, delete both nodes u and v from
    G and add them to C
end
```

Figure 17–4 Algorithm 2

from each edge chosen, and thus no node cover can be smaller than half the size of C. Thus Algorithm 2 has a relative error of at most 100 percent. This worst-case error can be actually achieved: simply think of a graph consisting of many disjoint edges.

What we have illustrated above, in terms of the NODE COVER problem and the two proposed heuristics, is the idea of evaluating a heuristic by analyzing its worst-case error bound. These concepts can be formalized as follows.

Definition 17.1

Let A be an optimization (minimization or maximization) problem with positive integral cost function c, and let $\mathcal{A}$ be an algorithm which, given an instance I of A, returns a feasible solution $f_{\mathcal{A}}(I)$; denote the optimal solution of I by $\hat{f}(I)$.† Then $\mathcal{A}$ is called an *ϵ-approximate algorithm* for A for some $\epsilon \geq 0$ if and only if

$$\frac{|c(f_{\mathcal{A}}(I)) - c(\hat{f}(I))|}{c(\hat{f}(I))} \leq \epsilon$$

for all instances I. □

For example, Algorithm 2 is a 1-approximate algorithm for node cover. Algorithm 1 is not an ϵ-approximate algorithm for *any* $\epsilon > 0$, because—as we indicated—its relative error will violate, for appropriate instances, all constant bounds. To describe the worst-case performance of algorithms such as Algorithm 1, we sometimes allow ϵ to be a function of the input. For instance, if n denotes the number of nodes in an instance of the NODE COVER problem, it can be shown that Algorithm 1 is a ln *n-approximate* algorithm (Problem 2). This means that for all graphs G with n nodes, the algorithm yields a set C satisfying

$$\frac{|C| - |\hat{C}|}{|\hat{C}|} \leq \ln n$$

where $\hat{C}$ is the optimal node cover.

†Certain approximation algorithms contain incompletely specified steps, such as the *break ties arbitrarily* clause in Algorithm 1, or the *select any edge* step of Algorithm 2. In such situations, $c(f_{\mathcal{A}}(I))$ is taken to be the *worst possible* cost resulting from $\mathcal{A}$ when applied to I.

17.2 Approximation Algorithms for the Traveling Salesman Problem

In this section we explore the possiblity of applying the ideas of the previous section to the traveling salesman problem (TSP). Our goal is to develop efficient algorithms that yield "good" approximate solutions of the TSP. Unfortunately, we are going to prove in Sec. 17.4 that this task for the general (that is, unrestricted) TSP is essentially as hopeless as that of solving it exactly. Nevertheless, we can show that reasonably successful strategies exist for a very natural special case of the problem.

Consider an $n \times n$ distance matrix $[d_{ij}]$ with positive real entries. As usual, we assume that $[d_{ij}]$ is symmetric—that is, $d_{ij} = d_{ji}$ for all i, j—and that $d_{jj} = 0$ for all j. We say that $[d_{ij}]$ *satisfies the triangle inequality* if

$$d_{ij} + d_{jk} \geq d_{ik} \qquad \text{for all } 1 \leq i, j, k \leq n$$

What the triangle inequality constraint essentially says is that going from city i to city k through city j cannot be cheaper than going directly from i to k (see Fig. 17–5(a) and (b)). This is very reasonable since the imposed visit to city j appears to be an additional constraint, which can only increase the cost. The triangle inequality is satisfied automatically, for example, whenever the distance matrix is *induced by a metric*—as it is in the important special case of *Euclidean*

$$\begin{bmatrix} 0 & 5 & 3 & 1 \\ 5 & 0 & \textcircled{7} & 2 \\ 3 & 7 & 0 & 4 \\ 1 & 2 & 4 & 0 \end{bmatrix} \qquad \begin{bmatrix} 0 & 3 & 3 & 1 \\ 3 & 0 & 6 & 2 \\ 3 & 6 & 0 & 4 \\ 1 & 2 & 4 & 0 \end{bmatrix}$$

(a) (b)

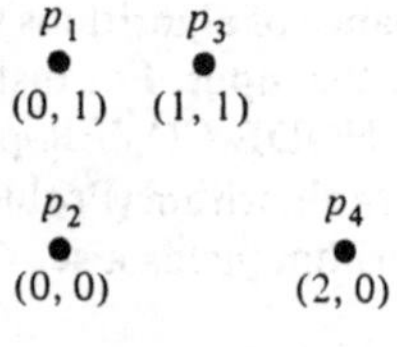

p_1 (0, 1) p_3 (1, 1)

p_2 (0, 0) p_4 (2, 0)

$$\begin{bmatrix} 0 & 1 & 1 & \sqrt{5} \\ 1 & 0 & \sqrt{2} & 2 \\ 1 & \sqrt{2} & 0 & \sqrt{2} \\ \sqrt{5} & 2 & \sqrt{2} & 0 \end{bmatrix}$$

(c) (d)

Figure 17–5 The matrix in (a) does *not* satisfy the triangle inequality because, for example, $d_{23} > d_{24} + d_{43}$. The matrix in (b) does. In fact, this matrix is the closure of the previous one. In (d) we show the Euclidean distance matrix arising from the four-point map shown in (c).

distance matrices, in which each city represents a point p_j on a 2-dimensional map with coordinates (x_j, y_j), and $d_{ij} = [(x_i - x_j)^2 + (y_i - y_j)^2]^{1/2}$ (see Fig. 17–5(c) and (d) and Problem 17 of Chapter 15).

Another important class of distance matrices that automatically satisfy the triangle inequality are *closure* matrices. We say that the matrix $[\bar{d}_{ij}]$ is the closure of $[d_{ij}]$ if $\bar{d}_{ij}$ is the length of the shortest path from i to j in the complete graph of n nodes $\{1, 2, \ldots, n\}$, where the length of the edge $[i, j]$ is d_{ij}. The closure of an $n \times n$ distance matrix can be calculated in $O(n^3)$ time by the Floyd-Warshall algorithm (Sec. 6.5). For example, in Fig. 17–5(b) we show the closure of the matrix in Fig. 17–5(a). The closure $[\bar{d}_{ij}]$ of any distance matrix $[d_{ij}]$ satisfies the triangle inequality because, if $\bar{d}_{ik} > \bar{d}_{ij} + \bar{d}_{jk}$, then $\bar{d}_{ik}$ is not the length of a shortest path from i to k.

The triangle inequality is also satisfied by distance matrices that model the cost structure of far more general situations. These include scheduling (in which case d_{ij} may stand for the start-up cost of job j, whenever job i was the last one processed), and routing (d_{ij} may incorporate, besides a nearly Euclidean distance, such cost components as fuel cost, personnel costs, waiting times, and so on). As a rule of thumb, whenever the entries of the distance matrix represent *costs*, the triangle inequality is automatically satisfied. The reason that these distance matrices satisfy the triangle inequality is because they are, in effect, closure matrices of positive cost matrices. This may fail to be the case if some of the entries are *gains*—that is, negative costs. For example, if each visit of city j results in some sort of "bonus," it may very well be that $d_{ij} + d_{jk} < d_{ik}$.

Definition 17.2

The *Triangle Inequality* (or *Metric*) *TSP* (abbreviated ΔTSP) is the TSP restricted to matrices satisfying the triangle inequality. If we further restrict our problem to Euclidean distance matrices, we have the *Euclidean TSP*. □

Theorem 17.1 *The (recognition version of) ΔTSP is NP-complete.*

Proof Recall the transformation from HAMILTON CIRCUIT to TSP. Given a graph (V, E) we construct an instance $([d_{ij}], |V|)$ of the $|V| \times |V|$ TSP with $d_{ij} = 1$ if $[v_i, v_j] \in E$, and 2 otherwise. It was immediate that this instance had a tour of cost $|V|$ or less if and only if G was Hamiltonian. Observe that any distance matrix with all entries equal to 1 or 2, such as $[d_{ij}]$, satisfies the triangle inequality. Thus HAMILTON CIRCUIT polynomially transforms to ΔTSP. □

We shall next present two approximation algorithms for ΔTSP. These heuristics construct an *Eulerian spanning graph* rather than a tour. We shall show that, under the assumption of triangle inequality, these solutions can be transformed to tours at no extra cost.

A *multigraph* (V, E)—that is, a graph with repetitions of edges allowed—is called *Eulerian* if it has a closed walk (called an *Eulerian walk*) in which each node appears *at least* once and each edge appears *exactly* once. For example, the multigraph of Fig. 17–6(a) is Eulerian, since the closed walk $[v_1, v_6, v_7, v_8, v_5, v_4, v_1, v_3, v_2, v_5, v_4, v_7, v_6, v_9, v_{10}, v_8, v_9, v_7, v_4, v_2, v_1]$ traverses all nodes,

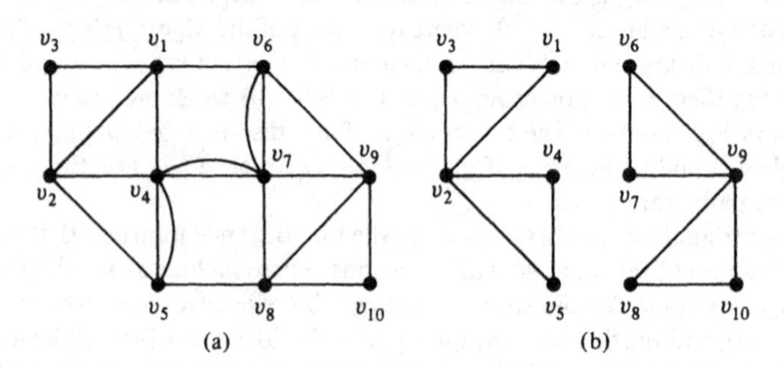

Figure 17–6

and each edge exactly once. The following simple result, due to Euler [Eu], is historically the first theorem in graph theory.

Theorem 17.2 *A multigraph $G = (V, E)$ is Eulerian if and only if*

(a) *G is connected, and*

(b) *All nodes in V have even degree.*

Proof That these two conditions are necessary is immediate. To show that they are sufficient, let us proceed by induction on the number of edges of G. The induction basis—a multigraph with one node and no edges—is trivial. Now, assume that G satisfies (a) and (b), and moreover, all multigraphs with fewer edges than G satisfying (a) and (b) are Eulerian. Choose a node v of G and start a walk on the edges of G, never traversing an edge twice, until v is met again. By (b), this will always be possible. Now, removing the edges of this walk from G, we have a number of connected components. Each component, however, satisfies both (a) and (b) and hence is Eulerian, by induction. It is easy to see that we can create an Eulerian walk for G by "appending" the Eulerian walks of the components to the original walk. □

This constructive proof of the sufficiency of the conditions suggests the recursive algorithm in Fig. 17–7 for finding an Eulerian walk in any multigraph (V, E) satisfying (a) and (b) in $O(|E|)$ time.

```
procedure Euler(v1)
(comment: it returns an Eulerian walk of the connected
          component of G containing v1)
 begin
 if v1 has no edges then return [v1] (comment: the empty walk)
   else
   begin
     starting from v1 create a walk of G, never visiting the
     same edge twice, until v1 is reached again;
     let [v1,v2,...,vn,v1] be this walk;
     delete [v1,v2],...,[vn,v1] from G;
     return [Euler(v1),Euler(v2),...,Euler(vn),v1]†
   end
 end
```

Figure 17–7

Example 17.1

Let us apply the procedure Euler to the graph of Fig. 17–6(a). Euler (v_1) may first return the walk $[v_1, v_6, v_7, v_4, v_7, v_8, v_5, v_4, v_1]$. We then call Euler ($v_1$), and, at that point, the graph is as shown in Fig. 17–6(b). The walk returned is, for example, $[v_1, v_3, v_2, v_4, v_5, v_2, v_1]$. We then call Euler($v_6$), which returns $[v_6, v_7, v_9, v_6]$. Euler(v_7) returns $[v_7]$, Euler(v_4) returns $[v_4]$, and so on, up to Euler(v_8), which returns $[v_8, v_9, v_{10}, v_8]$. The remaining calls of Euler return empty paths; the resulting Eulerian walk is $[v_1, v_3, v_2, v_4, v_5, v_2, v_1, v_6, v_7, v_9, v_6, v_7, v_4, v_7, v_8, v_9, v_{10}, v_8, v_5, v_4, v_1]$. □

Let $[d_{ij}]$ be an $n \times n$ distance matrix satisfying the triangle inequality. An *Eulerian spanning graph* is an Eulerian multigraph $G = (V, E)$ with $V = \{1, 2, \ldots, n\}$. The *cost* of G is $c(G) = \sum_{[i,j] \in E} d_{ij}$.

Theorem 17.3 *If $G = (V, E)$ is an Eulerian spanning graph, then we can find a tour τ of V with $c(\tau) \leq c(G)$ in $O(|E|)$ time.*

Proof By hypothesis G has an Eulerian walk w; because w visits all nodes at least once, we can write $w = [\alpha_0 i_1 \alpha_1 i_2 \cdots i_n \alpha_n]$, where $\tau = (i_1, \ldots, i_n)$ is a tour and $\alpha_0, \alpha_1, \ldots, \alpha_n$ are sequences (possibly empty) of integers in $\{1, \ldots, n\}$. (We say that τ *is embedded in* G). Now, the triangle inequality implies that $d_{ik} \leq d_{ij_1} + d_{j_1 j_2} + \cdots + d_{j_{m-1} j_m} + d_{j_m k}$ for *any* $m \geq 1$. Consequently, the total length of w—which is exactly $c(G)$—can be no smaller than $d_{i_1 i_2} + d_{i_2 i_3} + \cdots + d_{i_n i_1} = c(\tau)$. □

†Our convention is that, for example, a function call $z := F(G(x), H(y))$ means that G is called before H. This is important in cases, such as the present, in which functions have side-effects.

What Theorem 17.3 says, therefore, is that under the triangle inequality assumption, the problem of finding the shortest tour is *equivalent* to that of finding the shortest Eulerian spanning graph—the other direction is obvious, since a tour is a special case of an Eulerian spanning graph.

Consider, therefore, the algorithm in Fig. 17–8 for finding a short Eulerian spanning graph and one of the corresponding embedded tours.

1. Find the minimum spanning tree T under $[d_{ij}]$.
2. Create a multigraph G by using *two copies* of each edge of T.
3. Find an Eulerian walk of G and an embedded tour.

Figure 17–8 The tree algorithm.

Theorem 17.4 *The tree algorithm is a* 1*-approximate algorithm for* ΔTSP.

Proof First, we have to establish that G is Eulerian, so that Step 3 is possible. Because G contains a spanning tree T, it is connected. Furthermore, all degrees of G are even, since they are twice the corresponding degrees in T.

We now turn to the error bound. If $\hat{c}$ is the cost of the shortest tour, by Theorem 17.3 it suffices to show that $c(G) \leq 2 \cdot \hat{c}$. Now, $c(G) = 2 \cdot c(T)$, where $c(T)$ is the cost of the minimum spanning tree. Furthermore, $c(T) \leq \hat{c}$; this is because *any* tour (including the shortest) can be transformed to a tree simply by erasing an edge—and the shortest spanning tree is at least as short as the result. $\square$

Example 17.2

The tree algorithm is illustrated in the example of Fig. 17-9. The shortest spanning tree T is shown in (a), and G in (b). We find an Eulerian walk of G (Fig. 17–10(c)), and we pick one of the occurrences of each integer from $\{1, 2, \ldots, n\}$—by definition of an Eulerian walk, each integer occurs at least once.

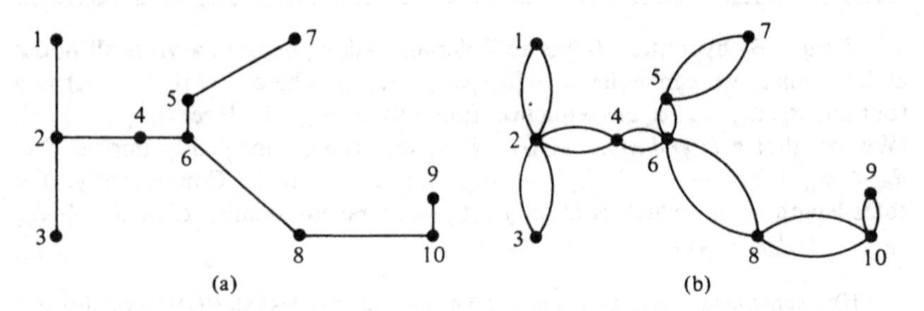

Figure 17–9 (a) A minimum spanning tree. (b) An Eulerian spanning graph. (c) An Eulerian walk (underlined: an embedded tour). (d) A 1-approximate tour. (e) The shortest tour.

$[\underline{1}, \underline{2}, \underline{3}, 2, \underline{4}, \underline{6}, \underline{5}, \underline{7}, 5, 6, \underline{8}, 10, \underline{9}, \underline{10}, 8, 6, 4, 2, 1]$

(c)

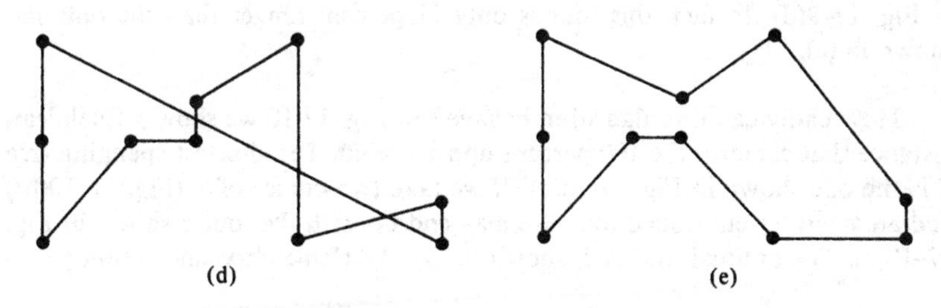

Figure 17–9 (*continued*)

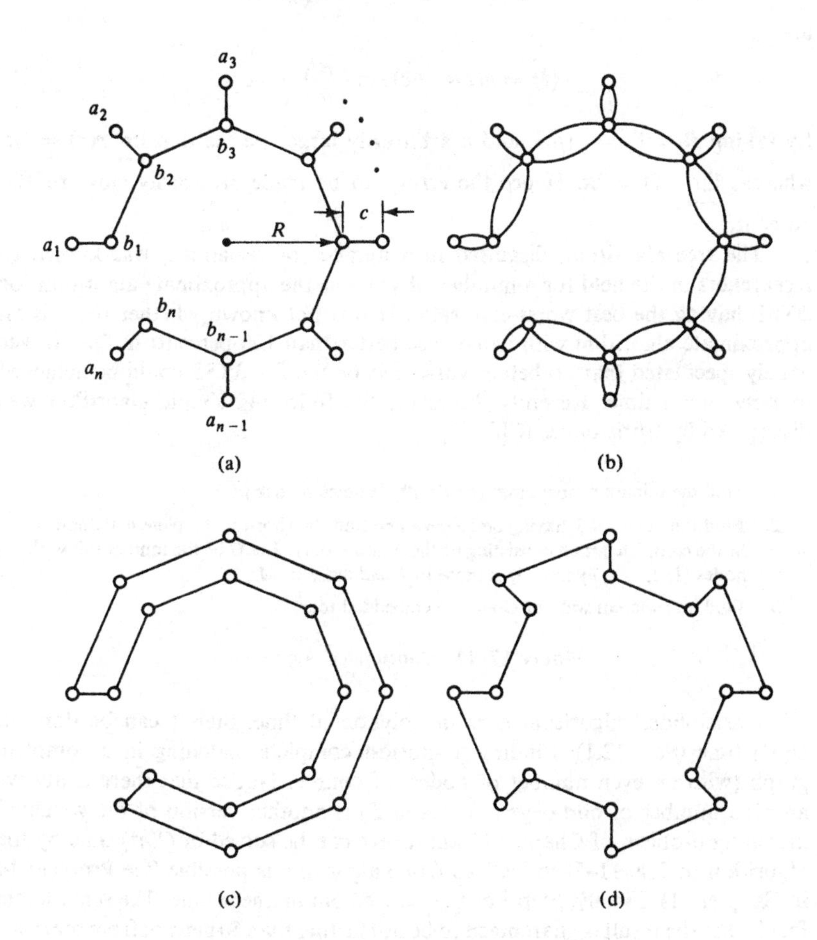

Figure 17–10

The resulting tour, guaranteed to be 1-approximate by Theorem 17.4, is shown in Fig. 17–9(d). In fact, this tour is only 11 percent longer than the optimal, shown in (e). □

How badly can this algorithm behave? In Fig. 17-10 we show a Euclidean instance that achieves the 100 percent upper bound. The shortest spanning tree T is the one shown in Fig. 17-10(a). If we take two copies of T (Fig. 17–10(b)) and an arbitrary embedded tour, we may end up with the tour τ shown in Fig. 17–10(c). The optimal tour is $\hat{\tau}$, shown in (d). An elementary calculation yields

$$c(\tau) = 2(n-1)(2R+c)\sin\left(\frac{\pi}{n}\right) + 2c$$

and

$$c(\hat{\tau}) = n(2R+c)\sin\left(\frac{\pi}{n}\right) + nc$$

By taking $R = 1$, $c = 1/n^2$, and n arbitrarily large, we see that $\lim_{n\to\infty} c(\tau) = 4\pi$, whereas $\lim_{n\to\infty} c(\hat{\tau}) = 2\pi$. Hence the error can be made arbitrarily close to 100 percent.

The tree algorithm, disguised in a number of variations, was known to researchers in the field for a number of years as the approximate algorithm for ΔTSP having the best worst-case ratio. It was not known whether there is an approximate algorithm with worst-case better than 100 percent; in fact, it was widely speculated that no better worst-case bound for ΔTSP could be achieved in polynomial time. Recently, however, the following simple algorithm was discovered by Christofides [Ch].

1. Find the minimum spanning tree T with distance matrix $[d_{ij}]$.
2. Find the nodes of T having *odd degree* and find the shortest complete matching M in the complete graph consisting of these nodes only. Let G be the multigraph with nodes $\{1, 2, \ldots, n\}$ and edges those in T *and* those in M.
3. Find an Eulerian walk of G and an embedded tour.

Figure 17–11 Christofides' Algorithm.

Christofides' algorithm runs in polynomial time. Step 1 can be done in $O(n^2)$ time (Sec. 12.1). Finding a shortest complete matching in a complete graph (with an even number of nodes, of course: Notice that there is always an even number of odd-degree nodes in T) is another version of the weighted matching problem of Chapter 11 and hence can be solved in $O(n^4)$ time by the algorithm in Fig. 11–5. In fact, an $O(n^3)$ algorithm is possible (see Problem 14 in Chapter 11). Finally, Step 3 can be carried out in linear time. The remarkable fact is that the result is guaranteed to be not farther than 50 percent from optimal:

Theorem 17.5 *Christofides' algorithm is a $\frac{1}{2}$-approximate algorithm for the* ΔTSP.

Proof The graph G constructed in Step 2 is Eulerian. To see this, notice that if a node had an even degree in T, it has the same degree in G. If it has an odd degree in T, it has one more edge, coming from the matching M, incident upon it. Furthermore, G is certainly connected, because it contains a spanning tree as a subgraph—namely, T.

To prove the $\frac{1}{2}$ error bound, recall that the graph G consists of T and M; hence the cost of the resulting tour τ satisfies

$$c(\tau) \leq c(G) = c(T) + c(M) \tag{17.1}$$

Now,

$$c(T) \leq c(\hat{\tau}) \tag{17.2}$$

where $\hat{\tau}$ is the shortest tour. Also, let $\{i_1, i_2, \ldots, i_{2m}\}$ be the set of odd-degree nodes in T, in the order that they appear in $\hat{\tau}$. In other words $\hat{\tau} = [\alpha_0 i_1 \alpha_1 i_2 \cdots \alpha_{2m-1} i_{2m} \alpha_{2m}]$, where the α's are (possibly empty) sequences of nodes from $\{1, 2, \ldots, n\}$. Consider the two matchings of the odd-degree nodes $M_1 = \{[i_1, i_2], [i_3, i_4], \ldots, [i_{2m-1}, i_{2m}]\}$ and $M_2 = \{[i_2, i_3], [i_4, i_5], \ldots, [i_{2m}, i_1]\}$. By the triangle inequality (see Fig. 17–12) $c(\hat{\tau}) \geq c(M_1) + c(M_2)$. However, M is the optimal matching, and so $c(\hat{\tau}) \geq 2c(M)$, or

$$c(M) \leq \tfrac{1}{2} c(\hat{\tau}) \tag{17.3}$$

Substituting (17.2) and (17.3) in (17.1), we get

$$c(\tau) \leq \tfrac{3}{2} c(\hat{\tau})$$

or

$$\frac{c(\tau) - c(\hat{\tau})}{c(\hat{\tau})} \leq \tfrac{1}{2}$$

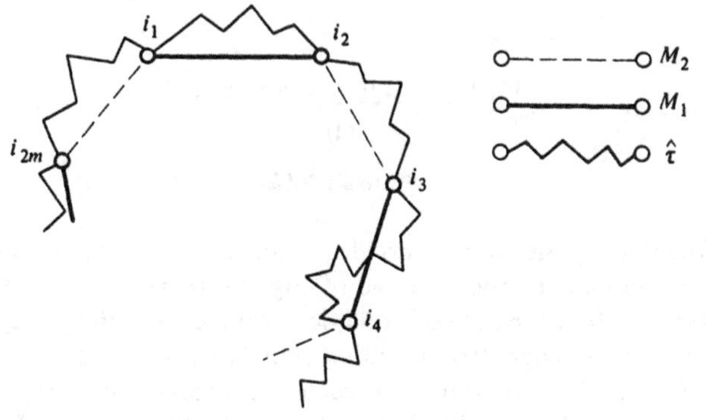

□

Figure 17–12

Example 17.2 (Continued)

In Fig. 17–13(a), we show the minimum spanning tree T, with the odd-degree nodes encircled, for the map examined in Fig. 17–9. Figure 17–13(b) shows the shortest matching M of these nodes, and (c) shows the graph G. An Eulerian walk of G is constructed in Fig. 17–13(d), and a corresponding tour is shown in (e). The tour constructed is only slightly better than the one constructed in Figure 17–9 using the tree algorithm: it is less than 8 percent off the optimum. □

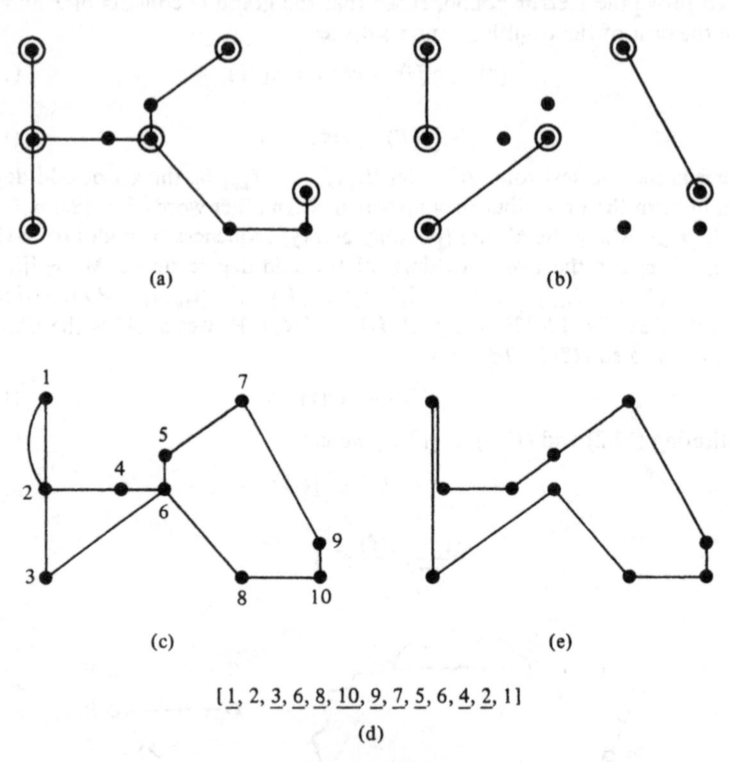

Figure 17–13

Like the tree algorithm, Christofides' algorithm can asymptotically achieve its worst-case bound. In the example of Fig. 17–14(a), we show the shortest spanning tree T. There are only two odd-degree nodes, and therefore the optimal matching is a single edge. The resulting Eulerian graph is a tour, and hence Step 3 is trivial. The tour thus constructed has total length $3n$, whereas the shortest tour (Fig. 17–14(b)) has length $2n + 1$. So the error can become arbitrarily close to $\frac{1}{2}$.

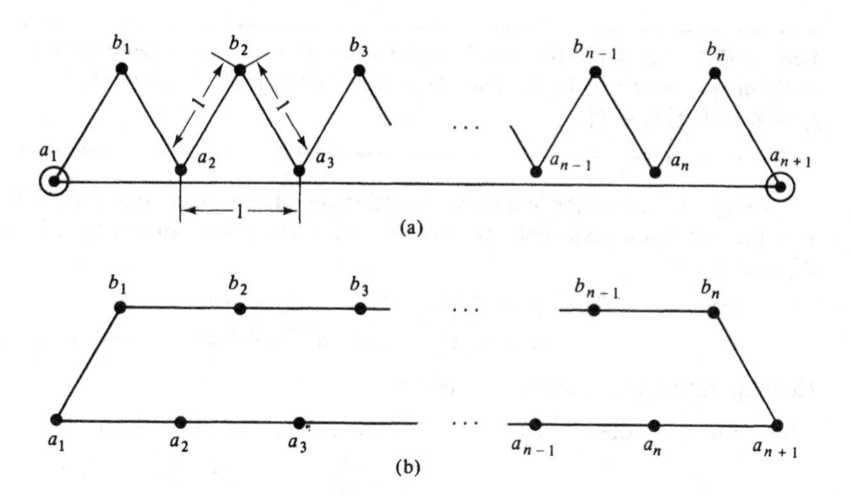

Figure 17–14

17.3 Approximation Schemes

The 0-1 KNAPSACK problem was defined in Chapter 15 as the following problem:

> Given positive integers $c_1, c_2, \ldots, c_n$ and K, is there a subset S of $\{1, 2, \ldots, n\}$ such that $\sum_{j \in S} c_j = K$?

The 0-1 KNAPSACK problem can be solved by a pseudo-polynomial algorithm which is a modification of the algorithm given in Sec. 16.2 for INTEGER KNAPSACK. Again we construct the digraph $G(c_1, c_2, \ldots, c_n; K) = (V, A)$ with $V = \{0, 1, \ldots, K\}$ and $A = A_1 \cup A_2 \cup \cdots \cup A_n$, where $A_j = \{(v, u) \in V^2 : u - v = c_j\}$. We then apply the algorithm in Fig. 17–15.

1. Mark the node 0.
2. For $j = 1, 2, \ldots, n$ do:
   For each marked node v mark the node u such that $(v, u) \in A_j$.
3. Conclude that the instance has a solution if and only if K is marked.

Figure 17–15 Algorithm DP-I

This algorithm is an instance of a very general class of methods, usually referred to as *dynamic programming* (see Sec. 18.6).

Lemma 17.1 *Let M_j be the set of marked nodes after the jth execution of Step 2 in the algorithm of Fig. 17–15. Then $M_j = \{v \in V$: there is a set $S \subseteq \{1, 2, \ldots, j\}$ such that $\sum_{i \in S} c_i = v\}$.*

Proof We prove the lemma by induction on j. It is trivially true for $j = 0$. For the induction step, consider the jth iteration, $j > 0$, and write $M_j = M_{j-1} \cup B_j$, where

$$\begin{aligned} B_j &= \{u: v \in M_{j-1} \quad \text{and} \quad (v, u) \in A_j\} \\ &= \{u: v \in M_{j-1} \quad \text{and} \quad u = v + c_j\} \end{aligned}$$

Consequently, by the induction hypothesis,

$$\begin{aligned} M_j &= \{v \in V: \text{there is a set } S' \subseteq \{1, 2, \ldots, j-1\} \text{ such that either} \\ &\qquad\qquad \sum_{i \in S'} c_i = v \quad \text{or} \quad \sum_{i \in S'} c_i + c_j = v\} \\ &= \{v \in V: \text{there is a set } S \subseteq \{1, 2, \ldots, j\} \text{ such that } \sum_{i \in S} c_i = v\}. \end{aligned}$$

□

Theorem 17.6 *The algorithm DP-I correctly solves* 0-1 *KNAPSACK in $O(nK)$ time.*

Proof The correctness follows directly from the lemma: K is marked if and only if there exists a subset S of $\{1, 2, \ldots, n\}$ such that $\sum_{j \in S} c_j = K$. For the time bound, observe that each execution of Step 2 takes $O(K)$ time, because at most K nodes are marked. □

Example 17.3

Consider the instance (11, 18, 24, 42, 15, 7; 56). Applying our algorithm, we construct the following M_j sets.

M_0: {0}

M_1: {0, 11}

M_2: {0, 11, 18, 29}

M_3: {0, 11, 18, 24, 29, 35, 42, 53}

M_4: {0, 11, 18, 24, 29, 35, 42, 53}

M_5: {0, 11, 15, 18, 24, 26, 29, 33, 35, 39, 42, 44, 50, 53}

M_6: {0, 7, 11, 15, 18, 22, 24, 25, 26, 29, 31, 33, 35, 36, 39, 40, 42, 44, 46, 49, 50, 51, 53}.

Thus the set M_j contains *all* possible sums of integers in $\{c_1, c_2, \ldots, c_j\}$ that do not exceed K. Hence, DP-I can be thought of as a process of constructing the subsets M_j instead of marking nodes in a graph, as shown in Fig. 17-16.

1. $M_0 = \{0\}$.
2. For $j = 1, 2, \ldots, n$ do:
   $M_j = \varnothing$;
   For each integer c in M_{j-1} add to M_j the integers c and $c + c_j$, if they do not exceed K and are not already there.
3. Conclude that the instance has a solution if and only if $K \in M_n$.

Figure 17–16 Algorithm DP-II

Consequently, the instance examined has no solution, because $56 \notin M_6$. □

Let us now introduce the following related *optimization* problem.

OPTIMIZATION 0-1 KNAPSACK
Given the integers $(w_1, \ldots, w_n; c_1, \ldots, c_n; K)$, maximize

$$\sum_{j=1}^{n} c_j x_j$$

subject to

$$\sum_{j=1}^{n} w_j x_j \leq K \qquad x_j = 0, 1.$$

Without loss of generality, we shall assume that $w_j \leq K, j = 1, \ldots, n$. Recall that in our formalism for optimization problems, these are defined in terms of two algorithms $\mathcal{A}_F$ (deciding whether a solution is feasible) and $\mathcal{A}_c$ (evaluating the cost of feasible solutions). An *instance* of an optimization problem is thus represented by a set S of parameters for $\mathcal{A}_F$—in our example of OPTIMIZATION 0-1 KNAPSACK, $S = \{w_1, w_2, \ldots, w_n, K\}$—and a set Q of parameters for $\mathcal{A}_c$, $\{c_1, \ldots, c_n\}$ in our example. It is important to notice at this point that OPTIMIZATION 0-1 KNAPSACK has the property that S and Q are *disjoint*. In other words, feasibility and cost are determined by totally different sets of parameters.

Can OPTIMIZATION 0-1 KNAPSACK be solved by a pseudopolynomial algorithm like DP-II (Fig. 17–16)? The idea, of course, would be to construct the set M_j of all integers expressible as a sum $\sum_{i=1}^{j} x_i c_i$ with $\sum_{i=1}^{j} x_i w_i \leq K$ and $x_i = 0$ or 1, and then select the largest integer in M_n. There is a difficulty, though. In the algorithm DP-II, M_j did not contain duplicate copies of integers; if two subsets of $\{1, 2, \ldots, j\}$ had the same sum of c_i's, they were indistinguishable. In this problem, however, two subsets S and S' of $\{1, 2, \ldots, j\}$ may have the same sum of c_i's but different sums of w_i's. The key observation here is that if

$$\sum_{i \in S} c_i = \sum_{i \in S'} c_i \quad \text{and} \quad \sum_{i \in S} w_i \leq \sum_{i \in S'} w_i$$

(in other words, S and S' are partial solutions with equal costs but different weights), then we can disregard S', the partial solution that has greater total weight. This is because, intuitively, if $S' \cup T$ is a feasible solution with $T \subseteq \{j+1, \ldots, n\}$, $S \cup T$ is also a feasible solution with the same costs.† This leads to the algorithm for OPTIMIZATION 0-1 KNAPSACK in Fig. 17–17.

1. Let $M_0 = \{(\varnothing, 0)\}$.
2. For $j = 1, 2, \ldots, n$ do Steps (a) through (c)
   (a) Let $M_j = \varnothing$.
   (b) For each element (S, c) of M_{j-1}, add to M_j the element (S, c), and also $(S \cup \{j\}, c + c_j)$ if $\sum_{i \in S} w_i + w_j \leq K$.
   (c) Examine M_j for pairs of elements (S, c) and (S', c) with the same second component. For each such pair, delete (S', c) if $\sum_{i \in S'} w_i \geq \sum_{i \in S} w_i$ and delete (S, c) otherwise.
3. The optimal solution is S, where (S, c) is the element of M_n having the largest second component.

Figure 17–17 Algorithm DP-III

Lemma 17.2 *Suppose that $(S, c) \in M_j$ at the end of the execution of DP-III. Then*

(a) $S \subseteq \{1, 2, \ldots, j\}$.

(b) $\sum_{i \in S} c_i = c$.

(c) $\sum_{i \in S} w_i \leq K$.

(d) *If $(S', c) \in M_j$, then $S' = S$.*

(e) *If $S' \subseteq \{1, 2, \ldots, j\}$ and $\sum_{i \in S'} c_i = c$, then $\sum_{i \in S} w_i \leq \sum_{i \in S'} w_i$.*

(f) *Furthermore, if $S \subseteq \{1, 2, \ldots, j\}$ with $\sum_{i \in S} w_i \leq K$ and $\sum_{i \in S} c_i = c$, then there is some $(S', c) \in M_j$.*

Proof The proof is by induction on j. All statements hold trivially for $j = 0$. For the induction step, consider some $j > 0$ and $(S, c) \in M_j$. There are two cases.

Case 1 $j \notin S$. Then (S, c) was transferred at Step 2(b) from M_{j-1} and hence (a), (b), (c), and (d) follow by the induction hypothesis.

Case 2 $j \in S$. Then $(S - \{j\}, c - c_j) \in M_{j-1}$. Again, (a), (b), (c), and (d) hold.

†We shall see further examples of such *dominance* relations among partial solutions of problems in Chapter 18 on branch-and-bound.

To show (e), assume that $S \neq S'$. We have three cases.

Case 1 $j \notin S, S'$. Then (S, c) and (S', c) were also in M_{j-1} and by the induction hypothesis, $S = S'$.

Case 2 $j \in S, S'$. Then $(S - \{j\}, c - c_j), (S' - \{j\}, c - c_j) \in M_{j-1}$, and hence, by induction, $S = S'$.

Case 3 $j \in S$ but $j \notin S'$, or vice-versa. Then (S', c) was deleted at Step 2(c), and (e) follows.

To prove (f), we use induction on max S. It holds for $S = \varnothing$; for the induction step, assume that $k = \max S$. Then in M_{k-1} there is, by the induction hypothesis, an element $(S', c - c_k)$. Thus, in Step 2(b) of the construction of M_k, the element $(S' \cup \{k\}, c)$ was added to M_k. Then either this pair is itself in M_j, or there is an element (S'', c) in M_j. In both cases, (f) holds. □

Theorem 17.7 *The algorithm DP-III solves OPTIMIZATION 0-1 KNAPSACK in $O(n^2c)$ time, where c is the value of the optimal cost.*

Proof Correctness follows directly from the lemma: Because the optimum is chosen by DP-III in Step 3 to be the first component S of the element of M_n with the largest second component c, it follows that S is feasible (by (c)), its cost is c (by (*b*)), and there is no better feasible solution (by (f)).

For the time bound, notice that the size of each set M_{j-1} is no greater than c, because there is at most one element in M_{j-1} with the same second component (by (*d*)). Each update operation of step 2(b) can be done in $O(n)$ time and must be repeated for all $O(c)$ elements of M_{j-1}. Step 2(c) also requires $O(nc)$ time, because it can be implemented concurrently with Step 2(b): We store all elements of M_j in an array of length c, indexing them by their second component, and every time that we attempt to insert a second element at the same location, we calculate and compare the sums of the w_i's, deleting the element that has the largest sum. The theorem follows. □

Example 17.4

Consider the following instance of OPTIMIZATION 0-1 KNAPSACK.

j	1	2	3	4	5
w_j	1	1	3	2	2
c_j	6	11	17	3	9

$K = 5$

Executing the DP-III algorithm results in the following sets:

$M_0 = \{(\varnothing, 0)\}$

$M_1 = \{(\varnothing, 0), (\{1\}, 6)\}$

$M_2 = \{(\varnothing, 0), (\{1\}, 6), (\{2\}, 11), (\{1, 2\}, 17)\}$

$M_3 = \{(\varnothing, 0), (\{1\}, 6), (\{2\}, 11), (\{1, 2\}, 17), (\{1, 3\}, 23), (\{2, 3\}, 28), (\{1, 2, 3\}, 34)\}.$

$M_4 = \{(\varnothing, 0), (\{4\}, 3), (\{1\}, 6), (\{1, 4\}, 9), (\{2\}, 11), (\{2, 4\}, 14), (\{1, 2\}, 17), (\{1, 2, 4\}, 20), (\{1, 3\}, 23), (\{2, 3\}, 28), (\{1, 2, 3\}, 34)\}.$

$M_5 = \{(\varnothing, 0), (\{4\}, 3), (\{1\}, 6), (\{5\}, 9), (\{2\}, 11), (\{4, 5\}, 12), (\{2, 4\}, 14), (\{1, 5\}, 15), (\{1, 2\}, 17), (\{1, 4, 5\}, 18), (\{2, 5\}, 20), (\{1, 3\}, 23), (\{1, 2, 5\}, 26), (\{2, 3\}, 28), (\{1, 2, 3\}, 34)\}.$

Hence the optimal subset is $\{1, 2, 3\}$ with $\sum_{j \in S} c_j = 34$. □

So far, we have been concerned with finding the *exact* optimum of the OPTIMIZATION 0-1 KNAPSACK problem. It turns out, however, that we can give up accuracy in exchange for efficiency. To illustrate this point, suppose that we wish to solve the following instance.

j	1	2	3	4	5	6	7	
w_j	4	1	2	3	2	1	2	$K = 10$
c_j	299	73	159	221	137	89	157	

If we apply DP-III to this problem, we finally conclude that the optimum is $S = \{1, 2, 3, 6, 7\}$ with $\sum_{j \in S} c_j = 777$; the algorithm, however, went through a tedious construction of a total of 91 pairs (S, c). A very straightforward idea for simplifying matters is to *ignore the last decimal digit of the parameters* c_i; the result is the following instance.

j	1	2	3	4	5	6	7	
w_j	4	1	2	3	2	1	2	$K = 10$
$\bar{c}_j$	290	70	150	220	130	80	150	

What this instance lacks in accuracy, though, it gains in efficiency. The DP-III algorithm yields the optimum $S' = \{1, 3, 4, 6\}$ with sum $\sum_{j \in S'} \bar{c}_j = 740$ (about 5 percent from optimal) after the construction of only 36 items. In larger problems, with more drastic truncations, the savings would be more impressive. Furthermore, this sum of 740 corresponds to an even better sum in the original problem—namely, $\sum_{j \in S'} c_j = 768$. In fact, we could have guessed beforehand that truncating would not result in too large an error. To see this, let S and S' be the optimal solutions of the original and the truncated version, respectively. We then have the following inequalities.

$$\sum_{j \in S} c_j \geq \sum_{j \in S'} c_j \geq \sum_{j \in S'} \bar{c}_j \geq \sum_{j \in S} \bar{c}_j \geq \sum_{j \in S} (c_j - 10) \geq \sum_{j \in S} c_j - n \cdot 10$$

So the error

$$\sum_{j \in S} c_j - \sum_{j \in S'} c_j$$

is bounded by $n \cdot 10$ (in our case, 70). In general, if we choose to truncate the last t decimal digits of the c_j's, we know that the deviation from optimality will not exceed $n \cdot 10^t$.

Let us now evaluate the gains in the complexity of the execution of DP-III resulting from our truncation idea. Let c_m be the largest coefficient among the c_j's. The time required by DP-III on the original problem is $O(n^3 c_m)$; on the version after a truncation of t digits, the bound becomes $O(n^3 c_m 10^{-t})$ because the c_j's have been, in effect, divided by 10^t.

It is now very interesting to observe that for every choice of t, the algorithm DP-III applied to the instance with the last t digits of the c_j's truncated is an ϵ-approximate algorithm, with $\epsilon = n10^t/c_m$. This is because

$$\frac{\sum_{j \in S} c_j - \sum_{j \in S'} c_j}{\sum_{j \in S} c_j} \leq \frac{n10^t}{c_m} = \epsilon$$

Consequently, for *every* given value of ϵ, we can design an ϵ-approximate algorithm for OPTIMIZATION 0-1 KNAPSACK, running in $O(n^4/\epsilon)$ time. The algorithm involves truncating the last

$$t = \left\lfloor \log_{10}\left(\frac{\epsilon c_m}{n}\right) \right\rfloor$$

digits and then applying DP-III. Such a favorable state of affairs naturally calls for a definition.

Definition 17.3

We say that an algorithm is a *polynomial-time approximation scheme* (PTAS) for an optimization problem A if, when supplied with an instance of A and an $\epsilon > 0$, it returns an ϵ-approximate solution within time which is bounded by a polynomial (depending on ϵ) in the length of the instance. □

What we have described is therefore a PTAS for OPTIMIZATION 0-1 KNAPSACK; the polynomial dependent on ϵ is $p_\epsilon(n) = (1/\epsilon) \cdot n^4$. Thus the bound is a polynomial both in n *and* $1/\epsilon$. This is quite fortunate, because Definition 17.3 allows polynomials of the form $p_\epsilon(n) = n^{1/\epsilon^2}$ or $n^{2^{1/\epsilon}}$, for example. To see the difference, just calculate $p_{0.1}(n)$ in all three cases! In Problem 4 we show actual examples of such PTAS's. To distinguish between these two kinds of schemes, we have the second part of Def. 17.3.

Definition 17.3 (Continued)

A PTAS is called a *fully polynomial-time approximation scheme* (FPTAS) if it operates within a bound that is polynomial both in the length of the instance *and* $1/\epsilon$. □

We can summarize this discussion as follows.

Theorem 17.8 *The algorithm DP-IV in Fig. 17–18 is an FPTAS for OPTIMIZATION* 0-1 *KNAPSACK.*

---

1. Let c_m be the largest of the c_j's.
2. Let $t = \lfloor \log_{10} (\epsilon c_m/n) \rfloor$.
3. For $j = 1, \ldots, n$ do $\bar{c}_j := \lfloor c_j/10^t \rfloor \cdot 10^t$.
4. Apply DP-III to the instance $(w_1, \ldots, w_n; \bar{c}_1, \bar{c}_2, \ldots, \bar{c}_n; K)$.

Figure 17–18 Algorithm DP-IV

Let us now try to pinpoint the features of the OPTIMIZATION 0-1 KNAPSACK problem that enabled us to turn the pseudopolynomial algorithm DP-III into the FPTAS DP-IV. An important property of this problem is that, as was pointed out before, in any instance the feasibility-checking parameters $S = \{w_1, \ldots, w_n, K\}$ are disjoint from the cost-evaluating parameters $Q = \{c_1, \ldots, c_n\}$. A key point is that the complexity of DP-III is not only bounded by a polynomial in $|I|$ and number(I), but in particular by a polynomial in $|I|$ and c_m, the largest among the c_j's. Another regularity property that makes Theorem 17.8 possible is the fact that the optimal cost c_I and c_m are polynomially related via $|I|$; that is, for two polynomials p_1 and p_2, we have

$$c_I \leq p_1(|I|, c_m) \quad \text{and} \quad c_m \leq p_2(|I|, c_I)$$

Finally, another important feature is that the cost of any fixed feasible solution is a linear functional of the c_j's. We leave it as an exercise to establish that DP-IV is indeed an instance of a far more general methodology for devising an FPTAS starting from a pseudopolynomial algorithm.

Theorem 17.9 *Suppose that all instances $I = (S, Q)$ of an optimization problem* A *are such that $Q = \{q_1, \ldots, q_n\}$ is a set of integers, and*

(*a*) *The optimal cost of the instance c_I satisfies, for some polynomials p_1 and p_2, $c_I \leq p_1(|I|, q)$, and $q \leq p_2(|I|, c_I)$, where q is the largest integer appearing in Q.*

(*b*) *For any given feasible solution f, the cost $c(f, Q)$ is a linear functional of $\{q_1, \ldots, q_n\}$.*

(*c*) *Problem* A *can be solved by an algorithm time-bounded by $p_3(|I|, q)$ where p_3 is a polynomial.*

Then A *has an FPTAS.*

We note here that Conditions (a) and (b), despite their technical appearance, hold true for many optimization problems.

17.4 Negative Results

For some combinatorial optimization problems the theory of *NP*-completeness can be applied to prove not only that they cannot be solved exactly by polynomial-time algorithms (unless $P = NP$), but also that they do not have ϵ-approximate algorithms, for various ranges of ϵ, again unless $P = NP$. In this section, we prove three representative results of this sort. Our first theorem concerns the general TSP.

Theorem 17.10 *Unless $P = NP$, there is no ϵ-approximate polynomial algorithm for the TSP for any $\epsilon > 0$.*

Proof Assume that there is an ϵ-approximate polynomial algorithm $\mathcal{A}_\epsilon$ for the TSP for some $\epsilon > 0$. We shall prove that we then have a polynomial algorithm $\mathcal{A}_{HC}$ that solves the HAMILTON CIRCUIT problem. Since this problem is *NP*-complete, the theorem will follow.

The algorithm $\mathcal{A}_{HC}$ works as follows: Given any graph $G = (V, E)$, it constructs a $|V|$-city instance of the TSP. The distance d_{ij} is taken to be 1 if $[i, j]$ is an edge of G; otherwise $d_{ij} = 2 + \epsilon|V|$. Then $\mathcal{A}_{HC}$ applies the assumed algorithm $\mathcal{A}_\epsilon$ to this instance. We *claim* that $\mathcal{A}_\epsilon$ will return a tour of cost $|V|$ if and only if G has a Hamilton circuit; this will imply that $\mathcal{A}_{HC}$ correctly solves the HAMILTON CIRCUIT problem in polynomial time. One direction of our claim is obvious. If $\mathcal{A}_\epsilon$ returns a tour of cost $|V|$, it follows that there is a tour using only distances 1, because the shortest distance in this instance of TSP is 1. Thus there is a Hamilton circuit in G.

For the other direction, assume that $\mathcal{A}_\epsilon$ returns a tour of length greater than $|V|$ but there is still a Hamilton circuit in G. The tour must have length at least $1 + (1 + \epsilon)|V|$, because the next longer distance after 1 is $2 + \epsilon|V|$. And, since G has a Hamilton circuit, the optimal tour has length $|V|$. Thus the tour returned by $\mathcal{A}_\epsilon$ is not ϵ-suboptimal, because

$$\frac{1 + (1 + \epsilon)|V| - |V|}{|V|} = \frac{1}{|V|} + \epsilon > \epsilon$$

This is a contradiction. □

Recall that for the ΔTSP, we do have a $\frac{1}{2}$-approximate algorithm (Sec. 16.2), which cleverly exploits the triangle inequality. Theorem 17.10 suggests that for the general problem there can be no ϵ-approximate algorithms—not even with $\epsilon = 1000$.

Our next result concerns the maximization version of the CLIQUE problem: Given a graph, find its largest complete subgraph. To prove the next theorem, we first need some graph-theoretic facts.

Definition 11.4

Let $G = (V, E)$ be a graph. The graph $G^2 = (V^2, E^2)$ is the graph with node set $V^2 = V \times V$ and set of edges $E^2 = \{[(v, u), (v', u')]$: either $v = v'$ and $[u, u'] \in E$, or $[v, v'] \in E\}$. For example, if G is the graph of Fig. 17–19(a), G^2 is shown in Fig. 17–19(b) (the sheaves of edges are complete bipartite graphs). □

The following lemma establishes a connection between this construction and the clique problem.

Lemma 17.3 *G has a clique of size k if and only if G^2 has a clique of size k^2.*

Proof Suppose that G has a clique $C = \{v_1, v_2, \ldots, v_k\}$. Then G^2 obviously has the clique $C^2 = \{(v, u): v, u \in C\}$ of size k^2.

For the other direction, suppose that G^2 has a clique C^2 of size k^2, but G has no clique of size k. Now, the set $D = \{v: \exists\, u \in V$ such that $(v, u) \in C^2\}$ is certainly a clique in G; hence $|D| \leq k - 1$. Therefore, one of the $|D|$ sets $F_v = \{u: (v, u) \in E\}$ for $v \in D$ must have k or more nodes. It is not hard to see that such sets are also cliques; this proves the lemma. □

Theorem 17.11 *If the clique problem has a polynomial-time ϵ-approximate algorithm for* any $1 > \epsilon > 0$, *then it has a polynomial ϵ-approximate algorithm for* all $1 > \epsilon > 0$.†

Proof Suppose that there is a polynomial ϵ-approximate algorithm $\mathcal{A}_\epsilon$ for the clique problem for some $\epsilon > 0$. We shall first construct, based on $\mathcal{A}_\epsilon$, a polynomial δ-approximate algorithm $\mathcal{A}_\delta$, where $\delta < \epsilon$. The algorithm $\mathcal{A}_\delta$

†Since clique is a maximization problem, only approximation ratios ϵ less than 1 are meaningful.

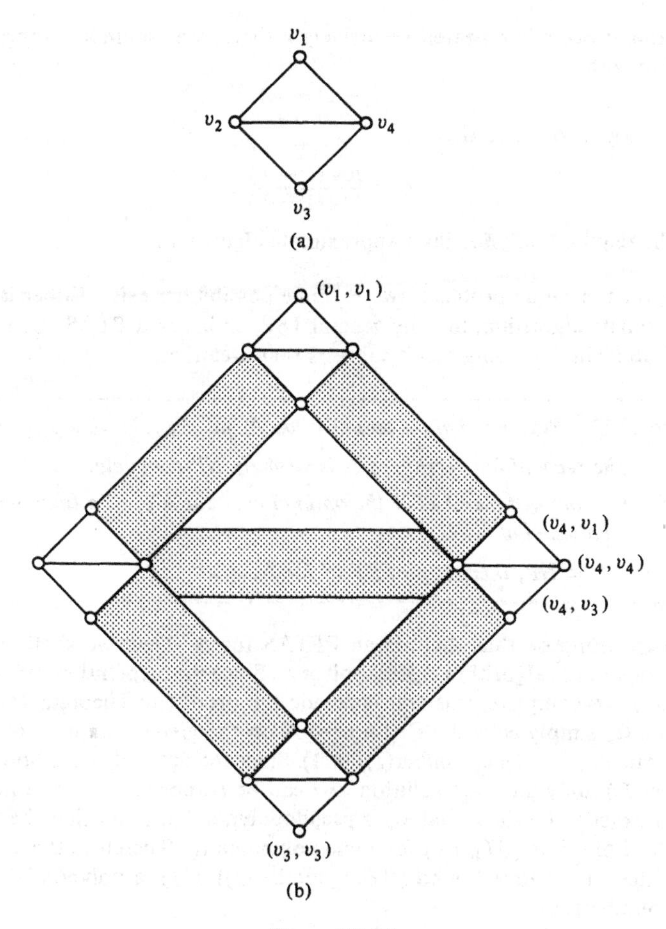

Figure 17–19

operates as follows: Given a graph G, it constructs G^2 and then applies $\mathcal{A}_\epsilon$ to it. From the clique C^2 of G^2, it selects the largest of the sets D and F_v for $v \in D$, as in the proof of the lemma. The resulting set, C, is a clique of G and satisfies $|C| \geq \lceil\sqrt{|C^2|}\rceil$. Now let k be the size of the largest clique in G; by the lemma, the largest clique in G^2 has k^2 nodes. Since $\mathcal{A}_\epsilon$ is ϵ-approximate, we have

$$\frac{k^2 - |C^2|}{k^2} \leq \epsilon \quad \text{or} \quad |C^2| \geq k^2(1 - \epsilon)$$

So

$$|C| \geq \sqrt{|C^2|} \geq k\sqrt{1 - \epsilon}$$

and hence $\mathcal{A}_\delta$ is indeed a δ-approximate algorithm, with $\delta = 1 - \sqrt{1 - \epsilon}$.

If this process is repeated recursively r times, we obtain a δ-approximate algorithm with

$$\delta = 1 - (1 - \epsilon)^{1/2^r}$$

Fixing r large enough so that

$$2^r > \frac{\log (1 - \epsilon)}{\log (1 - \delta)}$$

yields the required polynomial δ-approximate algorithm. □

So, for the clique problem, two extreme possibilities exist: Either it has no ϵ-approximate algorithm, like the general TSP, or it has a PTAS. Can it have an FPTAS? The following result suggests that it cannot.

Theorem 17.12 *Suppose that an optimization* A *has the following properties.*

(*a*) *The recognition version of* A *is strongly NP-complete.*

(*b*) *For any instance I of A the optimal cost obeys $c_I \leq p$ (number(I)) for a polynomial p(n).*

Then, unless $P = NP$, there is no FPTAS for A.

Proof Suppose that there is an FPTAS for A. Then we shall exhibit a pseudopolynomial algorithm $\mathcal{A}_\psi$ that solves A. Since the recognition version of A is strongly *NP*-complete, this will conclude the proof (by Theorem 16.4). The algorithm $\mathcal{A}_\psi$ simply calls the FPTAS for A on the given instance I of A with error parameter $\epsilon = 1/(p(\text{number}(I)) + 1)$. Since the optimal cost is bounded by $p(\text{number}(I))$, only an exact solution to I can be ϵ-approximate, and hence $\mathcal{A}_\psi$ solves A exactly. To show that $\mathcal{A}_\psi$ is pseudopolynomial, recall that the FPTAS operates within time $q(|I|, 1/\epsilon)$ for some polynomial q. Therefore the algorithm $\mathcal{A}_\psi$ operates within time bound $q(|I|, p(\text{number}(I)) + 1)$, a polynomial in both $|I|$ and number(I). □

PROBLEMS

1. We proved (Lemma 15.4) that NODE COVER, CLIQUE, and INDEPENDENT SET are all equivalent. Can you, then, transform Algorithm 2 of Section 17.1 into approximation algorithms for CLIQUE and INDEPENDENT SET? Why not? (*Note:* It is difficult to develop a theory of approximate algorithms that parallels that of *NP*-completeness. One reason is the sensitivity of the cost to even simple reductions, as exemplified in this problem.)

*2. Show that Algorithm 1 always returns a node cover of size at most ln n times the optimal for graphs with n nodes.

3. A *Hamilton walk* of a graph $G = (V, E)$ is a closed walk that visits each node at least once.
   (a) Show that the problem of finding the shortest Hamilton walk in a graph is *NP*-complete (that is, its recognition version is).
   (b) Give a $\frac{1}{2}$-approximate algorithm for this problem.
4. The following is an optimization version of the PARTITION problem (recall Corollary 1 to Theorem 15.8).

   Given n integers $c_1, \ldots, c_n$, find a partition of $\{1, 2, \ldots, n\}$ into two subsets S_1, S_2 that minimizes the quantity $\max\,(\sum_{j \in S_1} c_j, \sum_{j \in S_2} c_j)$.

   Consider the following heuristic for some fixed integer k:

   1. Choose the k largest c_j's.
   2. Find the optimal partition of these k integers (**comment**: by some exhaustive method).
   3. Complete this into a partition of $\{1, 2, \ldots, n\}$ by considering each of the remaining c_j's and adding it to the partition which at the time has the smallest sum.

   *(a) Prove that this $O(2^k + n)$ algorithm is $\dfrac{1}{2+k}$-approximate.
   (b) Based on (a), devise an approximation scheme for this problem. What is the complexity of your PTAS, in terms of n and $1/\epsilon$?
5. Prove Theorem 17.9.
6. Show that, unless $P = NP$, there can be no polynomial-time ϵ-approximate algorithm for MULTIPROCESSOR SCHEDULING for any $\epsilon < \frac{1}{3}$. (*Hint:* Recall the proof of Theorem 15.5.)
7. Describe a polynomial-time $\frac{1}{2}$-approximation algorithm for the wandering salesman problem (Problem 8 of Chapter 12) with the triangle inequality.

NOTES AND REFERENCES

Approximation for *NP*-complete problems is currently a very active area of research. Despite a wealth of impressive results, little order or unifying theory appears to exist. A snapshot of research in this area in 1976 can be found in

[GJ1] GAREY, M. R., and D. S. JOHNSON, "Approximation Algorithms for Combinatorial Problems: An Annotated Bibliography," pp. 41–52 in *Algorithms and Complexity: New Directions and Recent Results*, ed. J. F. Traub. New York: Academic Press, Inc., 1976.

Approximation algorithms for scheduling problems are surveyed by R. L. Graham in Chapter 5 of

[Co] COFFMAN, E. G., JR., ed., *Computer and Jobshop Scheduling Theory*. New York: Wiley-Interscience, 1976.

The solution to Problem 4 can be found there. Algorithm 1 (Sec. 17.1 and Problem 2) is analyzed in

[Jo] JOHNSON, D. S., "Approximation Algorithms for Combinatorial Problems," *JCSS*, 9 (1974), 256–78.

Algorithm 2 was discovered independently by Fanica Gavril and Mihalis Yannakakis. Christofides' algorithm is from

[Ch] CHRISTOFIDES, N., "Worst-case Analysis of a New Heuristic for the Traveling Salesman Problem," Technical Report, GSIA, Carnegie-Mellon Univ., 1976.

The tree algorithm was apparently known to researchers for some time. An elaborate version was published in

[RSL] ROSENKRANTZ, D. J., R. E. STEARNS, and P. M. LEWIS, "An Analysis of Several Heuristics for the Traveling Salesman Problem," *J. SIAM Comp.*, 6 (1977), 563–81.

For FPTAS's, see

[IK] IBARRA, O. H., and C. E. KIM, "Fast Approximation Algorithms for the Knapsack and Sum of Subset Problems," *J. ACM*, 22 (1975), 463–68.

[La] LAWLER, E. L., "Fast Approximation Schemes for Knapsack Problems," *Proc. 18th Ann. Symp. on Foundations of Computer Science*, IEEE Computer Soc. (1977), 206–13.

and

[Sa] SAHNI, S., "General Techniques for Combinatorial Approximation," *OR*, 25 (1977), 920–36.

Theorem 17.10 is due to

[SG] SAHNI, S., and T. GONZALEZ, "*P*-complete Approximation Problems," *J. ACM*, 23 (1976), 555–65.

Theorem 17.11 is mentioned in

[GJ2] GAREY, M. R., and D. S. JOHNSON, "The Complexity of Near-Optimal Graph Coloring," *J. ACM*, 23 (1976), 43–49.

Theorem 17.12 is from

[GJ3] GAREY, M. R., and D. S. JOHNSON, "Strong *NP*-completeness Results: Motivation, Examples, and Implications," *J. ACM*, 25 (1978), 499–508.

Theorem 17.2 is from

[Eu] EULER, L., "Solutio Problematis ad Geometriam Situs Pertinentis," *Commentarii Academiae Petropolitanae*, 8 (1736), 128–140 (in Latin).

18

Branch-and-Bound and Dynamic Programming

18.1 Branch-and-Bound for Integer Linear Programming

The branch-and-bound method is based on the idea of intelligently enumerating all the feasible points of a combinatorial optimization problem. The qualification *intelligently* is important here because, as should be clear by now, it is hopeless simply to look at all feasible solutions. Perhaps a more sophisticated way of describing the approach is to say that we try to construct a proof that a solution is optimal, based on successive partitioning of the solution space. The *branch* in branch-and-bound refers to this partitioning process; the *bound* refers to lower bounds that are used to construct a proof of optimality without exhaustive search. We shall develop the method in this section for ILP, and then put things in a more abstract framework.

Consider, then, the ILP

$$\begin{aligned} & \min z = c'x = c(x) \\ \textit{Problem 0} \qquad & Ax \leq b \\ & x \geq 0, \text{ integer} \end{aligned} \tag{18.1}$$

If we solve the LP relaxation, we obtain a solution x^0, which in general is not integer. The cost $c(x^0)$ of this solution is, however, a lower bound on the optimal cost $c(x^*)$ (where x^* is the optimal solution to Problem 0), and if x^0 were integer, we would in fact be done. In the cutting-plane algorithm, we would now add a constraint to the relaxed problem that does not exclude feasible solutions to (18.1). Here, however, we are going to split the problem into two subproblems by adding two *mutually exclusive* and *exhaustive* constraints. Suppose that component x_i^0 of x^0 is noninteger, for example. Then the two subproblems are

$$\text{Problem 1} \quad \begin{aligned} \min z &= c'x = c(x) \\ Ax &\leq b \\ x &\geq 0, \text{ integer} \\ x_i &\leq \lfloor x_i^0 \rfloor \end{aligned} \tag{18.2}$$

and

$$\text{Problem 2} \quad \begin{aligned} \min z &= c'x = c(x) \\ Ax &\leq b \\ x &\geq 0, \text{ integer} \\ x_i &\geq \lfloor x_i^0 \rfloor + 1 \end{aligned} \tag{18.3}$$

Example 18.1

A simple ILP is shown in Fig. 18–1(a); the solution is $x^* = (2, 1)$ and $c(x^*) = -(x_1 + x_2) = -3$. The initial relaxed problem has the solution $x^0 = (\frac{3}{2}, \frac{5}{2})$ with cost $c(x^0) = -4$. Figure 18–1(b) shows the two subproblems generated by choosing the noninteger component $x_1^0 = \frac{3}{2}$ and introducing the constraints

$$x_1 \leq 1 \quad \text{and} \quad x_1 \geq 2 \quad \square$$

The solution to the original problem must lie in the feasible region of one of these two problems, simply because one of

$$x_i^* \leq \lfloor x_i^0 \rfloor$$
$$x_i^* \geq \lfloor x_i^0 \rfloor + 1$$

must be true.

We now choose one of the subproblems, say Problem 1, which is after all an LP, and solve it. The solution x^1 will in general not be integer, and we may split Problem 1 into two subproblems just as we split Problem 0, creating Problems 3 and 4. We can visualize this process continuing indefinitely as a successively finer and finer subdivision of the feasible region, as shown in Fig. 18–2. Each subset in a given partition represents a subproblem i, with relaxed solution x^i and lower bound $z_i = c(x^i)$ on the cost of any solution in the partition.

We can also visualize this process as a tree, as shown in Fig. 18–3. The root represents the original feasible region and each node represents a subproblem.

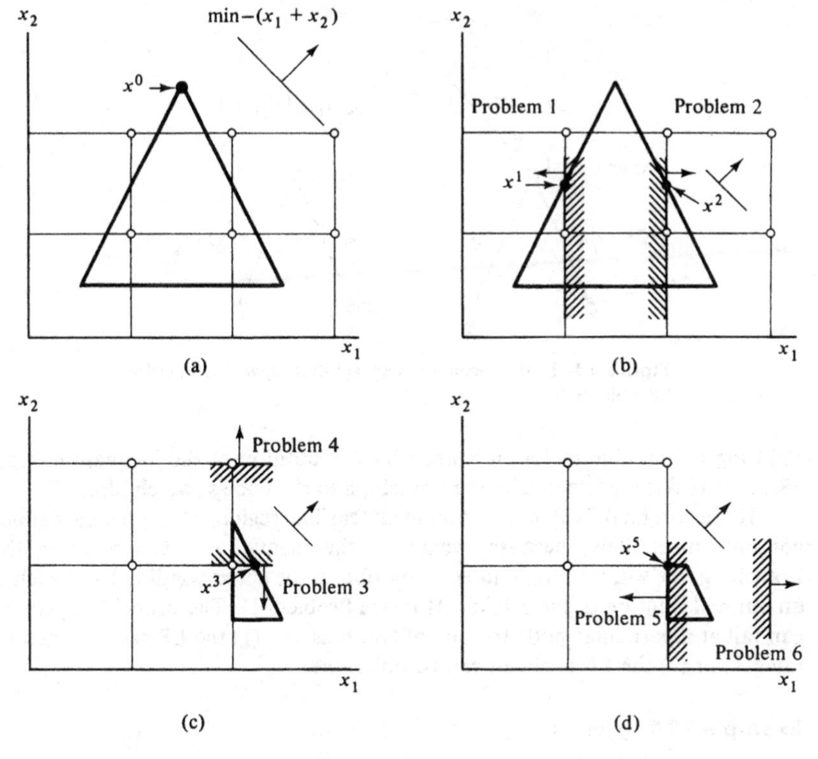

Figure 18–1 Stages in the solution of an ILP by branch-and-bound.

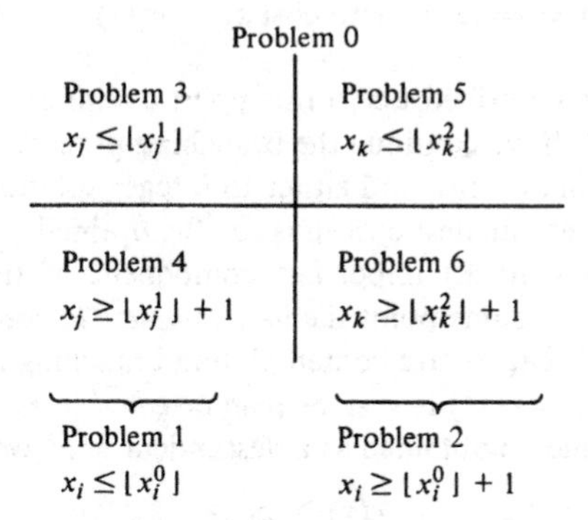

Figure 18–2 Successive subdivision of the feasible region of an ILP by addition of inequalities.

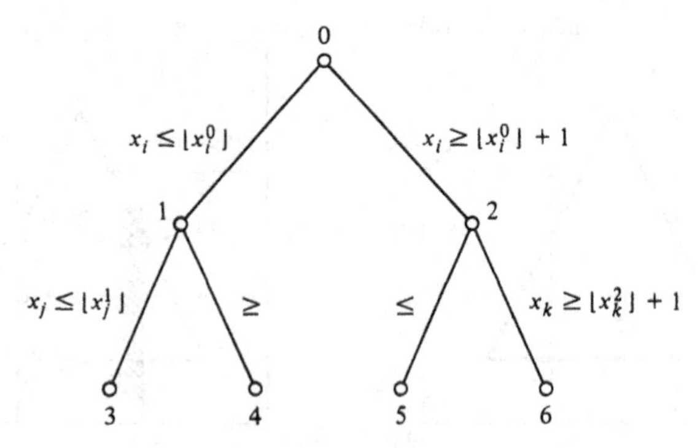

Figure 18–3 Representation of solution space subdivision by a binary tree.

Splitting the feasible region at a node by the addition of the inequalities Eqs. 18.2 and 18.3 is represented by the branching to the node's two children.

If the original ILP has a bounded feasible region, this process cannot continue indefinitely, because eventually the inequalities at a node in the branching tree will lead to an integer solution to the corresponding LP, which is an optimal solution to the original ILP (see Problem 1). The branching process can fail at a particular node for one of two reasons: (1) the LP solution can be integer; or (2) the LP problem can be infeasible.

Example 18.1 (Continued)

If we continue branching from Problem 2 in the example in Fig. 18–1, we obtain the branching tree shown in Fig. 18–4. Three leaves are reached in the right subtree; these leaves correspond to two infeasible LP's, and one LP with an integer solution $x^5 = (2, 1)$ with cost $z_5 = c(x^5) = -3$. □

What we have described up to this point comprises the *branching* part of branch-and-bound. If we continue the branching process until all the nodes are leaves of the tree and correspond either to integer solutions or infeasible LP's, then the leaf with the smallest cost must be the optimal solution to the original ILP. We come now to an important component of the branch-and-bound approach: Suppose at some point the *best complete integer solution* obtained so far has cost z_m and that we are contemplating branching from a node at which the lower bound $z_k = c(x^k)$ is *greater than or equal to* z_m. This means that any solution x that would be obtained as a descendent of x^k would have cost

$$c(x) \ge z_k \ge z_m$$

and hence we need not proceed with a branching from x^k. In such a case, we say that the node x^k has been *killed*, and refer to it (as one might guess) as *dead.*

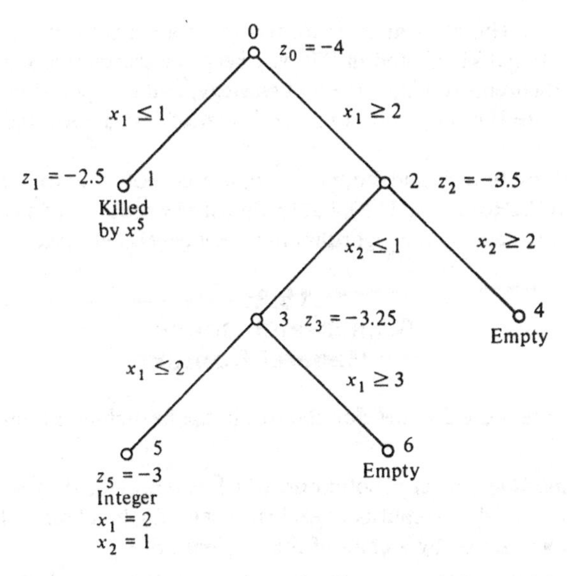

Figure 18–4 The binary tree leading to a solution to the problem.

(The term *fathomed* is also used.) The remaining nodes, from which branching is still possibly fruitful, are referred to as *live*.

Example 18.1 (Continued)

Referring again to Fig. 18–1 and 18–4, the node corresponding to Problem 1 has associated with it a lower bound of -2.5, which is greater than the solution cost of -3 associated with node 5. It is therefore killed by node 5, as shown. Since no live nodes remain, node 5 must represent the optimal solution. □

There are now still two important details in the algorithm that need to be specified: We must decide how to choose, at each branching step, which node to branch from; and we must decide how to choose which noninteger variable is to determine the added constraint. Dakin [Da] recommends branching in a depth-first manner to reduce the amount of storage needed for intermediate tableaux. If storage is not a determining factor, branching from the live node with the lowest lower bound might seem to be a reasonable heuristic. We shall discuss this problem later on in this chapter in a more general context; but little is known in general about this choice.

As for the second choice—the variable to add the constraint—Dakin [Da] reports that the best strategy is to find that constraint which leads to the largest increase in the lower bound z after performing one iteration of the dual simplex algorithm after that constraint is added, and to add either that constraint or its alternative [Da]. The motivation is to find the branch from a given node that is most likely to get killed, and in this way keep the search tree shallow. Again, there are no theorems to tell us the best strategy, and computational experience and intuition are the only guides to the design of fast algorithms of this type known at this time.

The idea of branch-and-bound is applicable not only to a problem formulated as an ILP (or mixed ILP), but to almost any problem of a combinatorial nature. We next develop the algorithm in a very general context.

18.2 Branch-and-Bound in a General Context

Two things were needed to develop the tree in the branch-and-bound algorithm for ILP.

1. *Branching* A set of solutions, which is represented by a node, can be partitioned into mutually exclusive sets. Each subset in the partition is represented by a child of the original node.
2. *Lower bounding* An algorithm is available for calculating a lower bound on the cost of any solution in a given subset.

No other properties of ILP were used. We may therefore formulate the method for any optimization problem in which (1) and (2) are available, whether or not the cost function or constraints are linear.

Figure 18–5 shows the basic algorithm. We use the set *activeset* to hold the live nodes at any point; the variable U is used to hold the cost of the best complete solution obtained at any given time (U is an *upper bound* on the optimal

```
begin
  activeset:={0}; (comment: "0" is the original problem)
  U:=∞;
  currentbest:=anything;
  while activeset is not empty do
    begin
      choose a branching node, node k ∈ activeset;
      remove node k from activeset;
      generate the children of node k, child i, i = 1, . . . , n_k,
      and the corresponding lower bounds, z_i;
      for i = 1, . . . ,n_k do
        begin
          if z_i ≥ U then kill child i
            else if child i is a complete solution then
                    U:=z_i, currentbest:=child i
                    else add child i to activeset
        end
    end
end
```

Figure 18–5 The basic branch-and-bound algorithm.

cost.) Notice that the branching process need not produce only two children of a given node, as in the ILP version, but any finite number.

Example 18.2 (The Shortest-Path Problem)

The shortest-path problem with nonnegative arc weights provides us with a transparent application of branch-and-bound, although we already have an efficient algorithm for its solution. Figure 18–6(a) shows an instance of the shortest-path problem. To solve this instance (Problem 0) by branch-and-bound we *branch* by choosing the next arc with which to continue the path. Thus a subset of feasible solutions corresponds to all paths from s to t that start by the choices already made. Fig. 18–6(b) shows a snapshot of the search tree that results when we branch from a node with the lowest lower bound at any point. The lower bound used is quite naturally the cumulative length of the partial path up to the particular point in the graph represented by the node in the search tree. For example, the path b-g-l brings us to a node with lower bound 10, the sum of the costs of the edges b, g, and l. Notice that in this example it is quite easy to compute at once the lower bounds for all the children of a node at which branching is taking place. In the ILP application, we needed to solve a linear program to obtain a lower bound, so we did not necessarily find both of the lower bounds immediately on branching, with the hope that we might avoid some of these calculations by killing later on.

Figure 18–6(c) shows the final search tree for this example. The first complete solution found has a cost of 8, and this turns out to be optimal. The killed nodes are indicated by bars below their lower bounds. □

Example 18.3 (The Traveling Salesman Problem)

A more realistic application of branch-and-bound is provided by an *NP*-complete problem like the TSP. There is more than one way to formulate a branching process for the TSP; the simplest, perhaps, is to partition the solution space into two sets at any point, according to whether a given edge is or is not in the tour. Little and others [LMSK] use this approach, together with a heuristic for the lower bound.

Another approach, attributed to Eastman [Ea], makes use of the fact that an efficient algorithm exists for the weighted bipartite matching, or assignment, problem (Chapter 11). This provides us with yet another example of one problem showing up as an easier subproblem of a more difficult one.

If we let the variable $x_{ij} = 1$ if edge $[i, j]$ is in a tour and zero otherwise, a tour for the n-city TSP with weights c_{ij} must satisfy

$$\min z = \sum_{i,j=1}^{n} c_{ij}x_{ij}$$

$$\sum_{i=1}^{n} x_{ij} = 1 \qquad j = 1, \ldots, n \tag{18.4}$$

$$\sum_{j=1}^{n} x_{ij} = 1 \qquad i = 1, \ldots, n$$

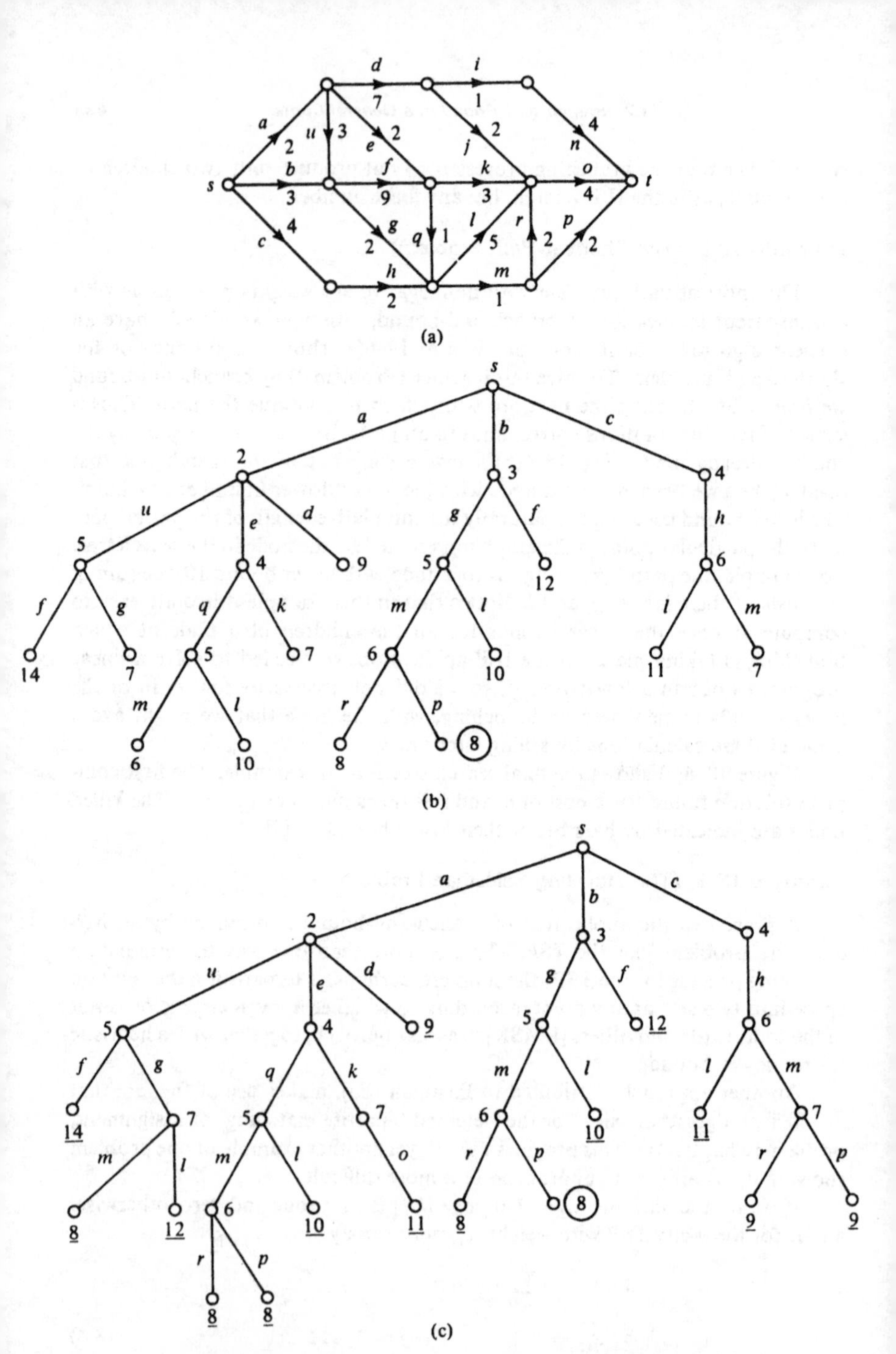

Figure 18–6 A shortest-path problem and its solution by branch-and-bound.

The two sets of equality constraints express the fact that exactly one edge enters and leaves each node. This formulation (the assignment problem of Sec. 11.2) is not sufficient to capture the difficulty of the TSP, since we have no way of ensuring that the solution is a single tour with precisely one cycle. Thus the constraints in (18.4) are necessary but not sufficient for the TSP, and a solution to (18.4) yields a value of z that is a lower bound on the cost of the TSP.

Furthermore, if the solution to (18.4) *is* a tour, then it solves our TSP. If it is not a tour, then the solution contains a cycle of length less than n, a *subtour* $[x_{12}, x_{23}, \ldots, x_{k1}]$ with

$$x_{12} = x_{23} = \ldots = x_{k1} = 1$$

Not all these variables can equal one in a solution to the TSP, so we can partition the solution space into k subsets by adding one of the constraints

$$x_{12} = 0$$
$$x_{23} = 0$$
$$\vdots$$
$$x_{k1} = 0$$

at a time. This yields k problems, each of which is also an assignment problem. (Just put a very high cost on the edge x_{ij} we wish to exclude from the solution.) The branching step is illustrated in Fig. 18–7. The solution to the assignment problem at any node, by the methods of Chapter 11, then provides the lower bound at that node. □

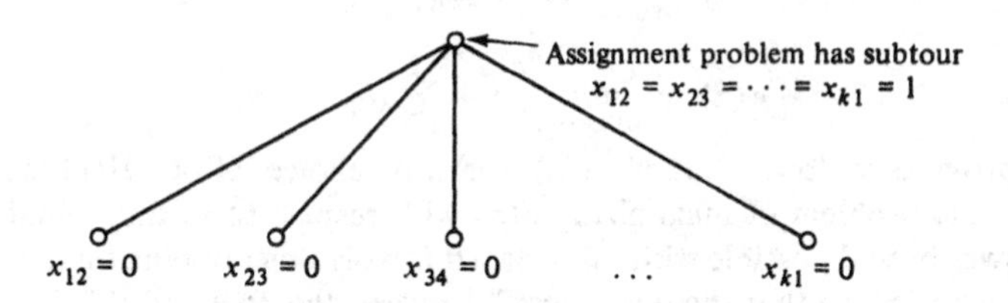

Figure 18–7 The branching step of branch-and-bound for the TSP.

Example 18.4 (A Spanning Tree Bound for the TSP)

Held and Karp [HK1, HK2] describe a very effective branch-and-bound algorithm for the TSP, based on a lower bound computed from related minimal spanning trees. We need the following idea.

Definition 18.1

Given a complete graph $G = (V, E)$ with distance matrix $[d_{ij}]$ and $n = |V|$ nodes, a *1-tree* is a graph formed by a tree on the node set $\{2, \ldots, n\}$, plus two edges incident on node 1. □

Now every tour is a 1-tree (but not vice-versa), so the minimal cost of a 1-tree is a lower bound on the cost of a tour. Furthermore, the cost of a minimal 1-tree is easy to compute (see Problem 2). If we branch in a branch-and-bound algorithm for the TSP by including and excluding sets of edges, the subproblems are also TSP problems, and the corresponding 1-tree problems give us lower bounds. Held and Karp do not rest at this point, but pursue much tighter lower bounds based on this idea.

Suppose that we transform a TSP by replacing its distance matrix by $[d_{ij} + \pi_i + \pi_j]$ for some numbers π_i. Then every tour has its cost increased by the amount

$$2\sum_{i=1}^{n} \pi_i$$

because every node is entered and exited exactly once. Thus the relative ranking of the costs of tours is unaffected by this transformation, and in particular an optimal tour remains optimal. On the other hand, the optimal 1-tree may change. If the degree of node i in a 1-tree is δ_i, then the cost of a 1-tree, after a transformation of the distance matrix by π, is

$$c + \sum_{i=1}^{n} \delta_i \pi_i$$

where c is the cost of the 1-tree with the original cost matrix $[d_{ij}]$. If c^* is the cost of an optimal tour, we therefore have

$$c^* + 2\sum_{i=1}^{n} \pi_i \geq \min_{\text{all 1-trees}} \left[c + \sum_{i=1}^{n} \delta_i \pi_i \right]$$

We can rewrite this as

$$c^* \geq w(\pi)$$

where

$$w(\pi) = \min_{\text{all 1-trees}} \left[c + \sum_{i=1}^{n} (\delta_i - 2)\pi_i \right]$$

This provides a lower bound $w(\pi)$ for any choice of π; Held and Karp pursue the problem of maximizing $w(\pi)$ with respect to π, thus obtaining the best lower bound possible with this idea. (It is worth mentioning that, in general, $c^* > \max_\pi w(\pi)$, so that there is a "gap" between the TSP and this corresponding lower-bounding problem.)

This bound obtained in [HK2] is so tight that the search trees for some fairly large problems (up to $n = 64$) can be exhibited in their entirety. This is a dramatic example of the importance of an effective lower bound in the branch-and-bound approach, since previous applications resulted in much larger search trees. □

18.3 Dominance Relations

Thus far, the only way a node can be eliminated for contention as the ancestor of an optimal solution—that is, killed—is by having its lower bound above the current upper bound. There is another way to kill a node, however: Consider the shortest-path problem in Example 18.2, for example. Suppose we branch to obtain the two nodes determined by edges a, e, q (with a lower bound of 5) and c, h (with a lower bound of 6). These two paths lead to the same point in the original graph, and we may say that the two paths in the branching tree have *merged*. There is no sense in pursuing the c, h path, because we have reached the same point with less cost via the a, e, q path. We may therefore kill the node corresponding to the c, h path, even before any complete solutions have been obtained. This is an example of a *dominance* relation; we say that the c, h node is *dominated* by the a, e, q node. One general way to define such a relation is as follows.

Definition 18.2

If we can show at any point that the best descendant of a node y is at least as good as the best descendant of node x, then we say *y dominates x*, and y can kill x. □

The existence of a practical algorithm for testing dominance depends very much on the particular problem at hand.

18.4 Branch-and-Bound Strategies

There are now many choices in how we implement a branch-and-bound algorithm for a given problem; we shall discuss some of them in this section.

First, there is the choice of the branching itself—there may be many schemes for partitioning the solution space, as in the TSP, or in the general ILP, for that matter.

Next, there is the lower-bound calculation. One often has a choice here between bounds that are relatively tight but require relatively large computation time and bounds that are not so tight but can be computed fast. A similar trade-off may exist in the choice of a dominance relation.

Third, there are many ways in which we can use the lower-bound and dominance relations. To explain, let $AS(x)$ be the *activeset* when node x is branched from; let $CH(x)$ be the set of children of x in the branching tree; and let the upper bound $U(x)$ be the best cost of a complete solution when x is branched from (see Fig. 18–5.). Then Fig. 18–8 shows four possible ways in which nodes can be killed:

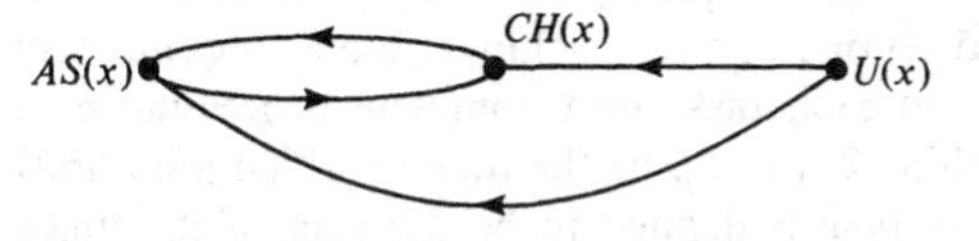

Figure 18–8 Possible ways in which nodes can be killed.

(a) A live node in $AS(x)$ can kill nodes in $CH(x)$.

(b) Nodes in $CH(x)$ can kill nodes in $AS(x)$.

(c), (d) The upper bound $U(x)$ can kill nodes in $AS(x)$ or $CH(x)$.

Again there is a trade-off in the choice of what is to be implemented: the time taken to test $CH(x)$ against $AS(x)$ may or may not be worthwhile in any particular problem.

Fourth, there is the choice at each branching step of which node to branch from. The usual alternatives are least-lower-bound-next, last-in-first-out, or first-in-first-out.

Still another choice must be made at the start of the algorithm. It is often practical to generate an initial solution by some heuristic construction, such as those described in Chapters 17 or 19. This gives us an initial upper bound $U < \infty$ and may be very useful for killing nodes early in the algorithm. As usual, however, we must trade off the time required for the heuristic against possible benefit.

Finally, we should mention that the branch-and-bound algorithm is often terminated before optimality is reached, either by design or necessity. In such a case we have a complete solution with cost U, and the lowest lower bound L of any live node provides a lower bound on the optimal cost. We are therefore within a ratio of $(U - L)/L$ of optimal.

It should be clear by now that the branch-and-bound idea is not one specific algorithm, but rather a very wide class. Its effective use is dependent on the design of a strategy for the particular problem at hand, and at this time is as much art as science. We conclude the discussion of branch-and-bound with an example of its application to a scheduling problem.

18.5 Application to a Flowshop Scheduling Problem

We now describe something of a "case history" of the application of branch-and-bound to a scheduling problem of some general interest—the two-machine flowshop scheduling problem with a sum-finishing-time criterion, defined as follows.

Definition 18.3

We are given a set of n jobs, J_i, $i = 1, \ldots, n$. Each job has two tasks, each to be performed on one of two machines. Job J_j requires a processing time τ_{ij} on machine i, and each task must complete processing on machine 1 before starting on machine 2. Let F_{ji} be the time at which job i finishes on machine j. The *sum finishing time* is defined to be the sum of the times that all the jobs finish processing on machine 2:

$$f = \sum_{i=1}^{n} F_{2i}$$

The sum-finishing-time problem (SFTP) is the problem of determining the order in which to assign the tasks to machines so f is minimum. □

Example 18.5

A common example of a situation in which a problem like SFTP may arise is a computer that executes one program at a time. We can identify the jobs with individual computer programs, machine 1 with the central processor, and machine 2 with the printer. We then assume that we are given a set of jobs with known execution and printing times and wish to schedule them so that the sum (or, equivalently, the average) finishing time is as small as possible. □

We now quote two important facts about SFTP. The first can be found in Conway, Maxwell, and Miller [CMM] and allows us to restrict our search for a single permutation that determines a complete schedule.

Theorem 18.1 *There is an optimal schedule for SFTP in which both machines process the jobs in the same order with no unnecessary idle time between jobs. (These are called* permutation *schedules.)*

The second, more recent, result is due to Garey, Johnson, and Sethi [GJS] and justifies the serious pursuit of a branch-and-bound algorithm.

Theorem 18.2 *The problem SFTP is NP-complete. (We mean, of course, the* yes-no *problem corresponding to SFTP with a solution of cost less than or equal to some L.)*

Example 18.6

Consider the following numerical example with 3 jobs.

τ_{ij}	Machine 1	Machine 2
Job 1	2	1
Job 2	3	1
Job 3	2	3

Figure 18–9 shows all 6 possible permutation schedules, among which the optimal schedule must lie, by Theorem 18.1. The unique optimum has cost 18. □

```
Machine 1   1 1 2 2 2 3 3
Machine 2       1     2   3 3 3           f = 19
                 ↑     ↑       ↑

Machine 1   1 1 3 3 2 2 2
Machine 2       1   3 3 3 2               f = 18
                 ↑       ↑ ↑

Machine 1   2 2 2 1 1 3 3
Machine 2         2   1   3 3 3           f = 20
                   ↑   ↑       ↑

Machine 1   2 2 2 3 3 1 1
Machine 2         2   3 3 3 1             f = 21
                   ↑       ↑ ↑

Machine 1   3 3 1 1 2 2 2
Machine 2       3 3 3 1   2               f = 19
                     ↑ ↑   ↑

Machine 1   3 3 2 2 2 1 1
Machine 2       3 3 3 2   1               f = 19
                     ↑ ↑   ↑
```

Figure 18–9 The six possible permutation schedules in Example 18-5, and their associated costs. The arrows indicate finishing times on machine 2.

This problem is, with the help of Theorem 18.1, a problem of finding one permutation of n objects, and the natural way to branch is to choose the first job to be scheduled at the first level of the branching tree, the second job at the next level, and so on. What we need next is a lower-bound function.

Ignall and Schrage [IS] describe a very effective lower bound, which we derive here. Suppose we are at a node at which the jobs in the set $M \subseteq \{1, \ldots, n\}$ have been scheduled, where $|M| = r$. Let t_k, $k = 1, \ldots, n$, be the index of the kth job under any schedule which is a descendant of the node under consideration. The cost of this schedule, which we wish to bound, is

$$f = \sum_{i \in M} F_{2i} + \sum_{i \notin M} F_{2i} \tag{18.5}$$

Now if every job could start its processing on machine 2 immediately after completing its processing on machine 1, the second sum in Eq. 18.5 would become

$$S_1 = \sum_{k=r+1}^{n} [F_{1t_r} + (n - k + 1)\tau_{1t_k} + \tau_{2t_k}] \tag{18.6}$$

(see Problem 3). If that is not possible, S_1 can only increase, so

$$\sum_{i \notin M} F_{2i} \geq S_1 \tag{18.7}$$

Similarly, if every job can start on machine 2 immediately after the preceding job finishes on machine 2, the second sum in Eq. 18.5 would become

$$S_2 = \sum_{k=r+1}^{n} [\max(F_{2t_r}, F_{1t_r} + \min_{i \notin M} \tau_{1i}) + (n - k + 1)\tau_{2t_k}] \tag{18.8}$$

Again, this is a lower bound:

$$\sum_{i \in M} F_{2i} \geq S_2 \tag{18.9}$$

Therefore

$$f \geq \sum_{i \in M} F_{2i} + \max(S_1, S_2) \tag{18.10}$$

The bound depends on the way the remaining jobs are scheduled, through t_k. This dependence can be eliminated by noting that S_1 is minimized by choosing t_k so that the tasks of length τ_{1t_k} are in ascending order, and that S_2 is minimized by choosing t_k so that the tasks of length τ_{2t_k} are likewise in ascending order. Call the resulting minimum values $\hat{S}_1$ and $\hat{S}_2$. Then

$$f \geq \sum_{i \in M} F_{2i} + \max(\hat{S}_1, \hat{S}_2) \tag{18.11}$$

is an easily computed lower bound.

Example 18.6 (Continued)

At the first branching step, Eq. 18.11 yields the lower bounds

$$f = \begin{cases} 18 & \text{if job 1 is scheduled first} \\ 20 & \text{if job 2 is scheduled first} \\ 18 & \text{if job 3 is scheduled first} \end{cases}$$

From the results in Figure 18-9, we see that the first two of these are as low as possible. Figure 18–10 shows a complete search tree in which the least lower bound is branched from first, from left to right in case of ties. The optimal solution (1, 3, 2) kills all the others when it is obtained. □

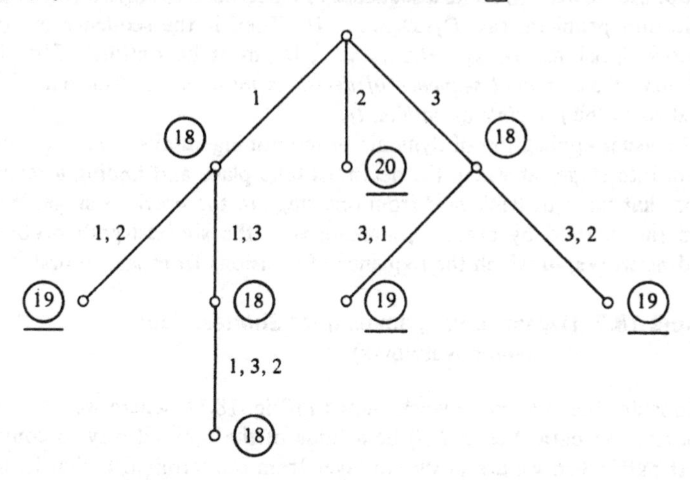

Figure 18–10 The search tree for Example 18.6.

Finally, we describe a natural dominance relation also given by Ignall and Schrage [IS]. Suppose we have two nodes t and u representing partial assignments of the same set of jobs, M. Let the kth scheduled job be t_k and u_k, $k = 1, \ldots, r$, under partial schedules t and u, respectively. Then if

$$F_{2t_r} \leq F_{2u_r} \tag{18.12}$$

(the set of jobs in M finishes no later on machine 2 under the partial schedule t), and if the accumulated cost under partial schedule t is no more than that under u,

$$\sum_{i \in M} F_{2i}\big|_{\text{schedule } t} \leq \sum_{i \in M} F_{2i}\big|_{\text{schedule } u} \tag{18.13}$$

then the best completion of schedule t is at least as good as the best completion of u. Equations 18.12 and 18.13 therefore define a dominance relation of t over u.

Example 18.6 (Continued)

Consider the nodes $t = (1, 2)$ and $u = (2, 1)$ (not generated in Fig. 18–10). Then t dominates u. On the other hand, $t = (1, 3)$ does not dominate $u = (3, 1)$. This instance is too small to show the real power of a dominance relation. □

18.6 Dynamic Programming

Dynamic programming is related to branch-and-bound in the sense that it performs an intelligent enumeration of all the feasible points of a problem, but it does so in a different way. The idea is to work backwards from the last decisions to the earlier ones.

Suppose we need to make a sequence of n decisions to solve a combinatorial optimization problem, say $D_1, D_2, \ldots, D_n$. Then if the sequence is optimal, the last k decisions, $D_{n-k+1}, D_{n-k+2}, \ldots, D_n$, must be optimal. That is, the *completion of an optimal sequence of decisions must be optimal.* This is often referred to as the *principle of optimality.*

The usual application of dynamic programming entails breaking down the problem into stages at which the decisions take place and finding a recurrence relation that takes us backward from one stage to the previous stage. We shall explain the method by example, starting with the shortest-path problem for layered networks, in which the sequence of decisions from last to first is clear.

Example 18.7 (Dynamic Programming for Shortest Path in Layered Networks)

Consider the layered network shown in Fig. 18.11, where we want to find the shortest s-t path. Let *table*(i) be a table of the optimal way to continue a shortest path when we are in the ith layer from the terminal t; that is, *table*(i) contains the best decision for each node in the layer i arcs from t. Thus the first

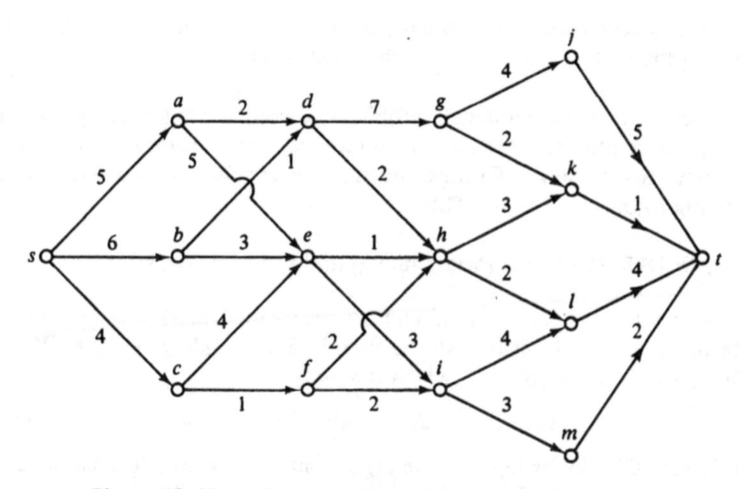

Figure 18–11 A shortest-path problem in a layered network.

table is simply

$$
table(1) = \begin{cases} \text{node} & j \quad k \quad l \quad m \\ \text{next node} & t \quad t \quad t \quad t \\ \text{total cost} & 5 \quad 1 \quad 4 \quad 2 \end{cases}
$$

Now consider the construction of *table*(2). At node g we can reach j or k. We know the cost of the optimal completion from j or k by *table*(1) and thus can find the best decision at node g by comparing the cost of arc (g, j) plus the cost of completion from j, with the cost of arc (g, k) plus the cost of completion from k. Continuing for nodes h and i, *table*(2) is

$$
table(2) = \begin{cases} \text{node} & g \quad h \quad i \\ \text{next node} & k \quad k \quad m \\ \text{total cost} & 3 \quad 4 \quad 5 \end{cases}
$$

Next we find

$$
table(3) = \begin{cases} \text{node} & d \quad e \quad f \\ \text{next node} & h \quad h \quad h \\ \text{total cost} & 6 \quad 5 \quad 6 \end{cases}
$$

$$
table(4) = \begin{cases} \text{node} & a \quad b \quad c \\ \text{next node} & d \quad d \quad f \\ \text{total cost} & 8 \quad 7 \quad 7 \end{cases}
$$

$$
table(5) = \begin{cases} \text{node} & s \\ \text{next node} & c \\ \text{total cost} & 11 \end{cases}
$$

We can now reconstruct an optimal path by backtracking through the tables, obtaining the path (s, c, f, h, k, t), with a cost of 11. □

Of course, in more difficult problems, dynamic programming runs into time and space problems, because the tables may grow in size at an exponential rate from stage to stage. To illustrate this, we conclude with a dynamic programming formulation of the TSP.

Example 18.8 (Dynamic Programming and the TSP [HK3])

Given a set $S \subseteq \{2, 3, \ldots, n\}$ and $k \in S$, we let $C(S, k)$ be the optimal cost of starting from city 1, visiting all the cities in S, and ending at city k. We begin by finding $C(S, k)$ for $|S| = 1$, which is simply

$$C(\{k\}, k) = d_{1k} \qquad \text{all } k = 2, \ldots, n \tag{18.14}$$

To calculate $C(S, k)$ for $|S| > 1$, we argue that the best way to accomplish our journey from 1 to all of S, ending at k, is to consider visiting m immediately before k, for all m, and looking up $C(S - \{k\}, m\}$ in our preceding table. Thus

$$C(S, k) = \min_{m \in S - \{k\}} [C(S - \{k\}, m) + d_{mk}] \tag{18.15}$$

This must be calculated for all sets S of a given size and for each possible city m in S. (We also must save the city m for which a minimum is achieved, so that we can reconstruct the optimal tour by backtracking.) If we count each value of $C(S, k)$ as one storage location, we need space equal to

$$\sum_{k=1}^{n-1} k\binom{n-1}{k} = (n-1)2^{n-2} = O(n2^n) \tag{18.16}$$

locations [HK3] and a number of additions and comparisons equal to

$$\sum_{k=2}^{n-1} k(k-1)\binom{n-1}{k} + (n-1) = (n-1)(n-2)2^{n-3} + (n-1) = O(n^2 2^n) \tag{18.17}$$

These are exponential functions of the problem size n, and may seem prohibitively large. But when we consider the fact that there are $(n - 1)!$ distinct tours in a naïve enumeration, we see that in fact this approach results in enormous savings. Since there is no algorithm known for the TSP that is better than exponential, the dynamic programming approach cannot be dismissed out of hand, although branch-and-bound algorithms have proven more effective in this application. □

As can be seen from the two examples above, dynamic programming is a very general idea and can demand varying amounts of ingenuity to find good ways of breaking down a problem into stages so that a convenient recurrence relation can be found. Some thought will show that some of the algorithms seen earlier in this book can be considered to be applications of dynamic programming (see Problem 7 and 8 and Section 17.3 on the 0-1 KNAPSACK problem).

PROBLEMS

1. Prove that the branch-and-bound algorithm when applied to ILP terminates at optimality within a number of steps bounded by an exponential in the problem size.
2. Describe an $O(n^2)$-time algorithm for finding a minimum 1-tree, given an $n \times n$ distance matrix.
3. Prove that Eqs. 18.6 and 18.8 are the claimed lower bounds.
4. Analyze the time and space complexities of the dynamic programming algorithm for shortest path in a layered network and compare them with the time and space complexities of Dijkstra's algorithm (Chapter 6) applied to the same problem.
5. Establish Eqs. 18.16 and 18.17, giving the space and time requirements of the dynamic programming algorithm for the TSP. What is the asymptotic savings in time over complete enumeration of $(n - 1)!$ tours?
6. In the branching step of the general branch-and-bound algorithm, is it necessary that the partition of the set of solutions be a *disjoint* partition?
7. Interpret Dijkstra's algorithm (Chapter 6) for shortest path (with nonnegative distances) as an application of dynamic programming. Define explicitly the definition of a stage and the recurrence relation and boundary conditions analogous to Eqs. 18.14 and 18.15.
8. Repeat Problem 7 for the Floyd-Warshall algorithm (See. 6.5).
9. We shall derive an $O(|V|^3)$ algorithm for shortest path with negative distances allowed, using dynamic programming. Consider an undirected graph $G = (V, E)$ with source node s and distance matrix $[d_{ij}]$.
   (a) Let the label $p_i(x)$ be the shortest length of any path from source s to node x, using i or fewer intermediate edges. Write the recurrence relation for $p_i(x)$ and its boundary conditions.
   (b) Show that if no $p_i(x)$ changes from stage i to $i + 1$, we have reached optimality and can stop.
   (c) Show that if we have not converged in the sense of Part (b) after $i = |V|$ stages, there is a negative-cost cycle. From this, prove that the algorithm takes $O(|V|^3)$ time.
   (d) Compare the worst-case behavior with that of the Floyd-Warshall algorithm.
10. Suppose we want to compute the product of n matrices, $A_1, A_2, \ldots, A_n$, where A_i has p_i rows and q_i columns. We assume, of course, that they are compatible;

that is, $q_i = p_{i+1}$, $i = 1, \ldots, n-1$. Devise an $O(n^3)$-time dynamic programming algorithm to find the order in which to multiply these matrices to minimize the total number of scalar multiplications, assuming that multiplying A_i and A_{i+1} takes $p_i q_i q_{i+1}$ scalar multiplications. How much space does your algorithm require?

11. Explain why Algorithm DP-I in Sec. 17.3 for 0-1 KNAPSACK is considered a dynamic programming algorithm.

12. Reformulate the dynamic programming algorithm for the TSP (Example 18.8) as a branch-and-bound algorithm with no upper and lower bounds, but with a dominance relation.

NOTES AND REFERENCES

A good review of branch-and-bound, up to 1966, with pertinent early references, is

[LW] LAWLER, E. L., and D. E. WOOD, "Branch-and-Bound Methods: A Survey," *OR*, 14 (1966), 699–719.

Lawler and Wood describe the approach of Little and others to the TSP

[LMSK] LITTLE, J. D. C., K. G. MURTY, D. W. SWEENY, and C. KAREL, "An Algorithm for the Traveling-Salesman Problem," *OR*, 11 (1963), 972–89

as well as Eastman's:

[Ea] EASTMAN, W. L., "Linear Programming with Pattern Constraints," Ph.D. Thesis, Report No. BL. 20, The Computation Laboratory, Harvard University, 1958.

The formulation of ILP in Sec. 18.1 is given in

[Da] DAKIN, R. J., "A Tree-Search Algorithm for Mixed Integer Programming Problems," *Comp. J.*, 8, no. 3 (1965), 250–55.

The formulation of branch-and-bound given here, as well as the flowshop examples, are after W. H. Kohler and K. Steiglitz in Chapter 6 of

[Co] COFFMAN, E. G., JR., ed., *Computer and Job-Shop Scheduling*. New York: Wiley-Interscience, 1976.

The results of some computational experiments with branch-and-bound can be found there, as well as a dynamic programming formulation of a scheduling problem that turns out to be equivalent to branch-and-bound.

The lower bound and dominance relation for the flowshop problem is from

[IS] IGNALL, E., and L. SCHRAGE, "Application of the Branch and Bound Technique to Some Flow-Shop Scheduling Problems," *OR*, 13, no. 3 (1965), 400–12.

Theorem 18.1 can be found in

[CMM] CONWAY, R. W., W. L. MAXWELL, and L. W. MILLER, *Theory of Scheduling*. Reading, Mass.: Addison-Wesley Publishing Co., Inc., 1967.

Theorem 18.2 is due to

[GJS] GAREY, M. R., D. S. JOHNSON, and R. SETHI, "The Complexity of Flowshop and Jobshop Scheduling," *Math. of Operations Res.*, 1 (1976), 117–29.

Held and Karp's work on 1-trees is from

[HK1] HELD, M., and R. M. KARP, "The Traveling Salesman Problem and Minimum Spanning Trees," *OR*, 18 (1970), 1138–62.

[HK2] ———, "The Traveling Salesman Problem and Minimum Spanning Trees: Part II," *Math. Prog.*, 1 (1971), 6–25.

The dynamic programming formulation of the TSP can be found in

[HK3] HELD, M., and R. M. KARP, "A Dynamic Programming Approach to Sequencing Problems," *J. SIAM*, 10, 1 (1962), 196–210.

as well as formulations of other combinatorial problems. Solutions to Problems 7–9 can be found in

[DL] DREYFUS, S. E., and A. M. LAW, *The Art and Theory of Dynamic Programming*. New York: Academic Press, Inc., 1977.

This book contains a comprehensive chapter on the shortest-path problem including some (constant-factor) improvements for the single-source and all-pairs problems. The algorithm in Problem 9 has an uncertain origin. In

[Dr] DREYFUS, S. E., "An Appraisal of Some Shortest-Path Algorithms," *OR*, 17 (1969), 395–412.

references to L. R. Ford, Jr., E. F. Moore, and R. E. Bellman are given, in 1956, 1957, and 1958, respectively.

19

Local Search

19.1 Introduction

Local search is based on what is perhaps the oldest optimization method—trial and error. The idea is so simple and natural, in fact, that it is surprising just how successful local search has proven on a variety of difficult combinatorial optimization problems. The best way to gain an appreciation of its power and the subtlety of its design questions is by example, and we shall therefore begin this chapter with a succession of case histories. We shall then present some theoretical aspects of local search, which relate it in spirit to the simplex algorithm.

We have already touched on the basic components of local search in Chapter 1. We now summarize the general algorithm. Given an instance (F, c) of an optimization problem, where F is the feasible set and c is the cost mapping, we choose a neighborhood

$$N: F \longrightarrow 2^F$$

which is searched at point $t \in F$ for improvements by the subroutine

$$\text{improve}(t) = \begin{cases} \text{any } s \in N(t) \text{ with } c(s) < c(t) \text{ if such an } s \text{ exists} \\ \text{"no" otherwise} \end{cases}$$

The general local search algorithm is shown in Fig. 19–1. We start at some initial feasible solution $t \in F$ and use subroutine improve to search for a better solution in its neighborhood. So long as an improved solution exists, we adopt it and repeat the neighborhood search from the new solution; when we reach a local optimum we stop.

```
procedure local search
begin
  t := some initial starting point in F;
  while improve(t) ≠ 'no' do
    t:=improve(t);
  return t
end
```

Figure 19–1 The general local search algorithm.

To apply this approach to a particular problem, we must make a number of choices. First, we must decide how to obtain an initial feasible solution. It is sometimes practical to execute local search from several different starting points and to choose the best result. In such cases, we must also decide how many starting points to try and how to distribute them.

Next, we must choose a "good" neighborhood for the problem at hand, and a method for searching it. This choice is usually guided by intuition, because very little theory is available as a guide. One can see a clear trade-off here, however, between small and large neighborhoods. A larger neighborhood would seem to hold promise of providing better local optima but will take longer to search, so we may expect that fewer of them can be found in a fixed amount of computer time. Do we generate fewer "stronger" local optima or more "weaker" ones?

These and similar questions are usually answered empirically, and the design of effective local search algorithms has been, and remains, very much an art.

19.2 Problem 1: The TSP

Recall from Chapter 1 that the *k-change* neighborhood ($k \geq 2$) for the TSP is defined at a tour f by

$$N_k(f) = \{g: g \in F \text{ and } g \text{ can be obtained from } f \text{ as follows: remove } k \text{ edges from the tour; then replace them with } k \text{ edges}\}$$

Two papers appeared in 1958 that used k-change local search for the TSP, and both combined the idea with enumerative methods similar to branch-and-bound to yield optimal solutions. Croes [Cr] used N_2, and he called a 2-change an "inversion." Bock [Bo] used N_3.

Two more papers applying local search to the TSP appeared in 1965.

Reiter and Sherman [RS] examined many different neighborhoods, but it was Lin [Li] who first convincingly demonstrated the power of the 3-change neighborhood N_3.

For example, Lin found empirically that a 3-opt tour for the 48-city problem of Held and Karp [HK1] has a probability of about 0.05 of being optimal, and hence a run from 100 random starts will yield the optimum with a probability of 0.99. One of Lin's important contributions is the emphasis on using many different randomized starting solutions; in the TSP it is effective to use starting tours that are completely random.

One might think that starting with solutions that are better on the average than a completely random tour (which is likely to be quite bad) would improve the quality of the local optima obtained. In the TSP using 3-opt, however, this does not seem to be the case, as was shown by extensive computer experiments performed by Peter Weiner [unpublished]. The explanation seems to be as follows: 3-change is powerful enough so that it will quickly improve a random tour, and by starting from completely random tours we get a wide sample of all local optima. However, in problems where we are forced to use relatively weaker neighborhoods, it may be crucial to use "good" starting solutions.

Another important contribution of Lin's paper is the demonstration that 3-opt solutions are much better than 2-opt, but that 4-opt solutions are not sufficiently better than 3-opt to justify the additional running time. Just why this is true is not completely clear.

Lin found empirically that the probability of a 3-opt tour being optimal was, for his class of examples, about $2^{-n/10}$, where n is the number of cities. Such an estimate enables one to decide how many runs from random starts to use to achieve a given probability of optimality. But unfortunately, there is not yet any theoretical justification for an empirical result of this form.

The results in Lin on the previously published problems of known difficulty were spectacular, and the general approach was so successful on the TSP that it stimulated the application of local search to a variety of other problems.

19.3 Problem 2: Minimum-Cost Survivable Networks [StWK]

The TSP is the archetype application of local search and has a number of convenient features that are not always present in other problems. For example, it is very easy to generate completely random feasible solutions, and it is very easy to test a candidate for feasibility. These properties are not shared by our second example—the problem of designing a network with prescribed connectivity and minimum cost. To define the problem we need a preliminary definition.

Definition 19.1

Given a connected undirected graph $G = (V, E)$ and two distinct vertices $i, j \in V$, a set of paths from i to j is said to be *vertex-disjoint* if no vertex other than i and j is on more than one of them. The *vertex-connectivity* between i and j is the maximum possible number of vertex-disjoint paths between them. □

Example 19.1

Figure 19–2 shows a graph in which the vertex connectivity between every distinct pair of vertices is 3. Such a graph is called *3-connected.* □

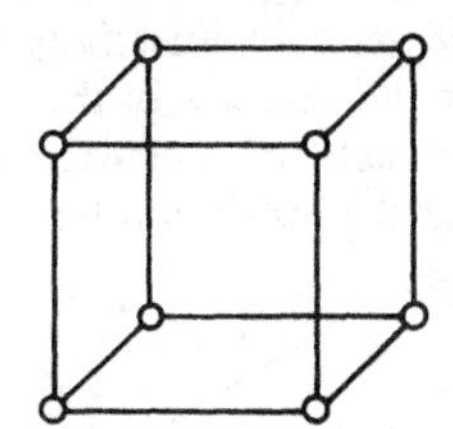

Figure 19–2 A 3-connected graph.

If the graph G represents a communication network, the vertex-connectivity between two vertices is a measure of the reliability of communication between the two vertices, because (as the reader was asked to show in Problem 8 of Chapter 6) it is equal to the minimum number of vertices that need to be removed from the network to disconnect the two vertices.

Given a cost matrix, then, it is an interesting question to ask for a network with prescribed connectivity between all pairs of vertices and minimum total cost. More formally, we define the following problem.

Definition 19.2

Given a cost matrix $[d_{ij}]$ and connectivity matrix $[s_{ij}]$, the *minimum-cost survivable network problem* (MCSN) is that of finding a graph $G = (V, E)$ with minimum total cost

$$\sum_{[i,j] \in E} d_{ij}$$

and vertex-connectivity r_{ij} between distinct vertices i, j satisfying

$$r_{ij} \geq s_{ij} \qquad \text{all } i, j, i \neq j \quad \square$$

The MCSN problem was attacked by local search in [StWK], and we describe those results next. The concept of *NP*-completeness was not yet invented in 1969, but in fact an appropriate version of this problem is *NP*-complete (recall Subsec. 16.3.1).

The first difficulty that arises is the calculation of the actual connectivity matrix $[r_{ij}]$ of a given graph. We state an upper bound on the complexity of calculating the point-to-point connectivity in the next theorem.

Theorem 19.1 *Given an undirected graph $G = (V, E)$ and two distinct vertices $i, j \in V$, we can find the vertex-connectivity r_{ij} in $O(|V|^{2.5})$ time.*

Proof Create a directed graph $G' = (V', E')$ replacing every vertex $v \in V$ by two vertices v' and $v'' \in V'$, as shown in Fig. 19–3. Next consider the flow network obtained by assigning a unit capacity to every arc of G'. Then the maximum i-j flow in this network is the vertex-connectivity r_{ij}, because the unit capacity arc from v' to v'' means that the vertex v in the original graph G can lie on at most one i-j path. The flow network corresponding to G' is simple, so by Theorem 9.4 we can calculate the maximum flow in $O(|V|^{2.5})$ time. □

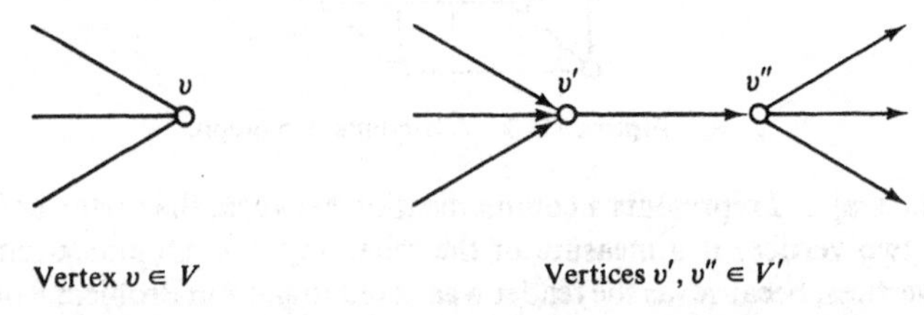

Figure 19–3 The construction in the proof of Theorem 19.1.

How many point-to-point connectivity calculations are required to check the feasibility requirement $r_{ij} \geq s_{ij}$, all $i, j, i \neq j$, for a given graph? As might be guessed, it is not necessary to calculate all $|V|(|V| - 1)/2$ vertex connectivities r_{ij}. In the case that s_{ij} is uniformly k, for example, it can be shown that $k(|V| - (k + 1)/2)$ is sufficient, which is $O(k|V|)$ (see Problem 13 and Problem 13 of Chapter 9). Thus the problem of verifying the feasibility of a graph is polynomial, but is considerably more complicated than in the TSP.

Next we come to the problem of constructing an initial feasible solution. We describe the method used in [StWK], which is an empirically effective greedy heuristic. The idea is based on the observation that the degree of vertex i must be at least $\max_j r_{ij}$. Define the *deficiency* of a vertex i in an undirected graph G as

$$\text{deficiency}(i) = \max_j r_{ij} - \text{degree}(i)$$

We shall add edges to the graph G until all the deficiencies are nonpositive. We then test the resulting graph for feasibility using the max-flow algorithm mentioned above.

In more detail, we begin with an array of size $|V|$ containing the deficiency of each vertex. At each stage we add an edge between a vertex with the largest deficiency and one with the next highest deficiency. Of all the vertices with next highest deficiency, we choose one that results in the smallest increase in cost; all other ties are resolved by choosing the earliest vertex in the array. Multiple edges are not allowed.

Example 19.2

Consider a problem with uniform connectivity requirement $s_{ij} = 3$ for all i, j, and the cost matrix determined by Euclidean distance in the plane between the eight vertices shown in Fig. 19–4(a). The initial array is

Vertex	1	2	3	4	5	6	7	8
Deficiency	3	3	3	3	3	3	3	3

The largest deficiency is of course 3, and the first edge chosen is (1, 2), since vertex 2 is, of all the vertices with deficiency 3, closest to vertex 1. Thereafter, the edges [3, 4], [5, 6], [7, 8], [1, 7], [2, 3], [4, 5], [6, 8], [1, 8], [2, 7], [3, 6], [4, 6], and [5, 7] are added in that order; the resulting graph is shown in Fig. 19–4(b). The reader can verify that this result does have vertex-connectivity

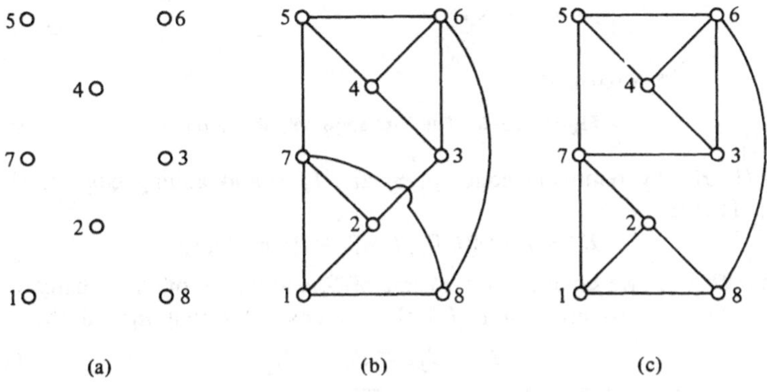

Figure 19–4 (a) The vertex placement for Example 19.2. (b) The result of the starting algorithm. (c) A favorable X-change that results in infeasibility.

3 between all pairs of vertices. Notice that the result has two vertices of degree 4; the numbering of the vertices can affect the number of edges in the answer. Furthermore, if the vertices are renumbered, the result may very well be infeasible. □

We mentioned before that it is desirable to randomize the starting algorithm; this is done in the present case by randomly ordering the vertices before the application of the algorithm. This method not only produces an assortment of starting feasible solutions, but these starting solutions tend to have low cost and a small number of edges. In contrast to the TSP, the MCSN problem was found to require starting solutions with relatively low cost. This is a consequence of the neighborhood used, which we describe next.

Definition 19.3

Let the set of graphs feasible in an instance of the MCSN problem be denoted by F. That is, F consists of all graphs with a given number of vertices and vertex-connectivity satisfying

$$r_{ij} \geq s_{ij} \qquad \text{all } i, j, i \neq j$$

Consider a graph $G = (V, E) \in F$ in which the edges $[i, m]$ and $[j, l]$ are *present* and the edges $[i, l]$ and $[j, m]$ are *absent* (see Figure 19–5). Define a new graph

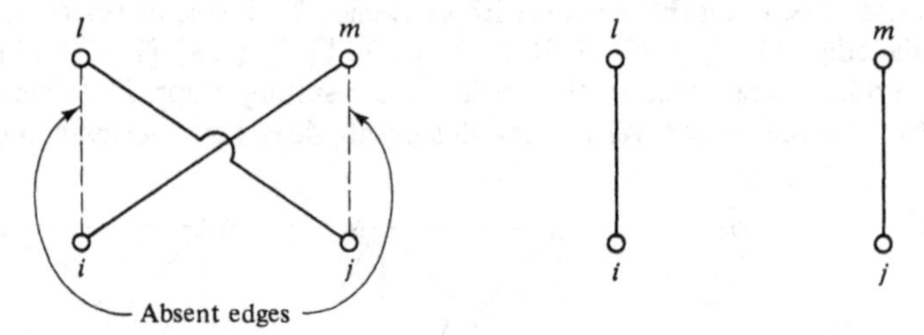

Figure 19–5 The *X*-change neighborhood.

$G' = (V, E')$ by removing edges $[i, m]$ and $[j, l]$ and adding edges $[i, l]$ and $[j, m]$. That is

$$E' = E \cup \{[i, l], [j, m]\} - \{[i, m], [j, l]\}$$

Then if $G' \in F$, we say it is an *X-change* of G, and the set of all X-changes of G defines the *X-change neighborhood.* If the new cost is less than the old, that is,

$$d_{il} + d_{jm} < d_{im} + d_{jl} \tag{19.1}$$

then the X-change is called *favorable.* □

The X-change neighborhood has the property that it preserves the number of edges and the degree of every vertex, which means that it is desirable to have an assortment of starting solutions, possibly with different numbers of edges and different vertex degrees. However, there is a certain difficulty in searching the X-change neighborhood. It is easy enough to search over all pairs of edges for a candidate pair $[i, m]$ and $[j, l]$ satisfying Eq. 19.1. But then the resulting graph G' may not be feasible. This is illustrated in Fig. 19–4(c), where the graph shown has been obtained from that in Fig. 19–4(b) by removing edges [2, 3] and [7, 8] and adding [2, 8] and [3, 7]. The result does not have vertex-connectivity 3; for example, there are only 2 vertex-disjoint 1-3 paths.

If we had to check the entire graph for feasibility after each favorable X-change candidate was discovered, the local search algorithm would be very slow, but it turns out that a complete check is not necessary (see Problem 13). In the case that all r_{ij} are uniformly k, in fact, it is necessary only to check two vertex-connectivities to test the feasibility of the new graph.

The local search algorithm specified by the starting algorithm and neighborhood defined above has been found effective on a number of problems, and the results for one simple problem are given next.

Example 19.3

A 7-vertex example with a uniform connectivity requirement of 3 (from [StWK]) is shown in Figure 19–6. One hundred local optima were generated with the following distribution of costs, where the cost matrix was obtained by taking the integer part of the Euclidean distances.

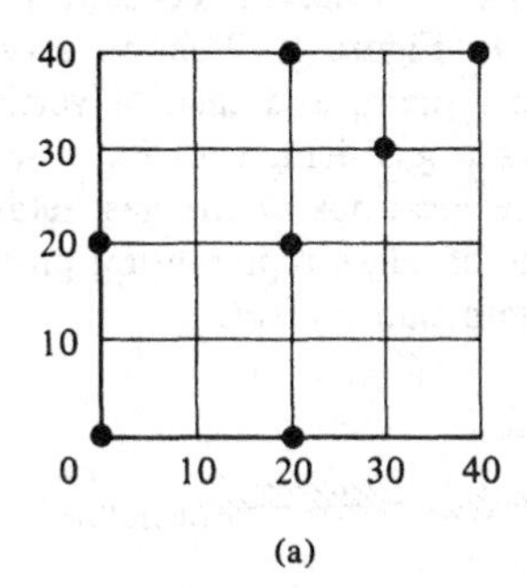

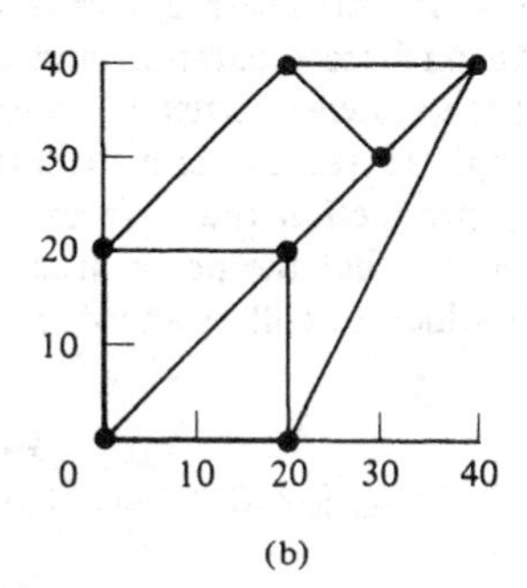

Figure 19–6 (a) A seven-node example [StWK]. (b) The optimal solution.

Cost	Number of Occurrences out of 100
242	39
245	9
250	20
251	7
253	6
258	1
265	9
270	9

The solution with cost 242 has been proven optimal by exhaustive search, so we see that the randomized local search algorithm has a probability of about 0.39 of yielding a global optimum. The probability of finding an optimal solution after 100 trials is therefore about

$$1 - (0.61)^{100} \cong 1 - 0.34 \times 10^{-21}$$

which, it can be argued, is greater than the probability that a computer will execute a 1-second program without an undetected error (see Problem 10).

We note also that there is a wide variety of local optima with costs close to optimal. These results are typical of local search, but for larger problems the probability of finding the global optimum decreases and the costs of local optima tend to spread out. □

In summary, the MCSN problem is one where the cost of testing feasibility is critical and the existence of efficient tests makes local search practical.

19.4 Problem 3: Topology of Offshore Natural-Gas Pipeline Systems [RFRSK]

We come now to a problem where it is not the feasibility but the *cost* calculation that is critical. The problem is to collect offshore gas reserves and deliver them to an onshore separation and compressor plant. Figure 19–7 shows a typical pipeline system: vertex 1 represents the onshore plant, and each of vertices 2 through 15 represents a drilling platform over a gas field, with its estimated daily production rate. We can assume that the locations of the gas fields are given and that the problem is to choose a tree of edges representing pipelines with which to collect and deliver the gas to the onshore plant.

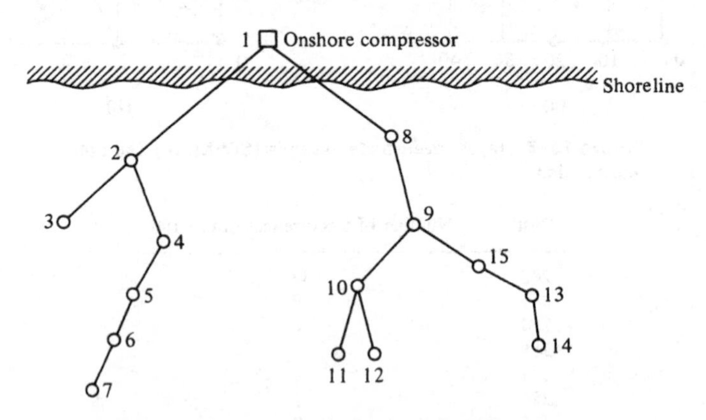

Figure 19–7 An offshore natural-gas pipeline system.

We next have to explain how the cost of a given tree is to be determined. Each edge of the tree represents a pipe with one of 7 standard diameters, ranging (in 1968) from a diameter of $10\frac{3}{4}$ inches and a cost of about $70,000 per mile to a diameter of 30 inches and a cost of about $310,000 per mile. Because we know the production rate at every vertex, given a tree we know the flow rate in every edge. This allows us to determine the pressure drop along an edge for each choice of diameter, by an empirical formula called the *panhandle* equation. The pipe diameters must then be chosen to minimize the cost subject to the following constraints:

1. The maximum pressure anywhere is not to exceed a given P_{max}.
2. The delivery pressure at the onshore plant must be at least a given P_{min}.
3. The pressure of the gas collected at each vertex is at least P_{min}.

The problem of choosing an optimal set of pipe diameters is a combinatorial optimization problem in its own right, which is solved by a method reminiscent of dynamic programming. This takes a reasonable but nontrivial amount of computer time for each topology, say about 1 second for a 20-vertex tree.

The problem that remains, then, is to choose a minimum-cost tree, where the cost of any particular tree must be determined by a complicated subroutine that takes about 1 second. If local search is to be used, we must be careful to choose a neighborhood that is sufficiently small—otherwise the running time will become exorbitantly large. The most natural neighborhood that comes to mind is the elementary tree transformation described in Example 1.5: Add in turn every possible edge to the tree and remove in turn each possible edge on the resulting cycle. This neighborhood is exact for the minimal spanning tree problem, but is of size $O(|V|^3)$, since we need to consider $|V|(|V| - 1)/2$ added edges, and each created cycle can be as long as $\Omega(|V|)$. However, it seems likely that if an improvement exists, there will be one that will be discovered by connecting a vertex to one that is geographically close. This motivates a restricted neighborhood called Δ-change, defined as follows: From each vertex x, find the three *closest* vertices y_1, y_2, and y_3 that are not adjacent to x in the tree. Then search the elementary tree transformations determined by the edges $[x, y_1]$, $[x, y_2]$, and $[x, y_3]$ (see Fig. 19–8). This neighborhood is of size $3k|V|$, where k is the average length of a cycle found in the elementary tree transformations, which is considerably smaller than $|V|$ because the edges added are short.

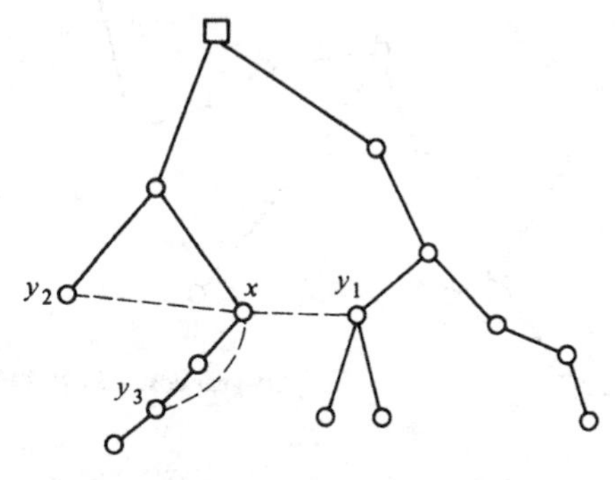

Figure 19–8 The three edges at vertex x determining a Δ-change are shown dotted.

As might be guessed, the starting routine is a heuristic construction meant to produce as good an initial feasible topology as possible, because the neighborhood Δ-change is not very powerful. Fig. 19–9 illustrates a real example from [RFRSK] showing the first successful Δ-change, which decreases the 20-year cost from $110,970,000 to $109,096,000. Figure 10–9(b) shows the local optimum, with a 20-year cost of $101,340,000 obtained after 12 successful Δ-

New

Old

Starting 20–year cost = $110,970,000

New 20–year cost = $109,096,000

(a)

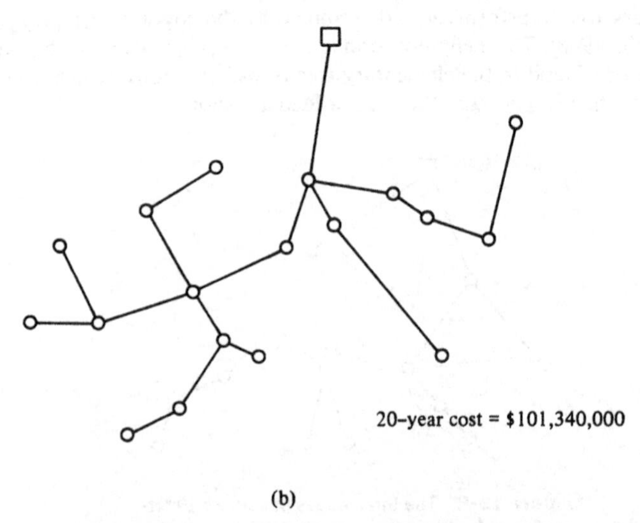

(b)

Figure 19–9 (a) The tree after the first successful Δ-change. (b) The local optimum after 12 successful Δ-changes.

changes. The potential savings of $9,630,000 over 20 years would certainly seem to justify a great deal of program development!

19.5 Problem 4: Uniform Graph Partitioning [KL]

Next we report on an application of local search to a problem related to circuit board wiring and program segmentation. This will lead to a description of the "variable-depth" method of Lin and Kernighan [LK].

We first define a simple version of the general problem:

Definition 19.4

Given a symmetric cost matrix $[d_{ij}]$ defined on the edges of a complete undirected graph $G = (V, E)$ with $|V| = 2n$ vertices, a partition $V = A \cup B$ such that $|A| = |B|$ is called a *uniform* partition. *The uniform graph partitioning* (UGP) problem is that of finding a uniform partition $V = A \cup B$ such that the cost

$$C(A, B) = \sum_{\substack{i \in A \\ j \in B}} d_{ij}$$

is minimum over all uniform partitions. □

The problem can be visualized as in Figure 19–10; we can think of the problem as that of dividing a circuit into two pieces of equal size so that the weight of the wires connecting the two pieces is as small as possible.

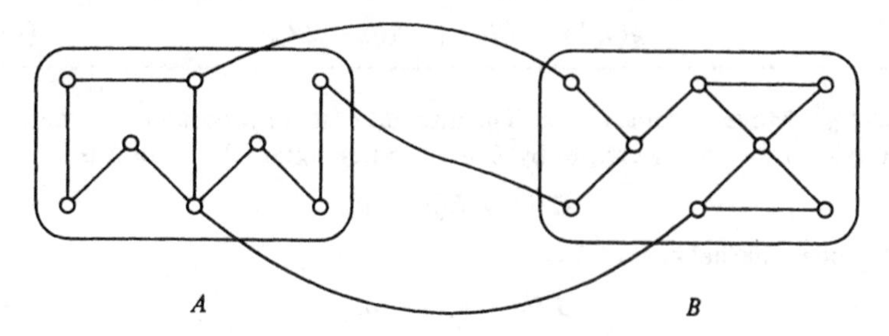

Figure 19–10 The uniform graph partitioning problem.

We now make a simple observation before describing a neighborhood. Suppose A^*, B^* is an optimal uniform partition and we are considering some partition A, B. Let X be those elements of A that are not in A^*—the "incorrect" elements—and let Y be similarly defined for B. Then $|X| = |Y|$, and

$$A^* = (A - X) \cup Y$$
$$B^* = (B - Y) \cup X \tag{19.2}$$

That is, we can obtain the optimal uniform partition by *interchanging* the elements in set X with those in Y. We therefore view the problem at any point in a local search algorithm as that of finding a sequence of favorable *swaps* of two elements.

Definition 19.5

Given a uniform partition A, B and elements $a \in A$ and $b \in B$, the operation of forming

$$\begin{aligned} A' &= (A - \{a\}) \cup \{b\} \\ B' &= (B - \{b\}) \cup \{a\} \end{aligned} \tag{19.3}$$

is called a *swap*. □

We next consider the problem of determining the effect that a swap has on the cost of a partition A, B. Let us define the *external cost* $E(a)$ associated with an element $a \in A$ by

$$E(a) = \sum_{i \in B} d_{ai} \tag{19.4}$$

and the *internal* cost by

$$I(a) = \sum_{j \in A} d_{aj} \tag{19.5}$$

(and similarly for elements of B). Let

$$D(v) = E(v) - I(v) \tag{19.6}$$

be the difference between external and internal cost for all $v \in V$.

Lemma 19.1 *The swap of a and b results in a reduction of cost (gain) of*

$$g(a, b) = D(a) + D(b) - 2d_{ab} \tag{19.7}$$

Proof Move a from A to B. The internal costs of a become external, and vice versa, so the cost decreases by $D(a)$. The new external cost of b is

$$E'(b) = E(b) - d_{ab} \tag{19.8}$$

and the new internal cost of b is

$$I'(b) = I(b) + d_{ab} \tag{19.9}$$

so its new difference is

$$D'(b) = E'(b) - I'(b) = D(b) - 2d_{ab} \tag{19.10}$$

When b is next moved from B to A, the cost decreases by $D'(b)$, which together with $D(a)$ gives Eq. 19.7. □

We can now very naturally define a neighborhood using the notion of swap.

Definition 19.6

The *swap neighborhood* N_s for the UGP problem is

$$N_s(A, B) = \{\text{all uniform partitions } A', B' \text{ that can be obtained from the uniform partition } A, B \text{ by a single swap}\} \quad \square$$

We can search for a favorable swap in $O(n^2)$ time by examining the gain $g(a, b)$ over all pairs $a \in A$, $b \in B$. When a swap is actually performed, the D's are updated and the search continues.

Kernighan and Lin [KL] report that locally optimal solutions for N_s for 0-1 distance matrices of size 32×32 are globally optimal about 10 percent of the time, and within 1 or 2 of globally optimal about 75 percent of the time.

Thus far things seem quite the same as in the previous examples, especially the TSP, because there are no special difficulties with testing feasibility or costing. One might be tempted to investigate the neighborhood defined by interchanging two elements of A with two of B, a neighborhood of size $O(n^4)$. However, Kernighan and Lin suggest the following intriguing idea, which opens new possibilities for the general approach of local search.

The idea is to replace the search for one favorable swap by a search for a favorable *sequence* of swaps, using the *costs* of the particular problem instance to guide the search. Thus a favorable sequence of k swaps is not found by examining a neighborhood containing all such sequences but is obtained sequentially as follows.

1. Calculate $D(v)$ for all elements $v \in V$.
2. Choose the pair a_1', b_1' so that the gain
$$g_1 = D(a_1') + D(b_1') - 2d_{a_1'b_1'} \tag{19.11}$$
is as large as possible (not necessarily positive).
3. Swap a_1' and b_1' and recompute the D values by
$$\begin{aligned} D'(x) &= D(x) + 2d_{xa'_1} - 2d_{xb'_1}, \quad & x \in A - \{a_1'\} \\ D'(y) &= D(y) + 2d_{yb'_1} - 2d_{ya'_1}, \quad & y \in B - \{b_1'\} \end{aligned} \tag{19.12}$$
(see Problem 3).
4. Repeat Steps 2 and 3, obtaining a sequence of swapped pairs a_2', b_2'; a_3', b_3'; . . . ; a_n', b_n'. Once a pair is swapped, it is no longer considered for swapping in Step 2.

In this way we obtain a sequence of gains $g_1, \ldots, g_n$, with corresponding swapped pairs, such that the result of interchanging the set $X = \{a_1', \ldots, a_k'\}$ with $Y = \{b_1', \ldots, b_k'\}$ is

$$G(k) = \sum_{i=1}^{k} g_i \tag{19.13}$$

Notice that when $k = n$, we are in effect interchanging all of A with all of B, so that $G(n) = 0$. We choose k so that $G(k)$ is maximum. If $G(k) \leq 0$, we stop.

If $G(k) > 0$, we interchange the corresponding sets X and Y and start the procedure over again from Step 1.

What we have accomplished is the following. Suppose $n = 12$ and the sequence of g_i produces the function G shown in Fig. 19–11, which peaks at

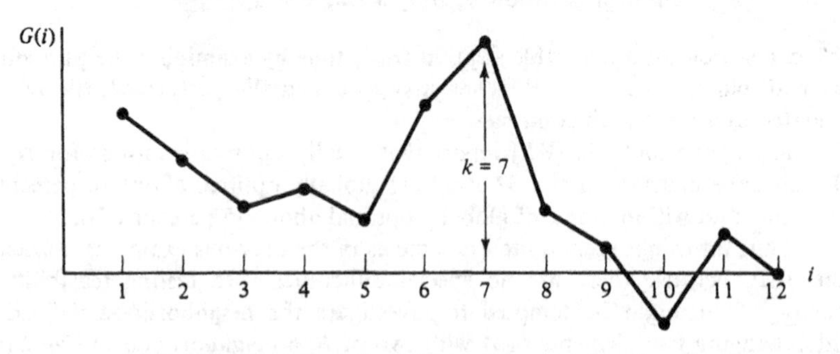

Figure 19–11 A hypothetical cumulative gain function G.

$k = 7$. We therefore find two sets of size $k = 7$ to interchange. But this interchange would have been difficult to find by searching for one swap at a time, because the second, third, and fifth swaps are actually *unfavorable* (notice from Fig. 19–11 that g_2, g_3, and g_5 are negative). We have therefore been able to take a "deep" stab, seven swaps away from where we were, without the necessity of exhausting all such sequences. The method, which we shall call *variable depth search*, will always find a single favorable swap if one exists, so the results are at least locally optimal with respect to the swap neighborhood N_s. But its real power depends on the use of the cost to find the best gain at each execution of Step 2. Figure 19–12 is a pictorial attempt to contrast local search with variable depth search.

The method is an empirical success on the UGP problem. Kernighan and Lin [KL] report that the probability of global optimality for problems of size

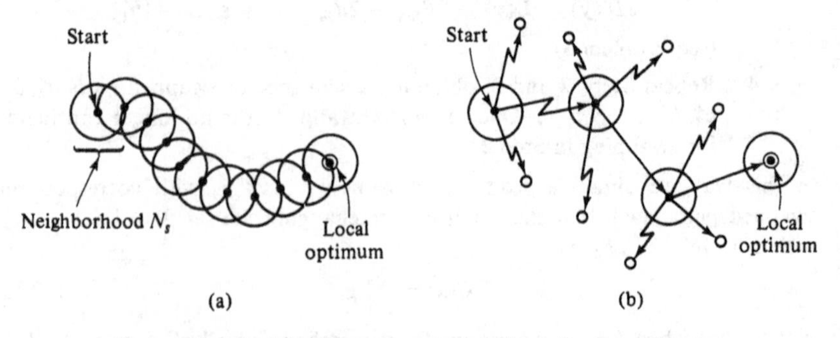

Figure 19–12 (a) Schematic representation of ordinary local search. (b) Schematic representation of variable depth search.

$n = 30$ is about 0.5 (as contrasted with the value of about 0.1 for ordinary local search mentioned above) and approximately $p = 2^{-n/30}$ for the problems studied by them. Furthermore the running time of the entire variable depth search algorithm is reported experimentally to be $O(n^{2.4})$ (see Problem 6).

The basic idea has also been applied with great success to the TSP [LK], where, however, additional complication is caused by the need to keep track of the feasibility of a sequence of edge exchanges. Lin and Kernighan [LK] discuss the algorithm in detail, together with several refinements that either improve the running time without sacrificing much power or improve the quality of the solutions without greatly increasing running time.

19.6 General Issues in Local Search

The previous four examples illustrate the diversity of problems to which local search has been applied. (Some other applications are mentioned in the Notes and References.) Each application has its own peculiarities and difficulties to be overcome, but certain patterns appear, which we shall now discuss.

The first issue that usually arises is the selection of a neighborhood or a class of neighborhoods, and this is tied to the notion of a "natural" perturbation of a feasible solution. This may be almost forced, like the swap perturbation in the graph partitioning problem. Or it may be somewhat less obvious, like the X-change perturbation in the survivable network problem. It may require some ingenuity to find, like "Δ-opt" in the multicommodity flow problem (see [Gr] and [GS]). Many times a perturbation has an "order" k associated with it, such as k-change for the TSP, and the resulting neighborhoods are of size $O(n^k)$.

What happens next, as a general rule, is that for a given range of problem sizes, a natural neighborhood of manageable size has a certain *strength*—that is, local optima produced by local search have a certain average quality. This strength seems to be strongly related to the correlation of the quality of the starting feasible solutions with that of the resulting local optima. A strong neighborhood seems to produce local optima whose quality is largely independent of the quality of starting solutions; weak neighborhoods seem to produce local optima whose quality is strongly correlated with that of the starts. For example, 3-change seems to be strong for TSP's of size up to at least 50 cities, so completely random initial tours are good starts because they enable us to sample the set of local optima widely. In contrast, X-change was found weaker [StWK] and requires a heuristic construction to produce starts of good average cost. Hence the relative strength of a neighborhood determines whether biased or completely random starts should be used.

Another question that arises is the manner in which the neighborhood is searched. The two extremes are *first-improvement*, in which a favorable change is accepted as soon as it is found with no further searching, and *steepest descent*, where the entire neighborhood is searched and a solution with the lowest cost is selected. It is often not worth the extra time to implement steepest descent, although it is dangerous to generalize about such matters. The great advantage of first-improvement is that only the final neighborhood necessarily needs to be searched completely, so that local optima can generally be found faster. The variable depth method of [KL] can be thought of as a kind of incomplete search strategy applied to an expanded version of the original neighborhood.

The order in which the neighborhood is searched is also a question. It is usually simplest to use some natural lexicographic ordering induced by indexing. The neighborhood can be ordered randomly, however, and this has the advantage of producing randomized local optima with the first-improvement policy, even from a single start. This strategy may be useful if starting solutions are difficult to obtain.

When a fixed ordering of the neighborhood is used with first-improvement, a new neighborhood search can restart at the beginning of the ordering or it can continue from the point where the last search left off. (Krone [Kr] calls these variations A and B, respectively.) We call the latter option *circular* searching, and a counter is used to determine when a local optimum has been found—that is, when we have gone around 360° at one solution. One can argue that circular searching is likely to be more efficient than the restart strategy, because changes that have been found most recently to be unfavorable in one neighborhood are more likely to remain unfavorable in the next. This needs to be tested experimentally in a given application.

When an approximate algorithm like local search is used for a minimization problem, an upper bound on the cost of an optimal solution is produced. It is often very desirable to have a lower bound as well, because that gives us a bound on the relative deviation of the approximate solution from optimality. A lower bound can often be obtained with relatively little effort. For example, in the minimum-cost survivable network problem, each vertex i must have degree at least $\delta_i = \max_j (s_{ij})$, and so we can do no better than connect it to its δ_i nearest neighbors $k_1, \ldots, k_{\delta_i}$. A lower bound is then

$$L = \sum_{i=1}^{|V|} \left(\sum_{j=1}^{\delta_i} d_{ik_j} \right) \tag{19.14}$$

As another example, Kernighan and Lin [KL] point out that the graph-partitioning problem can be lower-bounded by using a max-flow algorithm to find the value of a min-cut (see Problem 4).

We conclude this section by mentioning some elaborations and refinements that have been used in connection with local search. The first natural idea is to obtain a local optimum in the usual way and then to invest some effort in trying to improve it further. Krone [Kro] uses such a two-phase method for a flowshop scheduling problem. Phase I finds schedules that are locally optimal within the class of *permutation* schedules (see Sec. 18.5), and Phase II searches for changes that lie outside this class.

Kernighan and Lin [KL] also discuss a two-phase method for graph partitioning. They partition each of the locally optimal partition sets A and B into $A = A_1 \cup A_2$ and $B = B_1 \cup B_2$, using local search again, and then use the partition $A_1 \cup B_1, A_2 \cup B_2$ as a new starting solution. This is based on the empirical observation that when an $n \times n$ partition is locally but not globally optimal, an interchange of sets of size about $n/2$ is required to achieve global optimality.

A theme that was exploited by Lin [Li] in 1965 is the idea called *reduction.* This is based on the observation that in a particular problem some features will be common to all good solutions. These can be thought of as representing the "easy" parts of the problem. The strategy developed by Lin for the TSP is to obtain several local optima and then identify edges that are common to all of them. These are then fixed, thus reducing the time to find more local optima. This idea is developed further in [LK] and [GL].

As is common with heuristics, one can also argue for exactly the opposite idea. Once such common features are detected, they are *forbidden* rather than fixed. The justification is the following: If we fear that the global optimum is escaping us, this could be because our heuristic is "fooled" by these tempting features. Forbidding them could finally put us on the right track towards the global optimum. This is called *denial* in [SW].

Another idea, mentioned in [LK], is based on the fact that when many local optima are being found, there is likely to be a great deal of duplication among them. Furthermore, a large proportion of the search time is likely to be spent in checking out the optimality of each locally optimal solution by searching its neighborhood unsuccessfully. This can be avoided by keeping a dictionary of previously obtained local optima and their costs, and by checking current costs against locally optimal costs as the search proceeds. When a solution with locally optimal cost is encountered, it can be checked to see if it is identical to a previous local optimum. If it is, its checkout can be skipped. An efficient method for doing this cost testing must be used if time is to be saved overall (see Problem 5).

Having described some of the important practical considerations in applying local search, we turn next to some of its theoretical aspects.

19.7 The Geometry of Local Search

Local search is very closely related to the simplex algorithm; it can, in fact, be considered identical to it on an appropriately defined polytope. To pursue this idea we introduce the notion of a discrete linear subset problem.

Definition 19.7

Let $N = \{1, \ldots, n\}$ and let F be a set of subsets of N

$$F \subseteq 2^N$$

with the property that no set in F is properly contained in another. The *discrete linear subset* problem (DLS) $(F, \hat{d})$ is the combinatorial optimization problem with feasible set F and cost

$$c(f) = \sum_{i \in f} d_i \qquad f \in F$$

where $\hat{d} = (d_1, \ldots, d_n)$ is a given weight vector in R^n. □

Many important combinatorial optimization problems are DLS problems.

Example 19.4

The TSP is a DLS problem. Let n be the number of edges in a complete graph G, and let F contain exactly those sets of integers (that is, edges represented by integers) corresponding to tours of G. The cost vector $\hat{d}$ is simply a vector of edge weights, a vectorial representation of the distance matrix. □

Example 19.5

The MST problem is also a DLS problem, where now F contains exactly those sets of edges corresponding to spanning trees. □

Similarly, the shortest-path problem and some matroidal problems, as well as many others, can be formulated as DLS problems (see Problem 12). The DLS problems are closely related to the *subset systems* of Chapter 12.

A feasible point $f \in F$ can be thought of as a point in R^n with each coordinate i either 1 or zero depending on whether i is or is not in f, respectively. We shall use f to represent this n-vector as well as the set $f \in F$, with no danger of confusion. Thus, F can be thought of as a set of points on the unit cube in R^n with cost $\hat{d}'f$.

Now any point $f_0 \in F$ can be made uniquely optimal in F by the choice

$$d_i = \begin{cases} 0 & f_i = 1 \\ 1 & f_i = 0 \end{cases}$$

because this makes the cost of $f_0 = 0$ and the cost of all other f at least 1, by the assumption in Def. 19.7 that no feasible set is properly contained in another. This means that there is a hyperplane (namely $\hat{d}'\hat{x} = 0$) through the point $f_0 \in R^n$ with all the other points in F on one side ($\hat{d}'f > 0$). Hence no point f is a proper convex combination of others, and therefore the set F consists of vertices of a convex polytope in R^n.

Consider next the convex hull $CH(F)$, which is the polytope in the positive orthant of R^n which has as vertex set precisely the set F. Because the minimum value of the linear function $\hat{d}'f$ occurs at a vertex of $CH(F)$, the original DLS problem can be thought of as that of minimizing $\hat{d}'f$ over the polytope $CH(F)$: a linear programming problem. The polytope $CH(F)$ can in fact always be described by the intersection of a finite set of halfspaces; that is, by a set of inequalities

$$\hat{a}_i'\hat{x} \leq b_i \qquad i = 1, \ldots, m$$
$$\hat{x} \geq 0$$

and so the DLS problem can be written as what we shall call LP_1

$$\begin{aligned} &\min \hat{d}'\hat{x} \\ &\hat{A}\hat{x} \leq b \\ &\hat{x} \geq 0 \end{aligned} \tag{19.15}$$

with feasible set $F_1 \subseteq R^n$. (For a proof of the fact that a polytope can be so expressed, see [Ro, Theorem 19.1, p. 171]. This is one of those intuitively clear facts of geometry whose proof is far from trivial.)

It appears at first that we have accomplished a great deal—we have reduced any DLS problem, which might very well be *NP*-complete, to linear programming. We must look a little closer at the reduction, however, and ask how to obtain the matrix representation in (19.15) from the set F. No method suggests itself immediately, so we may begin to suspect that this is in general a difficult task. Confirmation of this is provided by a result in [KP], where it is shown that there is no polynomially concise characterization of the rows of the matrix A for any *NP*-complete DLS problem, unless $NP = co\text{–}NP$. Furthermore, it is shown in [Pa2] that for the TSP, determining whether two tours represent non-adjacent vertices of $CH(F)$ is itself an *NP*-complete problem.

We shall next pursue further the interpretation of DLS problems as LP's and derive two characterizations of the adjacency relation on $CH(F)$, originally given by Savage and his co-workers [Sa1, Sa2, WSB, SaWK]. In order to use what we know about vertices and adjacency from the simplex algorithm, we shall add slack variables to LP_1 in the usual way and move the problem into R^{n+m}, yielding the standard form LP_2 with feasible set $F_2 \subseteq R^{n+m}$:

$$\begin{aligned} &\min d'x \\ &Ax = b \\ &x \geq 0 \end{aligned} \tag{19.16}$$

where

$$d = (\hat{d} \,|\, 0_m)$$
$$A = [\hat{A} \,|\, I_m]$$
$$x = (\hat{x} \,|\, \epsilon_1, \ldots, \epsilon_m)$$

(Recall the notation of Subsec. 2.3.3.)

We define the transformation of a point $\hat{x} \in F_1$ to its corresponding point $x \in F_2$ as

$$\mu: \hat{x} \longrightarrow (\hat{x} \,|\, b - \hat{A}\hat{x}) = x$$

and the projection back from F_2 to F_1 as

$$\pi: x \longrightarrow (\text{first } n \text{ components of } x) = \hat{x}$$

Note that every point $x \in F_2$ can be written uniquely as $\mu\hat{x}$, where $\hat{x} \in F_1$, and in fact $\hat{x} = \pi x$. We next need a lemma that relates adjacency in F_1 and F_2.

Lemma 19.2 *Let $\hat{x}, \hat{y}$ be two distinct vertices in F_1. Then $\mu\hat{x}$ and $\mu\hat{y}$ are distinct vertices in F_2. Furthermore, $\hat{x}$ and $\hat{y}$ are adjacent in F_1 if and only if $\mu\hat{x}$ and $\mu\hat{y}$ are adjacent in F_2.*

Proof Suppose $\hat{x}$ is a vertex and $\mu\hat{x}$ is not. Then $\mu\hat{x}$ can be written as

$$\mu\hat{x} = \tfrac{1}{2}(\mu\hat{u} + \mu\hat{v})$$

where $\mu\hat{u}$ and $\mu\hat{v}$ are distinct points in F_2. Therefore

$$\hat{x} = \tfrac{1}{2}(\hat{u} + \hat{v})$$

so that $\hat{u} = \hat{v}$, which implies that $\mu\hat{u} = \mu\hat{v}$, a contradiction. The fact that $\mu\hat{x} \neq \mu\hat{y}$ if $\hat{x} \neq \hat{y}$ follows directly from the definition of μ.

We next show that if $\hat{x}$ and $\hat{y}$ are adjacent, so are $\mu\hat{x}$ and $\mu\hat{y}$. Suppose not; then a point $\mu\hat{z}$ on the line segment $L = [\mu\hat{x}, \mu\hat{y}]$ can be written

$$\mu\hat{z} = \tfrac{1}{2}(\mu\hat{u} + \mu\hat{v})$$

where $\mu\hat{u}$ and $\mu\hat{v}$ do not lie on L. Then

$$\hat{z} = \tfrac{1}{2}(\hat{u} + \hat{v})$$

where $\hat{u}$ and $\hat{v}$ do not lie on $\hat{L} = [\hat{x}, \hat{y}]$, which is a contradiction. The proof that $\hat{x}$ and $\hat{y}$ are adjacent if $\mu\hat{x}$ and $\mu\hat{y}$ are adjacent is similar. ☐

We now can establish the first characterization of adjacency.

Theorem 19.2 *Let a DLS problem generate the associated LP_1 in Eq. 19.15 in the manner described above. Then two distinct vertices $\hat{x}, \hat{y} \in F_1$ are adjacent if and only if there is a cost vector $\hat{d}$ such that $\hat{x}$ is the uniquely optimal vertex and $\hat{y}$ is a second-best vertex.*

Proof *Only if* Let $\hat{x}$ and $\hat{y}$ be two distinct and adjacent vertices of the convex polytope F_1. Because they are adjacent, there is a hyperplane that contains $\hat{x}$ and $\hat{y}$ and no other vertices and that *supports* the convex polytope: That is, all points in the polytope lie on one side of the hyperplane (see Sec. 2.3). This means there is a vector $\hat{d}$ such that

$$\hat{d}'\hat{x} = \hat{d}'\hat{y} = \hat{d}_0$$

and

$$\hat{d}'\hat{v} > \hat{d}_0 \qquad \text{for all vertices } \hat{v} \in F_1, \hat{v} \neq \hat{x}, \hat{y}$$

Let Q be defined by

$$Q = \min\,[\hat{d}'\hat{v} - \hat{d}_0] \qquad \text{over all vertices } \hat{v} \in F_1, \hat{v} \neq \hat{x}, \hat{y}$$

We shall perturb the hyperplane using the quantity Q so that $\hat{x}$ stays on it and

$\hat{y}$ becomes the "closest" vertex. To do this, choose a coordinate j where $y_j = 1$ and $x_j = 0$. (Such a coordinate must exist because of the noninclusion assumption in Def. 19.7.) Now increase d_j by $Q/2$. The cost of $\hat{x}$ is still $\hat{d}_0$, the cost of $\hat{y}$ becomes $\hat{d}_0 + Q/2$, and the cost of all other vertices is still at least $\hat{d}_0 + Q$.

If By Lemma 19.2, it suffices to establish the result for the images of $\hat{x}$ and $\hat{y}$ in F_2, the polytope corresponding to LP_2, which is a linear program in standard form.

Choose a cost vector $\hat{d}$ with the property described and place the simplex algorithm at vertex $y = \mu\hat{y}$, with cost vector $d = (\hat{d}|0)$. The vertex $x = \mu\hat{x}$ will be uniquely optimal in F_2 and y will be second-best. Suppose x and y are not adjacent. Because polytopal adjacency is an exact neighborhood for the simplex algorithm (Theorem 2.10) and because y has no polytopal neighbors with better cost, we conclude that y is optimal, which is a contradiction. □

We are now in a position to show that the adjacency neighborhood on $CH(F)$ is the *unique minimal exact neighborhood* for local search in the original DLS problem. In other words, the smallest local search neighborhood that guarantees optimality is precisely the one that would be searched by the simplex algorithm on $CH(F)$.

Theorem 19.3 *Polytopal adjacency on $CH(F)$ is the unique minimal exact neighborhood for local search over F.*

Proof Arguing again on F_2, polytopal adjacency is exact by the simplex algorithm; the problem is to show that every vertex adjacent to x on $CH(F)$ must be in every exact neighborhood at x. To this end, let t be a vertex adjacent to x but not in some exact neighborhood. By Theorem 19.2 there is a cost vector d that makes t uniquely optimal and x second-best. But the exact local search algorithm at x would proclaim x optimal, a contradiction. □

19.8 An Example of a Large Minimal Exact Neighborhood [PS1]

We shall next give a detailed example of a DLS problem—*job scheduling with deadlines*—in which we can actually exhibit the minimal exact neighborhood of a feasible solution. This will serve not only to fix the ideas of the previous section, but to introduce the question of how much time is required to search an exact neighborhood.

Definition 19.8

Let $N = \{1, 2, \ldots, n\}$ be a set of jobs, each with deadline D_j and execution time T_j. A subset $f \subseteq N$ of jobs is *feasible* if all the jobs in $N - \{f\}$ can be executed within their deadlines on a single processor, and $N - \{f\}$ is *maximal*;

that is, $N - \{f\}$ is not properly contained in another subset of N with the same property. An instance of the *job scheduling with deadlines* problem (JSD) is the DLS with this feasible set F and some cost d_j for $j \in N$. □

Notice that we have defined a feasible set to be a set of jobs *not* scheduled, in order to express the problem as a minimization problem. We thus want to find an $f \in F$ to minimize the cost

$$\sum_{j \in f} d_j$$

and we can think of d_j as the *penalty* incurred when job j does not meet its deadline. Notice also that the D_j and T_j are considered fixed; they are not read in as problem data but rather determine the feasible set F.

Now consider the JSD with its feasible set determined by fixing D_j and T_j at the following values, taking n odd:

$$D_j = \frac{(n-1)}{2} \qquad \text{all } i \in N$$

$$T_j = \begin{cases} \dfrac{(n-1)}{2} & j = 1 \\ 1 & j > 1 \end{cases}$$

If we schedule Job 1 there is no time for any other jobs to execute, so one feasible point is

$$f = N - \{1\}$$

If we do not schedule Job 1, there is room for precisely $(n-1)/2$ jobs with unit execution time, yielding the set of feasible points

$$F' = \left\{ s \subseteq N : 1 \in s \quad \text{and} \quad |s| = \frac{(n+1)}{2} \right\}$$

This exhausts all possible feasible points, so $F = \{f\} \cup F'$.

We shall next select any $s \in F'$ and exhibit a cost vector $\hat{d}(s)$ that renders s uniquely optimal and f second-best. By Theorem 19.2, this will show that f and s are adjacent. The instance that accomplishes this is

$$d_j(s) = \begin{cases} \dfrac{(n-3)}{2} & j = 1 \\ 0 & j \neq 1 \text{ and } j \in s \\ 1 & \text{otherwise} \end{cases}$$

The cost of s with this choice of $\hat{d}$ is $(n-3)/2$, the cost of f is $(n-1)/2$, and the cost of any other feasible point is at least $(n-1)/2$. Thus any s is in the minimal exact neighborhood of f, which is in fact precisely F'.

F' contains $\binom{n-1}{(n-1)/2}$ points. Using Stirling's formula, we can show that

$$|F'| \cong \frac{2^n}{\sqrt{2\pi n}} \tag{19.16}$$

We have thus found that the smallest exact neighborhood of f has an enormous number of points in it. This is a consequence of the choice of D_i and T_j but is a geometric fact that is independent of the cost vector $\hat{d}$, which we consider to be the input data.

If we envision a point-by-point enumeration of all of F', it would seem that exact local search from the point f must be very time consuming. The following result may come as a surprise.

Theorem 19.4 *For any instance of JSD with T_j, D_j as above and any cost $\hat{d}$, the minimal exact neighborhood of f, F', can be searched in $O(n)$ time.*

Proof Since $F = \{f\} \cup F'$, searching F' amounts to finding the optimal solution. This can be done by finding the $(n - 1)/2$ jobs in the set $N - \{1\}$ whose total cost is maximum and comparing that to d_1. The median algorithm mentioned in Problem 2 of Chapter 12 allows us to accomplish this in $O(n)$ time. □

Here we have an example where the search of a neighborhood is very strongly data-directed, along lines suggested by the variable-depth search in Sec. 19.5.

Savage and others [Sa1, Sa2, SWB, SaWK] show that the minimal exact neighborhood for the TSP is of exponential size (Problem 15), but the preceding example allows us to keep hope that perhaps there is an exact local search algorithm for the TSP with neighborhoods that can be searched fast, say in polynomial time. This would still not guarantee the existence of a polynomial-time algorithm for the TSP, because we still would not know how many improvements are needed to reach optimality. But it would at least provide an exact algorithm useful on small problems: exactly analogous to the simplex algorithm for LP, which also searches an exact neighborhood in polynomial time. The result in the next section destroys even this hope.

19.9 The Complexity of Exact Local Search for the TSP [PS1]

The goal of this section is to show that an exact neighborhood for the traveling salesman problem can be searched in polynomial time only if $P = NP$, a most unlikely circumstance. The problem we need is defined as follows.

RESTRICTED HAMILTON CIRCUIT (RHC)
Given a graph $G = (V, E)$ *and* a Hamilton path p of G, is there a Hamilton circuit c in G?

Thus we exhibit a Hamilton path (that is, a Hamilton circuit with one edge missing) and ask if there is a Hamilton circuit. It turns out that the Hamilton path does not help us make the decision.

Theorem 19.5 *The problem RHC is NP-complete.*

Proof The RHC problem is obviously in *NP*. We shall polynomially transform the ordinary Hamilton circuit problem (HC) to RHC with the help of the special-purpose subgraph shown in Fig. 19–13, which we call a *diamond.*

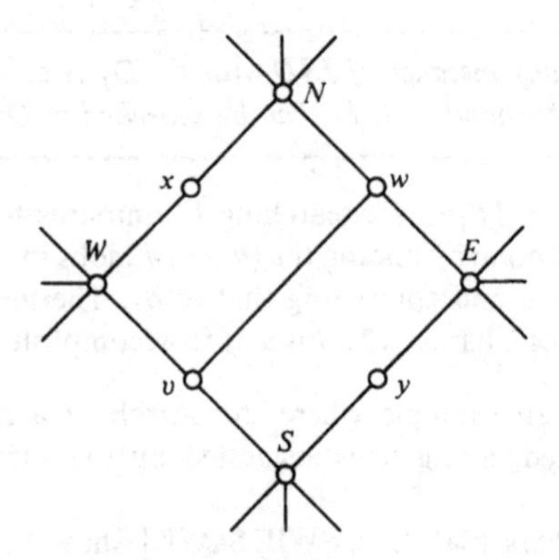

Figure 19–13 A diamond subgraph.

When this eight-vertex graph is a subgraph of a graph G, we shall assume that G touches the diamond only at the corner vertices N, S, E, and W. With this assumption we now show that if G has a Hamilton circuit c, the diamond D is traversed by c in exactly one of the two ways depicted in Fig. 19–14, which we call the *North-South* and *East-West* modes.

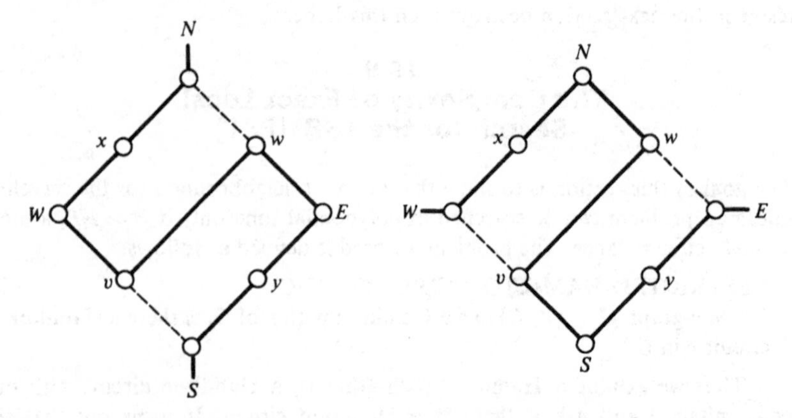

Figure 19–14 The North-South and East-West modes of traversing a diamond.

To see this, note that if a Hamilton circuit enters D at N, it must next go to x, or x would be stranded. It then reaches W, where it cannot leave D; for otherwise v and w could not be visited without stranding y. It then must go on to E and S, yielding a North-South traversal of D. The argument for an East-West traversal is similar.

Now let $G = (V, E)$ be an instance of the Hamilton circuit problem. We shall construct a new graph G' which always has a Hamilton path and which has a Hamilton circuit if and only if G has. The idea is to replace the nodes of G with diamonds and to put in a Hamilton path using the North-South modes. The graph G is then encoded into edges connecting east and west vertices of the diamonds.

In more detail, begin the construction of G' by placing $n = |V|$ diamonds D_i, with vertices N_i, S_i, E_i, and W_i, in a vertical column, and put in the edges $[S_i, N_{i+1}]$, $i = 1, \ldots, n-1$, as shown in Fig. 19–15(a). This ensures that G' has a Hamilton path p, namely, the North-South traversal of D_1, followed by $[S_1, N_2]$, followed by the North-South traversal of D_2, and so on.

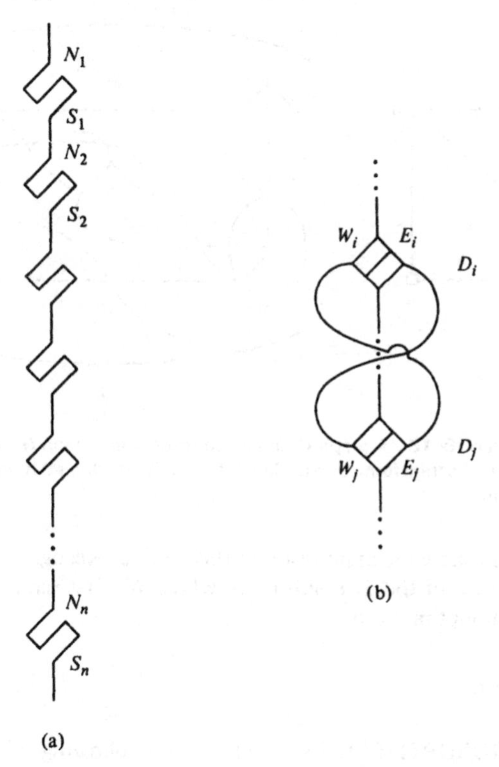

Figure 19–15 (a) The Hamilton path in G'. (b) The edges in G' caused by edge $[i, j]$ in G.

Next, for each edge $[v_i, v_j]$ of the original graph G, put edges $[W_i, E_j]$ and $[W_j, E_i]$ into G' (see Fig. 19–15(b)). If there is a Hamilton circuit in G, it can be mimicked in G' by visiting the diamonds in G' in the same order as the vertices in G and moving through each diamond in the East-West mode.

Conversely, if there is a Hamilton circuit in G', it cannot enter any diamond at a north or south vertex, for then it must traverse all the diamonds in the North-South mode and be stuck at N_1 or S_n. The Hamilton circuit must therefore traverse all the diamonds in the East-West mode, and the order in which the diamonds are visited determines a Hamilton circuit in G. □

Example 19.6

Figure 19–16 shows a simple graph G with no Hamilton circuit and the corresponding G', also with no Hamilton circuit, but with the Hamilton path p shown as a dotted line. □

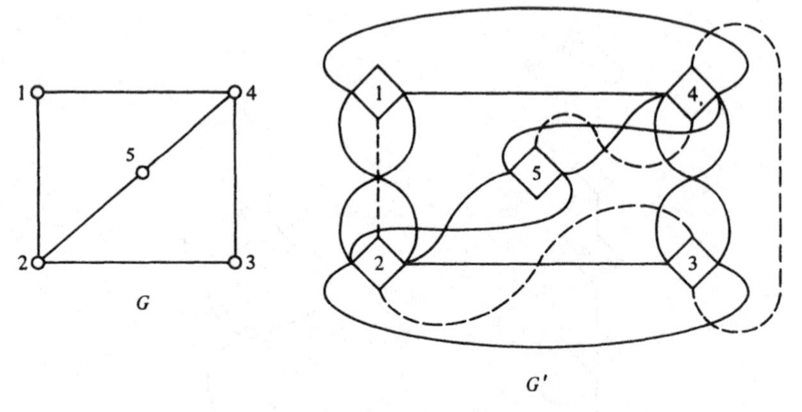

Figure 19–16 A graph G and its corresponding graph G', illustrating the transformation of Hamilton circuit to restricted Hamilton circuit.

We now can show the main result of this section: Recognizing a suboptimal tour in an instance of the TSP subsumes RHC. We first name the problem of recognizing a suboptimal tour.

Definition 19.9

The TSP SUBOPTIMALITY problem is the following: Given an instance of TSP and a tour f, is f suboptimal? □

Theorem 19.6 *TSP SUBOPTIMALITY is NP-complete.*

Proof The problem is in *NP*, because an optimal tour is a short certificate that shows f to be suboptimal. We now polynomially transform RHC to TSP SUBOPTIMALITY.

Let $G = (V, E)$ be the graph in an instance of RHC and let p be its Hamilton path. Construct an instance of TSP with $n = |V|$ cities and cost

$$d(i, j) = \begin{cases} 1 & \text{if } [i, j] \in E \\ 2 & \text{if } [i, j] \notin E \end{cases}$$

The Hamilton path p determines a tour f in the TSP with cost $(n - 1) + 2 = n + 1$. Suppose that f is suboptimal in TSP. Then there is a tour of cost n, which determines a Hamilton circuit in the original G of RHC.

Conversely, a Hamilton circuit in G determines a tour in the TSP with cost n, showing that f is suboptimal. □

Recall the subroutine improve(t) in our formulation of local search—with an exact neighborhood it returns an improved tour if t is suboptimal, so any instance of TSP SUBOPTIMALITY can be solved by one call of improve for any exact local search algorithm. We have therefore proved the following corollary.

Corollary *If P ≠ NP, no local search algorithm for the TSP having polynomial-time complexity per iteration can be exact.*

We thus end this chapter on a somewhat pessimistic note (see also Problem 14), abandoning hope that a local search algorithm with fast iterations—such as the simplex algorithm—can be found for the TSP, our prototypal *NP*-complete optimization problem. We should not forget, however, that local search can be a very effective heuristic for such problems. It is often, in fact, the best available.

PROBLEMS

*1. Show that the uniform graph partition problem of Sec. 19.5 is *NP*-complete.

2. Prove the following fact and discuss its relation to variable depth search [LK]. Let g_i, $i = 1, \ldots, k$ be a sequence of numbers, and let

$$G(j) = \sum_{i=1}^{j} g_i \qquad j = 1, \ldots, k$$

be their partial sums. Then if $G(k) > 0$, there is a cyclic permutation of the g_i: $g_r, g_{r+1}, \ldots, g_k, g_1, \ldots, g_{r-1}$, $1 \leq r \leq k$, with the property that all its partial sums are positive.

3. Prove that Eq. 19.12 gives the correct values for recomputing D after a swap.

4. Develop an algorithm for finding a lower bound in the uniform graph partition problem using the max-flow algorithm.

5. Suppose we want to check new feasible solutions during execution of a local search algorithm to see if they have already been proven to be locally optimal. Describe an efficient method for doing this.

6. To expedite the search for the next pair a_i', b_i' in the local search of a neighborhood by Eq. 19.11 in the uniform graph partitioning algorithm, Kernighan and Lin [KL] sort the D_a's and D_b's in descending order.
   (a) Explain how this speeds up this search.
   (b) Analyze the time complexity of each neighborhood search, and show it is $O(n^2 \log n)$.
   (c) The reported total time complexity is about $O(n^{2.4})$. What does this say about the number of passes needed to achieve local optimality?
   *(d) Compare the time complexities in (b) and (c) with the asymptotic time required to examine all possible pairs of equal-sized sets for interchange.

*7. [TS] The *diameter* of a graph is the longest shortest-path between any pair of nodes, and a *k-regular* graph G is one in which every node has the same degree. (In the directed case, indegree = outdegree = k.) Suppose we want to design a local search algorithm for designing a k-regular graph, directed or undirected, with n nodes and minimum diameter. Discuss the choice of (a) starting method; (b) neighborhood; and (c) cost criterion.

*8. [PS2] Use the diamond subgraph of Sec. 19.9 to construct n-city instances of the traveling salesman problem with the following properties: (a) there is a unique optimal tour, and (b) there are an exponential (in n) number of tours that are next best, have arbitrarily large cost, and cannot be improved by changing fewer than αn edges, where α is a fixed fraction.

9. [PS2] Consider the n-city traveling salesman problem in which the triangle inequality holds for the distance matrix. Let $c_0 > 0$ be the cost of an optimal tour, let c_s be the cost of a second-best tour, and define the *gap* g to be $(c_s - c_0)/c_0$. Prove that $g \leq 2/n$.

10. In a *Poisson process*, we think of time as being divided into tiny segments of length dt, and we assume that the probability of an event occuring in any segment is dt/τ, independently of any other segment.
    (a) Show that if we take the limit as dt approaches zero, the probability of an event *not* occuring in time T is $e^{-T/\tau}$.
    (b) Suppose we assume that a computer makes an undetected error with probability 10^{-6} in a year of operation, and that such errors occur as a Poisson

process. What is the probability that a computer executes a 1-second program without an undetected error?

(c) Suppose that a local search algorithm finds an optimal solution with probability 0.39. Under the assumptions of Part (b), how many independent trials are required so that we can place more confidence in the operation of the algorithm than in the operation of the computer?

11. (a) Establish Eq. 19.16, an estimate for the size of the set F'.

(b) Show that $|F'| > 2^{(n-1)/2}$.

12. Formulate the following problems as discrete linear subset problems.

(a) Shortest path.

(b) Min-cost flow.

(c) Maximum weight set in a matroid.

(d) Weighted matroid intersection.

(e) Weighted 3-matroid intersection.

***13.** [StWK] Suppose an undirected graph $G = (V, E)$ has actual connectivity matrix $[r_{ij}]$, and we want to check $r_{ij} \geq s_{ij}$ for all i, j, for a given requirement matrix $[s_{ij}]$.

(a) Prove the following: If

$$r_{am_i} \geq s_{ab}$$

$$r_{bm_i} \geq s_{ab}$$

for nodes $m_1, m_2, \ldots, m_{r_{ab}}$, distinct from a, b, and each other, then

$$r_{ab} \geq s_{ab}.$$

(b) Show that the result of Part (a) allows us to test the connectivity of a graph when $s_{ij} = k$ for all i, j with no more than $k(|V| - (k + 1)/2)$ tests.

(c) Show that if an X-change on a feasible network destroys feasibility by reducing r_{ab} below s_{ab}, then either

$$r_{im} < s_{ab}$$

or

$$r_{jl} < s_{ab}$$

where the X-change removed edges $[i, m]$ and $[j, l]$.

14. [PS1] Show that if $P \neq NP$ then no local search algorithm for the TSP having a polynomial time complexity per iteration can yield ϵ-approximate tours, for *any* fixed $\epsilon > 0$.

***15.** [WSB] Show that a tour of n cities has a number of adjacent tours in its minimal exact neighborhood that is exponential in n.

16. (a) Show that N_{n-3} is not an exact neighborhood for the TSP with n cities.

*(b) Show that N_{n-1} is an exact neighborhood.

NOTES AND REFERENCES

Section 19.2

[Bo] BOCK, F., "An Algorithm for Solving 'Traveling-Salesman' and Related Network Optimization Problems," presented at the fourteenth National Meeting of the Op. Res. Soc. of America, St. Louis, Missouri (Oct. 24, 1958).

[Cr] CROES, G. A., "A Method for Solving Traveling-Salesman Problems," *OR*, 6, no. 6 (November–December 1958), 791–812.

[RS] REITER, S., and G. SHERMAN, "Discrete Optimizing," *J. SIAM*, 13, no. 3 (Sept. 1965), 864–889.

[Li] LIN, S., "Computer Solutions of the Traveling Salesman Problem," *BSTJ*, 44, no. 10 (December 1965), 2245–69.

[HK1] HELD, M., and R. M. KARP, "A Dynamic Programming Approach to Sequencing Problems," *J. SIAM*, 10, no. 1 (March 1962), 196–210.

An early description of local search can be found in

[DFFN] DUNHAM, B., D. FRIDSHAL, R. FRIDSHAL, and J. H. NORTH, "Design by Natural Selection," IBM Res. Dept. RC-476, June 20, 1961.

Section 19.3

[StWK] STEIGLITZ, K., P. WEINER and D. J. KLEITMAN, "The Design of Minimal Cost Survivable Networks," *IEEE Trans. Cir. Theory*, CT–16, no. 4 (1969), 455–60.

Section 19.4

[RFRSK] ROTHFARB, B., H. FRANK, D. M. ROSENBAUM, K. STEIGLITZ, and D. J. KLEITMAN, "Optimal Design of Offshore Natural-Gas Pipeline Systems," *OR*, 18, no. 6 (November–December 1970), 992–1020.

Section 19.5

[KL] KERNIGHAN, B. W., and S. LIN, "An Efficient Heuristic Procedure for Partitioning Graphs," *BSTJ*, 49, no. 2 (February 1970), 291–307.

[LK] LIN, S., and B. W. KERNIGHAN, "An Effective Heuristic for the Traveling-Salesman Problem," *OR*, 21 (1973), 498–516.

Section 19.6

[Gr] GRATZER, F. J., "Computer Solution of Large Multicommodity Flow Problems," (unpublished Ph.D. dissertation, Princeton University, September 1970).

[GS] GRATZER, F. J., and K. STEIGLITZ, "A Heuristic Approach to Large Multicommodity Flow Problems," *Proc. Symp. Computer-Communications Networks and Teletraffic*, Microwave Res. Inst. Symp. Series, Vol. XXII, Polytechnic Press, N.Y. (1972), pp. 311–24.

[Kr] KRONE, M. J., "Heuristic Programming Applied to Scheduling Problems," (unpublished Ph.D. dissertation, Princeton University, September 1970).

[KS] KRONE, M. J., and K. STEIGLITZ, "Heuristic-Programming Solution of a Flowshop-Scheduling Problem," *OR*, 22, no. 3 (May-June 1974), 629–38.

[GL] GOLDSTEIN, A. J., and A. B. LESK, "Common Feature Techniques for Discrete Optimization," Comp. Sci. Tech. Report 27, Bell Tel. Labs. (March 1975).

[SW] STEIGLITZ, K., and P. WEINER, "Some Improved Algorithms for Computer Solution of the Travelling Salesman Problem," *Proc. Sixth Allerton Conf. on Circuit and System Theory*, Urbana, Illinois (1968), 814–21.

Section 19.7

[Ro] ROCKAFELLAR, R. T., *Convex Analysis*. Princeton, N.J.: Princeton Univ. Press, 1970.

[Pa2] PAPADIMITRIOU, C. H., "The Adjacency Relation on the Traveling Salesman Polytope is *NP*-Complete," *Math. Prog.*, 14 (1978), 312–24.

[Sa1] SAVAGE, S. L., "The Solution of Discrete Linear Optimization Problems by Neighborhood Search Techniques," (unpublished Ph.D. dissertation, Yale University, June 1973).

[Sa2] SAVAGE, S. L., "Some Theoretical Implications of Local Search," *Math. Prog.*, 10 (1976), 354–66.

[WSB] WEINER, P., S. L. SAVAGE, and A. BAGCHI, "Neighborhood Search Algorithms for Guaranteeing Optimal Traveling Salesman Tours Must be Inefficient," *J. Comp. Sys. Sci.*, 12 (1976), 25–35.

[SaWK] SAVAGE, S. L., P. WEINER, and M. J. KRONE, "Convergent Local Search," Res. Report 14, Dept. Comp. Sci., Yale University, New Haven, Conn. (1973).

[KP] KARP, R. M., and C. H. PAPADIMITRIOU, "On Linear Characterizations of Combinatorial Optimization Problems," *Proc. Twenty-First Annual Symp. on Foundations of Comp. Sci.*, IEEE (1980), 1–9.

Section 19.8

[Pa1] PAPADIMITRIOU, C. H., "The Complexity of Combinatorial Optimization Problems," (unpublished Ph.D. Thesis, Princeton University, August 1976).

[PS1] PAPADIMITRIOU, C. H., and K. STEIGLITZ, "On the Complexity of Local Search for the Traveling Salesman Problem," *J. SIAM Comp.*, 6, 1 (March 1977), 76–83.

FURTHER NOTES AND REFERENCES

Applications of local search to other problems have been described in the literature. The following references report favorable results in designing mechanical linkages, such as the printing-bed drive of a printing-press mechanism:

[LF] LEE, T. W., and F. FREUDENSTEIN, "Heuristic Combinatorial Optimization in the Kinematic Design of Mechanisms, Part I: Theory, Part 2: Applications," *Journal of Engineering for Industry*, 98, no. 4 (November 1976), 1277–84.

[LL] LEE, T. W., and N. LANGRANA, "Heuristic Approaches to the Inversed Dynamic Linkage Problems: With Special Application to Biomechanics," *Proc. Fifth World Congress on Theory of Machines and Mechanisms*, Paper No. USA-53, Montreal, Canada (July 1979).

The design of small-diameter networks, the subject of Problem 7, is described in

[TS] TOUEG, S., and K. STEIGLITZ, "The Design of Small-Diameter Networks by Local Search," *IEEE Trans. on Computers*, C-28, no. 7 (July 1979), 537–42.

Problems 8 and 9 are from

[PS2] PAPADIMITRIOU, C. H., and K. STEIGLITZ, "Some Examples of Difficult Traveling Salesman Problems," *OR*, 26, no. 3 (May–June 1978), 434–43.

The uniform graph partitioning problem in Problem 1 is a variation of minimum cut into bounded sets; for a proof of *NP*-completeness see

[GJS] GAREY, M. R., D. S. JOHNSON, and L. J. STOCKMEYER, "Some Simplified *NP*-Complete Graph Problems," *Theor. Comp. Sci.*, 1 (1976), 237–67.

Problem 16(b) is based on a graph-theoretic result first conjectured by Lin [Li] and proved in

[TH] THOMASON, A. G., "Hamiltonian Cycles and Uniquely Edge Colourable Graphs," *Annals of Discrete Math.*, 3 (1978), 259–68.

Index

B

C

H

I

J

K

L

M

Q

R

S

A CATALOG OF SELECTED

DOVER BOOKS

IN SCIENCE AND MATHEMATICS

Astronomy

BURNHAM'S CELESTIAL HANDBOOK, Robert Burnham, Jr. Thorough guide to the stars beyond our solar system. Exhaustive treatment. Alphabetical by constellation: Andromeda to Cetus in Vol. 1; Chamaeleon to Orion in Vol. 2; and Pavo to Vulpecula in Vol. 3. Hundreds of illustrations. Index in Vol. 3. 2,000pp. 6⅛ x 9¼.
Vol. I: 0-486-23567-X
Vol. II: 0-486-23568-8
Vol. III: 0-486-23673-0

EXPLORING THE MOON THROUGH BINOCULARS AND SMALL TELESCOPES, Ernest H. Cherrington, Jr. Informative, profusely illustrated guide to locating and identifying craters, rills, seas, mountains, other lunar features. Newly revised and updated with special section of new photos. Over 100 photos and diagrams. 240pp. 8¼ x 11. 0-486-24491-1

THE EXTRATERRESTRIAL LIFE DEBATE, 1750–1900, Michael J. Crowe. First detailed, scholarly study in English of the many ideas that developed from 1750 to 1900 regarding the existence of intelligent extraterrestrial life. Examines ideas of Kant, Herschel, Voltaire, Percival Lowell, many other scientists and thinkers. 16 illustrations. 704pp. 5⅜ x 8½. 0-486-40675-X

THEORIES OF THE WORLD FROM ANTIQUITY TO THE COPERNICAN REVOLUTION, Michael J. Crowe. Newly revised edition of an accessible, enlightening book re-creates the change from an earth-centered to a sun-centered conception of the solar system. 242pp. 5⅜ x 8½. 0-486-41444-2

ARISTARCHUS OF SAMOS: The Ancient Copernicus, Sir Thomas Heath. Heath's history of astronomy ranges from Homer and Hesiod to Aristarchus and includes quotes from numerous thinkers, compilers, and scholasticists from Thales and Anaximander through Pythagoras, Plato, Aristotle, and Heraclides. 34 figures. 448pp. 5⅜ x 8½.
0-486-43886-4

A COMPLETE MANUAL OF AMATEUR ASTRONOMY: TOOLS AND TECHNIQUES FOR ASTRONOMICAL OBSERVATIONS, P. Clay Sherrod with Thomas L. Koed. Concise, highly readable book discusses: selecting, setting up and maintaining a telescope; amateur studies of the sun; lunar topography and occultations; observations of Mars, Jupiter, Saturn, the minor planets and the stars; an introduction to photoelectric photometry; more. 1981 ed. 124 figures. 25 halftones. 37 tables. 335pp. 6½ x 9¼. 0-486-42820-8

AMATEUR ASTRONOMER'S HANDBOOK, J. B. Sidgwick. Timeless, comprehensive coverage of telescopes, mirrors, lenses, mountings, telescope drives, micrometers, spectroscopes, more. 189 illustrations. 576pp. 5⅝ x 8¼. (Available in U.S. only.)
0-486-24034-7

STAR LORE: Myths, Legends, and Facts, William Tyler Olcott. Captivating retellings of the origins and histories of ancient star groups include Pegasus, Ursa Major, Pleiades, signs of the zodiac, and other constellations. "Classic."—Sky & Telescope. 58 illustrations. 544pp. 5⅜ x 8½. 0-486-43581-4

Chemistry

THE SCEPTICAL CHYMIST: THE CLASSIC 1661 TEXT, Robert Boyle. Boyle defines the term "element," asserting that all natural phenomena can be explained by the motion and organization of primary particles. 1911 ed. viii+232pp. 5³/₈ x 8¹/₂.
0-486-42825-7

RADIOACTIVE SUBSTANCES, Marie Curie. Here is the celebrated scientist's doctoral thesis, the prelude to her receipt of the 1903 Nobel Prize. Curie discusses establishing atomic character of radioactivity found in compounds of uranium and thorium; extraction from pitchblende of polonium and radium; isolation of pure radium chloride; determination of atomic weight of radium; plus electric, photographic, luminous, heat, color effects of radioactivity. ii+94pp. 5³/₈ x 8¹/₂. 0-486-42550-9

CHEMICAL MAGIC, Leonard A. Ford. Second Edition, Revised by E. Winston Grundmeier. Over 100 unusual stunts demonstrating cold fire, dust explosions, much more. Text explains scientific principles and stresses safety precautions. 128pp. 5³/₈ x 8¹/₂. 0-486-67628-5

MOLECULAR THEORY OF CAPILLARITY, J. S. Rowlinson and B. Widom. History of surface phenomena offers critical and detailed examination and assessment of modern theories, focusing on statistical mechanics and application of results in mean-field approximation to model systems. 1989 edition. 352pp. 5³/₈ x 8¹/₂. 0-486-42544-4

CHEMICAL AND CATALYTIC REACTION ENGINEERING, James J. Carberry. Designed to offer background for managing chemical reactions, this text examines behavior of chemical reactions and reactors; fluid-fluid and fluid-solid reaction systems; heterogeneous catalysis and catalytic kinetics; more. 1976 edition. 672pp. 6¹/₈ x 9¹/₄. 0-486-41736-0 $31.95

ELEMENTS OF CHEMISTRY, Antoine Lavoisier. Monumental classic by founder of modern chemistry in remarkable reprint of rare 1790 Kerr translation. A must for every student of chemistry or the history of science. 539pp. 5³/₈ x 8¹/₂. 0-486-64624-6

MOLECULES AND RADIATION: An Introduction to Modern Molecular Spectroscopy. Second Edition, Jeffrey I. Steinfeld. This unified treatment introduces upper-level undergraduates and graduate students to the concepts and the methods of molecular spectroscopy and applications to quantum electronics, lasers, and related optical phenomena. 1985 edition. 512pp. 5³/₈ x 8¹/₂. 0-486-44152-0

A SHORT HISTORY OF CHEMISTRY, J. R. Partington. Classic exposition explores origins of chemistry, alchemy, early medical chemistry, nature of atmosphere, theory of valency, laws and structure of atomic theory, much more. 428pp. 5³/₈ x 8¹/₂. (Available in U.S. only.) 0-486-65977-1

GENERAL CHEMISTRY, Linus Pauling. Revised 3rd edition of classic first-year text by Nobel laureate. Atomic and molecular structure, quantum mechanics, statistical mechanics, thermodynamics correlated with descriptive chemistry. Problems. 992pp. 5³/₈ x 8¹/₂.
0-486-65622-5

ELECTRON CORRELATION IN MOLECULES, S. Wilson. This text addresses one of theoretical chemistry's central problems. Topics include molecular electronic structure, independent electron models, electron correlation, the linked diagram theorem, and related topics. 1984 edition. 304pp. 5³/₈ x 8¹/₂. 0-486-45879-2

Engineering

DE RE METALLICA, Georgius Agricola. The famous Hoover translation of greatest treatise on technological chemistry, engineering, geology, mining of early modern times (1556). All 289 original woodcuts. 638pp. $6^3/_4$ x 11. 0-486-60006-8

FUNDAMENTALS OF ASTRODYNAMICS, Roger Bate et al. Modern approach developed by U.S. Air Force Academy. Designed as a first course. Problems, exercises. Numerous illustrations. 455pp. $5^3/_8$ x $8^1/_2$. 0-486-60061-0

DYNAMICS OF FLUIDS IN POROUS MEDIA, Jacob Bear. For advanced students of ground water hydrology, soil mechanics and physics, drainage and irrigation engineering and more. 335 illustrations. Exercises, with answers. 784pp. $6^1/_8$ x $9^1/_4$. 0-486-65675-6

THEORY OF VISCOELASTICITY (SECOND EDITION), Richard M. Christensen. Complete consistent description of the linear theory of the viscoelastic behavior of materials. Problem-solving techniques discussed. 1982 edition. 29 figures. xiv+364pp. $6^1/_8$ x $9^1/_4$. 0-486-42880-X

MECHANICS, J. P. Den Hartog. A classic introductory text or refresher. Hundreds of applications and design problems illuminate fundamentals of trusses, loaded beams and cables, etc. 334 answered problems. 462pp. $5^3/_8$ x $8^1/_2$. 0-486-60754-2

MECHANICAL VIBRATIONS, J. P. Den Hartog. Classic textbook offers lucid explanations and illustrative models, applying theories of vibrations to a variety of practical industrial engineering problems. Numerous figures. 233 problems, solutions. Appendix. Index. Preface. 436pp. $5^3/_8$ x $8^1/_2$. 0-486-64785-4

STRENGTH OF MATERIALS, J. P. Den Hartog. Full, clear treatment of basic material (tension, torsion, bending, etc.) plus advanced material on engineering methods, applications. 350 answered problems. 323pp. $5^3/_8$ x $8^1/_2$. 0-486-60755-0

A HISTORY OF MECHANICS, René Dugas. Monumental study of mechanical principles from antiquity to quantum mechanics. Contributions of ancient Greeks, Galileo, Leonardo, Kepler, Lagrange, many others. 671pp. $5^3/_8$ x $8^1/_2$. 0-486-65632-2

STABILITY THEORY AND ITS APPLICATIONS TO STRUCTURAL MECHANICS, Clive L. Dym. Self-contained text focuses on Koiter postbuckling analyses, with mathematical notions of stability of motion. Basing minimum energy principles for static stability upon dynamic concepts of stability of motion, it develops asymptotic buckling and postbuckling analyses from potential energy considerations, with applications to columns, plates, and arches. 1974 ed. 208pp. $5^3/_8$ x $8^1/_2$. 0-486-42541-X

BASIC ELECTRICITY, U.S. Bureau of Naval Personnel. Originally a training course; best nontechnical coverage. Topics include batteries, circuits, conductors, AC and DC, inductance and capacitance, generators, motors, transformers, amplifiers, etc. Many questions with answers. 349 illustrations. 1969 edition. 448pp. $6^1/_2$ x $9^1/_4$. 0-486-20973-3

ROCKETS, Robert Goddard. Two of the most significant publications in the history of rocketry and jet propulsion: "A Method of Reaching Extreme Altitudes" (1919) and "Liquid Propellant Rocket Development" (1936). 128pp. 5³/₈ x 8¹/₂. 0-486-42537-1

STATISTICAL MECHANICS: PRINCIPLES AND APPLICATIONS, Terrell L. Hill. Standard text covers fundamentals of statistical mechanics, applications to fluctuation theory, imperfect gases, distribution functions, more. 448pp. 5³/₈ x 8¹/₂. 0-486-65390-0

ENGINEERING AND TECHNOLOGY 1650–1750: ILLUSTRATIONS AND TEXTS FROM ORIGINAL SOURCES, Martin Jensen. Highly readable text with more than 200 contemporary drawings and detailed engravings of engineering projects dealing with surveying, leveling, materials, hand tools, lifting equipment, transport and erection, piling, bailing, water supply, hydraulic engineering, and more. Among the specific projects outlined-transporting a 50-ton stone to the Louvre, erecting an obelisk, building timber locks, and dredging canals. 207pp. 8³/₈ x 11¹/₄. 0-486-42232-1

THE VARIATIONAL PRINCIPLES OF MECHANICS, Cornelius Lanczos. Graduate level coverage of calculus of variations, equations of motion, relativistic mechanics, more. First inexpensive paperbound edition of classic treatise. Index. Bibliography. 418pp. 5³/₈ x 8¹/₂. 0-486-65067-7

PROTECTION OF ELECTRONIC CIRCUITS FROM OVERVOLTAGES, Ronald B. Standler. Five-part treatment presents practical rules and strategies for circuits designed to protect electronic systems from damage by transient overvoltages. 1989 ed. xxiv+434pp. 6¹/₈ x 9¹/₄. 0-486-42552-5

ROTARY WING AERODYNAMICS, W. Z. Stepniewski. Clear, concise text covers aerodynamic phenomena of the rotor and offers guidelines for helicopter performance evaluation. Originally prepared for NASA. 537 figures. 640pp. 6¹/₈ x 9¹/₄. 0-486-64647-5

INTRODUCTION TO SPACE DYNAMICS, William Tyrrell Thomson. Comprehensive, classic introduction to space-flight engineering for advanced undergraduate and graduate students. Includes vector algebra, kinematics, transformation of coordinates. Bibliography. Index. 352pp. 5³/₈ x 8¹/₂. 0-486-65113-4

HISTORY OF STRENGTH OF MATERIALS, Stephen P. Timoshenko. Excellent historical survey of the strength of materials with many references to the theories of elasticity and structure. 245 figures. 452pp. 5³/₈ x 8¹/₂. 0-486-61187-6

ANALYTICAL FRACTURE MECHANICS, David J. Unger. Self-contained text supplements standard fracture mechanics texts by focusing on analytical methods for determining crack-tip stress and strain fields. 336pp. 6¹/₈ x 9¹/₄. 0-486-41737-9

STATISTICAL MECHANICS OF ELASTICITY, J. H. Weiner. Advanced, self-contained treatment illustrates general principles and elastic behavior of solids. Part 1, based on classical mechanics, studies thermoelastic behavior of crystalline and polymeric solids. Part 2, based on quantum mechanics, focuses on interatomic force laws, behavior of solids, and thermally activated processes. For students of physics and chemistry and for polymer physicists. 1983 ed. 96 figures. 496pp. 5³/₈ x 8¹/₂. 0-486-42260-7

Mathematics

FUNCTIONAL ANALYSIS (Second Corrected Edition), George Bachman and Lawrence Narici. Excellent treatment of subject geared toward students with background in linear algebra, advanced calculus, physics and engineering. Text covers introduction to inner-product spaces, normed, metric spaces, and topological spaces; complete orthonormal sets, the Hahn-Banach Theorem and its consequences, and many other related subjects. 1966 ed. 544pp. $6^1/_8$ x $9^1/_4$. 0-486-40251-7

DIFFERENTIAL MANIFOLDS, Antoni A. Kosinski. Introductory text for advanced undergraduates and graduate students presents systematic study of the topological structure of smooth manifolds, starting with elements of theory and concluding with method of surgery. 1993 edition. 288pp. $5^3/_8$ x $8^1/_2$. 0-486-46244-7

VECTOR AND TENSOR ANALYSIS WITH APPLICATIONS, A. I. Borisenko and I. E. Tarapov. Concise introduction. Worked-out problems, solutions, exercises. 257pp. $5^3/_8$ x $8^1/_4$. 0-486-63833-2

AN INTRODUCTION TO ORDINARY DIFFERENTIAL EQUATIONS, Earl A. Coddington. A thorough and systematic first course in elementary differential equations for undergraduates in mathematics and science, with many exercises and problems (with answers). Index. 304pp. $5^3/_8$ x $8^1/_2$. 0-486-65942-9

FOURIER SERIES AND ORTHOGONAL FUNCTIONS, Harry F. Davis. An incisive text combining theory and practical example to introduce Fourier series, orthogonal functions and applications of the Fourier method to boundary-value problems. 570 exercises. Answers and notes. 416pp. $5^3/_8$ x $8^1/_2$. 0-486-65973-9

COMPUTABILITY AND UNSOLVABILITY, Martin Davis. Classic graduate-level introduction to theory of computability, usually referred to as theory of recurrent functions. New preface and appendix. 288pp. $5^3/_8$ x $8^1/_2$. 0-486-61471-9

AN INTRODUCTION TO MATHEMATICAL ANALYSIS, Robert A. Rankin. Dealing chiefly with functions of a single real variable, this text by a distinguished educator introduces limits, continuity, differentiability, integration, convergence of infinite series, double series, and infinite products. 1963 edition. 624pp. $5^3/_8$ x $8^1/_2$. 0-486-46251-X

METHODS OF NUMERICAL INTEGRATION (SECOND EDITION), Philip J. Davis and Philip Rabinowitz. Requiring only a background in calculus, this text covers approximate integration over finite and infinite intervals, error analysis, approximate integration in two or more dimensions, and automatic integration. 1984 edition. 624pp. $5^3/_8$ x $8^1/_2$. 0-486-45339-1

INTRODUCTION TO LINEAR ALGEBRA AND DIFFERENTIAL EQUATIONS, John W. Dettman. Excellent text covers complex numbers, determinants, orthonormal bases, Laplace transforms, much more. Exercises with solutions. Undergraduate level. 416pp. $5^3/_8$ x $8^1/_2$. 0-486-65191-6

RIEMANN'S ZETA FUNCTION, H. M. Edwards. Superb, high-level study of landmark 1859 publication entitled "On the Number of Primes Less Than a Given Magnitude" traces developments in mathematical theory that it inspired. xiv+315pp. $5^3/_8$ x $8^1/_2$. 0-486-41740-9

CALCULUS OF VARIATIONS WITH APPLICATIONS, George M. Ewing. Applications-oriented introduction to variational theory develops insight and promotes understanding of specialized books, research papers. Suitable for advanced undergraduate/graduate students as primary, supplementary text. 352pp. 5^3/$_8$ x 8^1/$_2$.
0-486-64856-7

MATHEMATICIAN'S DELIGHT, W. W. Sawyer. "Recommended with confidence" by *The Times Literary Supplement,* this lively survey was written by a renowned teacher. It starts with arithmetic and algebra, gradually proceeding to trigonometry and calculus. 1943 edition. 240pp. 5^3/$_8$ x 8^1/$_2$. 0-486-46240-4

ADVANCED EUCLIDEAN GEOMETRY, Roger A. Johnson. This classic text explores the geometry of the triangle and the circle, concentrating on extensions of Euclidean theory, and examining in detail many relatively recent theorems. 1929 edition. 336pp. 5^3/$_8$ x 8^1/$_2$. 0-486-46237-4

COUNTEREXAMPLES IN ANALYSIS, Bernard R. Gelbaum and John M. H. Olmsted. These counterexamples deal mostly with the part of analysis known as "real variables." The first half covers the real number system, and the second half encompasses higher dimensions. 1962 edition. xxiv+198pp. 5^3/$_8$ x 8^1/$_2$. 0-486-42875-3

CATASTROPHE THEORY FOR SCIENTISTS AND ENGINEERS, Robert Gilmore. Advanced-level treatment describes mathematics of theory grounded in the work of Poincaré, R. Thom, other mathematicians. Also important applications to problems in mathematics, physics, chemistry and engineering. 1981 edition. References. 28 tables. 397 black-and-white illustrations. xvii + 666pp. 6^1/$_8$ x 9^1/$_4$. 0-486-67539-4

COMPLEX VARIABLES: Second Edition, Robert B. Ash and W. P. Novinger. Suitable for advanced undergraduates and graduate students, this newly revised treatment covers Cauchy theorem and its applications, analytic functions, and the prime number theorem. Numerous problems and solutions. 2004 edition. 224pp. 6^1/$_2$ x 9^1/$_4$. 0-486-46250-1

NUMERICAL METHODS FOR SCIENTISTS AND ENGINEERS, Richard Hamming. Classic text stresses frequency approach in coverage of algorithms, polynomial approximation, Fourier approximation, exponential approximation, other topics. Revised and enlarged 2nd edition. 721pp. 5^3/$_8$ x 8^1/$_2$. 0-486-65241-6

INTRODUCTION TO NUMERICAL ANALYSIS (2nd Edition), F. B. Hildebrand. Classic, fundamental treatment covers computation, approximation, interpolation, numerical differentiation and integration, other topics. 150 new problems. 669pp. 5^3/$_8$ x 8^1/$_2$.
0-486-65363-3

MARKOV PROCESSES AND POTENTIAL THEORY, Robert M. Blumental and Ronald K. Getoor. This graduate-level text explores the relationship between Markov processes and potential theory in terms of excessive functions, multiplicative functionals and subprocesses, additive functionals and their potentials, and dual processes. 1968 edition. 320pp. 5^3/$_8$ x 8^1/$_2$. 0-486-46263-3

ABSTRACT SETS AND FINITE ORDINALS: An Introduction to the Study of Set Theory, G. B. Keene. This text unites logical and philosophical aspects of set theory in a manner intelligible to mathematicians without training in formal logic and to logicians without a mathematical background. 1961 edition. 112pp. 5^3/$_8$ x 8^1/$_2$. 0-486-46249-8

INTRODUCTORY REAL ANALYSIS, A.N. Kolmogorov, S. V. Fomin. Translated by Richard A. Silverman. Self-contained, evenly paced introduction to real and functional analysis. Some 350 problems. 403pp. $5^{3}/_{8}$ x $8^{1}/_{2}$. 0-486-61226-0

APPLIED ANALYSIS, Cornelius Lanczos. Classic work on analysis and design of finite processes for approximating solution of analytical problems. Algebraic equations, matrices, harmonic analysis, quadrature methods, much more. 559pp. $5^{3}/_{8}$ x $8^{1}/_{2}$. 0-486-65656-X

AN INTRODUCTION TO ALGEBRAIC STRUCTURES, Joseph Landin. Superb self-contained text covers "abstract algebra": sets and numbers, theory of groups, theory of rings, much more. Numerous well-chosen examples, exercises. 247pp. $5^{3}/_{8}$ x $8^{1}/_{2}$. 0-486-65940-2

QUALITATIVE THEORY OF DIFFERENTIAL EQUATIONS, V. V. Nemytskii and V.V. Stepanov. Classic graduate-level text by two prominent Soviet mathematicians covers classical differential equations as well as topological dynamics and ergodic theory. Bibliographies. 523pp. $5^{3}/_{8}$ x $8^{1}/_{2}$. 0-486-65954-2

THEORY OF MATRICES, Sam Perlis. Outstanding text covering rank, nonsingularity and inverses in connection with the development of canonical matrices under the relation of equivalence, and without the intervention of determinants. Includes exercises. 237pp. $5^{3}/_{8}$ x $8^{1}/_{2}$. 0-486-66810-X

INTRODUCTION TO ANALYSIS, Maxwell Rosenlicht. Unusually clear, accessible coverage of set theory, real number system, metric spaces, continuous functions, Riemann integration, multiple integrals, more. Wide range of problems. Undergraduate level. Bibliography. 254pp. $5^{3}/_{8}$ x $8^{1}/_{2}$. 0-486-65038-3

MODERN NONLINEAR EQUATIONS, Thomas L. Saaty. Emphasizes practical solution of problems; covers seven types of equations. ". . . a welcome contribution to the existing literature. . . ."—*Math Reviews*. 490pp. $5^{3}/_{8}$ x $8^{1}/_{2}$. 0-486-64232-1

MATRICES AND LINEAR ALGEBRA, Hans Schneider and George Phillip Barker. Basic textbook covers theory of matrices and its applications to systems of linear equations and related topics such as determinants, eigenvalues and differential equations. Numerous exercises. 432pp. $5^{3}/_{8}$ x $8^{1}/_{2}$. 0-486-66014-1

LINEAR ALGEBRA, Georgi E. Shilov. Determinants, linear spaces, matrix algebras, similar topics. For advanced undergraduates, graduates. Silverman translation. 387pp. $5^{3}/_{8}$ x $8^{1}/_{2}$. 0-486-63518-X

MATHEMATICAL METHODS OF GAME AND ECONOMIC THEORY: Revised Edition, Jean-Pierre Aubin. This text begins with optimization theory and convex analysis, followed by topics in game theory and mathematical economics, and concluding with an introduction to nonlinear analysis and control theory. 1982 edition. 656pp. $6^{1}/_{8}$ x $9^{1}/_{4}$. 0-486-46265-X

SET THEORY AND LOGIC, Robert R. Stoll. Lucid introduction to unified theory of mathematical concepts. Set theory and logic seen as tools for conceptual understanding of real number system. 496pp. $5^{3}/_{8}$ x $8^{1}/_{4}$. 0-486-63829-4

Math—Decision Theory, Statistics, Probability

INTRODUCTION TO PROBABILITY, John E. Freund. Featured topics include permutations and factorials, probabilities and odds, frequency interpretation, mathematical expectation, decision-making, postulates of probability, rule of elimination, much more. Exercises with some solutions. Summary. 1973 edition. 247pp. $5^3/_8$ x $8^1/_2$.
0-486-67549-1

STATISTICAL AND INDUCTIVE PROBABILITIES, Hugues Leblanc. This treatment addresses a decades-old dispute among probability theorists, asserting that both statistical and inductive probabilities may be treated as sentence-theoretic measurements, and that the latter qualify as estimates of the former. 1962 edition. 160pp. $5^3/_8$ x $8^1/_2$.
0-486-44980-7

APPLIED MULTIVARIATE ANALYSIS: Using Bayesian and Frequentist Methods of Inference, Second Edition, S. James Press. This two-part treatment deals with foundations as well as models and applications. Topics include continuous multivariate distributions; regression and analysis of variance; factor analysis and latent structure analysis; and structuring multivariate populations. 1982 edition. 692pp. $5^3/_8$ x $8^1/_2$. 0-486-44236-5

LINEAR PROGRAMMING AND ECONOMIC ANALYSIS, Robert Dorfman, Paul A. Samuelson and Robert M. Solow. First comprehensive treatment of linear programming in standard economic analysis. Game theory, modern welfare economics, Leontief input-output, more. 525pp. $5^3/_8$ x $8^1/_2$. 0-486-65491-5

PROBABILITY: AN INTRODUCTION, Samuel Goldberg. Excellent basic text covers set theory, probability theory for finite sample spaces, binomial theorem, much more. 360 problems. Bibliographies. 322pp. $5^3/_8$ x $8^1/_2$. 0-486-65252-1

GAMES AND DECISIONS: INTRODUCTION AND CRITICAL SURVEY, R. Duncan Luce and Howard Raiffa. Superb nontechnical introduction to game theory, primarily applied to social sciences. Utility theory, zero-sum games, n-person games, decision-making, much more. Bibliography. 509pp. $5^3/_8$ x $8^1/_2$. 0-486-65943-7

INTRODUCTION TO THE THEORY OF GAMES, J. C. C. McKinsey. This comprehensive overview of the mathematical theory of games illustrates applications to situations involving conflicts of interest, including economic, social, political, and military contexts. Appropriate for advanced undergraduate and graduate courses; advanced calculus a prerequisite. 1952 ed. x+372pp. $5^3/_8$ x $8^1/_2$. 0-486-42811-7

FIFTY CHALLENGING PROBLEMS IN PROBABILITY WITH SOLUTIONS, Frederick Mosteller. Remarkable puzzlers, graded in difficulty, illustrate elementary and advanced aspects of probability. Detailed solutions. 88pp. $5^3/_8$ x $8^1/_2$. 0-486-65355-2

PROBABILITY THEORY: A CONCISE COURSE, Y. A. Rozanov. Highly readable, self-contained introduction covers combination of events, dependent events, Bernoulli trials, etc. 148pp. $5^3/_8$ x $8^1/_4$. 0-486-63544-9

THE STATISTICAL ANALYSIS OF EXPERIMENTAL DATA, John Mandel. First half of book presents fundamental mathematical definitions, concepts and facts while remaining half deals with statistics primarily as an interpretive tool. Well-written text, numerous worked examples with step-by-step presentation. Includes 116 tables. 448pp. $5^3/_8$ x $8^1/_2$. 0-486-64666-1

Math—Geometry and Topology

ELEMENTARY CONCEPTS OF TOPOLOGY, Paul Alexandroff. Elegant, intuitive approach to topology from set-theoretic topology to Betti groups; how concepts of topology are useful in math and physics. 25 figures. 57pp. 5³/₈ x 8¹/₂. 0-486-60747-X

A LONG WAY FROM EUCLID, Constance Reid. Lively guide by a prominent historian focuses on the role of Euclid's Elements in subsequent mathematical developments. Elementary algebra and plane geometry are sole prerequisites. 80 drawings. 1963 edition. 304pp. 5³/₈ x 8¹/₂. 0-486-43613-6

EXPERIMENTS IN TOPOLOGY, Stephen Barr. Classic, lively explanation of one of the byways of mathematics. Klein bottles, Moebius strips, projective planes, map coloring, problem of the Koenigsberg bridges, much more, described with clarity and wit. 43 figures. 210pp. 5³/₈ x 8¹/₂. 0-486-25933-1

THE GEOMETRY OF RENÉ DESCARTES, René Descartes. The great work founded analytical geometry. Original French text, Descartes's own diagrams, together with definitive Smith-Latham translation. 244pp. 5³/₈ x 8¹/₂. 0-486-60068-8

EUCLIDEAN GEOMETRY AND TRANSFORMATIONS, Clayton W. Dodge. This introduction to Euclidean geometry emphasizes transformations, particularly isometries and similarities. Suitable for undergraduate courses, it includes numerous examples, many with detailed answers. 1972 ed. viii+296pp. 6¹/₈ x 9¹/₄. 0-486-43476-1

EXCURSIONS IN GEOMETRY, C. Stanley Ogilvy. A straightedge, compass, and a little thought are all that's needed to discover the intellectual excitement of geometry. Harmonic division and Apollonian circles, inversive geometry, hexlet, Golden Section, more. 132 illustrations. 192pp. 5³/₈ x 8¹/₂. 0-486-26530-7

THE THIRTEEN BOOKS OF EUCLID'S ELEMENTS, translated with introduction and commentary by Sir Thomas L. Heath. Definitive edition. Textual and linguistic notes, mathematical analysis. 2,500 years of critical commentary. Unabridged. 1,414pp. 5³/₈ x 8¹/₂. Three-vol. set.
Vol. I: 0-486-60088-2 Vol. II: 0-486-60089-0 Vol. III: 0-486-60090-4

SPACE AND GEOMETRY: IN THE LIGHT OF PHYSIOLOGICAL, PSYCHOLOGICAL AND PHYSICAL INQUIRY, Ernst Mach. Three essays by an eminent philosopher and scientist explore the nature, origin, and development of our concepts of space, with a distinctness and precision suitable for undergraduate students and other readers. 1906 ed. vi+148pp. 5³/₈ x 8¹/₂. 0-486-43909-7

GEOMETRY OF COMPLEX NUMBERS, Hans Schwerdtfeger. Illuminating, widely praised book on analytic geometry of circles, the Moebius transformation, and two-dimensional non-Euclidean geometries. 200pp. 5⁵/₈ x 8¹/₄. 0-486-63830-8

DIFFERENTIAL GEOMETRY, Heinrich W. Guggenheimer. Local differential geometry as an application of advanced calculus and linear algebra. Curvature, transformation groups, surfaces, more. Exercises. 62 figures. 378pp. 5³/₈ x 8¹/₂. 0-486-63433-7

History of Math

THE WORKS OF ARCHIMEDES, Archimedes (T. L. Heath, ed.). Topics include the famous problems of the ratio of the areas of a cylinder and an inscribed sphere; the measurement of a circle; the properties of conoids, spheroids, and spirals; and the quadrature of the parabola. Informative introduction. clxxxvi+326pp. $5^3/_8$ x $8^1/_2$. 0-486-42084-1

A SHORT ACCOUNT OF THE HISTORY OF MATHEMATICS, W. W. Rouse Ball. One of clearest, most authoritative surveys from the Egyptians and Phoenicians through 19th-century figures such as Grassman, Galois, Riemann. Fourth edition. 522pp. $5^3/_8$ x $8^1/_2$. 0-486-20630-0

THE HISTORY OF THE CALCULUS AND ITS CONCEPTUAL DEVELOPMENT, Carl B. Boyer. Origins in antiquity, medieval contributions, work of Newton, Leibniz, rigorous formulation. Treatment is verbal. 346pp. $5^3/_8$ x $8^1/_2$. 0-486-60509-4

THE HISTORICAL ROOTS OF ELEMENTARY MATHEMATICS, Lucas N. H. Bunt, Phillip S. Jones, and Jack D. Bedient. Fundamental underpinnings of modern arithmetic, algebra, geometry and number systems derived from ancient civilizations. 320pp. $5^3/_8$ x $8^1/_2$. 0-486-25563-8

THE HISTORY OF THE CALCULUS AND ITS CONCEPTUAL DEVELOPMENT, Carl B. Boyer. Fluent description of the development of both the integral and differential calculus—its early beginnings in antiquity, medieval contributions, and a consideration of Newton and Leibniz. 368pp. $5^3/_8$ x $8^1/_2$. 0-486-60509-4

GAMES, GODS & GAMBLING: A HISTORY OF PROBABILITY AND STATISTICAL IDEAS, F. N. David. Episodes from the lives of Galileo, Fermat, Pascal, and others illustrate this fascinating account of the roots of mathematics. Features thought-provoking references to classics, archaeology, biography, poetry. 1962 edition. 304pp. $5^3/_8$ x $8^1/_2$. (Available in U.S. only.) 0-486-40023-9

OF MEN AND NUMBERS: THE STORY OF THE GREAT MATHEMATICIANS, Jane Muir. Fascinating accounts of the lives and accomplishments of history's greatest mathematical minds—Pythagoras, Descartes, Euler, Pascal, Cantor, many more. Anecdotal, illuminating. 30 diagrams. Bibliography. 256pp. $5^3/_8$ x $8^1/_2$. 0-486-28973-7

HISTORY OF MATHEMATICS, David E. Smith. Nontechnical survey from ancient Greece and Orient to late 19th century; evolution of arithmetic, geometry, trigonometry, calculating devices, algebra, the calculus. 362 illustrations. 1,355pp. $5^3/_8$ x $8^1/_2$. Two-vol. set. Vol. I: 0-486-20429-4 Vol. II: 0-486-20430-8

A CONCISE HISTORY OF MATHEMATICS, Dirk J. Struik. The best brief history of mathematics. Stresses origins and covers every major figure from ancient Near East to 19th century. 41 illustrations. 195pp. $5^3/_8$ x $8^1/_2$. 0-486-60255-9

Physics

OPTICAL RESONANCE AND TWO-LEVEL ATOMS, L. Allen and J. H. Eberly. Clear, comprehensive introduction to basic principles behind all quantum optical resonance phenomena. 53 illustrations. Preface. Index. 256pp. $5^3/_8 \times 8^1/_2$. 0-486-65533-4

QUANTUM THEORY, David Bohm. This advanced undergraduate-level text presents the quantum theory in terms of qualitative and imaginative concepts, followed by specific applications worked out in mathematical detail. Preface. Index. 655pp. $5^3/_8 \times 8^1/_2$. 0-486-65969-0

ATOMIC PHYSICS (8th EDITION), Max Born. Nobel laureate's lucid treatment of kinetic theory of gases, elementary particles, nuclear atom, wave-corpuscles, atomic structure and spectral lines, much more. Over 40 appendices, bibliography. 495pp. $5^3/_8 \times 8^1/_2$. 0-486-65984-4

A SOPHISTICATE'S PRIMER OF RELATIVITY, P. W. Bridgman. Geared toward readers already acquainted with special relativity, this book transcends the view of theory as a working tool to answer natural questions: What is a frame of reference? What is a "law of nature"? What is the role of the "observer"? Extensive treatment, written in terms accessible to those without a scientific background. 1983 ed. xlviii+172pp. $5^3/_8 \times 8^1/_2$. 0-486-42549-5

AN INTRODUCTION TO HAMILTONIAN OPTICS, H. A. Buchdahl. Detailed account of the Hamiltonian treatment of aberration theory in geometrical optics. Many classes of optical systems defined in terms of the symmetries they possess. Problems with detailed solutions. 1970 edition. xv + 360pp. $5^3/_8 \times 8^1/_2$. 0-486-67597-1

PRIMER OF QUANTUM MECHANICS, Marvin Chester. Introductory text examines the classical quantum bead on a track: its state and representations; operator eigenvalues; harmonic oscillator and bound bead in a symmetric force field; and bead in a spherical shell. Other topics include spin, matrices, and the structure of quantum mechanics; the simplest atom; indistinguishable particles; and stationary-state perturbation theory. 1992 ed. xiv+314pp. $6^1/_8 \times 9^1/_4$. 0-486-42878-8

LECTURES ON QUANTUM MECHANICS, Paul A. M. Dirac. Four concise, brilliant lectures on mathematical methods in quantum mechanics from Nobel Prize-winning quantum pioneer build on idea of visualizing quantum theory through the use of classical mechanics. 96pp. $5^3/_8 \times 8^1/_2$. 0-486-41713-1

THIRTY YEARS THAT SHOOK PHYSICS: THE STORY OF QUANTUM THEORY, George Gamow. Lucid, accessible introduction to influential theory of energy and matter. Careful explanations of Dirac's anti-particles, Bohr's model of the atom, much more. 12 plates. Numerous drawings. 240pp. $5^3/_8 \times 8^1/_2$. 0-486-24895-X

ELECTRONIC STRUCTURE AND THE PROPERTIES OF SOLIDS: THE PHYSICS OF THE CHEMICAL BOND, Walter A. Harrison. Innovative text offers basic understanding of the electronic structure of covalent and ionic solids, simple metals, transition metals and their compounds. Problems. 1980 edition. 582pp. $6^1/_8 \times 9^1/_4$. 0-486-66021-4

HYDRODYNAMIC AND HYDROMAGNETIC STABILITY, S. Chandrasekhar. Lucid examination of the Rayleigh-Benard problem; clear coverage of the theory of instabilities causing convection. 704pp. $5^3/_8$ x $8^1/_4$. 0-486-64071-X

INVESTIGATIONS ON THE THEORY OF THE BROWNIAN MOVEMENT, Albert Einstein. Five papers (1905–8) investigating dynamics of Brownian motion and evolving elementary theory. Notes by R. Fürth. 122pp. $5^3/_8$ x $8^1/_2$. 0-486-60304-0

THE PHYSICS OF WAVES, William C. Elmore and Mark A. Heald. Unique overview of classical wave theory. Acoustics, optics, electromagnetic radiation, more. Ideal as classroom text or for self-study. Problems. 477pp. $5^3/_8$ x $8^1/_2$. 0-486-64926-1

GRAVITY, George Gamow. Distinguished physicist and teacher takes reader-friendly look at three scientists whose work unlocked many of the mysteries behind the laws of physics: Galileo, Newton, and Einstein. Most of the book focuses on Newton's ideas, with a concluding chapter on post-Einsteinian speculations concerning the relationship between gravity and other physical phenomena. 160pp. $5^3/_8$ x $8^1/_2$. 0-486-42563-0

PHYSICAL PRINCIPLES OF THE QUANTUM THEORY, Werner Heisenberg. Nobel Laureate discusses quantum theory, uncertainty, wave mechanics, work of Dirac, Schroedinger, Compton, Wilson, Einstein, etc. 184pp. $5^3/_8$ x $8^1/_2$. 0-486-60113-7

ATOMIC SPECTRA AND ATOMIC STRUCTURE, Gerhard Herzberg. One of best introductions; especially for specialist in other fields. Treatment is physical rather than mathematical. 80 illustrations. 257pp. $5^3/_8$ x $8^1/_2$. 0-486-60115-3

AN INTRODUCTION TO STATISTICAL THERMODYNAMICS, Terrell L. Hill. Excellent basic text offers wide-ranging coverage of quantum statistical mechanics, systems of interacting molecules, quantum statistics, more. 523pp. $5^3/_8$ x $8^1/_2$. 0-486-65242-4

THEORETICAL PHYSICS, Georg Joos, with Ira M. Freeman. Classic overview covers essential math, mechanics, electromagnetic theory, thermodynamics, quantum mechanics, nuclear physics, other topics. First paperback edition. xxiii + 885pp. $5^3/_8$ x $8^1/_2$. 0-486-65227-0

PROBLEMS AND SOLUTIONS IN QUANTUM CHEMISTRY AND PHYSICS, Charles S. Johnson, Jr. and Lee G. Pedersen. Unusually varied problems, detailed solutions in coverage of quantum mechanics, wave mechanics, angular momentum, molecular spectroscopy, more. 280 problems plus 139 supplementary exercises. 430pp. $6^1/_2$ x $9^1/_4$. 0-486-65236-X

THEORETICAL SOLID STATE PHYSICS, Vol. 1: Perfect Lattices in Equilibrium; Vol. II: Non-Equilibrium and Disorder, William Jones and Norman H. March. Monumental reference work covers fundamental theory of equilibrium properties of perfect crystalline solids, non-equilibrium properties, defects and disordered systems. Appendices. Problems. Preface. Diagrams. Index. Bibliography. Total of 1,301pp. $5^3/_8$ x $8^1/_2$. Two volumes. Vol. I: 0-486-65015-4 Vol. II: 0-486-65016-2

WHAT IS RELATIVITY? L. D. Landau and G. B. Rumer. Written by a Nobel Prize physicist and his distinguished colleague, this compelling book explains the special theory of relativity to readers with no scientific background, using such familiar objects as trains, rulers, and clocks. 1960 ed. vi+72pp. $5^3/_8$ x $8^1/_2$. 0-486-42806-0

A TREATISE ON ELECTRICITY AND MAGNETISM, James Clerk Maxwell. Important foundation work of modern physics. Brings to final form Maxwell's theory of electromagnetism and rigorously derives his general equations of field theory. 1,084pp. $5^3/_8$ x $8^1/_2$. Two-vol. set. Vol. I: 0-486-60636-8 Vol. II: 0-486-60637-6

MATHEMATICS FOR PHYSICISTS, Philippe Dennery and Andre Krzywicki. Superb text provides math needed to understand today's more advanced topics in physics and engineering. Theory of functions of a complex variable, linear vector spaces, much more. Problems. 1967 edition. 400pp. $6^1/_2$ x $9^1/_4$. 0-486-69193-4

INTRODUCTION TO QUANTUM MECHANICS WITH APPLICATIONS TO CHEMISTRY, Linus Pauling & E. Bright Wilson, Jr. Classic undergraduate text by Nobel Prize winner applies quantum mechanics to chemical and physical problems. Numerous tables and figures enhance the text. Chapter bibliographies. Appendices. Index. 468pp. $5^3/_8$ x $8^1/_2$. 0-486-64871-0

METHODS OF THERMODYNAMICS, Howard Reiss. Outstanding text focuses on physical technique of thermodynamics, typical problem areas of understanding, and significance and use of thermodynamic potential. 1965 edition. 238pp. $5^3/_8$ x $8^1/_2$. 0-486-69445-3

THE ELECTROMAGNETIC FIELD, Albert Shadowitz. Comprehensive under- graduate text covers basics of electric and magnetic fields, builds up to electromagnetic theory. Also related topics, including relativity. Over 900 problems. 768pp. $5^3/_8$ x $8^1/_4$. 0-486-65660-8

GREAT EXPERIMENTS IN PHYSICS: FIRSTHAND ACCOUNTS FROM GALILEO TO EINSTEIN, Morris H. Shamos (ed.). 25 crucial discoveries: Newton's laws of motion, Chadwick's study of the neutron, Hertz on electromagnetic waves, more. Original accounts clearly annotated. 370pp. $5^3/_8$ x $8^1/_2$. 0-486-25346-5

EINSTEIN'S LEGACY, Julian Schwinger. A Nobel Laureate relates fascinating story of Einstein and development of relativity theory in well-illustrated, nontechnical volume. Subjects include meaning of time, paradoxes of space travel, gravity and its effect on light, non-Euclidean geometry and curving of space-time, impact of radio astronomy and space-age discoveries, and more. 189 b/w illustrations. xiv+250pp. $8^3/_8$ x $9^1/_4$. 0-486-41974-6

THE VARIATIONAL PRINCIPLES OF MECHANICS, Cornelius Lanczos. Philosophic, less formalistic approach to analytical mechanics offers model of clear, scholarly exposition at graduate level with coverage of basics, calculus of variations, principle of virtual work, equations of motion, more. 418pp. $5^3/_8$ x $8^1/_2$. 0-486-65067-7

Paperbound unless otherwise indicated. Available at your book dealer, online at www.doverpublications.com, or by writing to Dept. GI, Dover Publications, Inc., 31 East 2nd Street, Mineola, NY 11501. For current price information or for free catalogues (please indicate field of interest), write to Dover Publications or log on to www.doverpublications.com and see every Dover book in print. Dover publishes more than 400 books each year on science, elementary and advanced mathematics, biology, music, art, literary history, social sciences, and other areas.